Elastic Wave Propagation and Generation in Seismology

Seismology has complementary observational and theoretical components, and a thorough understanding of the observations requires a sound theoretical background. Seismological theory, however, can be a difficult mathematical subject and introductory books do not generally give students the tools they need to solve seismological problems by themselves. This book addresses these shortcomings by bridging the gap between introductory textbooks and advanced monographs. It provides the necessary mathematical machinery and demonstrates how to apply it.

The author's approach is to consider seismological phenomena as problems in applied mathematics. To this end, each problem is carefully formulated and its solution is derived in a step-by-step approach. Although some exposure to vector calculus and partial differential equations is expected, most of the mathematics needed is derived within the book. This includes Cartesian tensors, solution of 3-D scalar and vector wave equations, Green's functions, and continuum mechanics concepts. The book covers strain, stress, propagation of body and surface waves in simple models (half-spaces and the layer over a half-space), ray theory for P and S waves (including amplitude equations), near and far fields generated by moment tensor sources in infinite media, and attenuation and the mathematics of causality.

Numerous programs for the computation of reflection and transmission coefficients, for the generation of P- and S-wave radiation patterns, and for near- and far-field synthetic seismograms in infinite media are provided by the author on a dedicated website. The book also includes problems for students to work through, with solutions available on the associated website. This book will therefore find a receptive audience among advanced undergraduate and graduate students interested in developing a solid mathematical background to tackle more advanced topics in seismology. It will also form a useful reference volume for researchers wishing to brush up on the fundamentals.

JOSE PUJOL received a B.S. in Chemistry, from the Universidad Nacional del Sur, Bahia Blanca, Argentina, in 1968 and then went on to graduate studies in quantum chemistry at Uppsala University, Sweden, and Karlsruhe University, Germany. Following further graduate studies in petroleum exploration at the University of Buenos Aires, Argentina, he studied for an M.S. in geophysics, at the University of Alaska (1982) and a Ph.D. at the University of Wyoming (1985). He has been a faculty member at the University of Memphis since 1985 where he is currently an Associate Professor in the Center for Earthquake Research and Information. Professor Pujol's research interests include earthquake and exploration seismology, vertical seismic profiling, inverse problems, earthquake location and velocity inversion, and attenuation studies using borehole data. He is also an associate editor for the Seismological Society of America.

Elastic Wave Propagation
and Generation in
Seismology

Jose Pujol

CAMBRIDGE
UNIVERSITY PRESS

CAMBRIDGE
UNIVERSITY PRESS

Shaftesbury Road, Cambridge CB2 8EA, United Kingdom

One Liberty Plaza, 20th Floor, New York, NY 10006, USA

477 Williamstown Road, Port Melbourne, VIC 3207, Australia

314–321, 3rd Floor, Plot 3, Splendor Forum, Jasola District Centre, New Delhi – 110025, India

103 Penang Road, #05–06/07, Visioncrest Commercial, Singapore 238467

Cambridge University Press is part of Cambridge University Press & Assessment,
a department of the University of Cambridge.

We share the University's mission to contribute to society through the pursuit of
education, learning and research at the highest international levels of excellence.

www.cambridge.org
Information on this title: www.cambridge.org/9780521520461

First published 2003

A catalogue record for this publication is available from the British Library

Library of Congress Cataloging-in-Publication data
Pujol, Jose, 1945–
Elastic wave propagation and generation in seismology / Jose Pujol.
p.   cm.
Includes bibliographical references and index.
ISBN 0 521 81730 7 – ISBN 0 521 52046 0 (pbk.)
1. Elastic waves   I. Title
QE538.5.P85 2003
551.22–dc21   2002071477

ISBN   978-0-521-81730-1   Hardback
ISBN   978-0-521-52046-1   Paperback

In memory of my father
Jose A. Pujol

Contents

Contents

Preface

The study of the theory of elastic wave propagation and generation can be a daunting task because of its inherent mathematical complexity. The books on the subject currently available are either advanced or introductory. The advanced ones require a mathematical background and/or maturity generally beyond that of the average seismology student. The introductory ones, on the other hand, address advanced subjects but usually skip the more difficult mathematical derivations, with frequent references to the advanced books. What is needed is a text that goes through the complete derivations, so that readers have the opportunity to acquire the tools and training that will allow them to pose and solve problems at an intermediate level of difficulty and to approach the more advanced problems discussed in the literature. Of course, there is nothing new in this idea; there are hundreds of physics, mathematics, and engineering books that do just that, but unfortunately this does not apply to seismology. Consequently, the student in a seismology program without a strong quantitative or theoretical component, or the observational seismologist interested in a clear understanding of the analysis or processing techniques used, do not have an accessible treatment of the theory. A result of this situation is an ever widening gap between those who understand seismological theory and those who do not. At a time when more and more analysis and processing computer packages are available, it is important that their users have the knowledge required to use those packages as something more than black boxes.

This book was designed to fill the existing gap in the seismological literature. The guiding philosophy is to start with first principles and to move progressively to more advanced topics without recourse to "it can be proved" or references to Aki and Richards (1980). To fully benefit from the book the reader is expected to have had exposure to vector calculus and partial differential equations at an introductory level. Some knowledge of Fourier transforms is convenient, but except for a section in Chapter 6 and Chapter 11 they are little used. However, it is also expected that readers without this background will also profit from the book because of its explanatory material and examples.

The presentation of the material has been strongly influenced by the books of Ben-Menahem and Singh (1981) and Aki and Richards (1980), but has also benefited from those of Achenbach (1973), Burridge (1976), Eringen and Suhubi (1975), Hudson (1980), and Sokolnikoff (1956), among others. In fact, the selection of the sources for the different chapters and sections of the book was based on a "pick and choose" approach, with the overall goal of giving a presentation that is as simple as possible while at the same time retaining the inherent level of complexity of the individual topics. Again, this idea is not new, and is summarized in the following statement due to Einstein, "everything should be made as simple as possible, but not simpler" (quoted in Ben-Menahem and Singh's book). Because of its emphasis on fundamentals, the book does not deal with observations or with data analysis. However, the selection of topics was guided in part by the premise that they should be applicable to the analysis of observations. Brief chapter descriptions follow.

The first chapter is a self-contained introduction to Cartesian tensors. Tensors are essential for a thorough understanding of stress and strain. Of course, it is possible to introduce these two subjects without getting into the details of tensor analysis, but this approach does not have the conceptual clarity that tensors provide. In addition, the material developed in this chapter has direct application to the seismic moment tensor, discussed in Chapters 9 and 10. For completeness, a summary of results pertaining to vectors, assumed to be known, is included. This chapter also includes an introduction to dyadics, which in some cases constitute a convenient alternative to tensors and are found in some of the relevant literature.

Chapters 2 and 3 describe the strain and rotation tensors and the stress tensor, respectively. The presentation of the material is based on a continuum mechanics approach, which provides a conceptually clearer picture than other approaches and has wide applicability. For example, although the distinction between Lagrangian and Eulerian descriptions of motion is rarely made in seismology, the reader should be aware of them because they may be important in theoretical studies, as the book by Dahlen and Tromp (1998) demonstrates. Because of its importance in earthquake faulting studies, the Mohr circles for stress are discussed in detail. Chapter 4 introduces Hooke's law, which relates stress and strain, and certain energy relations that actually permit proving Hooke's law. The chapter also discusses several classic elastic parameters, and derives the elastic wave equation and introduces the P and S waves.

Chapter 5 deals with solutions to the scalar and vector wave equations and to the elastic wave equation in unbounded media. The treatment of the scalar equation is fairly conventional, but that of the vector equations is not. The basic idea in solving them is to find vector solutions, which in the case of the elastic wave equation immediately lead to the concept of P, SV, and SH wave motion. This approach, used by Ben-Menahem and Singh, bypasses the more conventional approach based on potentials. Because displacements, not potentials, are the observables, it makes sense to develop a theory based on them, particularly when no additional complexity is involved.

The P, SV, and SH vector solutions derived in Chapter 5 are used in Chapter 6, which covers body waves in simple models (half-spaces and a layer over a half-space). Because of their importance in applications, the different cases are discussed in detail. Two important problems, generally ignored in introductory books, also receive full attention. One is the change in waveform shapes that takes place for angles of incidence larger than the critical. The second problem is the amplification of ground motion caused by the presence of a surficial low-velocity layer, which is of importance in seismic risk studies.

Chapter 7 treats surface waves in simple models, including a model with continuous vertical variations in elastic properties, and presents a thorough analysis of dispersion. Unlike the customary discussion of dispersion in seismology books, which is limited to showing the existence of phase and group velocities, here I provide an example of a dispersive system that actually shows how the period of a wave changes as a function of time and position.

Chapter 8 deals with ray theory for the scalar wave equation and the elastic wave equation. In addition to a discussion of the kinematic aspects of the theory, including a proof of Fermat's principle, this chapter treats the very important problem of P and S amplitudes. This is done in the so-called ray-centered coordinate system, for which there are not readily available derivations. This coordinate system simplifies the computation of amplitudes and constitutes the basis of current advanced techniques for the computation of synthetic seismograms using ray theory.

Chapter 9 begins the discussion of seismic point sources in unbounded media. The simplest source is a force along one of the coordinate axes, but obtaining the corresponding solution requires considerable effort. Once this problem has been solved it is easy to find solutions to problems involving combinations of forces such as couples and combinations thereof, which in turn lead to the concept of the moment tensor. Chapter 10 specializes to the case of earthquakes caused by slip on a fault, which is shown to be equivalent to a double couple. After this major result it is more or less straightforward to find the moment tensor corresponding to a fault of arbitrary orientation. Although the Earth is clearly bounded and not homogeneous, the theory developed in these two chapters has had a major impact in seismology and still constitutes a major tool in the study of earthquakes, particularly in combination with ray theory.

Chapter 11 is on attenuation, a subject that is so vast and touches on so many aspects of seismology and physics that actually it deserves a book devoted exclusively to it. For this reason I restricted my coverage to just the most essential facts. One of these is the constraint imposed by causality, which plays a critical role in attenuation studies, and because it is not well covered in most seismology books it has been given prominence here. Causality has been studied by mathematicians, physicists and engineers, and a number of basic theorems rely on the theory of complex variables. Consequently, some of the basic results had to be quoted without even attempting to give a justification, but aside from that the treatment is self-contained. The measurement of attenuation using seismic data has received a large amount of attention in the literature, and for this reason it is not discussed here except for a brief description of the widely used spectral ratio method, and a little-known bias effect introduced when the method is applied to windowed data. As scattering may be a strong contributor to attenuation, this chapter closes with an example based on the effect of a finely layered medium on wave amplitudes and shapes.

The book ends with several appendices designed to provide background material. Appendix A introduces the theory of distributions, which is an essential tool in the study of partial differential equations (among other things). Dirac's delta is the best known example of a distribution, and is customarily introduced as something that is not a "normal" function but then is treated as such. This lack of consistency is unnecessary, as the most basic aspects of distribution theory are easy to grasp when presented with enough detail, as is done in this Appendix. Distributions are already part of the geophysical and seismological literature (e.g., Bourbié *et al.*, 1987; Dahlen and Tromp, 1998), and this Appendix will allow the reader to gain a basic understanding of the concepts discussed there. Appendix B discusses the Hilbert transform, which is a basic element in studies involving causality, and is needed to describe waves incident on a boundary at angles larger than the critical angle. A recipe for the numerical computation of the transform is also given. Appendix C derives the Green's function for the 3-D scalar wave equation. This Green's function is essential to solving the problems discussed in Chapter 9, but in spite of its importance it is usually quoted, rather than derived. The last two appendices derive in detail two fundamental equations given in Chapter 9 for the displacements caused by a single force and by an arbitrary moment tensor.

The book includes some brief historical notes. More often than not, scientific books present a finished product, without giving the readers a sense of the struggle that generally preceded the formalization of a theory. Being aware of this struggle should be part of the education of every future scientist because science rarely progresses in the neat and linear way it is usually presented. Seismology, as well as elasticity, upon which seismology is founded, also had their share of controversy, and it is instructive to learn how our present understanding of seismic theory has come about. In this context, an enlightening review of the history of seismology by Ben-Menahem (1995) is highly recommended.

An important component of the book are the problems. Going through them will help the reader solidify the concepts and techniques discussed in the text. All the problems have hints, so that solving them should not be difficult after the background material has been mastered. In addition, full solutions are provided on a dedicated website:

`http://publishing.cambridge.org/resources/0521817307`

Because of the obvious importance of computers in research and because going from the theory to a computer application is not always straightforward, Fortran and Matlab codes used to generate most of the figures in Chapters 6–10 have been provided on the same website.

The book is based on class notes developed for a graduate course I taught at the University of Memphis over more than ten years. The amount of material covered depended on the background of the students, but about 70% was always covered.

Acknowledgements

First and foremost, I thank my late father and my mother, whose hard work helped pave the way that led to this book. Secondly, two of my undergraduate professors at the Universidad Nacional del Sur, Argentina, the late E. Oklander and Dr A. Benedek, provided most of the mathematical tools that allowed me to write this book, which I hope also conveys their inspiring teaching philosophy. I am also very grateful to the students that took the class upon which this book is based for their positive response to my teaching approach and for their feedback. One of them, Mr K. Kim, helped with the proofreading of the first six chapters and Mr Y. Zhang with the proofreading of Chapters 9 and 10. The book also benefited from reviews of earlier versions of individual chapters by Drs N. Biswas, D. Boore, R. Burridge, G. Goldstein, J. Julia, C. Langston, T. Long, A. Roma, and R. Smalley. Of course, any remaining errors are my own and I will be very grateful if they are brought to my attention. Thanks are also due to Mr S. D. Hoskyns for his great help during the copyediting and typesetting, and to Mrs S. Early for carefully typing a preliminary version of several chapters. Last but not least, I am extremely grateful to my wife, Ursula, and our son, Sebastian, for their unfailing patience and understanding over the years.

1

Introduction to tensors and dyadics

1.1 Introduction

Tensors play a fundamental role in theoretical physics. The reason for this is that physical laws written in tensor form are independent of the coordinate system used (Morse and Feshbach, 1953). Before elaborating on this point, consider a simple example, based on Segel (1977). Newton's second law is $\mathbf{f} = m\mathbf{a}$, where $\mathbf{f}$ and $\mathbf{a}$ are vectors representing the force and acceleration of an object of mass m. This basic law does not have a coordinate system attached to it. To apply the law in a particular situation it will be convenient to select a coordinate system that simplifies the mathematics, but there is no question that any other system will be equally acceptable. Now consider an example from elasticity, discussed in Chapter 3. The stress vector $\mathbf{T}$ (force/area) across a surface element in an elastic solid is related to the vector $\mathbf{n}$ normal to the same surface via the stress tensor. The derivation of this relation is carried out using a tetrahedron with faces along the three coordinate planes in a Cartesian coordinate system. Therefore, it is reasonable to ask whether the same result would have been obtained if a different Cartesian coordinate system had been used, or if a spherical, or cylindrical, or any other curvilinear system, had been used. Take another example. The elastic wave equation will be derived in a Cartesian coordinate system. As discussed in Chapter 4, two equations will be found, one in component form and one in vector form in terms of a combination of gradient, divergence, and curl. Again, here there are some pertinent questions regarding coordinate systems. For example, can either of the two equations be applied in non-Cartesian coordinate systems? The reader may already know that only the latter equation is generally applicable, but may not be aware that there is a mathematical justification for that fact, namely, that the gradient, divergence, and curl are independent of the coordinate system (Morse and Feshbach, 1953). These questions are generally not discussed in introductory texts, with the consequence that the reader fails to grasp the deeper meaning of the concepts of

1

vectors and tensors. It is only when one realizes that physical entities (such as force, acceleration, stress tensor, and so on) and the relations among them have an existence independent of coordinate systems, that it is possible to appreciate that there is more to tensors than what is usually discussed. It is possible, however, to go through the basic principles of stress and strain without getting into the details of tensor analysis. Therefore, some parts of this chapter are not essential for the rest of the book.

Tensor analysis, in its broadest sense, is concerned with arbitrary curvilinear co-ordinates. A more restricted approach concentrates on *orthogonal curvilinear co-ordinates*, such as cylindrical and spherical coordinates. These coordinate systems have the property that the unit vectors at a given point in space are perpendicular (i.e. orthogonal) to each other. Finally, we have the rectangular Cartesian system, which is also orthogonal. The main difference between general orthogonal and Cartesian systems is that in the latter the unit vectors do not change as a function of position, while this is not true in the former. Unit vectors for the spherical system will be given in §9.9.1. The theory of tensors in non-Cartesian systems is exceed-ingly complicated, and for this reason we will limit our study to Cartesian tensors. However, some of the most important relations will be written using dyadics (see §1.6), which provide a symbolic representation of tensors independent of the coor-dinate system. It may be useful to note that there are oblique Cartesian coordinate systems of importance in crystallography, for example, but in the following we will consider the rectangular Cartesian systems only.

1.2 Summary of vector analysis

It is assumed that the reader is familiar with the material summarized in this section (see, e.g., Lass, 1950; Davis and Snider, 1991).

A vector is defined as a directed line segment, having both magnitude and direc-tion. The magnitude, or length, of a vector **a** will be represented by $|\mathbf{a}|$. The sum and the difference of two vectors, and the multiplication of a vector by a scalar (real number) are defined using geometric rules. Given two vectors **a** and **b**, two products between them have been defined.

Scalar, or dot, product:

$$\mathbf{a} \cdot \mathbf{b} = |\mathbf{a}||\mathbf{b}| \cos \alpha, \tag{1.2.1}$$

where α is the angle between the vectors.

Vector, or cross, product:

$$\mathbf{a} \times \mathbf{b} = (|\mathbf{a}||\mathbf{b}| \sin \alpha)\,\mathbf{n}, \tag{1.2.2}$$

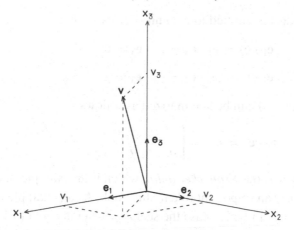

Fig. 1.1. Rectangular Cartesian coordinate system with unit vectors e_1, e_2, e_3, and decomposition of an arbitrary vector $\mathbf{v}$ into components v_1, v_2, v_3.

where α is as before, and $\mathbf{n}$ is a unit vector (its length is equal to 1) perpendicular to both $\mathbf{a}$ and $\mathbf{b}$ such that the three vectors form a right-handed system.

An important property of the vector product, derived using geometric arguments, is the distributive law

$$(\mathbf{a} + \mathbf{b}) \times \mathbf{c} = \mathbf{a} \times \mathbf{c} + \mathbf{b} \times \mathbf{c}. \tag{1.2.3}$$

By introducing a rectangular Cartesian coordinate system it is possible to write a vector in terms of three components. Let $\mathbf{e}_1 = (1, 0, 0)$, $\mathbf{e}_2 = (0, 1, 0)$, and $\mathbf{e}_3 = (0, 0, 1)$ be the three unit vectors along the x_1, x_2, and x_3 axes of Fig. 1.1. Then any vector $\mathbf{v}$ can be written as

$$\mathbf{v} = (v_1, v_2, v_3) = v_1\mathbf{e}_1 + v_2\mathbf{e}_2 + v_3\mathbf{e}_3 = \sum_{i=1}^{3} v_i\mathbf{e}_i. \tag{1.2.4}$$

The components v_1, v_2, and v_3 are the orthogonal projections of $\mathbf{v}$ in the directions of the three axes (Fig. 1.1).

Before proceeding, a few words concerning the notation are necessary. A vector will be denoted by a bold-face letter, while its components will be denoted by the same letter in italics with subindices (literal or numerical). A bold-face letter with a subindex represents a vector, not a vector component. The three unit vectors defined above are examples of the latter. If we want to write the kth component of the unit vector $\mathbf{e}_j$ we will write $(\mathbf{e}_j)_k$. For example, $(\mathbf{e}_2)_1 = 0$, $(\mathbf{e}_2)_2 = 1$, and $(\mathbf{e}_2)_3 = 0$. In addition, although vectors will usually be written in row form (e.g., as in (1.2.4)), when they are involved in matrix operations they should be considered as column vectors, i.e., as matrices of one column and three rows. For example, the matrix form of the scalar product $\mathbf{a} \cdot \mathbf{b}$ is $\mathbf{a}^T\mathbf{b}$, where T indicates transposition.

When the scalar product is applied to the unit vectors we find

$$\mathbf{e}_1 \cdot \mathbf{e}_2 = \mathbf{e}_1 \cdot \mathbf{e}_3 = \mathbf{e}_2 \cdot \mathbf{e}_3 = 0 \tag{1.2.5}$$

$$\mathbf{e}_1 \cdot \mathbf{e}_1 = \mathbf{e}_2 \cdot \mathbf{e}_2 = \mathbf{e}_3 \cdot \mathbf{e}_3 = 1. \tag{1.2.6}$$

Equations (1.2.5) and (1.2.6) can be summarized as follows:

$$\mathbf{e}_i \cdot \mathbf{e}_j = \delta_{ij} = \begin{cases} 1, & i = j \\ 0, & i \neq j. \end{cases} \tag{1.2.7}$$

The symbol δ_{jk} is known as the *Kronecker delta*, which is an example of a second-order tensor, and will play an important role in this book. As an example of (1.2.7), $\mathbf{e}_2 \cdot \mathbf{e}_k$ is zero unless $k = 2$, in which case the scalar product is equal to 1.

Next we derive an alternative expression for a vector $\mathbf{v}$. Using (1.2.4), the scalar product of $\mathbf{v}$ and $\mathbf{e}_i$ is

$$\mathbf{v} \cdot \mathbf{e}_i = \left(\sum_{k=1}^{3} v_k \mathbf{e}_k \right) \cdot \mathbf{e}_i = \sum_{k=1}^{3} v_k \mathbf{e}_k \cdot \mathbf{e}_i = \sum_{k=1}^{3} v_k (\mathbf{e}_k \cdot \mathbf{e}_i) = v_i. \tag{1.2.8}$$

Note that when applying (1.2.4) the subindex in the summation must be different from i. To obtain (1.2.8) the following were used: the distributive law of the scalar product, the law of the product by a scalar, and (1.2.7). Equation (1.2.8) shows that the ith component of $\mathbf{v}$ can be written as

$$v_i = \mathbf{v} \cdot \mathbf{e}_i. \tag{1.2.9}$$

When (1.2.9) is introduced in (1.2.4) we find

$$\mathbf{v} = \sum_{i=1}^{3} (\mathbf{v} \cdot \mathbf{e}_i) \mathbf{e}_i. \tag{1.2.10}$$

This expression will be used in the discussion of dyadics (see §1.6).

In terms of its components the length of the vector is given by

$$|\mathbf{v}| = \sqrt{v_1^2 + v_2^2 + v_3^2} = (\mathbf{v} \cdot \mathbf{v})^{1/2}. \tag{1.2.11}$$

Using purely geometric arguments it is found that the scalar and vector products can be written in component form as follows:

$$\mathbf{u} \cdot \mathbf{v} = u_1 v_1 + u_2 v_2 + u_3 v_3 \tag{1.2.12}$$

and

$$\mathbf{u} \times \mathbf{v} = (u_2 v_3 - u_3 v_2) \mathbf{e}_1 + (u_3 v_1 - u_1 v_3) \mathbf{e}_2 + (u_1 v_2 - u_2 v_1) \mathbf{e}_3. \tag{1.2.13}$$

The last expression is based on the use of (1.2.3).

Vectors, and vector operations such as the scalar and vector products, among others, are defined independently of any coordinate system. Vector relations derived without recourse to vector components will be valid when written in component form regardless of the coordinate system used. Of course, the same vector may (and generally will) have different components in different coordinate systems, but they will represent the same geometric entity. This is true for Cartesian and more general coordinate systems, such as spherical and cylindrical ones, but in the following we will consider the former only.

Now suppose that we want to define new vector entities based on operations on the components of other vectors. In view of the comments in §1.1 it is reasonable to expect that not every arbitrary definition will represent a vector, i.e., an entity intrinsically independent of the coordinate system used to represent the space. To see this consider the following example, which for simplicity refers to vectors in two-dimensional (2-D) space. Given a vector $\mathbf{u} = (u_1, u_2)$, define a new vector $\mathbf{v} = (u_1 + \lambda, u_2 + \lambda)$, where λ is a nonzero scalar. Does this definition result in a vector? To answer this question draw the vectors $\mathbf{u}$ and $\mathbf{v}$ (Fig. 1.2a), rotate the original coordinate axes, decompose $\mathbf{u}$ into its new components u_1' and u_2', add λ to each of them, and draw the new vector $\mathbf{v}' = (u_1' + \lambda, u_2' + \lambda)$. Clearly, $\mathbf{v}$ and $\mathbf{v}'$ are not the same geometric object. Therefore, our definition does not represent a vector.

Now consider the following definition: $\mathbf{v} = (\lambda u_1, \lambda u_2)$. After a rotation similar to the previous one we see that $\mathbf{v} = \mathbf{v}'$ (Fig. 1.2b), which is not surprising, as this definition corresponds to the multiplication of a vector by a scalar.

Let us look now at a more complicated example. Suppose that given two vectors $\mathbf{u}$ and $\mathbf{v}$ we want to define a third vector $\mathbf{w}$ as follows:

$$\mathbf{w} = (u_2 v_3 + u_3 v_2)\mathbf{e}_1 + (u_3 v_1 + u_1 v_3)\mathbf{e}_2 + (u_1 v_2 + u_2 v_1)\mathbf{e}_3. \qquad (1.2.14)$$

Note that the only difference with the vector product (see (1.2.13)) is the replacement of the minus signs by plus signs. As before, the question is whether this definition is independent of the coordinate system. In this case, however, finding an answer is not straightforward. What one should do is to compute the components w_1, w_2, w_3 in the original coordinate system, draw $\mathbf{w}$, perform a rotation of axes, find the new components of $\mathbf{u}$ and $\mathbf{v}$, compute w_1', w_2', and w_3', draw $\mathbf{w}'$ and compare it with $\mathbf{w}$. If it is found that the two vectors are different, then it is obvious that (1.2.14) does not define a vector. If the two vectors are equal it might be tempting to say that (1.2.14) does indeed define a vector, but this conclusion would not be correct because there may be other rotations for which $\mathbf{w}$ and $\mathbf{w}'$ are not equal.

These examples should convince the reader that establishing the vectorial character of an entity defined by its components requires a definition of a vector that will take this question into account automatically. Only then will it be possible to

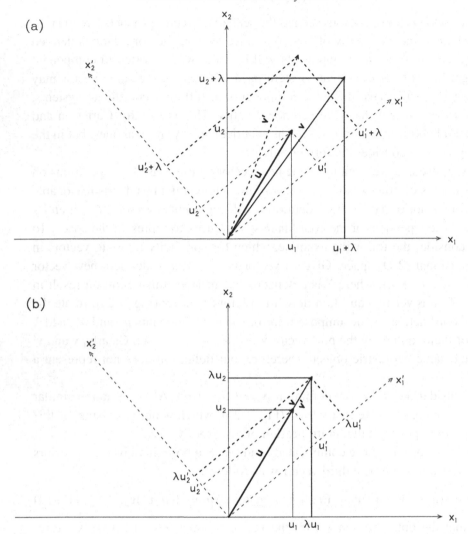

Fig. 1.2. (a) Vectors **v** and **v**' obtained from a vector **u** as follows. For **v**, add a constant λ to the components u_1 and u_2. For **v**', add a constant λ to the components u_1' and u_2' obtained by rotation of the axis. Because **v** and **v**' are not the same vector, we can conclude that the entity obtained by adding a constant to the components of a vector does constitute a vector under a rotation of coordinates. (b) Similar to the construction above, but with the constant λ multiplying the vector components. In this case **v** and **v**' coincide, which agrees with the fact that the operation defined is just the multiplication of a vector by a scalar. After Santalo (1969).

answer the previous question in a general way. However, before introducing the new definition it is necessary to study coordinate rotations in some more detail. This is done next.

1.3 Rotation of Cartesian coordinates. Definition of a vector

Let Ox_1, Ox_2, and Ox_3 represent a Cartesian coordinate system and Ox_1', Ox_2', Ox_3' another system obtained from the previous one by a rotation about their common origin O (Fig. 1.3). Let $\mathbf{e}_1$, $\mathbf{e}_2$, and $\mathbf{e}_3$ and $\mathbf{e}_1'$, $\mathbf{e}_2'$, and $\mathbf{e}_3'$ be the unit vectors along the three axes in the original and rotated systems. Finally, let a_{ij} denote the cosine of the angle between Ox_i' and Ox_j. The a_{ij}'s are known as direction cosines, and are related to $\mathbf{e}_i'$ and $\mathbf{e}_j$ by

$$\mathbf{e}_i' \cdot \mathbf{e}_j = a_{ij}. \qquad (1.3.1)$$

Given an arbitrary vector $\mathbf{v}$ with components v_1, v_2, and v_3 in the original system, we are interested in finding the components v_1', v_2', and v_3' in the rotated system. To find the relation between the two sets of components we will consider first the relation between the corresponding unit vectors. Using (1.3.1) $\mathbf{e}_i'$ can be written as

$$\mathbf{e}_i' = a_{i1}\mathbf{e}_1 + a_{i2}\mathbf{e}_2 + a_{i3}\mathbf{e}_3 = \sum_{j=1}^{3} a_{ij}\mathbf{e}_j \qquad (1.3.2)$$

(Problem 1.3a).

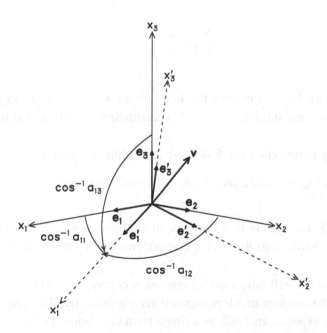

Fig. 1.3. Rotation of axes. Primed and unprimed quantities refer to the original and rotated coordinate systems, respectively. Both systems are rectangular Cartesian. The quantities a_{ij} indicate the scalar product $\mathbf{e}_i' \cdot \mathbf{e}_j$. The vector $\mathbf{v}$ exists independent of the coordinate system. Three relevant angles are shown.

Furthermore, in the original and rotated systems **v** can be written as

$$\mathbf{v} = \sum_{j=1}^{3} v_j \mathbf{e}_j \qquad (1.3.3)$$

and

$$\mathbf{v} = \sum_{i=1}^{3} v_i' \mathbf{e}_i'. \qquad (1.3.4)$$

Now introduce (1.3.2) in (1.3.4)

$$\mathbf{v} = \sum_{i=1}^{3} v_i' \sum_{j=1}^{3} a_{ij} \mathbf{e}_j \equiv \sum_{j=1}^{3} \left(\sum_{i=1}^{3} a_{ij} v_i' \right) \mathbf{e}_j. \qquad (1.3.5)$$

Since (1.3.3) and (1.3.5) represent the same vector, and the three unit vectors $\mathbf{e}_1$, $\mathbf{e}_1$, and $\mathbf{e}_3$ are independent of each other, we conclude that

$$v_j = \sum_{i=1}^{3} a_{ij} v_i'. \qquad (1.3.6)$$

If we write the $\mathbf{e}_j$s in terms of the $\mathbf{e}_i'$s and replace them in (1.3.3) we find that

$$v_i' = \sum_{j=1}^{3} a_{ij} v_j \qquad (1.3.7)$$

(Problem 1.3b).

Note that in (1.3.6) the sum is over the first subindex of a_{ij}, while in (1.3.7) the sum is over the second subindex of a_{ij}. This distinction is critical and must be respected.

Now we are ready to introduce the following definition of a vector:

three scalars are the components of a vector if under a rotation of coordinates they transform according to (1.3.7).

What this definition means is that if we want to define a vector by some set of rules, we have to verify that the vector components satisfy the transformation equations.

Before proceeding we will introduce a *summation convention* (due to Einstein) that will simplify the mathematical manipulations significantly. The convention applies to monomial expressions (such as a single term in an equation) and consists of dropping the sum symbol and summing over repeated indices.[1] This convention requires that the same index should appear no more than twice in the same term.

---

[1] In this book the convention will not be applied to uppercase indices

Repeated indices are known as *dummy indices*, while those that are not repeated are called *free indices*. Using this convention, we will write, for example,

$$\mathbf{v} = \sum_{j=1}^{3} v_j \mathbf{e}_j = v_j \mathbf{e}_j \tag{1.3.8}$$

$$v_j = \sum_{i=1}^{3} a_{ij} v_i' = a_{ij} v_i' \tag{1.3.9}$$

$$v_i' = \sum_{j=1}^{3} a_{ij} v_j = a_{ij} v_j. \tag{1.3.10}$$

It is important to have a clear idea of the difference between free and dummy indices. A particular dummy index can be changed at will as long as it is replaced (in its two occurrences) by some other index not equal to any other existing indices in the same term. Free indices, on the other hand, are fixed and cannot be changed inside a single term. However, a free index can be replaced by another as long as the change is effected in all the terms in an equation, and the new index is different from all the other indices in the equation. In (1.3.9) i is a dummy index and j is a free index, while in (1.3.10) their role is reversed. The examples below show legal and illegal index manipulations.

The following relations, derived from (1.3.9), are true

$$v_j = a_{ij} v_i' = a_{kj} v_k' = a_{lj} v_l' \tag{1.3.11}$$

because the repeated index i was replaced by a different repeated index (equal to k or l). However, it would not be correct to replace i by j because j is already present in the equation. If i were replaced by j we would have

$$v_j = a_{jj} v_j', \tag{1.3.12}$$

which would not be correct because the index j appears more than twice in the right-hand term, which is not allowed. Neither would it be correct to write

$$v_j = a_{ik} v_i' \tag{1.3.13}$$

because the free index j has been changed to k only in the right-hand term. On the other hand, (1.3.9) can be written as

$$v_k = a_{ik} v_i' \tag{1.3.14}$$

because the free index j has been replaced by k on both sides of the equation.

As (1.3.9) and (1.3.10) are of fundamental importance, it is necessary to pay attention to the fact that in the former the sum is over the first index of a_{ij} while in the latter the sum is over the second index of a_{ij}. Also note that (1.3.10) can be written as the product of a matrix and a vector:

$$\mathbf{v}' = \begin{pmatrix} v'_1 \\ v'_2 \\ v'_3 \end{pmatrix} = \begin{pmatrix} a_{11} & a_{12} & a_{13} \\ a_{21} & a_{22} & a_{23} \\ a_{31} & a_{32} & a_{33} \end{pmatrix} \begin{pmatrix} v_1 \\ v_2 \\ v_3 \end{pmatrix} \equiv \mathbf{A}\,\mathbf{v}, \qquad (1.3.15)$$

where $\mathbf{A}$ is the matrix with elements a_{ij}.

It is clear that (1.3.9) can be written as

$$\mathbf{v} = \mathbf{A}^{\mathrm{T}}\mathbf{v}', \qquad (1.3.16)$$

where the superscript T indicates transposition.

Now we will derive an important property of $\mathbf{A}$. By introducing (1.3.10) in (1.3.9) we obtain

$$v_j = a_{ij}a_{ik}v_k. \qquad (1.3.17)$$

Note that it was necessary to change the dummy index in (1.3.10) to satisfy the summation convention. Equation (1.3.17) implies that any of the three components of $\mathbf{v}$ is a combination of all three components. However, this cannot be generally true because $\mathbf{v}$ is an arbitrary vector. Therefore, the right-hand side of (1.3.17) must be equal to v_j, which in turn implies that the product $a_{ij}a_{ik}$ must be equal to unity when $j = k$, and equal to zero when $j \neq k$. This happens to be the definition of the Kronecker delta δ_{jk} introduced in (1.2.7), so that

$$a_{ij}a_{ik} = \delta_{jk}. \qquad (1.3.18)$$

If (1.3.9) is introduced in (1.3.10) we obtain

$$a_{ij}a_{kj} = \delta_{ik}. \qquad (1.3.19)$$

Setting $i = k$ in (1.3.19) and writing in full gives

$$1 = a_{i1}^2 + a_{i2}^2 + a_{i3}^2 = |\mathbf{e}'_i|^2; \qquad i = 1, 2, 3, \qquad (1.3.20)$$

where the equality on the right-hand side follows from (1.3.2).

When $i \neq k$, (1.3.19) gives

$$0 = a_{i1}a_{k1} + a_{i2}a_{k2} + a_{i3}a_{k3} = \mathbf{e}'_i \cdot \mathbf{e}'_k, \qquad (1.3.21)$$

where the equality on the right-hand side also follows from (1.3.2). Therefore, (1.3.19) summarizes the fact that the $\mathbf{e}'_j$s are unit vectors orthogonal to each other, while (1.3.18) does the same thing for the $\mathbf{e}_i$s. Any set of vectors having these properties is known as an orthonormal set.

In matrix form, (1.3.18) and (1.3.19) can be written as

$$\mathbf{A}^T\mathbf{A} = \mathbf{A}\mathbf{A}^T = \mathbf{I}, \tag{1.3.22}$$

where $\mathbf{I}$ is the identity matrix.

Equation (1.3.22) can be rewritten in the following useful way:

$$\mathbf{A}^T = \mathbf{A}^{-1}; \qquad (\mathbf{A}^T)^{-1} = \mathbf{A}, \tag{1.3.23}$$

where the superscript -1 indicates matrix inversion. From (1.3.22) we also find

$$|\mathbf{A}\mathbf{A}^T| = |\mathbf{A}||\mathbf{A}^T| = |\mathbf{A}|^2 = |\mathbf{I}| = 1, \tag{1.3.24}$$

where vertical bars indicate the determinant of a matrix.

Linear transformations with a matrix such that its determinant squared is equal to 1 are known as orthogonal transformations. When $|\mathbf{A}| = 1$, the transformation corresponds to a rotation. When $|\mathbf{A}| = -1$, the transformation involves the reflection of one coordinate axis in a coordinate plane. An example of reflection is the transformation that leaves the x_1 and x_2 axes unchanged and replaces the x_3 axis by $-x_3$. Reflections change the orientation of the space: if the original system is right-handed, then the new system is left-handed, and vice versa.

1.4 Cartesian tensors

In subsequent chapters the following three tensors will be introduced.

(1) The strain tensor ε_{ij}:

$$\varepsilon_{ij} = \frac{1}{2}\left(\frac{\partial u_i}{\partial x_j} + \frac{\partial u_j}{\partial x_i}\right); \qquad i, j = 1, 2, 3, \tag{1.4.1}$$

where the vector $\mathbf{u} = (u_1, u_2, u_3)$ is the displacement suffered by a particle inside a body when it is deformed.

(2) The stress tensor τ_{ij}:

$$T_i = \tau_{ij}n_j; \qquad i = 1, 2, 3, \tag{1.4.2}$$

where T_i and n_j indicate the components of the stress vector and normal vector referred to in §1.1.

(3) The elastic tensor c_{ijkl}, which relates stress to strain:

$$\tau_{ij} = c_{ijkl}\varepsilon_{kl}. \tag{1.4.3}$$

Let us list some of the differences between vectors and tensors. First, while a vector can be represented by a single symbol, such as $\mathbf{u}$, or by its components, such as u_j, a tensor can only be represented by its components (e.g., ε_{ij}), although the introduction of dyadics (see §1.6) will allow the representation of tensors by single symbols. Secondly, while vector components carry only one subindex, tensors carry two subindices or more. Thirdly, in the three-dimensional space we are considering, a vector has three components, while ε_{ij} and τ_{ij} have 3×3, or nine, components, and c_{ijkl} has 81 components ($3 \times 3 \times 3 \times 3$). Tensors ε_{ij} and τ_{ij} are known as second-order tensors, while c_{ijkl} is a fourth-order tensor, with the order of the tensor being given by the number of free indices. There are also differences among the tensors shown above. For example, ε_{ij} is defined in terms of operations (derivatives) on the components of a single vector, while τ_{ij} appears in a relation between two vectors. c_{ijkl}, on the other hand, relates two tensors.

Clearly, tensors offer more variety than vectors, and because they are defined in terms of components, the comments made in connection with vector components and the rotation of axes also apply to tensors. To motivate the following definition of a second-order tensor consider the relation represented by (1.4.2). For this relation to be independent of the coordinate system, upon a rotation of axes we must have

$$T'_l = \tau'_{lk} n'_k. \tag{1.4.4}$$

In other words, the functional form of the relation must remain the same after a change of coordinates. We want to find the relation between τ'_{lk} and τ_{ij} that satisfies (1.4.2) and (1.4.4). To do that multiply (1.4.2) by a_{li} and sum over i:

$$a_{li} T_i = a_{li} \tau_{ij} n_j. \tag{1.4.5}$$

Before proceeding rewrite T'_l and n_j using (1.3.10) with $\mathbf{v}$ replaced by $\mathbf{T}$ and (1.3.9) with $\mathbf{v}$ replaced by $\mathbf{n}$. This gives

$$T'_l = a_{li} T_i; \qquad n_j = a_{kj} n'_k. \tag{1.4.6a,b}$$

From (1.4.6a,b), (1.4.2), and (1.4.5) we find

$$T'_l = a_{li} T_i = a_{li} \tau_{ij} n_j = a_{li} \tau_{ij} a_{kj} n'_k = (a_{li} a_{kj} \tau_{ij}) n'_k. \tag{1.4.7}$$

Now subtracting (1.4.4) from (1.4.7) gives

$$0 = (\tau'_{lk} - a_{li} a_{kj} \tau_{ij}) n'_k. \tag{1.4.8}$$

As n_k is an arbitrary vector, the factor in parentheses in (1.4.8) must be equal to zero (Problem 1.7), so that

$$\tau'_{lk} = a_{li} a_{kj} \tau_{ij}. \tag{1.4.9}$$

Note that (1.4.9) does not depend on the physical nature of the quantities involved in (1.4.2). Only the functional relation matters. This result motivates the following definition.

Second-order tensor. Given nine quantities t_{ij}, they constitute the components of a second-order tensor if they transform according to

$$t'_{ij} = a_{il}a_{jk}t_{lk} \tag{1.4.10}$$

under a change of coordinates $v'_i = a_{ij}v_j$.

To write the tensor components in the unprimed system in terms of the components in the primed system, multiply (1.4.10) by $a_{im}a_{jn}$ and sum over i and j, and use the orthogonality relation (1.3.18):

$$a_{im}a_{jn}t'_{ij} = a_{im}a_{jn}a_{il}a_{jk}t_{lk} = a_{im}a_{il}a_{jn}a_{jk}t_{lk} = \delta_{lm}\delta_{kn}t_{lk} = t_{mn}. \tag{1.4.11}$$

Therefore,

$$t_{mn} = a_{im}a_{jn}t'_{ij}. \tag{1.4.12}$$

As (1.4.10) and (1.4.12) are similar, it is important to make sure that the arrangement of indices is strictly adhered to.

Equation (1.4.11) illustrates an important aspect of the Kronecker delta. For a given m and n, the expression $\delta_{lm}\delta_{kn}t_{lk}$ is a double sum over l and k, so that there are nine terms. However, since δ_{lm} and δ_{kn} are equal to zero except when $l = m$ and $k = n$, in which case the deltas are equal to one, the only nonzero term in the sum is t_{mn}. Therefore, the equality $\delta_{lm}\delta_{kn}t_{lk} = t_{mn}$ can be derived formally by replacing l and k in t_{lk} by m and n and by dropping the deltas.

The extension of (1.4.10) to higher-order tensors is straightforward.

Tensor of order n. Given 3^n quantities $t_{i_1i_2...i_n}$, they constitute the components of a tensor of order n if they transform according to

$$t'_{i_1i_2...i_n} = a_{i_1j_1}a_{i_2j_2} \cdots a_{i_nj_n}t_{j_1j_2...j_n} \tag{1.4.13}$$

under a change of coordinates $v'_i = a_{ij}v_j$. All the indices $i_1, i_2, \ldots$ and $j_1, j_2, \ldots$ can be 1, 2, 3.

The extension of (1.4.12) to tensors of order n is obvious. For example, for a third-order tensor we have the following relations:

$$t'_{ijk} = a_{il}a_{jm}a_{kn}t_{lmn}; \qquad t_{mnp} = a_{im}a_{jn}a_{kp}t'_{ijk}. \tag{1.4.14a,b}$$

The definition of tensors can be extended to include vectors and scalars, which can be considered as tensors of orders one and zero, respectively. The corresponding number of free indices are one and zero, but this does not mean that dummy indices are not allowed (such as in t_{jj} and $u_{i,i}$, introduced below).

An important consequence of definitions (1.4.10) and (1.4.13) is that if the components of a tensor are all equal to zero in a given coordinate system, they will be equal to zero in any other coordinate system.

In the following we will indicate vectors and tensors by their individual components. For example, u_i, u_j, and u_m, etc., represent the same vector $\mathbf{u}$, while t_{ij}, t_{ik}, t_{mn}, etc., represent the same tensor. As noted earlier, the introduction of dyadics will afford the representation of tensors by single symbols, but for the special case of a second-order tensor we can associate it with a 3×3 matrix. For example, the matrix $\mathbf{T}$ corresponding to the tensor t_{ij} is given by

$$\mathbf{T} = \begin{pmatrix} t_{11} & t_{12} & t_{13} \\ t_{21} & t_{22} & t_{23} \\ t_{31} & t_{32} & t_{33} \end{pmatrix}. \tag{1.4.15}$$

Introduction of $\mathbf{T}$ is convenient because (1.4.10), which can be written as

$$t'_{ij} = a_{il} t_{lk} a_{jk}, \tag{1.4.16}$$

has a simple expression in matrix form:

$$\mathbf{T'} = \mathbf{A}\mathbf{T}\mathbf{A}^{\mathrm{T}}, \tag{1.4.17}$$

where $\mathbf{A}$ is the matrix introduced in (1.3.15) and

$$\mathbf{T'} = \begin{pmatrix} t'_{11} & t'_{12} & t'_{13} \\ t'_{21} & t'_{22} & t'_{23} \\ t'_{31} & t'_{32} & t'_{33} \end{pmatrix}. \tag{1.4.18}$$

Equation (1.4.17) is very useful in actual computations.

To express $\mathbf{T}$ in terms of $\mathbf{T'}$, write (1.4.12) as

$$t_{mn} = a_{im} t'_{ij} a_{jn}, \tag{1.4.19}$$

which gives

$$\mathbf{T} = \mathbf{A}^{\mathrm{T}} \mathbf{T'} \mathbf{A}. \tag{1.4.20}$$

Otherwise, solve (1.4.17) for $\mathbf{T}$ by multiplying both sides by $(\mathbf{A}^{\mathrm{T}})^{-1}$ on the right and by $\mathbf{A}^{-1}$ on the left and use (1.3.23).

1.4.1 Tensor operations

(1) *Addition or subtraction of two tensors.* The result is a new tensor whose components are the sum or difference of the corresponding components. These two operations are defined for tensors of the same order. As an example, given the tensors t_{ij} and s_{ij}, their sum or difference is the tensor b_{ij} with components

$$b_{ij} = t_{ij} \pm s_{ij}. \tag{1.4.21}$$

Let us verify that b_{ij} is indeed a tensor. This requires writing the components of t_{ij} and s_{ij} in the primed system, adding or subtracting them together and verifying that (1.4.10) is satisfied. From (1.4.10) we obtain

$$t'_{ij} = a_{il}a_{jm}t_{lm} \qquad (1.4.22)$$

$$s'_{ij} = a_{il}a_{jm}s_{lm}. \qquad (1.4.23)$$

Adding or subtracting (1.4.22) and (1.4.23) gives

$$b'_{ij} = t'_{ij} \pm s'_{ij} = a_{il}a_{jm}(t_{lm} \pm s_{lm}) = a_{il}a_{jm}b_{lm}. \qquad (1.4.24)$$

Equation (1.4.24) shows that b_{ij} transforms according to (1.4.10) and is therefore a tensor.

(2) *Multiplication of a tensor by a scalar.* Each component of the tensor is multiplied by the scalar. The result of this operation is a new tensor. For example, multiplication of t_{ij} by a scalar λ gives the tensor b_{ij} with components

$$b_{ij} = \lambda t_{ij}. \qquad (1.4.25)$$

To show that b_{ij} is a tensor we proceed as before

$$b'_{ij} = \lambda t'_{ij} = \lambda a_{il}a_{jm}t_{lm} = a_{il}a_{jm}(\lambda t_{lm}) = a_{il}a_{jm}b_{lm}. \qquad (1.4.26)$$

(3) *Outer product of two tensors.* This gives a new tensor whose order is equal to the sum of the orders of the two tensors and whose components are obtained by multiplication of the components of the two tensors. From this definition it should be clear that the indices in one tensor cannot appear among the indices in the other tensor. As an example, consider the outer product of t_{ij} and u_k. The result is the tensor s_{ijk} given by

$$s_{ijk} = t_{ij}u_k. \qquad (1.4.27)$$

To show that s_{ijk} is a tensor proceed as before

$$s'_{ijk} = t'_{ij}u'_k = a_{il}a_{jm}t_{lm}a_{kn}u_n = a_{il}a_{jm}a_{kn}t_{lm}u_n = a_{il}a_{jm}a_{kn}s_{lmn}. \qquad (1.4.28)$$

As another example, the outer product of two vectors **a** and **b** is the tensor with components a_ib_j. This particular outer product will be considered again when discussing dyadics.

(4) *Contraction of indices.* In a tensor of order two or higher, set two indices equal to each other. As the two contracted indices become dummy indices, they indicate a sum over the repeated index. By contraction, the order of the tensor is reduced by two. For example, given t_{ij}, by contraction of i and j we obtain the scalar

$$t_{ii} = t_{11} + t_{22} + t_{33}, \qquad (1.4.29)$$

which is known as the *trace* of t_{ij}. Note that when t_{ij} is represented by a matrix,

t_{ii} corresponds to the sum of its diagonal elements, which is generally known as the trace of the matrix. Another example of contraction is the divergence of a vector, discussed in §1.4.5.

(5) *Inner product, or contraction, of two tensors.* Given two tensors, first form their outer product and then apply a contraction of indices using one index from each tensor. For example, the scalar product of two vectors **a** and **b** is equal to their inner product, given by $a_i b_i$. We will refer to the inner product as the contraction of two tensors. By extension, a product involving a_{ij}, as in (1.4.5) and (1.4.11), for example, will also be called a contraction.

1.4.2 Symmetric and anti-symmetric tensors

A second-order tensor t_{ij} is symmetric if

$$t_{ij} = t_{ji} \tag{1.4.30}$$

and anti-symmetric if

$$t_{ij} = -t_{ji}. \tag{1.4.31}$$

Any second-order tensor b_{ij} can be written as the following identity:

$$b_{ij} \equiv \frac{1}{2} b_{ij} + \frac{1}{2} b_{ij} + \frac{1}{2} b_{ji} - \frac{1}{2} b_{ji} = \frac{1}{2} (b_{ij} + b_{ji}) + \frac{1}{2} (b_{ij} - b_{ji}). \tag{1.4.32}$$

Clearly, the tensors in parentheses are symmetric and anti-symmetric, respectively. Therefore, b_{ij} can be written as

$$b_{ij} = s_{ij} + a_{ij} \tag{1.4.33}$$

with

$$s_{ij} = s_{ji} = \frac{1}{2} (b_{ij} + b_{ji}) \tag{1.4.34}$$

and

$$a_{ij} = -a_{ji} = \frac{1}{2} (b_{ij} - b_{ji}). \tag{1.4.35}$$

Examples of symmetric second-order tensors are the Kronecker delta, the strain tensor (as can be seen from (1.4.1)), and the stress tensor, as will be shown in the next chapter.

For higher-order tensors, symmetry and anti-symmetry are referred to pairs of indices. A tensor is completely symmetric (anti-symmetric) if it is symmetric (anti-symmetric) for all pairs of indices. For example, if t_{ijk} is completely symmetric, then

$$t_{ijk} = t_{jik} = t_{ikj} = t_{kji} = t_{kij} = t_{jki}. \tag{1.4.36}$$

If t_{ijk} is completely anti-symmetric, then

$$t_{ijk} = -t_{jik} = -t_{ikj} = t_{kij} = -t_{kji} = t_{jki}. \qquad (1.4.37)$$

The permutation symbol, introduced in §1.4.4, is an example of a completely anti-symmetric entity.

The elastic tensor in (1.4.3) has symmetry properties different from those described above. They will be described in detail in Chapter 4.

1.4.3 Differentiation of tensors

Let t_{ij} be a function of the coordinates x_i ($i = 1, 2, 3$). From (1.4.10) we know that

$$t'_{ij} = a_{ik}a_{jl}t_{kl}. \qquad (1.4.38)$$

Now differentiate both sides of (1.4.38) with respect to x'_m,

$$\frac{\partial t'_{ij}}{\partial x'_m} = a_{ik}a_{jl}\frac{\partial t_{kl}}{\partial x_s}\frac{\partial x_s}{\partial x'_m}. \qquad (1.4.39)$$

Note that on the right-hand side we used the chain rule of differentiation and that there is an implied sum over the index s. Also note that since

$$x_s = a_{ms}x'_m \qquad (1.4.40)$$

(Problem 1.8) then

$$\frac{\partial x_s}{\partial x'_m} = a_{ms}. \qquad (1.4.41)$$

Using (1.4.41) and introducing the notation

$$\frac{\partial t'_{ij}}{\partial x'_m} \equiv t'_{ij,m}; \qquad \frac{\partial t_{kl}}{\partial x_s} \equiv t_{kl,s} \qquad (1.4.42)$$

equation (1.4.39) becomes

$$t'_{ij,m} = a_{ik}a_{jl}a_{ms}t_{kl,s}. \qquad (1.4.43)$$

This shows that $t_{kl,s}$ is a third-order tensor.

Applying the same arguments to higher-order tensors shows that first-order differentiation generates a new tensor with the order increased by one. It must be emphasized, however, that in general curvilinear coordinates this differentiation does not generate a tensor.

1.4.4 The permutation symbol

This is indicated by ϵ_{ijk} and is defined by

$$\epsilon_{ijk} = \begin{cases} 0, & \text{if any two indices are repeated} \\ 1, & \text{if } ijk \text{ is an even permutation of 123} \\ -1, & \text{if } ijk \text{ is an odd permutation of 123.} \end{cases} \qquad (1.4.44)$$

A permutation is even (odd) if the number of exchanges of i, j, k required to order them as 123 is even (odd). For example, to go from 213 to 123 only one exchange is needed, so the permutation is odd. On the other hand, to go from 231 to 123, two exchanges are needed: $231 \rightarrow 213 \rightarrow 123$, and the permutation is even. After considering all the possible combinations we find

$$\epsilon_{123} = \epsilon_{231} = \epsilon_{312} = 1 \qquad (1.4.45)$$

$$\epsilon_{132} = \epsilon_{321} = \epsilon_{213} = -1. \qquad (1.4.46)$$

The permutation symbol is also known as the alternating or Levi-Civita symbol. The definition (1.4.44) is general in the sense that it can be extended to more than three subindices. The following equivalent definition can only be used with three subindices, but is more convenient for practical uses:

$$\epsilon_{ijk} = \begin{cases} 0, & \text{if any two indices are repeated} \\ 1, & \text{if } ijk \text{ are in cyclic order} \\ -1, & \text{if } ijk \text{ are not in cyclic order.} \end{cases} \qquad (1.4.47)$$

Three different indices ijk are in cyclic order if they are equal to 123 or 231 or 312. They are not in cyclic order if they are equal to 132 or 321 or 213 (see Fig. 1.4).

The permutation symbol will play an important role in the applications below, so its properties must be understood well. When using ϵ_{ijk} in equations, the values of ijk are generally not specified. Moreover, it may happen that the same equation includes factors such as ϵ_{jik} and ϵ_{kij}. In cases like that it is necessary to express

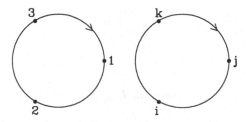

Fig. 1.4. Diagrams used to find out whether or not a combination of the integers 1, 2, 3, or any three indices i, j, k, are in cyclic order (indicated by the arrow). For example, the combination 312 is in cyclic order, while 213 is not. For arbitrary indices i, j, k, if the combination ikj is assumed to be in cyclic order, then combinations such as kji and jik will also be in cyclic order, but kij and ijk will not.

one of them in terms of the other. To do that assume that jik is in cyclic order (Fig. 1.4) and by inspection find out whether kij is in the same order or not. As it is not, $\epsilon_{kij} = -\epsilon_{jik}$.

Is the permutation symbol a tensor? Almost. If the determinant of the transformation matrix $\mathbf{A}$ is equal to 1, then the components of ϵ_{ijk} transform according to (1.4.14a). However, if the determinant is equal to -1, the components of ϵ_{ijk} transform according to -1 times the right-hand side of (1.4.14a). Entities with this type of transformation law are known as *pseudo tensors*. Another well-known example is the vector product, which produces a *pseudo vector*, rather than a vector.

Another important aspect of ϵ_{ijk} is that its components are independent of the coordinate system (see §1.4.7). A proof of the tensorial character of ϵ_{ijk} can be found in Goodbody (1982) and McConnell (1957).

1.4.5 Applications and examples

In the following it will be assumed that the scalars, vectors, and tensors are functions of x_1, x_2, and x_3, as needed, and that the required derivatives exist.

(1) By contraction of the second-order tensor $u_{i,j}$ we obtain the scalar $u_{i,i}$. When this expression is written in full it is clear that it corresponds to the *divergence* of $\mathbf{u}$:

$$u_{i,i} = \frac{\partial u_1}{\partial x_1} + \frac{\partial u_2}{\partial x_2} + \frac{\partial u_3}{\partial x_3} = \text{div } \mathbf{u} = \nabla \cdot \mathbf{u}. \tag{1.4.48}$$

In the last term we introduced the vector operator ∇ (nabla or del), which can be written as

$$\nabla = \left(\frac{\partial}{\partial x_1}, \frac{\partial}{\partial x_2}, \frac{\partial}{\partial x_3} \right). \tag{1.4.49}$$

This definition of ∇ is valid in Cartesian coordinate systems only.

(2) The derivative of the scalar function $f(x_1, x_2, x_3)$ with respect to x_i is the ith component of the *gradient* of f:

$$\frac{\partial f}{\partial x_i} = (\nabla f)_i = f_{,i}. \tag{1.4.50}$$

(3) The sum of second derivatives $f_{,ii}$ of a scalar function f is the *Laplacian* of f:

$$f_{,ii} = \frac{\partial^2 f}{\partial x_1^2} + \frac{\partial^2 f}{\partial x_2^2} + \frac{\partial^2 f}{\partial x_3^2} = \nabla^2 f. \tag{1.4.51}$$

(4) The second derivatives of the vector components u_i are $u_{i,jk}$. By contraction we obtain $u_{i,jj}$, which corresponds to the ith component of the Laplacian of the vector $\mathbf{u}$:

$$\nabla^2 \mathbf{u} = \left(\nabla^2 u_1, \nabla^2 u_2, \nabla^2 u_3\right) = \left(u_{1,jj}, u_{2,jj}, u_{3,jj}\right). \tag{1.4.52}$$

Again, this definition applies to Cartesian coordinates only. In general orthogonal coordinate systems the *Laplacian of a vector* is defined by

$$\nabla^2 \mathbf{u} = \nabla(\nabla \cdot \mathbf{u}) - \nabla \times \nabla \times \mathbf{u}, \tag{1.4.53}$$

where the appropriate expressions for the gradient, divergence, and curl should be used (e.g., Morse and Feshbach, 1953; Ben-Menahem and Singh, 1981). For Cartesian coordinates (1.4.52) and (1.4.53) lead to the same expression (Problem 1.9).

(5) Show that the Kronecker delta is a second-order tensor. Let us apply the transformation law (1.4.10):

$$\delta'_{ij} = a_{il}a_{jk}\delta_{lk} = a_{il}a_{jl} = \delta_{ij}. \tag{1.4.54}$$

This shows that δ_{ij} is a tensor. Note that in the term to the right of the first equality there is a double sum over l and k (nine terms) but because of the definition of the delta, the only nonzero terms are those with $l = k$ (three terms), in which case delta has a value of one. The last equality comes from (1.3.19).

(6) Let $\mathbf{B}$ be the 3×3 matrix with elements b_{ij}. Then, the determinant of $\mathbf{B}$ is given by

$$|\mathbf{B}| \equiv \begin{vmatrix} b_{11} & b_{12} & b_{13} \\ b_{21} & b_{22} & b_{23} \\ b_{31} & b_{32} & b_{33} \end{vmatrix} = \epsilon_{ijk}b_{1i}b_{2j}b_{3k} = \epsilon_{ijk}b_{i1}b_{j2}b_{k3}. \tag{1.4.55}$$

The two expressions in terms of ϵ_{ijk} correspond to the expansion of the determinant by rows and columns, respectively. It is straightforward to verify that (1.4.55) is correct. In fact, those familiar with determinants will recognize that (1.4.55) is the definition of a determinant.

(7) Vector product of $\mathbf{u}$ and $\mathbf{v}$. The ith component is given by

$$(\mathbf{u} \times \mathbf{v})_i = \epsilon_{ijk}u_j v_k. \tag{1.4.56}$$

(8) *Curl of a vector.* The ith component is given by

$$(\text{curl } \mathbf{u})_i = (\nabla \times \mathbf{u})_i = \epsilon_{ijk}\frac{\partial}{\partial x_j}u_k = \epsilon_{ijk}u_{k,j}. \tag{1.4.57}$$

(9) Let S_{kl} and A_{ij} be a symmetric and an anti-symmetric tensor, respectively. Then,

$$S_{ij}A_{ij} = 0. \qquad (1.4.58)$$

Proof

$$S_{ij}A_{ij} = -S_{ij}A_{ji} = -S_{ji}A_{ij} = -S_{ij}A_{ij} = 0. \qquad (1.4.59)$$

In the first equality the asymmetry of A_{ij} was used (see (1.4.31)). To get the second equality the dummy indices i and j were interchanged everywhere. The third equality results from the symmetry of S_{ij}. The last equality is obtained because $S_{ij}A_{ij}$ is a scalar, say α, which implies that $\alpha = -\alpha$, which in turn implies that α must be zero. $\qquad \square$

(10) If B_{ij} is a symmetric tensor, then

$$\epsilon_{ijk}B_{jk} = 0. \qquad (1.4.60)$$

The proof is similar to the one given above.

(11) Prove that

$$\nabla \times \nabla\phi = \mathbf{0}, \qquad (1.4.61)$$

where ϕ is a scalar function and $\mathbf{0}$ is the zero vector.

Proof Let $\nabla\phi = \mathbf{u}$. Then $u_k = \phi_{,k}$ and

$$(\nabla \times \nabla\phi)_i = (\nabla \times \mathbf{u})_i = \epsilon_{ijk}u_{k,j} = \epsilon_{ijk}\left(\phi_{,k}\right)_{,j} = \epsilon_{ijk}\phi_{,kj} = \epsilon_{ijk}\phi_{,jk} = 0. \qquad (1.4.62)$$

The last equality results from the fact that the order of differentiation is irrelevant, so that $\phi_{,kj}$ is symmetric in k and j. Therefore, using (1.4.60) we obtain (1.4.62). $\qquad \square$

(12) Prove that

$$\nabla \cdot \nabla \times \mathbf{u} = 0. \qquad (1.4.63)$$

Proof Let $\nabla \times \mathbf{u} = \mathbf{v}$. Then, $v_i = \epsilon_{ijk}u_{k,j}$ and

$$\nabla \cdot \nabla \times \mathbf{u} = \nabla \cdot \mathbf{v} = v_{i,i} = \left(\epsilon_{ijk}u_{k,j}\right)_{,i} = \epsilon_{ijk}u_{k,ji} = 0. \qquad (1.4.64)$$

Because ϵ_{ijk} does not depend on x_i, its derivatives vanish. The last equality follows from the fact that $u_{k,ji}$ is symmetric in i and j and (1.4.60). $\qquad \square$

(13) A relation between the permutation symbol and the Kronecker delta:

$$\epsilon_{ijr}\epsilon_{pqr} = \delta_{ip}\delta_{jq} - \delta_{iq}\delta_{jp} \qquad (1.4.65)$$

(Goodbody, 1982; Segel, 1977). To remember (1.4.65), note that the indices of the first delta are the first indices of the permutation symbols, the indices of the second delta are the second indices of the permutation symbols, while for the third and fourth deltas the indices are the first and second, and second and first, respectively. The following mnemonic rule may be helpful: *first–first, second–second, first–second, second–first*. This relation is extremely useful, as is shown below.

(14) Prove that

$$\nabla \times (\mathbf{a} \times \mathbf{b}) = \mathbf{b} \cdot \nabla \mathbf{a} + \mathbf{a}(\nabla \cdot \mathbf{b}) - \mathbf{b}(\nabla \cdot \mathbf{a}) - \mathbf{a} \cdot \nabla \mathbf{b}. \qquad (1.4.66)$$

Proof Let $\mathbf{v} = \mathbf{a} \times \mathbf{b}$. Then $v_q = \epsilon_{qjk} a_j b_k$ and

$$(\nabla \times (\mathbf{a} \times \mathbf{b}))_l = (\nabla \times \mathbf{v})_l = \epsilon_{lpq} v_{q,p} = \epsilon_{lpq} \epsilon_{qjk} \left(a_j b_k \right)_{,p}$$

$$= \epsilon_{lpq} \epsilon_{jkq} \left(a_{j,p} b_k + a_j b_{k,p} \right)$$

$$= \left(\delta_{lj} \delta_{pk} - \delta_{lk} \delta_{pj} \right) \left(a_{j,p} b_k + a_j b_{k,p} \right)$$

$$= \delta_{lj} \delta_{pk} a_{j,p} b_k + \delta_{lj} \delta_{pk} a_j b_{k,p} - \delta_{lk} \delta_{pj} a_{j,p} b_k - \delta_{lk} \delta_{pj} a_j b_{k,p}$$

$$= a_{l,p} b_p + a_l b_{p,p} - a_{p,p} b_l - a_p b_{l,p}$$

$$= (\mathbf{b} \cdot \nabla \mathbf{a})_l + a_l (\nabla \cdot \mathbf{b}) - b_l (\nabla \cdot \mathbf{a}) - (\mathbf{a} \cdot \nabla \mathbf{b})_l. \qquad (1.4.67)$$

$\qquad\qquad\qquad\qquad\qquad\qquad\qquad\qquad\qquad\qquad\qquad\qquad\qquad\qquad\qquad \square$

This example shows the power of indicial notation. The conventional proof would have taken much longer. It must be noted however, that some care is needed when manipulating the indices. In addition to choosing the dummy indices in such a way that they repeat only once, it is also important to recognize that l is a free index, so it cannot be replaced by any other letter. Therefore, the effect of δ_{lj} and δ_{lk} is to replace j by l, and k by l, respectively. Doing the opposite would be incorrect. For the other deltas their two subindices are dummy indices, so either of them can replace the other. Finally, to prove (1.4.66) it is necessary to write (1.4.67) as shown on the right-hand side of that equation. $\nabla \cdot \mathbf{a}$ and $\nabla \cdot \mathbf{b}$ are scalars, so the two terms that include them are easy to write. The other two terms, $a_{l,p} b_p$ and $a_p b_{l,p}$ must be interpreted as the scalar product of a tensor and a vector. In both cases the tensor is the *gradient of a vector*, which we have already encountered in (1.4.1), where the strain tensor was defined.

(15) If t_i is a vector and v_{jk} is an arbitrary tensor, then the relation

$$u_{ijk} v_{jk} = t_i \qquad (1.4.68)$$

implies that u_{ijk} is a tensor. The following derivation is based on Santalo (1969). The starting point is

$$u'_{ijk} v'_{jk} = t'_i. \qquad (1.4.69)$$

Next, write t'_i and v'_{jk} in terms of the components in the unprimed system and use (1.4.68). This gives

$$u'_{ijk}a_{jl}a_{ks}v_{ls} = a_{im}t_m = a_{im}u_{mls}v_{ls}, \tag{1.4.70}$$

which in turn implies

$$\left(u'_{ijk}a_{jl}a_{ks} - a_{im}u_{mls}\right)v_{ls} = 0. \tag{1.4.71}$$

Because v_{ls} is arbitrary, we can take all the components but one equal to zero. This means that the factor in parentheses in (1.4.71) must be equal to zero. When all the possible tensors with only one nonzero components are considered we find that the factor in parentheses must always be zero (e.g., Hunter, 1976), which means that

$$u'_{ijk}a_{jl}a_{ks} = a_{im}u_{mls}. \tag{1.4.72}$$

To show that u_{ijk} is a tensor we must verify that it satisfies (1.4.14a). To do that we will contract (1.4.72) with a_{ql} and a_{ps} and then use the orthogonality relation (1.3.19). This gives

$$u'_{ijk}\delta_{jq}\delta_{kq} = u'_{iqp} = a_{im}a_{ql}a_{ps}u_{mls}. \tag{1.4.73}$$

Therefore, u_{ijk} is a tensor.

This result is a special case of the so-called *quotient theorem*, which states that an entity is a tensor if its contraction with an arbitrary tensor gives a tensor (e.g., Goodbody, 1982). Similar arguments will be used in Chapters 4 and 9 to show that entities of interest are tensors.

1.4.6 Diagonalization of a symmetric second-order tensor

Problems involving symmetric second-order tensors (such as the stress or strain tensors) generally become simpler when they are studied in a coordinate system in which the tensors have nonzero elements only when the two indices are equal to each other. In other words, if t_{ij} is symmetric, we want to find a new coordinate system in which t'_{11}, t'_{22}, and t'_{33} are the only nonzero components. When this is achieved, it is said that the tensor has been diagonalized, in analogy with the diagonalization of matrices. However, because we are interested in determining the transformation that diagonalizes the tensor, we will derive our results starting with the transformation law for tensors. The following analysis is based on Lindberg (1983) and Santalo (1969).

From (1.4.10) we know that

$$t'_{ij} = a_{ik}a_{jl}t_{kl}. \tag{1.4.74}$$

At this point the tensor t_{kl} is not required to be symmetric. In (1.4.74) t'_{ij} and the components a_{jl} of **A** (see (1.3.15)) are unknown and must be determined. To do that we start by contracting (see §1.4.1) both sides of (1.4.74) with a_{jm} and then using the orthogonality properties of a_{ij} and the properties of the Kronecker delta discussed in connection with (1.4.11)

$$a_{jm}t'_{ij} = a_{jm}a_{ik}a_{jl}t_{kl} = a_{ik}\delta_{lm}t_{kl} = a_{ik}t_{km}. \qquad (1.4.75)$$

Writing (1.4.75) in full we have

$$a_{ik}t_{km} = a_{jm}t'_{ij} = a_{1m}t'_{i1} + a_{2m}t'_{i2} + a_{3m}t'_{i3}. \qquad (1.4.76)$$

Now let $i = 1, 2, 3$ and recall that only t'_{11}, t'_{22}, and t'_{33} are different from zero. Then

$$a_{1k}t_{km} = a_{1m}t'_{11} \qquad (1.4.77a)$$

$$a_{2k}t_{km} = a_{2m}t'_{22} \qquad (1.4.77b)$$

$$a_{3k}t_{km} = a_{3m}t'_{33}. \qquad (1.4.77c)$$

Let

$$u_k = a_{1k}; \quad v_k = a_{2k}; \quad w_k = a_{3k}; \qquad k = 1, 2, 3. \qquad (1.4.78)$$

u_k, v_k, and w_k can be considered as components of three vectors **u**, **v**, and **w**, respectively. With these definitions (1.4.77a–c) become

$$u_k t_{km} = u_m t'_{11} \qquad (1.4.79a)$$

$$v_k t_{km} = v_m t'_{22} \qquad (1.4.79b)$$

$$w_k t_{km} = w_m t'_{33}. \qquad (1.4.79c)$$

These three equations have the same form

$$z_k t_{km} = \lambda z_m; \qquad m = 1, 2, 3, \qquad (1.4.80)$$

where z is either u, v, or w, and λ is either t'_{11}, t'_{22}, or t'_{33}.

Equation (1.4.80) will be rewritten using the fact that $z_m \equiv \delta_{km}z_k$

$$z_k t_{km} - \delta_{km}\lambda z_k = 0; \qquad m = 1, 2, 3, \qquad (1.4.81)$$

which in turn can be written as

$$(t_{km} - \delta_{km}\lambda)z_k = 0; \qquad m = 1, 2, 3. \qquad (1.4.82)$$

Equation (1.4.82) represents a system of three linear equations in the three unknowns z_1, z_2, and z_3. Furthermore, this system is homogeneous because the right-hand side of each equation is equal to zero. Therefore, for a nonzero solution the determinant of the system must be equal to zero:

$$|t_{km} - \delta_{km}\lambda| = 0. \tag{1.4.83}$$

Writing the determinant in full and expanding we obtain

$$\begin{vmatrix} t_{11} - \lambda & t_{12} & t_{13} \\ t_{21} & t_{22} - \lambda & t_{23} \\ t_{31} & t_{32} & t_{33} - \lambda \end{vmatrix} = -\lambda^3 + A\lambda^2 - B\lambda + C = 0, \tag{1.4.84}$$

where

$$A = t_{ii}, \qquad C = |t_{km}| \tag{1.4.85}$$

and, for symmetric tensors,

$$B = t_{11}t_{22} - t_{12}^2 + t_{22}t_{33} - t_{23}^2 + t_{11}t_{33} - t_{13}^2. \tag{1.4.86}$$

The quantities A, B, and C do not depend on the coordinate system used and are called invariants of t_{ij}. As noted earlier, t_{ii} is known as the trace of t_{ij} (see (1.4.29)).

Equation (1.4.84) is of third order in λ, so it has three roots, λ_1, λ_2, λ_3. These roots are called the *eigenvalues*[2] or *proper values* of t_{ij}, and the three corresponding vectors u_k, v_k, and w_k are called the *eigenvectors* or *proper vectors*. In the general case of a nonsymmetric tensor, the eigenvalues may be real or complex, but for the case of interest to us, symmetric tensors with real components, the eigenvalues are real (Problem 1.14), although not necessarily all different. When they are different, the three eigenvectors are unique and orthogonal to each other (Problem 1.14). If one of the roots is repeated, only the eigenvector corresponding to the distinct eigenvalue is unique. The two other eigenvectors can be chosen so as to form an orthogonal set. When the three eigenvalues are equal to each other (triple root), the tensor is a multiple of the Kronecker delta, and every vector is an eigenvector. A detailed general discussion of these matters can be found in Goodbody (1982). Finally, we note that for symmetric tensors (1.4.80) can be written as

$$t_{mk}z_k = \lambda z_m; \qquad m = 1, 2, 3 \tag{1.4.87}$$

or

$$\mathbf{T}\mathbf{z} = \lambda \mathbf{z}, \tag{1.4.88}$$

where $\mathbf{T}$ is the matrix representation of t_{ij} and $\mathbf{z}$ is the vector with components z_k.

---

[2] Eigen is the German word for proper, in the sense of belonging.

1.4.6.1 Example

Let

$$t_{ij} = \begin{pmatrix} 1 & -1 & -1 \\ -1 & 1 & -1 \\ -1 & -1 & 1 \end{pmatrix}. \tag{1.4.89}$$

To determine the eigenvalues solve

$$\begin{vmatrix} 1-\lambda & -1 & -1 \\ -1 & 1-\lambda & -1 \\ -1 & -1 & 1-\lambda \end{vmatrix} = (1-\lambda)^3 - 3(1-\lambda) - 2 = 0. \tag{1.4.90}$$

The three roots are $\lambda_1 = -1$, and $\lambda_2 = \lambda_3 = 2$.

Now use λ_1 with (1.4.79a), which written in full gives

$$u_1 t_{11} + u_2 t_{21} + u_3 t_{31} = \lambda_1 u_1 \tag{1.4.91a}$$

$$u_1 t_{12} + u_2 t_{22} + u_3 t_{32} = \lambda_1 u_2 \tag{1.4.91b}$$

$$u_1 t_{13} + u_2 t_{23} + u_3 t_{33} = \lambda_1 u_3. \tag{1.4.91c}$$

Because t_{ij} is symmetric, this equation can be written in matrix form as

$$\begin{pmatrix} t_{11}-\lambda_1 & t_{12} & t_{13} \\ t_{21} & t_{22}-\lambda_1 & t_{23} \\ t_{31} & t_{32} & t_{33}-\lambda_1 \end{pmatrix} \begin{pmatrix} u_1 \\ u_2 \\ u_3 \end{pmatrix} = \mathbf{0}. \tag{1.4.92}$$

Replacing t_{ij} by its components and performing the matrix multiplication gives

$$2u_1 - u_2 - u_3 = 0 \tag{1.4.93a}$$

$$-u_1 + 2u_2 - u_3 = 0 \tag{1.4.93b}$$

$$-u_1 - u_2 + 2u_3 = 0. \tag{1.4.93c}$$

Because of the way (1.4.92) was derived, we know that at least one of the three equations above must be a linear combination of the remaining equations (in this example, the last one is equal to the sum of the first two after a change of sign). Therefore, one of the three equations is redundant and can be ignored. Furthermore, two of the three u_i will depend on the remaining one. Therefore, we can write

$$u_2 + u_3 = 2u_1 \tag{1.4.94a}$$

$$2u_2 - u_3 = u_1. \tag{1.4.94b}$$

Solving (1.4.94) for u_2 and u_3 gives $u_2 = u_1$ and $u_3 = u_1$, so that $\mathbf{u} = (u_1, u_1, u_1)$. To find u_1 recall that $\mathbf{u}$ gives the first row of the transformation matrix A (see

(1.4.78)), so that the elements of **u** are constrained by the orthonormality conditions (1.3.19). In particular, from

$$a_{i1}^2 + a_{i2}^2 + a_{i3}^2 = 1 \qquad (1.4.95)$$

(see (1.3.20)), setting $i = 1$ gives

$$a_{11}^2 + a_{12}^2 + a_{13}^2 \equiv u_1^2 + u_2^2 + u_3^2 = |\mathbf{u}|^2 = 3u_1^2 = 1. \qquad (1.4.96)$$

Using (1.4.96) we find that

$$\mathbf{u} = \frac{1}{\sqrt{3}}(1, 1, 1). \qquad (1.4.97)$$

To find the components of the second eigenvector (**v**), (1.4.92) must be used with λ_1 replaced by λ_2. In this case the same equation

$$v_1 + v_2 + v_3 = 0 \qquad (1.4.98)$$

occurs three times. This means that at least one of the components (say v_1) has to be written in terms of the other two. This gives $\mathbf{v} = (-v_2 - v_3, v_2, v_3)$. Note that $\mathbf{u} \cdot \mathbf{v} = 0$ regardless of the values of v_2 and v_3. In other words, **v** is in a plane perpendicular to **u**. Because there are no further constraints on both v_2 and v_3, one can choose the value of one of them, say $v_3 = 0$. Then $\mathbf{v} = (-v_2, v_2, 0)$. To constrain the value of v_2 use (1.4.95) with $i = 2$, which gives

$$a_{21}^2 + a_{22}^2 + a_{23}^2 \equiv v_1^2 + v_2^2 + v_3^2 = 2v_2^2 = 1, \qquad (1.4.99)$$

so that

$$\mathbf{v} = \frac{1}{\sqrt{2}}(-1, 1, 0). \qquad (1.4.100)$$

If the three λ_i were all different, then the third eigenvector would be determined by using (1.4.92) with λ_1 replaced by λ_3. For this example, however, this cannot be done because two of the λ_i are equal to each other. In this case the eigenvector can be found under the condition that it be in the plane perpendicular to **u**, as discussed above, and different from **v**. In particular, we can choose

$$\mathbf{w} = \mathbf{u} \times \mathbf{v} = \frac{1}{\sqrt{6}}(-1, -1, 2). \qquad (1.4.101)$$

If (1.4.95) is written for $i = 3$ we find

$$a_{31}^2 + a_{32}^2 + a_{33}^2 \equiv w_1^2 + w_2^2 + w_3^2 = 1. \qquad (1.4.102)$$

This condition is already satisfied by **w**.

As **u**, **v**, and **w** are the first, second, and third rows of the matrix of the transformation (see (1.4.78)), we can write

$$A = \frac{1}{\sqrt{6}} \begin{pmatrix} \sqrt{2} & \sqrt{2} & \sqrt{2} \\ -\sqrt{3} & \sqrt{3} & 0 \\ -1 & -1 & 2 \end{pmatrix}. \tag{1.4.103}$$

Using (1.4.17) it is easy to verify that by a rotation of coordinates with matrix **A** the tensor t_{ij} becomes diagonal with $t'_{11} = -1$, $t'_{22} = 2$, $t'_{33} = 2$ (Problem 1.15). Also note that $t_{ii} = t'_{ii} = 3$, so that the trace of the tensor remains invariant after the rotation, as indicated after (1.4.86) (see also Problem 1.16).

1.4.7 Isotropic tensors

A tensor is *isotropic* if it has the same components in any coordinate system. The trivial case of a tensor with all of its components equal to zero, which is isotropic, is excluded. Isotropic tensors are important in applications (e.g., elasticity). The following are some relevant results (Goodbody, 1982; Jeffreys and Jeffreys, 1956; Segel, 1977).

(1) There are no isotropic vectors or pseudo vectors.
(2) The only isotropic tensor of second order is $\lambda \delta_{ij}$, where λ is a scalar.
(3) In 3-D space there are no third-order isotropic tensors, and the only isotropic pseudo tensor is $\lambda \epsilon_{ijk}$, with λ a scalar.
(4) The only isotropic tensor of fourth order is

$$c_{ijkl} = \lambda \delta_{ij} \delta_{kl} + \mu \delta_{ik} \delta_{jl} + \nu \delta_{il} \delta_{jk}, \tag{1.4.104}$$

where λ, μ, and ν are scalars. Equation (1.4.104) plays a critical role in the theory of wave propagation in isotropic media.

1.4.8 Vector associated with a second-order anti-symmetric tensor

If W_{ij} is an anti-symmetric tensor, then

$$W_{ij} = -W_{ji} \tag{1.4.105}$$

and

$$W_{JJ} = 0; \qquad J = 1, 2, 3. \tag{1.4.106}$$

There is no summation over uppercase indices. Therefore, the tensor can be described by just three independent components, which can be considered as components of some vector **w** $= (w_1, w_2, w_3)$. We will call **w** the *vector associated with* W_{ij}. Clearly, each w_i is one of W_{12}, W_{13}, or W_{23} (taken as the three independent components of W_{ij}). The question is how to relate the two sets of components. This can be done with the help of the permutation symbol:

$$W_{ij} = \epsilon_{ijk} w_k. \tag{1.4.107}$$

The right-hand side of this expression is zero when $i = j$ and changes sign when i and j are interchanged, thus satisfying the anti-symmetry properties of W_{ij}.

To find w_k in terms of W_{ij} let us write (1.4.107) in full for i and j equal to 1 and 2, and 2 and 1:

$$W_{12} = \epsilon_{123} w_3 = w_3 \tag{1.4.108}$$

$$W_{21} = \epsilon_{213} w_3 = -w_3. \tag{1.4.109}$$

Subtraction of (1.4.109) from (1.4.108) gives

$$w_3 = (W_{12} - W_{21})/2 = W_{12}. \tag{1.4.110}$$

Similar expressions are obtained for the other combinations of i and j:

$$w_1 = (W_{23} - W_{32})/2 = W_{23} \tag{1.4.111}$$

$$w_2 = (W_{31} - W_{13})/2 = -W_{13}. \tag{1.4.112}$$

The left-hand side equalities in (1.4.110)–(1.4.112) can be summarized using the permutation symbol:

$$w_i = \frac{1}{2}\epsilon_{ijk} W_{jk} \tag{1.4.113}$$

(see Problems 1.17–1.19).

1.4.9 Divergence or Gauss' theorem

Let $\mathbf{v}(\mathbf{x})$ be a vector field over a volume V bounded by a closed surface S having outward normal $\mathbf{n}(\mathbf{x})$ and let $\mathbf{v}$ and $\nabla \cdot \mathbf{v}$ be continuous. Then the divergence theorem states that

$$\int_V \nabla \cdot \mathbf{v}\, dV = \int_S \mathbf{v} \cdot \mathbf{n}\, dS \tag{1.4.114}$$

or, in component form,

$$\int_V v_{i,i}\, dV = \int_S v_i n_i\, dS. \tag{1.4.115}$$

The extension of this important theorem to tensors is straightforward. If $T_{ijk...}$ represents an arbitrary tensor field, then

$$\int_V T_{ijk...,p}\, dV = \int_S T_{ijk...} n_p\, dS \tag{1.4.116}$$

(Chou and Pagano, 1967).

1.5 Infinitesimal rotations

These rotations are important in the analysis of strain and therefore they will be analyzed in detail. This section is based on Sokolnikoff (1956).

To introduce the idea of infinitesimal rotation note that the identity matrix (with elements δ_{ij}) represents a rotation of axes that leaves the components of an arbitrary vector unchanged:

$$x_i' = \delta_{ij} x_j = x_i. \tag{1.5.1}$$

An infinitesimal rotation of axes is represented by a matrix whose elements ω_{ij} differ slightly from those of the identity matrix. In other words,

$$\omega_{ij} = \delta_{ij} + \alpha_{ij}; \qquad |\alpha_{ij}| \ll 1. \tag{1.5.2}$$

Because of the condition $|\alpha_{ij}| \ll 1$, infinitesimal rotations have the important property that they commute. This can be seen by considering the effect of an infinitesimal rotation $\delta_{ij} + \beta_{ij}$ on a vector $\mathbf{v}$ obtained by applying another infinitesimal rotation $\delta_{ij} + \alpha_{ij}$ to a vector $\mathbf{u}$. The result is another vector $\mathbf{w}$ with component i given by

$$w_i = (\delta_{ij} + \beta_{ij})v_j = (\delta_{ij} + \beta_{ij})(\delta_{jk} + \alpha_{jk})u_k = (\delta_{ik} + \beta_{ik} + \alpha_{ik} + \beta_{ij}\alpha_{jk})u_k$$

$$= (\delta_{ik} + \beta_{ik} + \alpha_{ik})u_k. \tag{1.5.3}$$

In the final step, the second-order terms $\beta_{ij}\alpha_{jk}$ have been neglected. Since $\beta_{ik} + \alpha_{ik} = \alpha_{ik} + \beta_{ik}$, (1.5.3) shows that the order in which the rotations are applied is irrelevant, and thus that the infinitesimal rotations commute. This property is not shared by finite rotations.

Next, we show that if $\omega_{ij} = \delta_{ij} + \alpha_{ij}$ is an infinitesimal rotation, then $\alpha_{ij} = -\alpha_{ji}$. To see this use (1.3.23) to find the elements of the inverse rotation matrix

$$(\omega^{-1})_{ij} = (\omega^{\mathrm{T}})_{ij} = \delta_{ji} + \alpha_{ji}. \tag{1.5.4}$$

Now use the fact that the product of a matrix and its inverse is equal to the identity matrix:

$$\delta_{ij} = \omega_{ik}(\omega^{-1})_{kj} = (\delta_{ik} + \alpha_{ik})(\delta_{jk} + \alpha_{jk}) = \delta_{ij} + \alpha_{ij} + \alpha_{ji}. \tag{1.5.5}$$

As before, second-order terms have been neglected. This equation is satisfied when

$$\alpha_{ji} = -\alpha_{ij}. \tag{1.5.6}$$

Therefore, α_{ij} is anti-symmetric. This, in turn, implies that the diagonal elements of α_{ij} and ω_{ij} are equal to zero and one, respectively. This result is consistent with the fact that the elements of any rotation matrix are direction cosines, which cannot be larger than one.

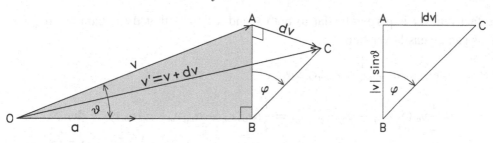

Fig. 1.5. Left: geometry for the infinitesimal rotation of a vector $\mathbf{v}$ about a vector $\mathbf{a}$. The angle between the two vectors is ϑ. The rotated vector is $\mathbf{v}'$. The angle φ is the rotation angle (grossly exaggerated). Right: a detailed view of the triangle ACB on the left. After Hunter (1976).

Now let us consider the vector $\mathbf{v}'$ obtained by the infinitesimal rotation of an arbitrary vector $\mathbf{v}$:

$$v_i' = (\delta_{ij} + \alpha_{ij})v_j = v_i + \alpha_{ij}v_j = v_i + dv_i \qquad (1.5.7)$$

with

$$dv_i = \alpha_{ij}v_j. \qquad (1.5.8)$$

Therefore,

$$\mathbf{v}' = \mathbf{v} + \mathbf{dv}. \qquad (1.5.9)$$

Because rotations preserve vector length (see Problem 1.6), equation (1.5.9) shows that $|\mathbf{dv}|$ has to be much smaller that $|\mathbf{v}|$, so that, to first order, $|\mathbf{v}| = |\mathbf{v}'|$.

To determine the angle between $\mathbf{dv}$ and $\mathbf{v}$ compute their scalar product:

$$\mathbf{dv} \cdot \mathbf{v} = dv_i\, v_i = \alpha_{ij}v_j v_i = \alpha_{ij}v_i v_j = 0, \qquad (1.5.10)$$

with the last equality following from the anti-symmetry of α_{ij} and the symmetry of $v_i v_j$. Therefore, $\mathbf{dv}$ lies in a plane perpendicular to $\mathbf{v}$.

Now write α_{ij} in terms of its associated vector $\mathbf{a}$ (see §1.4.8) and form the scalar product between $\mathbf{dv}$ and $\mathbf{a}$:

$$\mathbf{dv} \cdot \mathbf{a} = dv_i a_i = \alpha_{ij}v_j a_i = \epsilon_{ijk}a_k v_j a_i = 0. \qquad (1.5.11)$$

The last equality derives from the fact that ϵ_{ijk} and $a_k a_i$ are anti-symmetric and symmetric in i and k, respectively. This result shows that $\mathbf{dv}$ is also perpendicular to $\mathbf{a}$, which, in turn, shows that for an infinitesimal rotation the rotation axis is along the vector associated with the anti-symmetric component of the rotation (Fig. 1.5).

The fact that $\mathbf{dv}$ is perpendicular to both $\mathbf{v}$ and $\mathbf{a}$ means that $\mathbf{dv}$ is parallel to $\mathbf{v} \times \mathbf{a}$, as can be easily verified

$$(\mathbf{v} \times \mathbf{a})_l = \epsilon_{lmn} v_m a_n = \frac{1}{2} v_m \epsilon_{lmn} \epsilon_{npq} \alpha_{pq}$$

$$= \frac{1}{2} v_m (\delta_{lp}\delta_{mq} - \delta_{lq}\delta_{mp}) \alpha_{pq} = \frac{1}{2}(\alpha_{lm} v_m - \alpha_{ml} v_m) = \alpha_{lm} v_m = dv_l.$$

$$(1.5.12)$$

Therefore,

$$\mathbf{dv} = \mathbf{v} \times \mathbf{a} \qquad\qquad (1.5.13)$$

and

$$|\mathbf{dv}| = |\mathbf{v}||\mathbf{a}| \sin \vartheta \qquad\qquad (1.5.14)$$

where ϑ is the angle between $\mathbf{v}$ and $\mathbf{a}$.

Finally, we will find the angle of rotation, which is the angle φ between the planes containing $\mathbf{a}$ and $\mathbf{v}$ and $\mathbf{a}$ and $\mathbf{v}'$. The appropriate geometry is shown in Fig. 1.5, from which it follows that

$$\varphi \approx \tan \varphi = \frac{\overline{CA}}{\overline{AB}} = \frac{|\mathbf{dv}|}{|\mathbf{v}| \sin \vartheta} = |\mathbf{a}|. \qquad\qquad (1.5.15)$$

In conclusion, if $\omega_{ij} = \delta_{ji} + \alpha_{ji}$ is an infinitesimal rotation and $\mathbf{a}$ is the vector associated with α_{ji}, then ω_{ij} is equivalent to a rotation of angle $|\mathbf{a}|$ about an axis along $\mathbf{a}$.

1.6 Dyads and dyadics

As noted before, while vectors can be represented by either single symbols or by their components, tensors are defined only by their components. To represent tensors by single symbols Gibbs (Wilson, 1901) introduced dyads and dyadics. The use of dyadics may be convenient when applied to simple problems because it has all the advantages of a symbolic notation. Once the rules that govern the operations between symbols have been established, mathematical derivations are simplified, and the relations between physical entities may become clearer as they are not masked by the subindices that accompany the same relations written in component form. However, when higher-order tensors are involved, the symbolic notation is less convenient because more symbols, and rules for them, are needed to denote new operations between tensors. An additional advantage of dyadics is that relations involving them are independent of the coordinate system, which makes them convenient when we want to write the relations in non-Cartesian systems. It must be noted, however, that this independence of the coordinate system must be demonstrated. For example, we know that the elastic wave equation written in vector form is general, although it will be derived using a Cartesian system, because, as has been noted in §1.1, the gradient, divergence and curl are independent of the coordinate system. On the other hand, the Cartesian expressions for the strain and stress tensors cannot be used in other systems.

Dyadics are heavily used by Ben-Menahem and Singh (1981) and are discussed in detail in Chou and Pagano (1967), Goodbody (1982), Morse and Feshbach (1953), and Nadeau (1964), among others. A number of worked examples are given in Mase (1970). The following discussion is introductory and its purpose is to help the reader get acquainted with the most basic aspects of dyadics, which will be indicated with calligraphic uppercase letters.

1.6.1 Dyads

A *dyad* is an entity obtained from the juxtaposition of two vectors. If **u** and **v** are any two vectors, they generate two dyads, **uv** and **vu**, in general different from each other. The first vector of the dyad is called the antecedent of the dyad, and the second one the consequent. Dyads are defined in terms of their operations on vectors. Let **a**, **b**, and **c** be arbitrary vectors, and let $\mathcal{D}$ be the dyad **ab**. The scalar product of $\mathcal{D}$ and **c** is given by

$$\mathcal{D} \cdot \mathbf{c} = (\mathbf{ab}) \cdot \mathbf{c} = \mathbf{a}(\mathbf{b} \cdot \mathbf{c}) \equiv (\mathbf{b} \cdot \mathbf{c})\mathbf{a}. \tag{1.6.1}$$

The factor in parentheses is the standard scalar product between two vectors, so that it is a scalar. Therefore, $\mathcal{D} \cdot \mathbf{c}$ is a vector.

Given two dyadics $\mathcal{S}$ and $\mathcal{T}$, their sum or difference are new dyads such that

$$(\mathcal{S} \pm \mathcal{T}) \cdot \mathbf{d} = \mathcal{S} \cdot \mathbf{d} \pm \mathcal{T} \cdot \mathbf{d}, \tag{1.6.2}$$

where **d** is an arbitrary vector.

The following are some of the relations satisfied by dyads:

$$\mathbf{a}(\mathbf{b} + \mathbf{c}) = \mathbf{ab} + \mathbf{ac} \tag{1.6.3}$$

$$(\mathbf{a} + \mathbf{b})\mathbf{c} = \mathbf{ac} + \mathbf{bc} \tag{1.6.4}$$

$$(\mathbf{a} + \mathbf{b})(\mathbf{c} + \mathbf{d}) = \mathbf{ac} + \mathbf{ad} + \mathbf{bc} + \mathbf{bd} \tag{1.6.5}$$

$$(\lambda + \mu)\mathbf{ab} = \lambda\mathbf{ab} + \mu\mathbf{ab} \tag{1.6.6}$$

$$(\lambda\mathbf{a})\mathbf{b} = \mathbf{a}(\lambda\mathbf{b}) = \lambda\mathbf{ab}, \tag{1.6.7}$$

where λ and μ are scalars.

It is straightforward to verify these relations. The first one will be used as an example. Because $\mathbf{b} + \mathbf{c}$ is a vector, the left-hand side of (1.6.3) is a dyad. Now compute the scalar product of this dyad with an arbitrary vector $\mathbf{d}$, and apply the properties of the scalar product between vectors:

$$[\mathbf{a}(\mathbf{b} + \mathbf{c})] \cdot \mathbf{d} = \mathbf{a}[(\mathbf{b} + \mathbf{c}) \cdot \mathbf{d}] = \mathbf{a}(\mathbf{b} \cdot \mathbf{d} + \mathbf{c} \cdot \mathbf{d}) = \mathbf{a}(\mathbf{b} \cdot \mathbf{d}) + \mathbf{a}(\mathbf{c} \cdot \mathbf{d})$$

$$= (\mathbf{ab}) \cdot \mathbf{d} + (\mathbf{ac}) \cdot \mathbf{d} = (\mathbf{ab} + \mathbf{ac}) \cdot \mathbf{d}.$$

In the first and last steps, the definitions (1.6.1) and (1.6.2), respectively, were used.

1.6.2 Dyadics

A *dyadic* can be interpreted as a symbolic representation of a second-order tensor. When a dyadic is multiplied scalarly with a vector the result is another vector. If $\mathcal{T}$ and $\mathbf{u}$ are an arbitrary dyadic and vector, then

$$\mathcal{T} \cdot \mathbf{u} = \mathbf{v}, \tag{1.6.8}$$

where $\mathbf{v}$ is a vector generally different from $\mathbf{u}$. In the following we will introduce a number a definitions and properties of dyadics.

(1) *Nonion form and components of a dyadic.* Before discussing the most general case let us consider the dyadic obtained by adding dyads:

$$\mathcal{D} = \sum_{i=1}^{n} \mathbf{u}_i \mathbf{v}_i, \tag{1.6.9}$$

where n is arbitrary finite integer, and the subindex i identifies a vector, not a vector component. The summation symbol is used explicitly to make clear the meaning of the following equations. Now use (1.2.10) to write $\mathbf{u}_i$ and $\mathbf{v}_i$ as

$$\mathbf{u}_i = \sum_{k=1}^{3} (\mathbf{u}_i \cdot \mathbf{e}_k) \mathbf{e}_k \tag{1.6.10}$$

and

$$\mathbf{v}_i = \sum_{l=1}^{3} (\mathbf{v}_i \cdot \mathbf{e}_l) \mathbf{e}_l. \tag{1.6.11}$$

Introducing (1.6.10) and (1.6.11) in (1.6.9) gives

$$\mathcal{D} = \sum_{i=1}^{n} \left[\sum_{k=1}^{3} (\mathbf{u}_i \cdot \mathbf{e}_k) \mathbf{e}_k \right] \left[\sum_{l=1}^{3} (\mathbf{v}_i \cdot \mathbf{e}_l) \mathbf{e}_l \right]. \tag{1.6.12}$$

This expression can be rewritten taking into account that the two factors in parentheses are scalars, interchanging the order of summation, and using (1.6.7). What cannot be done is to change the order of the vectors in the dyad $\mathbf{e}_k\mathbf{e}_l$:

$$\mathcal{D} = \sum_{k,l=1}^{3} \left[\sum_{i=1}^{n} (\mathbf{u}_i \cdot \mathbf{e}_k)(\mathbf{v}_i \cdot \mathbf{e}_l) \right] \mathbf{e}_k\mathbf{e}_l. \tag{1.6.13}$$

Replacing the scalar in parentheses by d_{kl} and returning to the summation convention we finally obtain

$$\mathcal{D} = \sum_{k,l=1}^{3} d_{kl}\mathbf{e}_k\mathbf{e}_l = d_{kl}\mathbf{e}_k\mathbf{e}_l. \tag{1.6.14}$$

This expression shows that the dyadic $\mathcal{D}$ can be written as the sum of nine terms. Equation (1.6.14) is known as the *nonion form* of the dyadic.

The components of $\mathcal{D}$ are obtained as follows. Using (1.6.14) and (1.2.7) we can write

$$\mathbf{e}_m \cdot \mathcal{D} = \mathbf{e}_m \cdot (d_{kl}\mathbf{e}_k\mathbf{e}_l) = d_{kl} \left(\mathbf{e}_m \cdot \mathbf{e}_k \right) \mathbf{e}_l = d_{kl}\delta_{mk}\mathbf{e}_l = d_{ml}\mathbf{e}_l \tag{1.6.15}$$

and

$$\mathbf{e}_m \cdot \mathcal{D} \cdot \mathbf{e}_n = d_{ml}\mathbf{e}_l \cdot \mathbf{e}_n = d_{mn}. \tag{1.6.16}$$

The scalar d_{mn} is the mn component of the dyadic $\mathcal{D}$. As an example let $\mathcal{D} = \mathbf{ab}$. In this case

$$\mathcal{D} = (a_i\mathbf{e}_i)(b_j\mathbf{e}_j) = a_ib_j\mathbf{e}_i\mathbf{e}_j \tag{1.6.17}$$

and the components of $\mathcal{D}$ are given by

$$d_{pq} = a_ib_j\mathbf{e}_p \cdot \mathbf{e}_i\mathbf{e}_j \cdot \mathbf{e}_q = a_pb_q. \tag{1.6.18}$$

Therefore, the dyad $\mathbf{ab}$ corresponds to the outer product of $\mathbf{a}$ and $\mathbf{b}$ introduced in §1.4.1.

As a second example consider the dyadic $\mathcal{D} = \mathbf{ab} - \mathbf{ba}$. From (1.6.18) it is straightforward to find that its components are of the form $a_ib_j - b_ia_j$. Written in matrix form we have

$$\begin{pmatrix} 0 & a_1b_2 - b_1a_2 & a_1b_3 - b_1a_3 \\ a_2b_1 - b_2a_1 & 0 & a_2b_3 - b_2a_3 \\ a_3b_1 - b_3a_1 & a_3b_2 - b_3a_2 & 0 \end{pmatrix}. \tag{1.6.19}$$

As discussed in Problem 1.20, the elements of this matrix are directly related to the elements of $\mathbf{a} \times \mathbf{b}$.

To get the components of a general dyadic $\mathcal{T}$ let the vector $\mathbf{u}$ in (1.6.8) be one of the unit vectors, say $\mathbf{e}_q$. In this case the resulting vector will be written as

$$T \cdot \mathbf{e}_q = t_{1q}\mathbf{e}_1 + t_{2q}\mathbf{e}_2 + t_{3q}\mathbf{e}_3 = t_{kq}\mathbf{e}_k. \qquad (1.6.20)$$

The scalar product of $\mathbf{e}_p$ with the vector in (1.6.20) gives

$$\mathbf{e}_p \cdot T \cdot \mathbf{e}_q = t_{kq}\mathbf{e}_p \cdot \mathbf{e}_k = t_{pq}. \qquad (1.6.21)$$

The right-hand side of (1.6.21) is the pq component of T. This equation should be compared with (1.2.9), which gives the component of a vector.

Equation (1.6.21) shows that T can be written in nonion form

$$T = t_{ij}\mathbf{e}_i\mathbf{e}_j \qquad (1.6.22)$$

as can be verified by finding the components of T:

$$\mathbf{e}_p \cdot T \cdot \mathbf{e}_q = t_{ij}\mathbf{e}_p \cdot \mathbf{e}_i\mathbf{e}_j \cdot \mathbf{e}_q = t_{pq}. \qquad (1.6.23)$$

(2) *Components of the scalar product of a dyadic and a vector.* Given a dyadic T and a vector $\mathbf{v}$, the components of their scalar products are determined from

$$T \cdot \mathbf{v} = t_{ij}\mathbf{e}_i\mathbf{e}_j \cdot \mathbf{v} = t_{ij}\mathbf{e}_i(\mathbf{e}_j \cdot \mathbf{v}) = t_{ij}\mathbf{e}_i v_j = (t_{ij}v_j)\mathbf{e}_i, \qquad (1.6.24)$$

which means that the ith component is given by

$$(T \cdot \mathbf{v})_i = t_{ij}v_j. \qquad (1.6.25)$$

The right-hand side of (1.6.25) can be interpreted as the product of a matrix and a vector.

(3) *The unit dyadic or idempotent.* Is defined by the property that for any given vector $\mathbf{v}$,

$$\mathcal{I} \cdot \mathbf{v} = \mathbf{v} \cdot \mathcal{I} = \mathbf{v}. \qquad (1.6.26)$$

To find an expression for $\mathcal{I}$, write $\mathbf{v}$ as follows:

$$\mathbf{v} = (\mathbf{v} \cdot \mathbf{e}_1)\mathbf{e}_1 + (\mathbf{v} \cdot \mathbf{e}_2)\mathbf{e}_2 + (\mathbf{v} \cdot \mathbf{e}_3)\mathbf{e}_3 = \mathbf{v} \cdot (\mathbf{e}_1\mathbf{e}_1 + \mathbf{e}_2\mathbf{e}_2 + \mathbf{e}_3\mathbf{e}_3). \qquad (1.6.27)$$

Comparison of (1.6.26) and (1.6.27) gives

$$\mathcal{I} = \mathbf{e}_1\mathbf{e}_1 + \mathbf{e}_2\mathbf{e}_2 + \mathbf{e}_3\mathbf{e}_3. \qquad (1.6.28)$$

The components of $\mathcal{I}$ are obtained using (1.6.26)

$$\mathbf{e}_i \cdot \mathcal{I} \cdot \mathbf{e}_j = \mathbf{e}_i \cdot (\mathcal{I} \cdot \mathbf{e}_j) = \mathbf{e}_i \cdot \mathbf{e}_j = \delta_{ij} \qquad (1.6.29)$$

or from (1.6.28). Note that $\mathcal{I}$ and the identity matrix have the same components.

(4) *Components of the sum or subtraction of dyadics.* Given two dyadics S and T, the components of their sum or difference are given by

$$\mathbf{e}_i \cdot (S \pm T) \cdot \mathbf{e}_j = \mathbf{e}_i \cdot S \cdot \mathbf{e}_j \pm \mathbf{e}_i \cdot T \cdot \mathbf{e}_j = s_{ij} \pm t_{ij}. \qquad (1.6.30)$$

(5) *The conjugate of a dyadic.* Given the dyadic $T = t_{ij}\mathbf{e}_i\mathbf{e}_j$, its conjugate, denoted by T_c, is the dyadic

$$T_c = t_{ji}\mathbf{e}_i\mathbf{e}_j = t_{ij}\mathbf{e}_j\mathbf{e}_i. \qquad (1.6.31)$$

The relation between T and T_c is similar to the relation between a matrix and its transpose.

Using the definition (1.6.25) we can show that

$$T_c \cdot \mathbf{v} = \mathbf{v} \cdot T \qquad (1.6.32)$$

(Problem 1.22)

(6) *Symmetric and anti-symmetric dyadics.* A dyadic T is symmetric if

$$T = T_c \qquad (1.6.33)$$

and anti-symmetric if

$$T = -T_c. \qquad (1.6.34)$$

These definitions are analogous to the definitions of symmetric and anti-symmetric second-order tensors given in (1.4.30) and (1.4.31). An example of an anti-symmetric dyadic is $\mathbf{ab} - \mathbf{ba}$.

Given an arbitrary dyadic T, it can be decomposed as the sum of two dyadics, one symmetric and one anti-symmetric:

$$T = \frac{1}{2}(T + T_c) + \frac{1}{2}(T - T_c). \qquad (1.6.35)$$

The first dyadic on the right-hand side is symmetric and the second one is anti-symmetric. Equation (1.6.35) is analogous to (1.4.33)–(1.4.35) for tensors.

Any symmetric dyadic S has six independent components, but by an appropriate rotation of coordinates it is possible to write it in terms of just three components:

$$S = \lambda_1\mathbf{e}_1'\mathbf{e}_1' + \lambda_2\mathbf{e}_2'\mathbf{e}_2' + \lambda_3\mathbf{e}_3'\mathbf{e}_3' \qquad (1.6.36)$$

(Nadeau, 1964). This is the dyadic counterpart of the diagonalization of tensors discussed in §1.4.6. λ_i and $\mathbf{e}_i'$ are the eigenvalues and eigenvectors of S.

Problems

1.1   Give a simple example of a transformation of coordinates for which $|A| = -1$ and show its graphical representation.

1.2   Using the definition of direction cosines, derive the expression for the matrices corresponding to clockwise and counterclockwise rotations of an angle α about the x_2 axis. Draw figures showing all the angles involved.

1.3   (a) Verify (1.3.2).
       (b) Verify (1.3.7).

1.4   Refer to Fig. 1.2. The vector $\mathbf{u}$ has components u_1 and u_2 equal to 0.3 and 0.5, respectively. The axes have been rotated by $40°$. The value of λ is 0.25 for (a) and 1.3 for (b). Find the coordinates of $\mathbf{v}$ and $\mathbf{v}'$ with respect to the unprimed axes for both cases.

1.5   Use index notation to prove the following relations.

(a) $\nabla \cdot (\mathbf{a} \times \mathbf{b}) = \mathbf{b} \cdot (\nabla \times \mathbf{a}) - \mathbf{a} \cdot (\nabla \times \mathbf{b})$
(b) $\nabla \cdot (f\mathbf{a}) = (\nabla f) \cdot \mathbf{a} + f(\nabla \cdot \mathbf{a})$, where f is a scalar function
(c) $\nabla \times (f\mathbf{a}) = (\nabla f) \times \mathbf{a} + f(\nabla \times \mathbf{a})$
(d) $\nabla \times \mathbf{r} = 0$, where $\mathbf{r} = (x_1, x_2, x_3)$
(e) $(\mathbf{a} \cdot \nabla)\mathbf{r} = \mathbf{a}$
(f) $\nabla|\mathbf{r}| = \mathbf{r}/|\mathbf{r}|$.

1.6   Show that orthogonal transformations preserve vector length.

1.7   Show that when n_k is an arbitrary vector the relation

$$(\tau'_{lk} - a_{li}a_{kj}\tau_{ij})n'_k = 0$$

implies that the factor in parentheses must be zero.

1.8   Why is (1.4.40) true?

1.9   Verify that for Cartesian coordinates

$$\nabla^2\mathbf{u} = \nabla(\nabla \cdot \mathbf{u}) - \nabla \times \nabla \times \mathbf{u}.$$

1.10   Show that

$$\epsilon_{ijk}\alpha_{jk} = 0$$

implies that α_{jk} is symmetric.

1.11   Let $\mathbf{B}$ indicate an arbitrary 3×3 matrix having elements b_{ij} and determinant $|\mathbf{B}|$. Using indicial notation show that:

(a) if two rows (or columns) of $\mathbf{B}$ are interchanged, the determinant becomes $-|\mathbf{B}|$;
(b) if $\mathbf{B}$ has two equal rows (or columns), then $|\mathbf{B}| = 0$;
(c) $\epsilon_{lmn}|\mathbf{B}| = \epsilon_{ijk}b_{il}b_{jm}b_{kn} = \epsilon_{ijk}b_{li}b_{mj}b_{nk}$.

1.12 Show that

$$\epsilon_{lmn}a_{kn} = \epsilon_{ijk}a_{il}a_{jm}$$

(after Jeffreys and Jeffreys, 1956).

1.13 Given two vectors **u** and **v**, show that their vector product is a vector (i.e., show that (1.3.10) is satisfied).

1.14 Let T_{ij} be a symmetric second-order tensor with real components.

    (a) Show that the eigenvalues of T_{ij} are real.

    (b) Determine under which conditions the eigenvectors of T_{ij} constitute an orthogonal set.

1.15 Verify that after a rotation of axes with rotation matrix given by (1.4.103) the tensor (1.4.89) becomes diagonal.

1.16 (a) Show that the trace of a second-order tensor is invariant under a rotation of axes.

    (b) Show that if **v** is an eigenvector of the tensor t_{ij} with eigenvalue λ, then after a rotation of axes the vector $\mathbf{v'}$ is an eigenvector of t'_{ij} with eigenvalue λ.

1.17 Verify that (1.4.107) and (1.4.113) are consistent.

1.18 Verify that the w_i given by (1.4.113) is a vector (i.e., show that (1.3.10) is satisfied).

1.19 Let W_{ij} be an anti-symmetric tensor.

    (a) Show that W_{ij} has one real eigenvalue, equal to 0, and two imaginary eigenvalues, equal to $\pm i|\mathbf{w}|$, where **w** is the vector associated with W_{ij} (see §1.4.8).

    (b) Verify that the eigenvector of W_{ij} corresponding to the zero eigenvalue is **w**.

1.20 Verify that the vector associated with the anti-symmetric tensor $a_ib_j - b_ia_j$ is $\mathbf{a} \times \mathbf{b}$.

1.21 Rewrite the matrix of Problem 1.2 for the clockwise rotation assuming that it is infinitesimal. How small should α be (in degrees) if we want to approximate the finite rotation matrix with an infinitesimal one and the approximations involved should not exceed 1% in absolute value.

1.22 Verify that

$$\mathcal{T}_c \cdot \mathbf{v} = \mathbf{v} \cdot \mathcal{T}.$$

2

Deformation. Strain and rotation tensors

2.1 Introduction

Elasticity theory, which lies at the core of seismology, is most generally studied as a branch of continuum mechanics. Although this general approach is not required in introductory courses, it is adopted here for two reasons. First, it affords a number of insights either not provided, or not as easily derivable, when a more restricted approach is used. Secondly, it is essential to solve advanced seismology problems, as can be seen from the discussion in Box 8.5 of Aki and Richards (1980), and references therein, or from the book by Dahlen and Tromp (1998).

Continuum mechanics studies the deformation and motion of bodies ignoring the discrete nature of matter (molecules, atoms, subatomic particles), and "confines itself to relations among gross phenomena, neglecting the structure of the material on a smaller scale" (Truesdell and Noll, 1965, p. 5). In this phenomenological approach a number of basic concepts (among them, stress, strain, and motion) and principles (see Chapter 3) are used to derive general relations applicable, in principle, to all types of media (e.g., different types of solids and fluids). These relations, however, are not enough to solve specific problems. To do that, additional relations, known as *constitutive equations*, are needed. These equations, in turn, are used to define ideal materials, such as perfect fluids, viscous fluids, and perfect elastic bodies (Truesdell and Toupin, 1960). Therefore, it can be argued that although continuum mechanics does not take into consideration the real nature of matter, the latter ultimately expresses itself through the different constitutive laws, and the values that the parameters required to specify a particular material can take.

For elastic solids, for example, the constitutive equation is the relation (1.4.3) between stress and strain. The result of combining this equation with general continuum mechanics relations is the elastic wave equation, but before this equation can be used to solve particular problems it is necessary to specify c_{ijkl}. In the simplest case (isotropic media) two parameters are sufficient, but in the most

40

general case 21 parameters are required. In either case, these parameters depend on the type of media under consideration. In the case of the Earth, in particular, they span a wide range, corresponding to the presence of different types of rocks and minerals, with physical properties that depend on factors such as chemical composition and structure, temperature and pressure.

2.2 Description of motion. Lagrangian and Eulerian points of view

Strain and rotation are two manifestations of *deformation*. When a body undergoes deformation, the relative position of the particles of the body change. To see this, consider the following experiment. On the surface of an inflated balloon mark four points corresponding to the corners of a square. Then squeeze the balloon in the vicinity of the points. The result of this experiment is a change in the relative positions of the four points. Changes in the angles at the corners of the square are also possible. Furthermore, if the squeezing is performed slowly, it will also be observed that the positions of the four points change continuously as time progresses. If we want to describe the change in position of one of the four points mathematically, we have to find a function that describes the evolution of the coordinates of the point. This function will depend on the original location of the point and on the time of observation. Alternatively, we may be interested in finding the original position of one of the points from the knowledge of the position of the point some time after the deformation. The following discussion addresses these questions from a general point of view, although we will concentrate on small deformations because they constitute a basic element in the theory of wave propagation in the Earth.

Figure 2.1 shows a volume V_o inside a body before deformation begins, and an external coordinate system that remains fixed during the deformation. All the positions described below are referred to this system. The use of upper- and lowercase symbols distinguishes undeformed and deformed states; it does not mean that two different coordinate systems are used. $V(t)$ is the volume at time $t_o + t$ occupied by the particles initially in the volume V_o occupied at time t_o.

Let $\mathbf{R} = (X_1, X_2, X_3)$ indicate the vector position corresponding to a particle in the undeformed body, and $\mathbf{r} = (x_1, x_2, x_3)$ the vector position in the deformed body corresponding to the particle that originally was at $\mathbf{R}$ (Fig. 2.1). Vector $\mathbf{R}$ uniquely labels the particles of the body, while vector $\mathbf{r}$ describes the motion of the particles. We will use uppercase letters when referring to particles. Since $\mathbf{R}$ denotes a generic particle initially inside a volume V_o, the vector

$$\mathbf{r} = \mathbf{r}(\mathbf{R}, t) \qquad (2.2.1)$$

represents the motion (deformation) of all the particles that where in V_o before the deformation began.

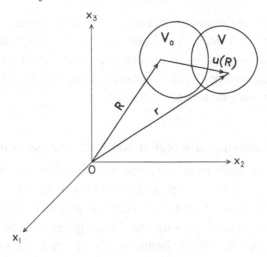

Fig. 2.1. Geometry of deformation. V_o represents a given volume of material that after deformation becomes a new volume V. Vectors $\mathbf{R}$ and $\mathbf{r}$ represent the position of a particle before and after deformation, respectively. The displacement vector $\mathbf{u}(\mathbf{R})$ is the difference between $\mathbf{r}$ and $\mathbf{R}$, and is a function of position in the undeformed state.

In component form (2.2.1) can be written as

$$r_i = x_i = x_i(X_1, X_2, X_3, t); \qquad i = 1, 2, 3. \tag{2.2.2}$$

Furthermore, it is a well-known result from calculus that given $\mathbf{r} = \mathbf{r}(\mathbf{R}, t)$ and assuming that the Jacobian

$$J = \frac{\partial(x_1, x_2, x_3)}{\partial(X_1, X_2, X_3)} = \begin{vmatrix} \dfrac{\partial x_1}{\partial X_1} & \dfrac{\partial x_1}{\partial X_2} & \dfrac{\partial x_1}{\partial X_3} \\[2mm] \dfrac{\partial x_2}{\partial X_1} & \dfrac{\partial x_2}{\partial X_2} & \dfrac{\partial x_2}{\partial X_3} \\[2mm] \dfrac{\partial x_3}{\partial X_1} & \dfrac{\partial x_3}{\partial X_2} & \dfrac{\partial x_3}{\partial X_3} \end{vmatrix} \tag{2.2.3}$$

is different from zero, it is possible to write $\mathbf{R}$ in terms of $\mathbf{r}$:

$$\mathbf{R} = \mathbf{R}(\mathbf{r}, t), \tag{2.2.4}$$

which in component form is written as

$$R_i = X_i = X_i(x_1, x_2, x_3, t); \qquad i = 1, 2, 3. \tag{2.2.5}$$

Equation (2.2.1) corresponds to the so-called *Lagrangian* (or *material*) *description* of motion, while (2.2.4) corresponds to the *Eulerian* (or *spatial*) *description*.

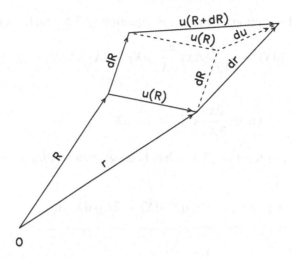

Fig. 2.2. The full description of deformation requires consideration of the changes experienced by the vector dR between two adjacent points. After deformation dR becomes dr. (After Sokolnikoff, 1956.)

We will chose as reference time $t_o = 0$, so that

$$\mathbf{r}(\mathbf{R}, t_o) \equiv \mathbf{R} \qquad (2.2.6)$$

and $\mathbf{r}(\mathbf{R}, t)$ is the current position of the particle that was at $\mathbf{R}$ at time $t = 0$.

In the Lagrangian description we follow the motion of a specified particle $\mathbf{R}$, while in the Eulerian description we are interested in the particle that occupies a given point $\mathbf{r}$ in space at a given time t.

2.3 Finite strain tensors

Let dR and dr be vector line elements corresponding to $\mathbf{R}$ and $\mathbf{r}$

$$\mathbf{dR} = (dX_1, dX_2, dX_3) \qquad (2.3.1)$$

$$\mathbf{dr} = (dx_1, dx_2, dx_3) \qquad (2.3.2)$$

(Fig. 2.2). Vectors dR and dr represent the same two close points before and after deformation, respectively. Their lengths, given by

$$dS = |\mathbf{dR}| = (\mathbf{dR} \cdot \mathbf{dR})^{1/2} = (dX_i \, dX_i)^{1/2} \qquad (2.3.3)$$

and

$$ds = |\mathbf{dr}| = (\mathbf{dr} \cdot \mathbf{dr})^{1/2} = (dx_i \, dx_i)^{1/2} \qquad (2.3.4)$$

will be used to quantify the deformation. Consider the difference

$$(ds)^2 - (dS)^2 = dx_i \, dx_i - dX_k \, dX_k. \qquad (2.3.5)$$

Using the Lagrangian description (see (2.2.1)), equation (2.3.5) can be rewritten as

$$(ds)^2 - (dS)^2 = \frac{\partial x_i}{\partial X_k}\, dX_k \frac{\partial x_i}{\partial X_l}\, dX_l - dX_k\, dX_l\, \delta_{kl}, \qquad (2.3.6)$$

where the chain rule

$$dx_i = \frac{\partial x_i}{\partial X_j}\, dX_j \equiv x_{i,j}\, dX_j \qquad (2.3.7)$$

has been used, and the last term of (2.3.6) has been written so that the two terms can be combined into one:

$$(ds)^2 - (dS)^2 = (x_{i,k} x_{i,l} - \delta_{kl})\, dX_k\, dX_l \equiv 2L_{kl}\, dX_k\, dX_l, \qquad (2.3.8)$$

where

$$L_{kl} = \frac{1}{2}(x_{i,k} x_{i,l} - \delta_{kl}) \qquad (2.3.9)$$

is known as the Green (or Lagrangian) finite strain tensor.

In the Eulerian description (see (2.2.4)) the equivalent result is

$$(ds)^2 - (dS)^2 = (\delta_{kl} - X_{i,k} X_{i,l})\, dx_k\, dx_l \equiv 2E_{kl}\, dx_k\, dx_l, \qquad (2.3.10)$$

where

$$E_{kl} = \frac{1}{2}(\delta_{kl} - X_{i,k} X_{i,l}) \qquad (2.3.11)$$

is known as the Eulerian (or Almansi) finite strain tensor.

We now introduce the *displacement vector* **u**

$$\mathbf{u}(\mathbf{R}, t) = \mathbf{r} - \mathbf{R} \qquad (2.3.12)$$

(Fig. 2.1). An example of displacement is the ground motion felt in an earthquake. In component form (2.3.12) is written as

$$u_i = x_i - X_i. \qquad (2.3.13)$$

Therefore

$$x_i = u_i + X_i \qquad (2.3.14)$$

so that

$$x_{i,k} = \frac{\partial x_i}{\partial X_k} = u_{i,k} + \delta_{ik} \qquad (2.3.15)$$

and

$$L_{kl} = \frac{1}{2}[(u_{i,k} + \delta_{ik})(u_{i,l} + \delta_{il}) - \delta_{kl}] = \frac{1}{2}(u_{l,k} + u_{k,l} + u_{i,k} u_{i,l}). \qquad (2.3.16)$$

Alternatively, it is possible to write

$$X_i = x_i - u_i \qquad (2.3.17)$$

so that

$$X_{i,k} = \frac{\partial X_i}{\partial x_k} = \delta_{ik} - u_{i,k} \qquad (2.3.18)$$

and

$$E_{kl} = \frac{1}{2}(u_{l,k} + u_{k,l} - u_{i,k}u_{i,l}). \qquad (2.3.19)$$

2.4 The infinitesimal strain tensor

Let us assume that the deformation is so small that the products of derivatives in (2.3.16) and (2.3.19) can be neglected. In this case we introduce the *infinitesimal strain tensor*

$$\varepsilon_{kl} = \frac{1}{2}(u_{k,l} + u_{l,k}). \qquad (2.4.1)$$

The assumption of *small deformations* has the effect of making the distinction between Lagrangian and Eulerian descriptions unnecessary, so that

$$\frac{\partial}{\partial x_l} = \frac{\partial}{\partial X_l} \qquad (2.4.2)$$

to first order (Hudson 1980, Segel, 1977). However, we will keep the distinction in the following discussions, as it makes clear what quantities are involved in the deformation.

In the following, reference to the strain tensor will mean the tensor ε_{kl}. That this entity is, in fact, a tensor is shown below. From its definition it is clear that ε_{kl} is symmetric, with diagonal elements $\varepsilon_{JJ} = u_{J,J}$ (no summation over uppercase indices). These elements are known as *normal strains*, while $\varepsilon_{i,j}$ ($i \neq j$) are known as *shearing strains*. Because ε_{kl} is symmetric, it can be diagonalized as discussed in §1.4.6. The eigenvectors of ε_{kl} are known as the *principal directions of strain* and the eigenvalues as the *principal strains*.

Equation (2.4.1) shows how to compute the strain tensor given a differentiable displacement field $\mathbf{u}(\mathbf{R}, t)$. On the other hand, equation (2.4.1) can also be viewed as a system of six equations in three unknowns, u_1, u_2, u_3. Because the number of equations is greater than the number of unknowns, in general (2.4.1) will not have a solution for arbitrary values of the strain components. This raises the following question: what restrictions must be placed on $\varepsilon_{kl}(\mathbf{R}, t)$ to ensure that $\mathbf{u}(\mathbf{R}, t)$ is a single-valued continuous field? (Sokolnikoff, 1956). The answer is that the strain components must satisfy the following equation:

$$\varepsilon_{ij,kl} + \varepsilon_{kl,ij} - \varepsilon_{ik,jl} - \varepsilon_{jl,ik} = 0. \tag{2.4.3}$$

The system (2.4.3) represents $3^4 = 81$ equations, but because some of them are satisfied identically and others are repeated because of the symmetry of ε_{jl}, equation (2.4.3) reduces to six equations, known as *compatibility equations* or *conditions* (Sokolnikoff, 1956; Segel, 1977). Two representative equations are

$$2\varepsilon_{23,23} = \varepsilon_{22,33} + \varepsilon_{33,22} \tag{2.4.4}$$

$$\varepsilon_{33,12} = -\varepsilon_{12,33} + \varepsilon_{23,13} + \varepsilon_{31,23} \tag{2.4.5}$$

(Problem 2.1).

2.4.1 Geometric meaning of ε_{ij}

Here it will be shown that the diagonal and off-diagonal elements of ε_{ij} are related to changes in lengths and angles, respectively, and that the trace of ε_{ij} is related to a change in volume.

We will consider the diagonal elements first. For small deformations, from (2.3.8), (2.3.16), and (2.4.1) we obtain

$$(ds)^2 - (dS)^2 = 2\varepsilon_{kl}\, dX_k\, dX_l. \tag{2.4.6}$$

Furthermore, $ds \approx dS$ (see (2.4.18) below), so that

$$(ds)^2 - (dS)^2 = (ds - dS)(ds + dS) \approx 2\, dS(ds - dS). \tag{2.4.7}$$

Introducing this approximation in (2.4.6) and dividing by $(dS)^2$ gives

$$\frac{ds - dS}{dS} = \varepsilon_{kl}\frac{dX_k}{dS}\frac{dX_l}{dS}. \tag{2.4.8}$$

The left-hand side of (2.4.8) is the change in length per unit length of the original line element dS. In addition, since dX_i is the ith element of $d\mathbf{R}$ and dS is the length of $d\mathbf{R}$, dX_i/dS represents the direction cosines of the line element.

Now assume that the line element is along the X_1 axis, so that

$$\frac{dX_1}{dS} = 1, \qquad \frac{dX_2}{dS} = 0, \qquad \frac{dX_3}{dS} = 0 \tag{2.4.9}$$

and

$$\frac{ds - dS}{dS} = \varepsilon_{11}. \tag{2.4.10}$$

Similarly, if the line element is along the X_2 or X_3 axes, the fractional change is equal to ε_{22} or ε_{33}. Therefore the diagonal elements of the strain tensor, known as *normal strains*, represent the fractional changes in the lengths of the line elements that prior to the deformation were along the coordinate axes.

To analyze the geometrical meaning of the off-diagonal elements, let us consider two line elements $d\mathbf{R}^{(1)} = (dS_1, 0, 0)$ and $d\mathbf{R}^{(2)} = (0, dS_2, 0)$ along the X_1 and X_2 coordinate axes, respectively, and let $d\mathbf{r}^{(1)}$ and $d\mathbf{r}^{(2)}$ be the corresponding line elements in the deformed state. The components of the line elements before and after deformation are related by (2.3.7)

$$dx_i^{(1)} = x_{i,k}\, dX_k^{(1)} = x_{i,1}\, dS_1; \qquad i = 1, 2, 3 \tag{2.4.11}$$

and

$$dx_i^{(2)} = x_{i,k}\, dX_k^{(2)} = x_{i,2}\, dS_2; \qquad i = 1, 2, 3. \tag{2.4.12}$$

Furthermore, using (2.3.15) we find

$$d\mathbf{r}^{(1)} = \left(dx_1^{(1)}, dx_2^{(1)}, dx_3^{(1)} \right) = (1 + u_{1,1}, u_{2,1}, u_{3,1})\, dS_1 \tag{2.4.13}$$

and

$$d\mathbf{r}^{(2)} = \left(dx_1^{(2)}, dx_2^{(2)}, dx_3^{(2)} \right) = (u_{1,2}, 1 + u_{2,2}, u_{3,2})\, dS_2. \tag{2.4.14}$$

Neglecting second-order terms (infinitesimal deformation), the length of $d\mathbf{r}^{(1)}$ is given by

$$ds_1 = (d\mathbf{r}^{(1)} \cdot d\mathbf{r}^{(1)})^{1/2} = (1 + 2u_{1,1})^{1/2}\, dS_1 = (1 + u_{1,1})\, dS_1, \tag{2.4.15}$$

where in the last step only the first two terms in the series expansion of the square root have been retained (Problem 2.2). Similarly,

$$ds_2 = (1 + u_{2,2})\, dS_2. \tag{2.4.16}$$

Note that (2.4.15) can be rewritten as

$$\frac{ds_1 - dS_1}{dS_1} = u_{1,1} = \varepsilon_{11}, \tag{2.4.17}$$

which is the same as (2.4.10).

Also note that (2.4.15) and (2.4.16) show that

$$ds_J \approx dS_J; \qquad J = 1, 2 \tag{2.4.18}$$

provided that $1 \gg |u_{J,J}|$ (no summation over J).

Now consider the scalar product between $d\mathbf{r}^{(1)}$ and $d\mathbf{r}^{(2)}$. Using (2.4.13), (2.4.14), and (2.4.18) and neglecting second-order terms we find

$$d\mathbf{r}^{(1)} \cdot d\mathbf{r}^{(2)} = ds_1\, ds_2 \cos\theta \approx (u_{1,2} + u_{2,1})\, dS_1\, dS_2 \approx 2\varepsilon_{12}\, ds_1\, ds_2, \tag{2.4.19}$$

where θ is the angle between the two line elements (Fig. 2.3). Furthermore, let

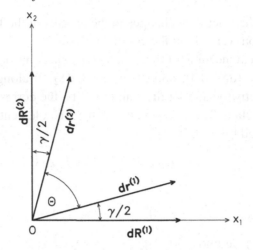

Fig. 2.3. Geometry for the interpretation of the off-diagonal elements of the strain tensor. Two line elements originally along coordinate axes, and thus at right angles with respect to each other, become line elements forming an angle θ.

$\gamma = \pi/2 - \theta$. Since the angle between the two line elements in the undeformed state was $\pi/2$, γ represents the change in the angle between the line elements caused by the deformation. Introducing γ in (2.4.19) and using the approximation $\sin \gamma \approx \gamma$, valid for small deformations, we find

$$\varepsilon_{12} = \frac{1}{2}\gamma. \qquad (2.4.20)$$

Therefore, ε_{12} represents one-half the change in the angle between two line elements that were originally along the X_1 and X_2 axes. Similar interpretations apply to the other nondiagonal elements.

Finally, we will consider an infinitesimal parallelepiped with sides originally along the coordinate axes and lengths equal to dS_1, dS_2, and dS_3. After deformation the corresponding lengths become ds_1, ds_2, and ds_3, respectively. Using (2.4.15) and (2.4.16) and a similar relation for ds_3, the volume of the deformed body is given by

$$ds_1\, ds_2\, ds_3 = (1 + u_{i,i})\, dS_1\, dS_2\, dS_3. \qquad (2.4.21)$$

If V_o and $V_o + dV$ are the volumes of the small elements before and after deformation, then (2.4.21) can be written as

$$\frac{dV}{V_o} = u_{i,i} = \nabla \cdot \mathbf{u} = \varepsilon_{ii}. \qquad (2.4.22)$$

The last equality follows from (2.4.1). This relationship between the fractional change in volume and the divergence of the displacement vector $\mathbf{u}$ is independent of the coordinate system used to derive it, because $u_{i,i}$ is the trace of ε_{ij}, which is a tensor invariant.

2.4.2 *Proof that ε_{ij} is a tensor*

To prove that ε_{ij} is a tensor it is necessary to show that under a rotation of axis the components transform according to

$$\varepsilon'_{kl} = a_{ki}a_{lj}\varepsilon_{ij} \tag{2.4.23}$$

(see (1.4.9)). To prove (2.4.23) let us translate the coordinate axes in such a way that the new origin is at the end of **R**. Then dX_i can be replaced by X_i, where X_i indicates (temporarily) local coordinates measured with respect to the new origin. Then, using (2.4.7), equation (2.4.6) can be written as

$$k \equiv dS(ds - dS) = \varepsilon_{ij}X_iX_j. \tag{2.4.24}$$

Because k is a quantity related to the length of vectors, it is independent of the coordinate system used to describe the deformation. Equation (2.4.24) represents a quadratic function, and is known as the *strain quadric*. After a rotation of the local coordinates the following relations are satisfied:

$$X'_i = a_{ij}X_j \tag{2.4.25}$$

$$X_i = a_{ji}X'_j. \tag{2.4.26}$$

In the rotated system the relation equivalent to (2.4.24) is

$$k = \varepsilon'_{ij}X'_iX'_j. \tag{2.4.27}$$

Therefore, from (2.4.24) and (2.4.27) we obtain

$$\varepsilon_{ij}X_iX_j = \varepsilon'_{ij}X'_iX'_j. \tag{2.4.28}$$

Introducing (2.4.26) in (2.4.28) gives

$$\varepsilon_{ij}a_{ki}X'_ka_{lj}X'_l = \varepsilon'_{kl}X'_kX'_l, \tag{2.4.29}$$

which can be rewritten as

$$(a_{ki}a_{lj}\varepsilon_{ij} - \varepsilon'_{kl})X'_kX'_l = 0. \tag{2.4.30}$$

This equation is valid for arbitrary $X'_kX'_l$, which implies that the term in parentheses has to be equal to zero. Therefore,

$$\varepsilon'_{kl} = a_{ki}a_{lj}\varepsilon_{ij}, \tag{2.4.31}$$

which shows that ε_{ij} is a tensor (Sokolnikoff, 1956).

2.5 The rotation tensor

The strain tensor was introduced by analyzing the change in length of line elements, but it does not represent the whole effect of the deformation. To see this refer to Fig. 2.2. The displacement vector **u** is a function of the coordinates of the point in the body being considered. To make this dependence explicit we will write **u(R)**. In general, the displacement will be different at different points. Therefore, the difference **u(R + dR) − u(R)** fully describes the deformation in the vicinity of **R**. Expanding this difference in a Taylor series under the assumption of small deformations we obtain

$$\mathrm{d}u_i = u_i(\mathbf{R} + \mathrm{d}\mathbf{R}) - u_i(\mathbf{R}) = \frac{\partial u_i}{\partial X_j}\mathrm{d}X_j = u_{i,j}\,\mathrm{d}X_j. \tag{2.5.1}$$

Because of the small-deformation hypothesis the higher-order terms in $\mathrm{d}X_j$ can be neglected.

We know from §1.4.3 that $u_{i,j}$ is a tensor, which can be written as the sum of two tensors, one symmetric and one anti-symmetric, as shown in §1.4.2. Adding and subtracting $u_{j,i}\,\mathrm{d}X_j/2$ to the last term of (2.5.1) we obtain

$$\mathrm{d}u_i = \frac{1}{2}(u_{i,j} + u_{j,i})\,\mathrm{d}X_j + \frac{1}{2}(u_{i,j} - u_{j,i})\,\mathrm{d}X_j = (\varepsilon_{ij} + \omega_{ij})\,\mathrm{d}X_j, \tag{2.5.2}$$

where

$$\omega_{ij} = \frac{1}{2}(u_{i,j} - u_{j,i}). \tag{2.5.3}$$

The tensor ω_{ij} is anti-symmetric, and in view of §1.5 it is likely to be related to some infinitesimal rotation. To show that this is indeed the case note that

$$\mathrm{d}\mathbf{r} = \mathrm{d}\mathbf{R} + \mathrm{d}\mathbf{u} \tag{2.5.4}$$

(see Fig. 2.2) or

$$\mathrm{d}x_i = \mathrm{d}X_i + \mathrm{d}u_i. \tag{2.5.5}$$

Then using (2.5.2) and (2.5.5) and writing $\mathrm{d}X_i = \mathrm{d}X_j\,\delta_{ij}$, we find

$$\mathrm{d}x_i = \mathrm{d}X_i + (\varepsilon_{ij} + \omega_{ij})\,\mathrm{d}X_j = [\varepsilon_{ij} + (\delta_{ij} + \omega_{ij})]\,\mathrm{d}X_j. \tag{2.5.6}$$

From §1.5 we know that $(\delta_{ij} + \omega_{ij})$ represents an infinitesimal rotation. For this reason ω_{ij} is known as the *rotation tensor*. Therefore, the deformation of the line element d**R** consists of two terms, one involving the strain tensor already described, and the other being an infinitesimal rotation of d**R**. It is important to recognize that this is a local rotation, associated with a particular d**R**, not a whole-body rotation.

The vector w_k associated with the anti-symmetric tensor ω_{ij} is

$$w_i = \frac{1}{2}\epsilon_{ijk}\omega_{jk} = \frac{1}{4}(\epsilon_{ijk}u_{j,k} - \epsilon_{ijk}u_{k,j}) = \frac{1}{4}(\epsilon_{ijk}u_{j,k} - \epsilon_{ikj}u_{j,k})$$

$$= -\frac{1}{2}\epsilon_{ikj}u_{j,k} = -\frac{1}{2}(\nabla \times \mathbf{u})_i \qquad (2.5.7)$$

(see (1.4.113)). Here we have used the following facts: $\epsilon_{ijk}u_{k,j} = \epsilon_{ikj}u_{j,k}$ because j and k are dummy indices, and $\epsilon_{ijk} = -\epsilon_{ikj}$.

Equations (2.5.6) and (2.5.7) show that the term $\delta_{ij} + \omega_{ij}$ corresponds to the rotation of the vector element $d\mathbf{R}$ through a small angle $\frac{1}{2}|\nabla \times \mathbf{u}|$ about an axis parallel to $\nabla \times \mathbf{u}$.

Finally, from (2.5.2) and (1.4.107), the contribution of the rotation tensor to du_i can be written as

$$\omega_{ij}\, dX_j = \epsilon_{ijk}w_k\, dX_j = (d\mathbf{R} \times \mathbf{w})_i. \qquad (2.5.8)$$

2.6 Dyadic form of the strain and rotation tensors

To write the strain and rotation tensors in dyadic form first note that

$$u_{i,j} = \frac{\partial}{\partial X_j}u_i = (\nabla\mathbf{u})_{ji}. \qquad (2.6.1)$$

Now we will introduce the dyadic $\mathbf{u}\nabla$, which corresponds to the conjugate of $\nabla\mathbf{u}$. In indicial form,

$$(\mathbf{u}\nabla)_{ji} = (\nabla\mathbf{u})_{ij} = u_{j,i}. \qquad (2.6.2)$$

The dyadic $\mathbf{u}\nabla$ is known as the *displacement gradient*. Using these two dyadics and (2.4.1), the strain tensor can be written in dyadic form as

$$\mathcal{E} = \frac{1}{2}(\mathbf{u}\nabla + \nabla\mathbf{u}). \qquad (2.6.3)$$

When $\mathcal{E}$ is diagonalized, it can be written as

$$\mathcal{E} = \epsilon_1\mathbf{e}_1'\mathbf{e}_1' + \epsilon_2\mathbf{e}_2'\mathbf{e}_2' + \epsilon_3\mathbf{e}_3'\mathbf{e}_3', \qquad (2.6.4)$$

where ϵ_1, ϵ_2, and ϵ_3 and $\mathbf{e}_1'$, $\mathbf{e}_2'$, and $\mathbf{e}_3'$ are the principal strains and principal strain directions of $\mathcal{E}$.

Equation (2.5.1) can be rewritten as

$$du_i = u_{i,j}\, dX_j = (\mathbf{u}\nabla)_{ij}\, dX_j = (\mathbf{u}\nabla \cdot d\mathbf{R})_i \qquad (2.6.5)$$

so that

$$d\mathbf{u} = \mathbf{u}\nabla \cdot d\mathbf{R}. \qquad (2.6.6)$$

In matrix form (2.6.6) is written as

$$\begin{pmatrix} \mathrm{d}u_1 \\ \mathrm{d}u_2 \\ \mathrm{d}u_3 \end{pmatrix} = \begin{pmatrix} u_{1,1} & u_{1,2} & u_{1,3} \\ u_{2,1} & u_{2,2} & u_{2,3} \\ u_{3,1} & u_{3,2} & u_{3,3} \end{pmatrix} \begin{pmatrix} \mathrm{d}X_1 \\ \mathrm{d}X_2 \\ \mathrm{d}X_3 \end{pmatrix}. \tag{2.6.7}$$

By adding and subtracting $\frac{1}{2}\nabla \mathbf{u}$ to the right-hand side of (2.6.6) we find

$$\mathrm{d}\mathbf{u} = \left[\frac{1}{2}(\mathbf{u}\nabla + \nabla\mathbf{u}) + \frac{1}{2}(\mathbf{u}\nabla - \nabla\mathbf{u}) \right] \cdot \mathrm{d}\mathbf{R}. \tag{2.6.8}$$

Note that

$$\frac{1}{2}(\mathbf{u}\nabla - \nabla\mathbf{u})_{ij} = \omega_{ij} \tag{2.6.9}$$

with ω_{ij} as in (2.5.3). Therefore, the rotation tensor can be written in dyadic form as

$$\Omega = \frac{1}{2}(\mathbf{u}\nabla - \nabla\mathbf{u}). \tag{2.6.10}$$

Finally, introducing (2.6.8) in (2.5.4) and using $\mathrm{d}\mathbf{R} \equiv \mathcal{I} \cdot \mathrm{d}\mathbf{R}$, where $\mathcal{I}$ is the unit dyadic, we obtain

$$\mathrm{d}\mathbf{r} = \left[\mathcal{I} + \frac{1}{2}(\mathbf{u}\nabla - \nabla\mathbf{u}) + \frac{1}{2}(\mathbf{u}\nabla + \nabla\mathbf{u}) \right] \cdot \mathrm{d}\mathbf{R} \equiv (\mathcal{R} + \mathcal{E}) \cdot \mathrm{d}\mathbf{R}, \tag{2.6.11}$$

where $\mathcal{R}$ is the dyadic

$$\mathcal{R} = \mathcal{I} + \Omega. \tag{2.6.12}$$

Equations (2.6.3) and (2.6.10) can be used with any orthogonal curvilinear coordinate system. This requires the expression for the components of $\nabla\mathbf{u}$ and $\mathbf{u}\nabla$. The corresponding results for the cylindrical and spherical coordinate systems can be found in Chou and Pagano (1967), Auld (1990), and Ben-Menahem and Singh (1981).

2.7 Examples of simple strain fields

In the following examples we assume small deformations, which imply approximations of the form $\tan\alpha \approx \alpha$.

(1) *Dilatation.* Is defined by

$$\mathbf{r} = \alpha\mathbf{R}, \qquad \alpha > 1 \tag{2.7.1}$$

or

$$x_i = \alpha X_i. \tag{2.7.2}$$

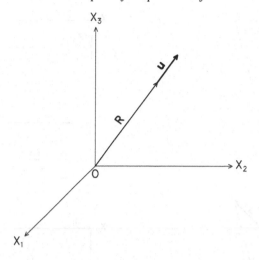

Fig. 2.4. Example of a simple strain field: dilatation. The vector **u** is in the direction of **R**, and the vector **r** (not shown) is equal to **R**+**u**. In the case of a contraction **u** points towards the origin.

The displacement is given by

$$\mathbf{u} = \mathbf{r} - \mathbf{R} = (\alpha - 1)\mathbf{R} \tag{2.7.3}$$

(Fig. 2.4) or

$$u_i = x_i - X_i = (\alpha - 1)X_i. \tag{2.7.4}$$

The strain tensor is given by

$$\varepsilon_{ij} = \frac{1}{2}(u_{i,j} + u_{j,i}) = \frac{\alpha - 1}{2}(X_{i,j} + X_{j,i}) = \frac{\alpha - 1}{2}(\delta_{ij} + \delta_{ji}) = (\alpha - 1)\delta_{ij} \tag{2.7.5}$$

and the rotation tensor by

$$\omega_{ij} = \frac{1}{2}(u_{i,j} - u_{j,i}) = 0 \tag{2.7.6}$$

so that there is no rotation involved in the deformation. Since δ_{ij} is an isotropic tensor, ε_{ij} is also isotropic and its components are the same in any coordinate system. In matrix form we have

$$\mathcal{E} = \begin{pmatrix} \alpha - 1 & 0 & 0 \\ 0 & \alpha - 1 & 0 \\ 0 & 0 & \alpha - 1 \end{pmatrix}. \tag{2.7.7}$$

The fractional volume change is given by

$$\nabla \cdot \mathbf{u} = u_{i,i} = 3(\alpha - 1). \tag{2.7.8}$$

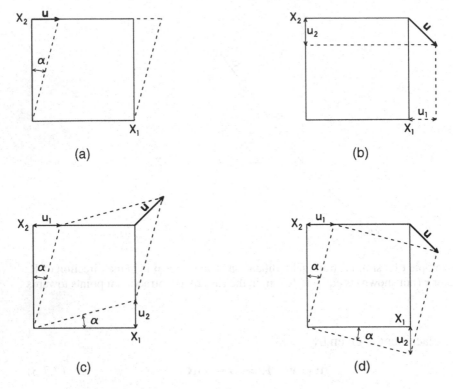

Fig. 2.5. Additional examples of simple strain fields: (a) simple shear; (b) pure shear; (c) another pure shear; (d) pure rotation.

If $\alpha < 1$, the strain field represents a contraction.

(2) *Simple shear.*[1] The deformation is shown in Fig. 2.5(a), and is defined by

$$\mathbf{u} = \alpha X_2 \mathbf{e}_1 \qquad (2.7.9)$$

or

$$u_1 = \alpha X_2; \qquad u_2 = 0; \qquad u_3 = 0. \qquad (2.7.10)$$

The components of the strain and rotation tensors are given by

$$\varepsilon_{12} = \varepsilon_{21} = \frac{\alpha}{2}; \qquad \varepsilon_{ij} = 0, \quad ij \neq 12, 21 \qquad (2.7.11)$$

$$\omega_{12} = -\omega_{21} = \frac{\alpha}{2}; \qquad \omega_{ij} = 0, \quad ij \neq 12, 21. \qquad (2.7.12)$$

---

[1] Some authors call it pure shear and vice versa.

In matrix form (2.7.11) and (2.7.12) become

$$\mathcal{E} = \begin{pmatrix} 0 & \alpha/2 & 0 \\ \alpha/2 & 0 & 0 \\ 0 & 0 & 0 \end{pmatrix}; \qquad \Omega = \begin{pmatrix} 0 & \alpha/2 & 0 \\ -\alpha/2 & 0 & 0 \\ 0 & 0 & 0 \end{pmatrix}. \qquad (2.7.13)$$

As $\nabla \cdot \mathbf{u} = 0$, simple shear does not involve a volume change.

Inspection of $\mathcal{R}$ shows that it corresponds to a rotation of angle $\alpha/2$ about the X_3 axis. This can also be found using (2.5.7),

$$w_i = -\frac{1}{2}(\nabla \times \mathbf{u})_i = \frac{\alpha}{2}\delta_{i3} \qquad (2.7.14)$$

or

$$\mathbf{w} = \frac{\alpha}{2}\mathbf{e}_3 \qquad (2.7.15)$$

(Problem 2.3).

Figure 2.5(a) shows the geometric relations involved. For small deformations, the angle subtended by u_1 is also small, and can be replaced by its tangent, given by $u_1/X_2 = \alpha$. Therefore, half of the change in angle generated by the deformation is caused by the strain tensor, while the other half is caused by the rotation tensor. An example of this deformation is the shearing of a deck of cards.

(3) *Pure shear.* The deformation is shown in Fig. 2.5(b), and is defined by

$$\mathbf{u} = \alpha(X_1\mathbf{e}_1 - X_2\mathbf{e}_2) \qquad (2.7.16)$$

or

$$u_1 = \alpha X_1; \qquad u_2 - -\alpha X_2; \qquad u_3 = 0. \qquad (2.7.17)$$

The components of the strain tensor are given by

$$\varepsilon_{11} = \alpha; \qquad \varepsilon_{22} = -\alpha; \qquad \varepsilon_{ij} = 0, \quad ij \neq 11, 22, \qquad (2.7.18)$$

while the components of the rotation tensor are all zero, so that there is no rotation involved. Also in this case $\nabla \cdot \mathbf{u} = 0$. In matrix form (2.7.18) becomes

$$\mathcal{E} = \begin{pmatrix} \alpha & 0 & 0 \\ 0 & -\alpha & 0 \\ 0 & 0 & 0 \end{pmatrix}. \qquad (2.7.19)$$

Note that the strain tensor is already in diagonal form.

(4) *Another pure shear.* The deformation is shown in Fig. 2.5(c), and is defined by

$$\mathbf{u} = \alpha(X_2\mathbf{e}_1 + X_1\mathbf{e}_2) \qquad (2.7.20)$$

or

$$u_1 = \alpha X_2; \qquad u_2 = \alpha X_1; \qquad u_3 = 0. \qquad (2.7.21)$$

The components of the strain tensor are given by

$$\varepsilon_{12} = \varepsilon_{21} = \alpha; \qquad \varepsilon_{ij} = 0, \quad ij \neq 12, 21. \tag{2.7.22}$$

In matrix form,

$$\mathcal{E} = \begin{pmatrix} 0 & \alpha & 0 \\ \alpha & 0 & 0 \\ 0 & 0 & 0 \end{pmatrix}. \tag{2.7.23}$$

An interesting property of $\mathcal{E}$ is that after a rotation of axes of $45°$ about the X_3 axis it becomes

$$\mathcal{E}' = \begin{pmatrix} \alpha & 0 & 0 \\ 0 & -\alpha & 0 \\ 0 & 0 & 0 \end{pmatrix}, \tag{2.7.24}$$

which is equal to the $\mathcal{E}$ in (2.7.19) (Problem 2.4).

For this deformation all the components of the rotation tensor are equal to zero and $\nabla \cdot \mathbf{u} = 0$.

(5) *Pure rotation.* The deformation is shown in Fig. 2.5(d), and defined by

$$\mathbf{u} = \alpha(X_2\mathbf{e}_1 - X_1\mathbf{e}_2) \tag{2.7.25}$$

or

$$u_1 = \alpha X_2; \qquad u_2 = -\alpha X_1; \qquad u_3 = 0. \tag{2.7.26}$$

The components of the strain tensor are all zero, while the components of the rotation tensor are

$$\omega_{12} = -\omega_{21} = \alpha; \qquad \omega_{ij} = 0, \quad ij \neq 12, 21, \tag{2.7.27}$$

which has the matrix form

$$\Omega = \begin{pmatrix} 0 & \alpha & 0 \\ -\alpha & 0 & 0 \\ 0 & 0 & 0 \end{pmatrix}. \tag{2.7.28}$$

Also in this case $\nabla \cdot \mathbf{u} = 0$.

Problems

2.1 Using the definition of strain tensor, show that (2.4.4) and (2.4.5) are satisfied.

2.2 Verify (2.4.13) and (2.4.15).

2.3 Verify (2.7.14) and (2.7.15).

2.4 Verify that (2.7.24) is obtained by diagonalizing (2.7.23).

2.5 Let ρ_o and ρ be the density of the mass in the volumes V_o and V before and after the deformation (during which the mass is conserved, i.e., is neither gained nor lost). Show that to first order

$$\frac{\rho - \rho_o}{\rho_o} = -\nabla \cdot \mathbf{u}.$$

2.6 Verify that

$$\Omega = \mathcal{I} \times \left(\frac{1}{2}\nabla \times \mathbf{u}\right),$$

where Ω is given by (2.6.10). Use the following definition:

$$\mathbf{a}\mathbf{u} \times \mathbf{v} = \mathbf{a}(\mathbf{u} \times \mathbf{v}),$$

where $\mathbf{a}$, $\mathbf{u}$, and $\mathbf{v}$ are arbitrary vectors. (After Ben-Menahem and Singh, 1981.)

2.7 Consider the *quadratic form*

$$f = \mathbf{x}^T\mathbf{T}\mathbf{x},$$

where $\mathbf{x}$ is a nonzero arbitrary vector and $\mathbf{T}$ is the matrix representation of a second-order tensor t_{ij}. If $f > 0$, $\mathbf{T}$ (and by extension t_{ij}) is said to be *positive definite* (e.g., Noble and Daniel, 1977).

Show that for positive-definite matrices the eigenvalues are positive. The importance of this result is that if $\mathbf{x}$ represents the variables x_1, x_2, x_3, then the quadratic form is the equation of an ellipsoid (e.g., Noble and Daniel, 1977).

2.8 (a) Show that

$$d\mathbf{R} = d\mathbf{r} \cdot (\mathcal{I} - \nabla\mathbf{u}),$$

where the differentiation is with respect to $\mathbf{x}$ (Eulerian description). Here and in the following, higher-order terms will be neglected.

   (b) Consider the set of points on the surface a sphere of radius $|d\mathbf{R}| = D$ centered at point P. Show that

$$D^2 = d\mathbf{R} \cdot d\mathbf{R} = d\mathbf{r} \cdot (\mathcal{I} - 2\mathcal{E}) \cdot d\mathbf{r}.$$

As D is positive, the dyadic in parentheses must be positive definite (see Problem 2.7).

   (c) Assume that $\mathcal{E}$ has been diagonalized. In the rotated system $d\mathbf{r}$ becomes

$$d\mathbf{r}' = dx_1'\,\mathbf{e}_1' + dx_2'\,\mathbf{e}_2' + dx_3'\,\mathbf{e}_3'.$$

Show that in the rotated system the equation in (b) becomes

$$(1 - 2\epsilon_1)(dx_1')^2 + (1 - 2\epsilon_2)(dx_2')^2 + (1 - 2\epsilon_3)(dx_3')^2 = D^2.$$

This is the equation of an ellipsoid in dx_i', known as the *material strain ellipsoid* (e.g., Eringen 1967).

(d) Let V_o and V be the volumes of the sphere of radius D and of the strain ellipsoid. Show that

$$V = \frac{4\pi}{3} D^3 (1 + \epsilon_1 + \epsilon_2 + \epsilon_3)$$

and

$$\frac{V - V_o}{V_o} = \nabla \cdot \mathbf{u}.$$

(After Ben-Menahem and Singh, 1981.)

3

The stress tensor

3.1 Introduction

The development of the theory of elasticity took about two centuries, beginning with Galileo in the 1600s (e.g., Love, 1927; Timoshenko, 1953). The most difficult problem was to gain an understanding of the forces involved in an elastic body. This problem was addressed by assuming the existence of attractive and repulsive forces between the molecules of a body. The most successful of the theories based on this assumption was that of Navier, who in 1821 presented the equations of motion for an elastic isotropic solid (Hudson, 1980; Timoshenko, 1953). Navier's results were essentially correct, but because of the molecular assumptions made, only one elastic constant was required, as opposed to the two that characterize an isotropic solid (see §4.6). Interestingly, the results based on the simple molecular theory used by the earlier researchers can be obtained by setting the ratio of P- to S-wave velocities equal to $\sqrt{3}$ in the more general results derived later. Navier's work attracted the attention of the famous mathematician Cauchy, who in 1822 introduced the concept of stress as we know it today. Instead of considering intermolecular forces, Cauchy introduced the idea of pressure on surfaces internal to the body, with the pressure not perpendicular to the surface, as it would be in the case of hydrostatic pressure. This led to the concept of stress, which is much more complicated than that of strain, and which requires additional continuum mechanics concepts for a full study. The relevant results are summarized below and in several problems. Very readable presentations of this material are provided by Atkin and Fox (1980), Hunter (1976) and Mase (1970).

3.2 Additional continuum mechanics concepts

Let us use the Eulerian description of motion introduced in §2.2, and let $p(\mathbf{r}, t)$ indicate the value of some property of the medium (e.g., pressure, temperature, velocity) at a given point $\mathbf{r}$ at time t. As t varies, different particles (identified by

59

different values of **R**) occupy the same spatial point **r**. Now let us concentrate on a single particle **R**. Using

$$\mathbf{r} = \mathbf{r}(\mathbf{R}, t) \tag{3.2.1}$$

we find

$$P(\mathbf{R}, t) = p(\mathbf{r}(\mathbf{R}, t), t). \tag{3.2.2}$$

Note that, in general, P and p will have different functional forms.[1]

When a body is in motion, the description of the time rate of change of a given property depends on how the motion is described. To motivate the definitions below, consider the following situation (based on Bird *et al.*, 1960). Assume that we are interested in measuring the time rate of change of some property (such as the temperature) of a river as a function of position and time. We can do at least two things. One is to conduct the measurements at a point that remains fixed with respect to the shoreline. This point will have a position given by **r** (in some coordinate system). The *local time rate of change* obtained in this way is the partial derivative of $p(\mathbf{r}, t)$ with respect to t, indicated by $\partial p/\partial t$.

A second thing we can do is to measure the property from a canoe that floats along the river. The canoe (which is representative of a particle in a continuous medium) is identified by the vector **R** (having the same origin as the vector **r** of the fixed point referred to above). The time rate of change determined from these measurements is known as the *material derivative* of P. More specifically, the material time derivative of a quantity P, indicated by DP/Dt, is the time rate of change of P as would be recorded by an observer moving with the particle identified by **R** and is written as

$$\frac{DP}{Dt} = \left.\frac{\partial P(\mathbf{R}, t)}{\partial t}\right|_{\mathbf{R}\text{ fixed}}. \tag{3.2.3}$$

The quantity P represents any scalar, vector, or tensor property of the medium.

If P is expressed in terms of **r** (see (3.2.2)) the material derivative becomes

$$\frac{Dp}{Dt} = \left.\frac{\partial p(\mathbf{r}, t)}{\partial t}\right|_{\mathbf{r}\text{ fixed}} + \left.\frac{\partial p(\mathbf{r}, t)}{\partial x_k}\frac{\partial x_k}{\partial t}\right|_{\mathbf{R}\text{ fixed}}. \tag{3.2.4}$$

Here (3.2.1) and the chain rule of partial derivatives have been used. With two exceptions, in the following the subscripts **r** and **R** and the label 'fixed' will be dropped to simplify the notation.

The first term on the right-hand side of (3.2.4) is the local time rate of change

---

[1] This can be seen with a simple example. Let $f(x, y) = x^2 + y^2$ and introduce the change of variables $x = x(X, Y) = X + \lambda Y$, $y = y(X, Y) = Y$. Then $F(X, Y) = f(x(X, Y), y(X, Y)) = X^2 + (1 + \lambda^2)Y^2 + 2\lambda XY$. Thus, f and F have different functional forms.

defined above. The second term is known as the convective time rate of change, and arises from the motion of the particles in the medium. The material derivative is also known as the substantial derivative.

The *velocity of a particle* is defined as the material time rate of change of the position vector of the particle

$$\mathbf{v} = \frac{D\mathbf{r}}{Dt} = \frac{\partial \mathbf{r}(\mathbf{R}, t)}{\partial t}\bigg|_{\mathbf{R}} \tag{3.2.5}$$

or, in component form,

$$v_k = \frac{\partial x_k}{\partial t}. \tag{3.2.6}$$

As defined, $\mathbf{v}$ is a function of a particular particle (identified by $\mathbf{R}$) and t, which is a material description. Therefore, we should have used $\mathbf{V}$ instead of $\mathbf{v}$, but this distinction is not always made explicitly. With this caveat we can use (2.2.4) and write

$$\mathbf{v} = \mathbf{V}(\mathbf{R}, t) = \mathbf{V}\big(\mathbf{R}(\mathbf{r}, t), t\big) = \mathbf{v}(\mathbf{r}, t), \tag{3.2.7}$$

where $\mathbf{v}(\mathbf{r}, t)$ represents the velocity field in the spatial description. In this description, $\mathbf{v}$ is known for all points in the medium. The particle $\mathbf{R}$ that happens to be at a point $\mathbf{r}$ at a given time t will have velocity $\mathbf{v}(\mathbf{r}, t)$ (Eringen, 1967).

A comparison of (3.2.4) and (3.2.6) shows that the second term on the right-hand side of (3.2.4) is the scalar product of $(\nabla p)_k$ and v_k. Therefore, equation (3.2.4) can be rewritten as

$$\frac{Dp}{Dt} = \frac{\partial p}{\partial t} + (\mathbf{v} \cdot \nabla)p. \tag{3.2.8}$$

Expressing $\mathbf{r}$ in terms of $\mathbf{u}$ (see (2.3.12)) equation (3.2.5) gives

$$\mathbf{v} = \frac{D(\mathbf{u} + \mathbf{R})}{Dt} = \frac{\partial (\mathbf{u}(\mathbf{R}, t) + \mathbf{R})}{\partial t}\bigg|_{\mathbf{R}} = \frac{\partial \mathbf{u}(\mathbf{R}, t)}{\partial t}. \tag{3.2.9}$$

Because $\mathbf{R}$ is independent of time, $\partial \mathbf{R}/\partial t = 0$.

If $\mathbf{u}$ is given in the spatial description, then v_k is obtained using (3.2.4):

$$v_k = \frac{\partial u_k}{\partial t} + \frac{\partial u_k}{\partial x_l}\frac{\partial x_l}{\partial t} = \frac{\partial u_k}{\partial t} + (\mathbf{v} \cdot \nabla)u_k \tag{3.2.10}$$

or, in vector form,

$$\mathbf{v} = \frac{\partial \mathbf{u}}{\partial t} + (\mathbf{v} \cdot \nabla)\mathbf{u}. \tag{3.2.11}$$

Note that the velocity is given in implicit form.

The *acceleration of a particle* is the material time rate of change of the velocity of a particle:

$$\mathbf{a} = \frac{D\mathbf{v}}{Dt}.$$

(3.2.12)

In the Eulerian description we have

$$a_k = \frac{\partial v_k}{\partial t} + \frac{\partial v_k}{\partial x_l}\frac{\partial x_l}{\partial t} = \frac{\partial v_k}{\partial t} + (\mathbf{v} \cdot \nabla)v_k$$

(3.2.13)

or, in vector form,

$$\mathbf{a} = \frac{\partial \mathbf{v}}{\partial t} + (\mathbf{v} \cdot \nabla)\mathbf{v}.$$

(3.2.14)

These definitions of velocity and acceleration will become clear with the example below.

To complete this section it is necessary to add a number of basic definitions and principles, as follows.

Mass. The mass m of a volume V of a body having variable density ρ is given by

$$m = \int_V \rho \, dV.$$

(3.2.15)

Linear momentum:

$$\mathbf{P} = \int_V \rho \mathbf{v} \, dV.$$

(3.2.16)

Angular momentum:

$$\mathbf{M} = \int_V \mathbf{r} \times \rho \mathbf{v} \, dV.$$

(3.2.17)

The last two definitions can be viewed as extensions of similar concepts in classical mechanics.

Conservation of mass:

$$\frac{dm}{dt} = 0.$$

(3.2.18)

Balance of linear momentum:

$$\frac{d\mathbf{P}}{dt} = \text{sum of forces applied to the body.}$$

(3.2.19)

Balance of angular momentum:

$$\frac{d\mathbf{M}}{dt} = \text{sum of torques applied about the origin.}$$

(3.2.20)

The last two principles have classical mechanics counterparts, and are due to Euler, another famous mathematician of the eighteenth century. The balance of linear momentum, in particular, is the equivalent of Newton's second law. It is important to note, however, that these principles should be viewed as axioms justified by the usefulness of the theories based on them (Atkin and Fox, 1980).

3.2.1 Example

Consider the motion given by

$$x_1 = X_1 + atX_2 \tag{3.2.21a}$$

$$x_2 = X_2 \tag{3.2.21b}$$

$$x_3 = (1 + bt)X_3, \tag{3.2.21c}$$

where a and b are constants. Because $\mathbf{r}$ is given as a function of $\mathbf{R}$, equation (3.2.21) corresponds to the Lagrangian description (see §2.2). The corresponding velocity and acceleration are given by

$$\mathbf{v} = \left(\frac{\partial x_1}{\partial t}, \frac{\partial x_2}{\partial t}, \frac{\partial x_3}{\partial t}\right) = (aX_2, 0, bX_3) \tag{3.2.22}$$

$$\mathbf{a} = \left(\frac{\partial^2 x_1}{\partial t^2}, \frac{\partial^2 x_2}{\partial t^2}, \frac{\partial^2 x_3}{\partial t^2}\right) = (0, 0, 0). \tag{3.2.23}$$

To compute the velocity and acceleration in the Eulerian description, first solve (3.2.21) for $\mathbf{R}$:

$$X_1 = x_1 - atx_2 \tag{3.2.24a}$$

$$X_2 = x_2 \tag{3.2.24b}$$

$$X_3 = \frac{x_3}{1 + bt}. \tag{3.2.24c}$$

The velocity can be obtained by straight replacement of (3.2.24) in (3.2.22), which gives

$$\mathbf{v} = \left(ax_2, 0, \frac{bx_3}{1 + bt}\right) \tag{3.2.25}$$

or can be determined using the definition (3.2.10), as follows. First, write the components of the displacement vector,

$$u_1 = x_1 - X_1 = atx_2 \tag{3.2.26a}$$

$$u_2 = x_2 - X_2 = 0 \tag{3.2.26b}$$

$$u_3 = x_3 - X_3 = \frac{btx_3}{1+bt} \tag{3.2.26c}$$

and then apply the definition keeping in mind that $\mathbf{r}$ must be fixed and use

$$u_{1,1} = u_{1,3} = 0; \qquad u_{1,2} = at \tag{3.2.27a}$$

$$u_{2,1} = u_{2,2} = u_{2,3} = 0 \tag{3.2.27b}$$

$$u_{3,1} = u_{3,2} = 0; \qquad u_{3,3} = \frac{bt}{1+bt}. \tag{3.2.27c}$$

This gives

$$v_1 = ax_2 + atv_2 \tag{3.2.28a}$$

$$v_2 = 0 \tag{3.2.28b}$$

$$v_3 = \frac{bx_3}{1+bt} - \frac{b^2tx_3}{(1+bt)^2} + \frac{btv_3}{1+bt}. \tag{3.2.28c}$$

Equations (3.2.28) represent a system of three equations in the three unknowns v_1, v_2, v_3. Introducing (3.2.28b) in (3.2.28a) and solving (3.2.28c) for v_3 gives

$$\mathbf{v} = \left(ax_2, 0, \frac{bx_3}{1+bt} \right), \tag{3.2.29}$$

which is equal to (3.2.25).

To determine the acceleration use (3.2.13) and (3.2.29)

$$\mathbf{a} = \left(0, 0, -\frac{b^2x_3}{(1+bt)^2} \right) + \left(ax_2\frac{\partial}{\partial x_1} + \frac{bx_3}{1+bt}\frac{\partial}{\partial x_3} \right)\left(ax_2, 0, \frac{bx_3}{1+bt} \right)$$

$$= (0, 0, 0). \tag{3.2.30}$$

3.3 The stress vector

In continuum mechanics two different types of forces are recognized, *body forces*, which act at a distance within a body or between bodies, and *surface* (or *contact*) *forces*, which only depend on the surface of contact of either two bodies in contact or any two portions of a body separated by an imaginary surface. Examples of body forces are the gravitational forces, which, as noted, may be the result of the action of the particles within a body, or may originate in another body. The effect of gravitation is generally ignored in wave propagation studies that do not involve the whole Earth. An example of a contact force is the hydrostatic pressure on the surface of a body immersed in a fluid. Other forces, such as the magnetic force, and force distributions, such as surface or volume distributions of couples (i.e., pairs of

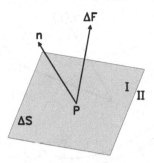

Fig. 3.1. Geometry for the definition of the stress vector. ΔS represent a planar surface element within a body which divides it into two media, indicated by I and II. ΔF is the force exerted by medium I on medium II. The normal to ΔS is **n**, which is usually at an angle with respect to ΔF.

forces in opposite directions) are possible, but they are not required in elasticity studies (e.g., Burridge, 1976).

To introduce the concept of stress vector we follow Love (1927) and Ben-Menahem and Singh (1981). Consider any plane surface ΔS within a body and a point P on the surface. Let **n** be one of the two possible normals to ΔS (Fig. 3.1). The surface divides the body into two portions, to be called media I and II, where medium I contains **n**. Now assume that medium I exerts a force on medium II across the surface ΔS, and that this force is equivalent to a force ΔF acting at P plus a couple ΔC about some axis. Furthermore, assume that as the surface is continuously contracted around P, both ΔF and ΔC go to zero, with the direction of ΔF reaching some limiting direction. Finally, assume that the ratio $\Delta C/\Delta S$ goes to zero while the ratio $\Delta F/\Delta S$ has a finite limit, known as the *stress vector* or *traction*, which can be written as

$$\mathbf{T(n)} = \lim_{\Delta S \to 0} \frac{\Delta F}{\Delta S} = \frac{d\mathbf{F}}{dS}, \tag{3.3.1}$$

where $\mathbf{T(n)}$ is the stress vector at P associated with the normal **n**. Note that changing **n** will change $\mathbf{T(n)}$, and that, in general, $\mathbf{T(n)}$ may depend on the coordinates of the point P and the time t, but to simplify the notation we will not use them explicitly. As **T** is a force per unit area, it has the dimensions of pressure. The projection of **T** on **n** is given by $\mathbf{T} \cdot \mathbf{n}$. If this projection is positive, it corresponds to a tension; if negative to a compression. For the case of hydrostatic pressure, the force and the normal are in opposite directions. For solids, they are usually in different directions. From (3.3.1) we also see that when **T** is given as a function of position, then the force across any infinitesimal surface element dS will be equal to $\mathbf{T}\,dS$.

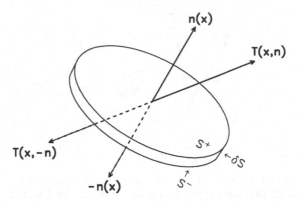

Fig. 3.2. Disk-shaped volume element used to show that $\mathbf{T}(\mathbf{x}, \mathbf{n}) = -\mathbf{T}(\mathbf{x}, -\mathbf{n})$. Here the dependence of $\mathbf{T}$ on position was made explicit. S^+ and S^- indicate the upper and lower surfaces of the disk and δS indicates the lateral surface.

Before proceeding two comments are in order. First, the assumptions that lead to (3.3.1) are justified by the success of the theory of elastic materials (Hudson, 1980). Secondly, we have implicitly used an Eulerian approach, i.e., $\mathbf{n}$ and ΔS are referred to the deformed state. When the undeformed state is used as a reference, the analysis of stress is considerably more complicated. A major difference is that in that case the stress tensor, known as the *Piola–Kirchhoff stress tensor*, is not symmetric (e.g., Aki and Richards, 1980; Atkin and Fox, 1980; Dahlen and Tromp, 1998). These questions, however, can be ignored in most wave propagation applications.

Now we will use the principle of linear momentum (3.2.19) to show that $\mathbf{T}(-\mathbf{n}) = -\mathbf{T}(\mathbf{n})$. Let S be the surface of a body with volume V, and let $\mathbf{f}$ be the body force per unit mass, so that $\rho\mathbf{f}$ is a force per unit volume. To apply (3.2.19) to an arbitrary body of volume V and surface S it is necessary to find the total force on the body, which is equal to the sum of the contributions of the body and surface forces, represented by the integrals below. Then, from (3.2.16) and (3.2.19) we find

$$\frac{\mathrm{d}}{\mathrm{d}t}\int_V \rho\mathbf{v}\,\mathrm{d}V = \int_S \mathbf{T}\,\mathrm{d}S + \int_V \rho\mathbf{f}\,\mathrm{d}V. \tag{3.3.2}$$

Another result we need is

$$\frac{\mathrm{d}}{\mathrm{d}t}\int_V \rho\mathbf{v}\,\mathrm{d}V = \int_V \rho\frac{\mathrm{D}\mathbf{v}}{\mathrm{D}t}\,\mathrm{d}V \tag{3.3.3}$$

(Problem 3.5).

Equations (3.3.2) and (3.3.3) will be applied to a disk-shaped volume having a thickness that will be allowed to go to zero (Fig. 3.2). Under these conditions the volume integrals and the integral over the surface δS go to to zero (Problem 3.6), and (3.3.2) becomes

$$\int_{S^+} \mathbf{T}(\mathbf{n}) \, dS^+ + \int_{S^-} \mathbf{T}(-\mathbf{n}) \, dS^- = 0, \qquad (3.3.4)$$

where S^+ and S^- are the two surfaces of the disk, and $\mathbf{n}$ is the normal to S^+ (Fig. 3.2). As the thickness of the disk goes to zero, S^- approaches S^+, and because S^+ is arbitrary, (3.3.4) implies that

$$\mathbf{T}(-\mathbf{n}) = -\mathbf{T}(\mathbf{n}) \qquad (3.3.5)$$

as long as $\mathbf{T}$ is continuous on S^+ (Hudson, 1980). Equation (3.3.5) is similar to Newton's law of action and reaction.

3.4 The stress tensor

Here we will find a functional relationship between $\mathbf{T}(\mathbf{n})$ and $\mathbf{n}$, which automatically introduces the stress tensor. Consider an infinitesimal tetrahedron with three faces along the coordinate planes (Fig. 3.3). $\mathbf{T}(\mathbf{n})$ is the stress tensor across the plane ABC having a normal $\mathbf{n}$. dS_i is the surface normal to coordinate axis x_i. Therefore, the normal to dS_i is $-\mathbf{e}_i$ and the normal to dS_n is $\mathbf{n}$. Note that all the normals are directed outwards.

Now apply (3.3.2) and (3.3.3) to the tetrahedron. Because the volume integrals go to zero faster than the surface integral as the volume of the tetrahedron goes to zero, we have to consider surface forces only (Problem 3.7). Let S be the surface

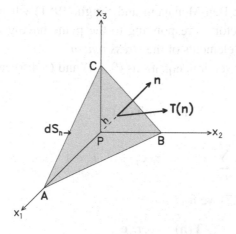

Fig. 3.3. Tetrahedron used to introduce the stress tensor. The surface dS_n corresponds to the face ACB. The vector $\mathbf{n}$ is the normal to dS_n and h is the distance from P to dS_n. The faces BPC, APC, and APB are labeled dS_1, dS_2, dS_3, respectively. The outer normal to dS_i is the vector $-\mathbf{e}_i$.

of the tetrahedron, so that $S = dS_n + dS_1 + dS_2 + dS_3$. Therefore,

$$0 = \int_S \mathbf{T} \, dS = \mathbf{T}(\mathbf{n}) \, dS_n + \mathbf{T}(-\mathbf{e}_1) \, dS_1 + \mathbf{T}(-\mathbf{e}_2) \, dS_2 + \mathbf{T}(-\mathbf{e}_3) \, dS_3. \quad (3.4.1)$$

Strictly speaking, on the right-hand side of (3.4.1) we should have written the integrals over the faces of the tetrahedron, but because they are assumed to be infinitesimal surfaces, each integral can be replaced by the product of the stress tensor at point P (see Fig. 3.3) and the surface area (Atkin and Fox, 1980). Then, using (3.3.5) we obtain

$$\mathbf{T}(\mathbf{n}) \, dS_n = \mathbf{T}(\mathbf{e}_1) \, dS_1 + \mathbf{T}(\mathbf{e}_2) \, dS_2 + \mathbf{T}(\mathbf{e}_3) \, dS_3. \quad (3.4.2)$$

Equation (3.4.2) can be simplified even further because

$$dS_i = (\mathbf{n} \cdot \mathbf{e}_i) \, dS_n = n_i \, dS_n; \qquad i = 1, 2, 3 \quad (3.4.3)$$

(Problem 3.7). Therefore

$$\mathbf{T}(\mathbf{n}) = n_1 \mathbf{T}(\mathbf{e}_1) + n_2 \mathbf{T}(\mathbf{e}_2) + n_3 \mathbf{T}(\mathbf{e}_3) = \sum_{i=1}^{3} n_i \mathbf{T}(\mathbf{e}_i). \quad (3.4.4)$$

The vectors $\mathbf{T}(\mathbf{e}_i)$ can be written in terms of the unit vectors:

$$\mathbf{T}(\mathbf{e}_1) = \tau_{11}\mathbf{e}_1 + \tau_{12}\mathbf{e}_2 + \tau_{13}\mathbf{e}_3$$

$$\mathbf{T}(\mathbf{e}_2) = \tau_{21}\mathbf{e}_1 + \tau_{22}\mathbf{e}_2 + \tau_{23}\mathbf{e}_3 \quad (3.4.5)$$

$$\mathbf{T}(\mathbf{e}_3) = \tau_{31}\mathbf{e}_1 + \tau_{32}\mathbf{e}_2 + \tau_{33}\mathbf{e}_3$$

(e.g., Atkin and Fox, 1980; Ben-Menahem and Singh, 1981) where τ_{ij} is the x_j component of the stress vector corresponding to the plane having $\mathbf{e}_i$ as a normal (Fig. 3.4). τ_{ij} constitute the elements of the *stress tensor*.

Using the summation convention, equations (3.4.4) and (3.4.5) can be rewritten as

$$\mathbf{T}(\mathbf{n}) = n_i \mathbf{T}(\mathbf{e}_i) \quad (3.4.6)$$

$$\mathbf{T}(\mathbf{e}_i) = \sum_{j=1}^{3} \tau_{ij}\mathbf{e}_j = \tau_{ij}\mathbf{e}_j; \qquad i = 1, 2, 3. \quad (3.4.7)$$

Then, from (3.4.6) and (3.4.7) we find

$$\mathbf{T}(\mathbf{n}) = n_i \tau_{ij}\mathbf{e}_j. \quad (3.4.8)$$

The vector $\mathbf{T}(\mathbf{n})$ can also be written as

$$\mathbf{T}(\mathbf{n}) = T_1(\mathbf{n})\mathbf{e}_1 + T_2(\mathbf{n})\mathbf{e}_2 + T_3(\mathbf{n})\mathbf{e}_3 = T_j(\mathbf{n})\mathbf{e}_j \quad (3.4.9)$$

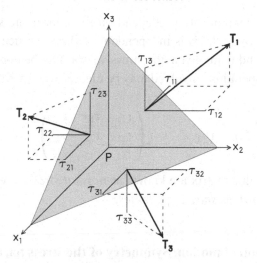

Fig. 3.4. The components of the stress tensor τ_{ij} are the components of the stress vectors $\mathbf{T}(\mathbf{e}_1)$, $\mathbf{T}(\mathbf{e}_2)$, $\mathbf{T}(\mathbf{e}_3)$ (see (3.4.5)). To simplify the figure the three vectors have been written as $\mathbf{T}_1$, $\mathbf{T}_2$, $\mathbf{T}_3$. (After Sokolnikoff, 1956.)

so that comparison with (3.4.8) shows that

$$T_j(\mathbf{n}) = n_i \tau_{ij}; \qquad j = 1, 2, 3. \tag{3.4.10}$$

As for the case of the strain tensor, one should show that τ_{ij} is a tensor, but here this is not necessary because (3.4.10) was used in §1.4 to motivate the definition of a second-order tensor.

Now, using

$$n_i = \mathbf{n} \cdot \mathbf{e}_i \tag{3.4.11}$$

equation (3.4.6) can be rewritten as

$$\mathbf{T}(\mathbf{n}) = (\mathbf{n} \cdot \mathbf{e}_i)\mathbf{T}(\mathbf{e}_i) = \mathbf{n} \cdot \big(\mathbf{e}_i \mathbf{T}(\mathbf{e}_i)\big). \tag{3.4.12}$$

The factor in parentheses is known as the *stress dyadic*:

$$\mathcal{T} = \mathbf{e}_i \mathbf{T}(\mathbf{e}_i), \tag{3.4.13}$$

which can be written in expanded form as follows:

$$\mathcal{T} = \mathbf{e}_1 \mathbf{T}(\mathbf{e}_1) + \mathbf{e}_2 \mathbf{T}(\mathbf{e}_2) + \mathbf{e}_3 \mathbf{T}(\mathbf{e}_3) = \tau_{11}\mathbf{e}_1\mathbf{e}_1 + \tau_{12}\mathbf{e}_1\mathbf{e}_2 + \tau_{13}\mathbf{e}_1\mathbf{e}_3$$

$$+ \tau_{21}\mathbf{e}_2\mathbf{e}_1 + \tau_{22}\mathbf{e}_2\mathbf{e}_2 + \tau_{23}\mathbf{e}_2\mathbf{e}_3 + \tau_{31}\mathbf{e}_3\mathbf{e}_1 + \tau_{32}\mathbf{e}_3\mathbf{e}_2 + \tau_{33}\mathbf{e}_3\mathbf{e}_3. \tag{3.4.14}$$

In terms of the stress dyadic, the stress vector can be written as

$$\mathbf{T}(\mathbf{n}) = \mathbf{n} \cdot \mathcal{T}. \tag{3.4.15}$$

Note that although the components τ_{ij} depend on the coordinate system used, the relationship indicated by (3.4.15) is independent of the reference frame. Also note that when τ_{ij}, or T, and $\mathbf{n}$ are given, the stress vector may be computed using (3.4.10) or (3.4.15). The operations involved may be clearer when written in matrix form:

$$(T_1, T_2, T_3) = (n_1, n_2, n_3) \begin{pmatrix} \tau_{11} & \tau_{12} & \tau_{13} \\ \tau_{21} & \tau_{22} & \tau_{23} \\ \tau_{31} & \tau_{32} & \tau_{33} \end{pmatrix}. \qquad (3.4.16)$$

The diagonal elements of this matrix are known as *normal stresses*, while the off-diagonal elements are known as *shearing stresses*.

3.5 The equation of motion. Symmetry of the stress tensor

Here we will combine several results from the previous sections into the so-called equation of motion, which constitutes one of the most important results of this chapter. In Chapter 4 we will see that this equation leads to the elastic wave equation and in this section it will be used to prove that the stress tensor τ_{ij} is symmetric. From (3.3.2), (3.3.3), and (3.4.10), and Gauss' theorem for tensors (see §1.4.9) we find

$$\int_S n_j \tau_{ji} \, dS + \int_V \rho f_i \, dV = \int_V (\tau_{ji,j} + \rho f_i) \, dV = \int_V \rho \frac{Dv_i}{Dt} \, dV, \qquad (3.5.1)$$

where $\tau_{ji,j}$ is the divergence of τ_{ji}, which is a vector. A good discussion of $\tau_{ji,j}$ and its expression in cylindrical and spherical coordinates is provided by Auld (1990).

The last equality in (3.5.1) can be written as

$$\int_V \left(\tau_{ji,j} + \rho f_i - \rho \frac{Dv_i}{Dt} \right) dV = 0. \qquad (3.5.2)$$

Since (3.5.2) is valid for any arbitrary volume V inside the body, and continuity of the integrand is assumed, we obtain

$$\tau_{ji,j} + \rho f_i = \rho \frac{Dv_i}{Dt}. \qquad (3.5.3)$$

The proof that (3.5.3) follows (3.5.2) is by contradiction. If the integrand were different from zero (say positive) at some point, by continuity it would be positive in a neighborhood of that point. If V is chosen within that neighborhood the integral would be different from zero. Equation (3.5.3) is known as *Euler's equation of motion*. In dyadic form (3.5.3) becomes

$$\nabla \cdot T + \rho \mathbf{f} = \rho \frac{D\mathbf{v}}{Dt}. \qquad (3.5.4)$$

The symmetry of the stress tensor will be proved using the principle of angular momentum, which should be written as follows when body forces are present:

$$\frac{d}{dt}\int_V (\mathbf{r} \times \rho\mathbf{v})\,dV = \int_S \mathbf{r} \times \mathbf{T}\,dS + \int_V (\mathbf{r} \times \rho\mathbf{f})\,dV. \qquad (3.5.5)$$

Writing in component form, collecting terms in one side, introducing τ_{ij}, and using Gauss' theorem and

$$\frac{d}{dt}\int_V \rho\phi\,dV = \int_V \rho\frac{D\phi}{Dt}\,dV, \qquad (3.5.6)$$

where ϕ is any scalar, vector, or tensor property (Problem 3.5) we obtain

$$0 = \int_V \epsilon_{ijk}\left[(x_j\tau_{rk})_{,r} + \rho x_j f_k - \rho\frac{D}{Dt}(x_j v_k)\right]dV$$

$$= \int_V \epsilon_{ijk}\left[(\tau_{jk} + x_j\tau_{rk,r}) + \rho x_j f_k - \rho v_k v_j - \rho x_j\frac{Dv_k}{Dt}\right]dV, \qquad (3.5.7)$$

where the relation $x_{j,r} = \delta_{jr}$ was used. Now using (3.5.3) and the symmetry of $v_k v_j$, equation (3.5.7) becomes

$$\int_V \epsilon_{ijk}\tau_{jk}\,dV = 0. \qquad (3.5.8)$$

Because the integration volume is arbitrary, if the integrand in (3.5.8) is continuous it must to be equal to zero:

$$\epsilon_{ijk}\tau_{jk} = 0, \qquad (3.5.9)$$

which means that τ_{jk} is symmetric,

$$\tau_{jk} = \tau_{kj} \qquad (3.5.10)$$

(see Problem 1.10).

Equation (3.5.10) can be used to rewrite (3.4.10) as

$$T_j(\mathbf{n}) = \tau_{ji}n_i; \qquad j = 1, 2, 3, \qquad (3.5.11)$$

so that the matrix equation (3.4.16) becomes

$$\begin{pmatrix} T_1 \\ T_2 \\ T_3 \end{pmatrix} = \begin{pmatrix} \tau_{11} & \tau_{12} & \tau_{13} \\ \tau_{21} & \tau_{22} & \tau_{23} \\ \tau_{31} & \tau_{32} & \tau_{33} \end{pmatrix}\begin{pmatrix} n_1 \\ n_2 \\ n_3 \end{pmatrix}. \qquad (3.5.12)$$

Equation (3.5.11) can also be written in dyadic form:

$$\mathbf{T} = \mathcal{T}\cdot\mathbf{n}. \qquad (3.5.13)$$

3.6 Principal directions of stress

Since the stress tensor is real and symmetric, the vectors $\mathbf{e}_i$ can be rotated so that $\tau_{ij} = 0$ if $i \neq j$, i.e., only τ_{11}, τ_{22}, and τ_{33} may differ from zero (see §1.4.6). We will use τ_1, τ_2, and τ_3 for the three diagonal elements. Confusion should not arise by the use of the same symbol with a different number of subindices because the two sets of quantities will never be used in the same equation. Let $\mathbf{a}_1$, $\mathbf{a}_2$, and $\mathbf{a}_3$ be the unit vectors in the rotated system. In this system

$$\mathcal{T} = \tau_1 \mathbf{a}_1 \mathbf{a}_1 + \tau_2 \mathbf{a}_2 \mathbf{a}_2 + \tau_3 \mathbf{a}_3 \mathbf{a}_3. \tag{3.6.1}$$

In matrix form $\mathcal{T}$ is written as

$$\begin{pmatrix} \tau_1 & 0 & 0 \\ 0 & \tau_2 & 0 \\ 0 & 0 & \tau_3 \end{pmatrix}. \tag{3.6.2}$$

In the new (rotated) system the stress vectors are perpendicular to the coordinate planes and the shear stresses vanish on these planes. The directions $\mathbf{a}_1$, $\mathbf{a}_2$, and $\mathbf{a}_3$ are known as *principal directions of stress*, and the planes normal to the principal directions of stress are known as *principal planes of stress*. In the rotated system the stress vector has a simple expression:

$$\begin{pmatrix} T_1 \\ T_2 \\ T_3 \end{pmatrix} = \begin{pmatrix} \tau_1 & 0 & 0 \\ 0 & \tau_2 & 0 \\ 0 & 0 & \tau_3 \end{pmatrix} \begin{pmatrix} n_1 \\ n_2 \\ n_3 \end{pmatrix} = \begin{pmatrix} n_1 \tau_1 \\ n_2 \tau_2 \\ n_3 \tau_3 \end{pmatrix}. \tag{3.6.3}$$

3.7 Isotropic and deviatoric components of the stress tensor

The stress tensor τ_{ij} will be written as the sum of two tensors, one isotropic and the other with zero trace:

$$\tau_{ij} \equiv P\delta_{ij} + (\tau_{ij} - P\delta_{ij}) = P\delta_{ij} + \sigma_{ij}, \tag{3.7.1}$$

where P is a scalar and

$$\sigma_{ij} = \tau_{ij} - P\delta_{ij}. \tag{3.7.2}$$

Because τ_{ij} and δ_{ij} are symmetric, so is σ_{ij}. To determine P set the trace of σ_{ij} equal to zero:

$$0 = \sigma_{ii} = \tau_{ii} - P\delta_{ii} = \tau_{ii} - 3P. \tag{3.7.3}$$

Therefore,

$$P = \frac{\tau_{ii}}{3} = \frac{\tau_{11} + \tau_{22} + \tau_{33}}{3}. \tag{3.7.4}$$

In matrix form (3.7.1) is written as

$$
\begin{pmatrix} \tau_{11} & \tau_{12} & \tau_{13} \\ \tau_{21} & \tau_{22} & \tau_{23} \\ \tau_{31} & \tau_{32} & \tau_{33} \end{pmatrix} = \begin{pmatrix} P & 0 & 0 \\ 0 & P & 0 \\ 0 & 0 & P \end{pmatrix} + \begin{pmatrix} \tau_{11} - P & \tau_{12} & \tau_{13} \\ \tau_{21} & \tau_{22} - P & \tau_{23} \\ \tau_{31} & \tau_{32} & \tau_{33} - P \end{pmatrix}.
$$

$$(3.7.5)$$

In dyadic form the equivalent relation is

$$
\mathcal{T} = P\mathcal{I} + (\mathcal{T} - P\mathcal{I}) = P\mathcal{I} + \Sigma, \tag{3.7.6}
$$

where

$$
\Sigma = \mathcal{T} - P\mathcal{I}. \tag{3.7.7}
$$

The tensor $P\delta_{ij}$ is known as the *isotropic* or *hydrostatic*, or *spherical part* of the stress tensor. In a fluid, the stress tensor is

$$
\tau_{ij} = -P\delta_{ij}, \tag{3.7.8}
$$

where P is the *hydrostatic pressure* (a positive number).

The tensor σ_{ij} is known as the *deviatoric part* of the stress tensor. To find the principal directions of σ_{ij} use (1.4.87) and (3.7.1), and let v_i and λ indicate the eigenvector and the eigenvalue, respectively. This gives

$$
\tau_{ij}v_j = \lambda v_i = P\delta_{ij}v_j + \sigma_{ij}v_j, \tag{3.7.9}
$$

which implies that

$$
\sigma_{ij}v_j = (\lambda - P)v_i. \tag{3.7.10}
$$

Therefore, τ_{ij} and σ_{ij} have the same principal directions, while the eigenvalues of σ_{ij} are given by $\lambda - P$, with λ equal to τ_1, τ_2, τ_3. Because in the rotated system

$$
P = \frac{\tau_1 + \tau_2 + \tau_3}{3} \tag{3.7.11}
$$

the eigenvalues of σ_{ij} are

$$
\lambda_1 = \frac{1}{3}(2\tau_1 - \tau_2 - \tau_3); \qquad \lambda_2 = \frac{1}{3}(2\tau_2 - \tau_1 - \tau_3) \qquad \lambda_3 = \frac{1}{3}(2\tau_3 - \tau_1 - \tau_2).
$$

$$(3.7.12)$$

3.8 Normal and shearing stress vectors

Here we will decompose the stress vector $\mathbf{T}(\mathbf{n})$ into two parts, one, $\mathbf{T}^N$, in the direction of $\mathbf{n}$ and the other, $\mathbf{T}^S$, perpendicular to $\mathbf{n}$ (Fig. 3.5). $\mathbf{T}^N$ is known as the *normal stress*, and $\mathbf{T}^S$ as the *shearing stress*. The three vectors are on the same plane. From Fig. 3.5 we see that

$$
|\mathbf{T}^N| = \big||\mathbf{T}|\cos\alpha\big| = |\mathbf{T}\cdot\mathbf{n}|. \tag{3.8.1}
$$

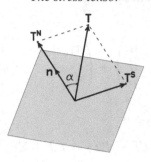

Fig. 3.5. Decomposition of the stress vector **T** into normal (**T**^{N}) and shearing (**T**^{S}) components.

Therefore,

$$\mathbf{T}^N = (\mathbf{T} \cdot \mathbf{n})\mathbf{n}. \tag{3.8.2}$$

On the other hand,

$$\mathbf{T} = \mathbf{T}^N + \mathbf{T}^S \tag{3.8.3}$$

so that

$$\mathbf{T}^S = \mathbf{T} - \mathbf{T}^N \tag{3.8.4}$$

and

$$|\mathbf{T}^S| = \big||\mathbf{T}| \sin\alpha\big| = |\mathbf{n} \times \mathbf{T}|. \tag{3.8.5}$$

Equation (3.8.5) does not mean that $\mathbf{T}^S$ is equal to $\mathbf{n} \times \mathbf{T}$ because the latter is perpendicular to both $\mathbf{n}$ and the direction of $\mathbf{T}^S$. However, $\mathbf{n} \times \mathbf{T} \times \mathbf{n}$ is in the direction of $\mathbf{T}^S$ and also has the right absolute value (given in (3.8.5)), so that

$$\mathbf{T}^S = \mathbf{n} \times \mathbf{T} \times \mathbf{n}. \tag{3.8.6}$$

To verify that this expression is correct first rewrite (3.8.4) in indicial form using (3.8.2),

$$T_k^S = T_k - T_k^N = n_l \tau_{lk} - n_k(\mathbf{T} \cdot \mathbf{n}) = n_l \tau_{lk} - n_k n_l \tau_{li} n_i = n_l(\tau_{lk} - n_k n_i \tau_{li}), \tag{3.8.7}$$

where (3.5.11) has been used twice. Next write (3.8.6) in indicial form,

$$(\mathbf{n} \times \mathbf{T} \times \mathbf{n})_k = \epsilon_{kij}(\mathbf{n} \times \mathbf{T})_i n_j = \epsilon_{kij}\epsilon_{ipq} n_p T_q n_j = -\epsilon_{ikj}\epsilon_{ipq} n_p n_j n_l \tau_{lq}$$

$$= -(\delta_{kp}\delta_{jq} - \delta_{kq}\delta_{jp}) n_p n_j n_l \tau_{lq} = n_j n_j n_l \tau_{lk} - n_k n_j n_l \tau_{lj}$$

$$= n_l(\tau_{lk} - n_k n_i \tau_{li}), \tag{3.8.8}$$

because $n_j n_j = |\mathbf{n}|^2 = 1$. Since the right-hand sides of (3.8.7) and (3.8.8) are equal to each other, equation (3.8.6) is indeed correct.

3.9 Stationary values and directions of the normal and shearing stress vectors

Given a stress tensor τ_{ij} and a point P within the stressed medium, there are an infinite number of stress vectors, each of them associated with one of the infinitely many plane surface elements through P. Each surface element has an associated normal vector $\mathbf{n}$, and now the question we want to ask is whether there are special directions along which the absolute values of the normal and shearing stress vectors have maximum and/or minimum values. Therefore, the problem is to find the vector or vectors $\mathbf{n}$ that render $\mathbf{T}^N$ and $\mathbf{T}^S$ stationary. The only restriction on $\mathbf{n}$ is that its absolute value is equal to 1. There will be no restrictions on τ_{ij} except for two special cases discussed at the end of this section. The analysis will be carried out in the principal axes coordinate system, as in this system all the equations are simpler. Working in this preferred system does not introduce any restriction; it is always possible to go back to the original coordinate system by applying appropriate rotations. This section is based on Sokolnikoff (1956).

Let

$$\mathbf{n} = n_1\mathbf{a}_1 + n_2\mathbf{a}_2 + n_3\mathbf{a}_3 \tag{3.9.1}$$

be the unit normal to a plane through some point P inside the medium. Then, using (3.6.3) and (3.8.1)

$$\mathbf{T} \cdot \mathbf{n} \equiv \tau_N = (n_1\tau_1, n_2\tau_2, n_3\tau_3) \cdot (n_1, n_2, n_3) = \tau_1 n_1^2 + \tau_2 n_2^2 + \tau_3 n_3^2. \tag{3.9.2}$$

Now use the relation

$$|\mathbf{n}|^2 = n_1^2 + n_2^2 + n_3^2 = 1 \tag{3.9.3}$$

to solve for n_1^2, which, after substitution in (3.9.2) gives

$$\tau_N = \tau_1(1 - n_2^2 - n_3^2) + \tau_2 n_2^2 + \tau_3 n_3^2. \tag{3.9.4}$$

We are interested in determining the maximum or minimum values of τ_N. To do that we will determine the values of n_2 and n_3 for which τ_N is stationary. In other words, we are interested in the values of n_2 and n_3 that satisfy the following conditions:

$$\frac{\partial \tau_N}{\partial n_2} = 0; \qquad \frac{\partial \tau_N}{\partial n_3} = 0. \tag{3.9.5}$$

This gives

$$\frac{\partial \tau_N}{\partial n_2} = -2n_2\tau_1 + 2n_2\tau_2 = 0 \tag{3.9.6}$$

$$\frac{\partial \tau_N}{\partial n_3} = -2n_3\tau_1 + 2n_3\tau_3 = 0. \tag{3.9.7}$$

Equations (3.9.6) and (3.9.7) imply $n_2 = n_3 = 0$, so that

$$\mathbf{n} = (\pm 1, 0, 0); \qquad \tau_N = \tau_1. \tag{3.9.8}$$

Solving (3.9.3) for n_2 and n_3, and repeating the previous steps gives

$$\mathbf{n} = (0, \pm 1, 0); \qquad \tau_N = \tau_2 \tag{3.9.9}$$

and

$$\mathbf{n} = (0, 0, \pm 1); \qquad \tau_N = \tau_3. \tag{3.9.10}$$

Equations (3.9.8)–(3.9.10) show that the principal directions are those along which the normal stresses are stationary.

Now we will determine the stationary values of the shearing stresses. From (3.8.3) we know that

$$|\mathbf{T}|^2 = |\mathbf{T}^N|^2 + |\mathbf{T}^S|^2. \tag{3.9.11}$$

Therefore, using (3.6.3) and (3.9.2),

$$|\mathbf{T}^S|^2 \equiv \tau_S^2 = |\mathbf{T}|^2 - |\mathbf{T}^N|^2 = \tau_1^2 n_1^2 + \tau_2^2 n_2^2 + \tau_3^2 n_3^2 - (\tau_1 n_1^2 + \tau_2 n_2^2 + \tau_3 n_3^2)^2. \tag{3.9.12}$$

Equation (3.9.12) is subject to the condition that $|\mathbf{n}^2| = n_i n_i = 1$. To find the stationary values of τ_S^2 the method of Lagrange multipliers will be used. To do that let

$$\phi = n_i n_i - 1 \tag{3.9.13}$$

and determine the stationary values of the new function

$$F = \tau_S^2 + \lambda \phi, \tag{3.9.14}$$

where λ is an unknown scalar to be determined. Now set $\partial F / \partial n_i = 0$ and use $|\mathbf{n}| = 1$. This gives

$$\frac{\partial F}{\partial n_i} = 2n_i \left[\tau_i^2 - 2(\tau_1 n_1^2 + \tau_2 n_2^2 + \tau_3 n_3^2) \tau_i + \lambda \right] = 0; \qquad i = 1, 2, 3. \tag{3.9.15}$$

We will solve (3.9.15) by examining a number of possibilities. To do that note that (3.9.15) can be written as

$$n_i f(n_i, \tau_i, \lambda) = 0, \tag{3.9.16}$$

with f representing the expression within brackets. An obvious solution is $\mathbf{n} = (0, 0, 0)$, but it does not satisfy the constraint $|\mathbf{n}^2| = 1$.

Let us investigate the possibility that two of the n_i are equal to zero. Assume that $\mathbf{n} = (\pm 1, 0, 0)$. Replacing this expression in (3.9.15) and (3.9.12) gives $\lambda = \tau_1^2$ and $\tau_S^2 = 0$. The latter is an expected result because $\mathbf{n} = (\pm 1, 0, 0)$ is normal to one of the principal planes of stress, on which τ_S is known to be zero. Similar results are obtained when $\mathbf{n} = (0, \pm 1, 0)$ or $\mathbf{n} = (0, 0, \pm 1)$.

Another possibility is that only one of the n_i is equal to zero. Let us assume that $n_1 = 0$. In this case we find

$$n_2^2 + n_3^2 = 1 \tag{3.9.17}$$

$$\tau_2^2 - 2(\tau_2 n_2^2 + \tau_3 n_3^2)\tau_2 + \lambda = 0 \tag{3.9.18}$$

$$\tau_3^2 - 2(\tau_2 n_2^2 + \tau_3 n_3^2)\tau_3 + \lambda = 0. \tag{3.9.19}$$

Equation (3.9.17) results from the constraint on $|\mathbf{n}|$, and (3.9.18) and (3.9.19) from the fact that the function f must be zero when $n_i \neq 0$.

Equations (3.9.17)–(3.9.19) represent a system of three equations and three unknowns (n_2^2, n_3^2, and λ). The solution is

$$n_2^2 = n_3^2 = \frac{1}{2}; \qquad \lambda = \tau_2 \tau_3. \tag{3.9.20}$$

Then

$$\mathbf{n} = \left(0, \pm\frac{1}{\sqrt{2}}, \pm\frac{1}{\sqrt{2}} \right) \tag{3.9.21}$$

and from (3.9.12)

$$\tau_S^2 = \frac{1}{2}(\tau_2^2 + \tau_3^2) - \frac{1}{4}(\tau_2 + \tau_3)^2 = \frac{1}{4}\tau_2^2 + \frac{1}{4}\tau_3^2 - \frac{1}{2}\tau_2\tau_3 = \frac{1}{4}(\tau_2 - \tau_3)^2 \tag{3.9.22}$$

so that

$$\tau_S = \pm\frac{1}{2}(\tau_2 - \tau_3) \tag{3.9.23}$$

and

$$|\tau_S| = \frac{1}{2}|\tau_2 - \tau_3|. \tag{3.9.24}$$

Since τ_S and τ_S^2 are not functions of $\mathbf{n}$, their derivatives with respect to n_i are equal to zero. Therefore, $\mathbf{n}$ given by (3.9.21) represents a stationary direction.

If instead of $n_1 = 0$ we had chosen $n_2 = 0$ or $n_3 = 0$, the corresponding results would have been:

$$\mathbf{n} = \left(\pm\frac{1}{\sqrt{2}}, 0, \pm\frac{1}{\sqrt{2}} \right); \qquad |\tau_S| = \frac{1}{2}|\tau_1 - \tau_3|. \tag{3.9.25}$$

$$\mathbf{n} = \left(\pm\frac{1}{\sqrt{2}}, \pm\frac{1}{\sqrt{2}}, 0 \right); \qquad |\tau_S| = \frac{1}{2}|\tau_1 - \tau_2|. \tag{3.9.26}$$

If now we assume that $\tau_1 > \tau_2 > \tau_3$, then $(\tau_1 - \tau_3)/2$ is the greatest shearing stress at P. This stress is one-half of the difference between the greatest and least

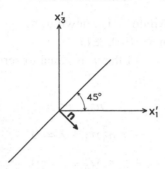

Fig. 3.6. For a given stress tensor with principal values τ_1, τ_2, τ_3, the greatest shearing stress acts on planes that contain the x_2' axis and bisect the x_1' and x_3' axes. The line at 45° is in the (x_1', x_3') plane and represents one of the planes.

principal stresses. It acts on planes that contain $\mathbf{a}_2$ and that bisect the right angles between the directions of these two principal stresses (Fig. 3.6).

Now let us go back to (3.9.16). We want to investigate the possibility that the three n_i are different from zero. As we will see, this occurs for some special values of τ_1, τ_2, and τ_3. When the three n_i are all nonzero we can write

$$\tau_1 n_1^2 + \tau_2 n_2^2 + \tau_3 n_3^2 = \frac{1}{2} \frac{(\lambda + \tau_1^2)}{\tau_1} \tag{3.9.27}$$

$$\tau_1 n_1^2 + \tau_2 n_2^2 + \tau_3 n_3^2 = \frac{1}{2} \frac{(\lambda + \tau_2^2)}{\tau_2} \tag{3.9.28}$$

$$\tau_1 n_1^2 + \tau_2 n_2^2 + \tau_3 n_3^2 = \frac{1}{2} \frac{(\lambda + \tau_3^2)}{\tau_3}. \tag{3.9.29}$$

The left-hand sides of (3.9.27)–(3.9.29) are equal to each other, so that there is no solution if the three τ_i are all different. However, if $\tau_1 = \tau_2 = \tau_3 = \sqrt{\lambda}$, then the three equations are satisfied identically. Also note that for this selection of τ_i, $\tau_S \equiv 0$ (see (3.9.12)). Furthermore, because there are no constraints on the components of $\mathbf{n}$, any direction is perpendicular to some surface on which shearing stresses are zero. Also note that in the rotated system the stress tensor is proportional to δ_{ij}, so that the tensor is isotropic and has the same components in any coordinate system.

Another solution exists when two of the τ_i are equal to each other. For example, if $\tau_1 = \tau_2$, equations (3.9.27) and (3.9.28) can be written as

$$\tau_1 (n_1^2 + n_2^2) + \tau_3 n_3^2 = \frac{1}{2} \frac{(\lambda + \tau_1^2)}{\tau_1} = \frac{1}{2} \left(\frac{\lambda}{\tau_1} + \tau_1 \right) \tag{3.9.30}$$

and (3.9.29) as

$$\tau_1(n_1^2 + n_2^2) + \tau_3 n_3^2 = \frac{1}{2}\frac{(\lambda + \tau_3^2)}{\tau_3} = \frac{1}{2}\left(\frac{\lambda}{\tau_3} + \tau_3\right). \tag{3.9.31}$$

It can be seen by inspection that the right-hand sides of (3.9.30) and (3.9.31) will be equal if $\lambda = \tau_1 \tau_3$. Introducing this expression in (3.9.30) gives

$$\tau_1(n_1^2 + n_2^2) + \tau_3 n_3^2 = \frac{1}{2}\tau_1 + \frac{1}{2}\tau_3. \tag{3.9.32}$$

Equation (3.9.32) is satisfied if

$$n_1^2 + n_2^2 = \frac{1}{2}; \qquad n_3^2 = \frac{1}{2}. \tag{3.9.33a,b}$$

The solution given by (3.9.33) satisfies the constraint $|\mathbf{n}| = 1$. Note that (3.9.33a) represents an infinite number of directions. This implies that there are an infinite number of directions $\mathbf{n}$ normal to a surface on which the shearing stress is a maximum. Introducing (3.9.33) and $\tau_1 = \tau_2$ in (3.9.12) gives

$$|\tau_S| = \frac{1}{2}|\tau_1 - \tau_3|. \tag{3.9.34}$$

3.10 Mohr's circles for stress

Here we will show that once the principal values of the stress tensor have been found, the normal and shear stress vectors can be determined graphically using a simple geometrical construction based on three circles with radii related to the differences of principal values. These circles are very important in rock fracture and earthquake faulting studies (e.g. Ramsey, 1967; Scholz, 1990; Yeats *et al.*, 1997) and this is why they are studied here. The starting point is the following relations:

$$n_1^2 + n_2^2 + n_3^2 = 1 \tag{3.10.1}$$

$$\tau_N = \tau_1 n_1^2 + \tau_2 n_2^2 + \tau_3 n_3^2 \tag{3.10.2}$$

$$|\mathbf{T}|^2 = \tau_N^2 + \tau_S^2 = n_1^2 \tau_1^2 + n_2^2 \tau_2^2 + n_3^2 \tau_3^2. \tag{3.10.3}$$

Next, we will solve this system of equations for n_1^2, n_2^2, and n_3^2. This will be done through the following steps, based on Sokolnikoff (1956).

Solve (3.10.1) for n_1^2,

$$n_1^2 = 1 - n_2^2 - n_3^2. \tag{3.10.4}$$

Introduce (3.10.4) in (3.10.2),

$$\tau_N = \tau_1 + n_2^2(\tau_2 - \tau_1) + n_3^2(\tau_3 - \tau_1). \tag{3.10.5}$$

This implies that

$$n_3^2 = \frac{\tau_N - \tau_1 - n_2^2(\tau_2 - \tau_1)}{\tau_3 - \tau_1}. \tag{3.10.6}$$

Introducing (3.10.4) in (3.10.3) and using (3.10.6) gives

$$\tau_N^2 + \tau_S^2 = \tau_1^2 + n_2^2(\tau_2^2 - \tau_1^2) + n_3^2(\tau_3^2 - \tau_1^2)$$

$$= \tau_1^2 + n_2^2(\tau_2^2 - \tau_1^2) + (\tau_3 + \tau_1)\left[\tau_N - \tau_1 - n_2^2(\tau_2 - \tau_1)\right]$$

$$= \tau_1^2 + n_2^2(\tau_2^2 - \tau_1^2) + (\tau_3 + \tau_1)(\tau_N - \tau_1) - n_2^2(\tau_2 - \tau_1)(\tau_3 + \tau_1)$$

$$= n_2^2(\tau_2 - \tau_1)[\tau_2 + \tau_1 - (\tau_3 + \tau_1)] + \tau_3\tau_N - \tau_3\tau_1 + \tau_1\tau_N. \tag{3.10.7}$$

After rearrangement of terms (3.10.7) becomes

$$n_2^2(\tau_2 - \tau_1)(\tau_2 - \tau_3) = \tau_N^2 + \tau_S^2 - \tau_3\tau_N + \tau_3\tau_1 - \tau_1\tau_N$$

$$= \tau_N(\tau_N - \tau_3) - \tau_1(\tau_N - \tau_3) + \tau_S^2$$

$$= (\tau_N - \tau_1)(\tau_N - \tau_3) + \tau_S^2. \tag{3.10.8}$$

Therefore,

$$n_2^2 = \frac{(\tau_N - \tau_1)(\tau_N - \tau_3) + \tau_S^2}{(\tau_2 - \tau_1)(\tau_2 - \tau_3)}. \tag{3.10.9}$$

The equivalent relations for n_1^2 and n_3^2 are

$$n_1^2 = \frac{(\tau_N - \tau_2)(\tau_N - \tau_3) + \tau_S^2}{(\tau_1 - \tau_2)(\tau_1 - \tau_3)} \tag{3.10.10}$$

$$n_3^2 = \frac{(\tau_N - \tau_1)(\tau_N - \tau_2) + \tau_S^2}{(\tau_3 - \tau_1)(\tau_3 - \tau_2)}. \tag{3.10.11}$$

Equations (3.10.9)–(3.10.11) are the basis for the Mohr circles of stress in the (τ_N, τ_S) plane.

Now consider (3.10.10). Because $n_1^2 \geq 0$ and we have chosen $\tau_1 > \tau_2 > \tau_3$, which implies that the denominator is positive, we can write

$$(\tau_N - \tau_2)(\tau_N - \tau_3) + \tau_S^2 \geq 0. \tag{3.10.12}$$

Operating on the left-hand side of (3.10.12) gives

$$\tau_N^2 - \tau_N(\tau_2 + \tau_3) + \tau_2\tau_3 + \tau_S^2 = \left[\tau_N - \frac{1}{2}(\tau_2 + \tau_3)\right]^2 - \frac{1}{4}(\tau_2 + \tau_3)^2 + \tau_2\tau_3 + \tau_S^2$$

$$= \left[\tau_N - \frac{1}{2}(\tau_2 + \tau_3)\right]^2 - \left(\frac{\tau_2 - \tau_3}{2}\right)^2 + \tau_S^2 \geq 0.$$

$$\tag{3.10.13}$$

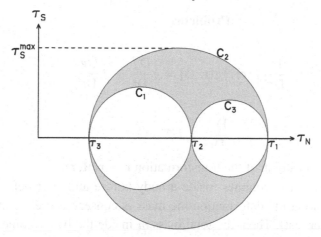

Fig. 3.7. Mohr's circles for stress. For a given stress tensor with principal values τ_1, τ_2, τ_3, the range of values that the normal (τ_N) and shearing (τ_S) components of the stress vector can take is given by the shaded area.

This implies that

$$\left[\tau_N - \frac{1}{2}(\tau_2 + \tau_3)\right]^2 + \tau_S^2 \geq \left(\frac{\tau_2 - \tau_3}{2}\right)^2. \qquad (3.10.14)$$

Equation (3.10.14) represents a circle with center at $[(\tau_2 + \tau_3)/2, 0]$ and radius $(\tau_2 - \tau_3)/2$ (Fig. 3.7, circle C_1). The $\geq$ sign indicates points on or outside C_1.

From (3.10.9), since $n_2^2 \geq 0$ and the denominator is negative we find

$$(\tau_N - \tau_3)(\tau_N - \tau_1) + \tau_S^2 \leq 0, \qquad (3.10.15)$$

which can be rewritten as

$$\left[\tau_N - \frac{1}{2}(\tau_1 + \tau_3)\right]^2 + \tau_S^2 \leq \left(\frac{\tau_1 - \tau_3}{2}\right)^2. \qquad (3.10.16)$$

Equation (3.10.16) represents the points inside or on the circle C_2 in Fig. 3.7.

From (3.10.11), since $n_3^2 \geq 0$ and the denominator is positive we obtain

$$\left[\tau_N - \frac{1}{2}(\tau_1 + \tau_2)\right]^2 + \tau_S^2 \geq \left(\frac{\tau_1 - \tau_2}{2}\right)^2. \qquad (3.10.17)$$

Equation (3.10.17) represents the points on or outside the circle C_3 in Fig. 3.7.

For a given stress tensor, the range of values that τ_N and τ_S can take is given by the shaded areas between the circles (Fig. 3.7). Note that τ_S cannot exceed $(\tau_1 - \tau_3)/2$.

Problems

3.1   Verify that

$$\frac{D}{Dt}[f(\mathbf{r}, t)g(\mathbf{r}, t)] = g\frac{Df}{Dt} + f\frac{Dg}{Dt}.$$

3.2   Prove that

$$\frac{DJ}{Dt} = J(\nabla \cdot \mathbf{v}),$$

where J is the Jacobian of the transformation $\mathbf{r} = \mathbf{r}(\mathbf{R}, t)$.

3.3   Let V_o and V be the volumes inside a body before and after deformation (see §2.2). During the deformation the mass is conserved (i.e., there is no mass gained or lost). Then, the total mass m inside the two volumes is the same. Mathematically this is expressed as follows:

$$m = \int_{V_o} \rho_o(\mathbf{R}, 0) \, dV_o = \int_V \rho(\mathbf{r}, t) \, dV.$$

Use this result to show that

$$\rho(\mathbf{r}, t)J = \rho_o(\mathbf{R}, 0).$$

3.4   Verify the following equations:

(a)

$$\frac{D\rho}{Dt} + \rho\nabla \cdot \mathbf{v} = 0$$

(b)

$$\frac{\partial\rho}{\partial t} + \nabla \cdot (\rho\mathbf{v}) = 0,$$

where ρ and $\mathbf{v}$ are functions of $\mathbf{x}$ and t. Each of these equations is known as an *equation of continuity*, and is a consequence of the conservation of mass.

3.5   Prove that

$$\frac{d}{dt}\int_V \rho(\mathbf{r}, t)\phi(\mathbf{r}, t) \, dV = \int_V \rho(\mathbf{r}, t)\frac{D\phi(\mathbf{r}, t)}{Dt} \, dV,$$

where ϕ is any scalar, vector, or tensor quantity. In particular, ϕ may be the velocity $\mathbf{v}$.

3.6   Refer to Fig. 3.2. Introduce (3.3.3) in (3.3.2) and then show that the volume integrals and the integral over the lateral surface δS go to zero as the thickness of the disk goes to zero.

3.7   Refer to Fig. 3.3.

(a) Show that the volume of the tetrahedron is given by

$$V = \frac{1}{3} h \, dS_n.$$

(b) Show that the integral of body forces goes to zero faster than the integral of surface forces as h (and V) tend to zero.

(c) Show that $dS_i = (\mathbf{n} \cdot \mathbf{e}_i) \, dS_n = n_i \, dS_n$.

3.8   Consider the stress tensor τ_{ij} given by

$$\begin{pmatrix} 1 & 0 & 0 \\ 0 & -1 & 1 \\ 0 & 1 & -1 \end{pmatrix}.$$

(a) Determine the stress vector acting on the plane which intersects the x_1, x_2, and x_3 axes at 3, 1, and 1, respectively.

(b) Compute $\mathbf{T}^N$ and $\mathbf{T}^S$, and verify that $\mathbf{T}^S$ lies in the plane determined in (a).

(c) Draw the Mohr circles for this tensor.

4
Linear elasticity – the elastic wave equation

4.1 Introduction

In the previous chapter we introduced the idea of small deformation, which allowed us to neglect the distinction between the Lagrangian and the Eulerian description. Now we will apply the small-deformation hypothesis to the equation of motion (3.5.3). The resulting equation will include spatial derivatives of the stress tensor, the acceleration of the displacement, and body forces. The displacement, in turn, is related to the strain tensor via (2.4.1). Therefore, we have two systems of equations, one for stress and displacement and one for strain and displacement. Within the approximations that have been introduced, these equations are valid for any continuous medium. To apply them to a specific type of medium (e.g., solid, viscous fluid) it is necessary to establish a general relation (known as a constitutive equation) between stress and strain.

In the case of solids, when a body is subjected to external forces it becomes deformed (strained), and internal stresses are generated within the body. The relation between stress and strain depends on the nature of the deformation and other external factors, such as the temperature. If the deformation is such that the deformed body returns to its original state after the force that caused the deformation is removed, then the deformation is said to be *elastic*. If this is not the case, i.e., if part of the deformation remains, the deformation is known as *plastic*. Clearly, whether a deformation is elastic or not depends on the magnitude of the force applied to the body, and the nature of the body. It is a matter of fact that a given force applied to say steel will produce a deformation different from that produced when applied to a completely different material, rubber, for example. In the following we will be concerned with *linear elastic solids*, for which a linear relation between stress and strain exists (generalized Hooke's law). In addition we will show that Hooke's law is a consequence of the more restrictive condition that a strain energy density exists, in which case the solid is known as *hyperelastic*. After a discussion of these general

84

questions we specialize to the case of the isotropic elastic solid, which includes a consideration of a number of important parameters, such as Poisson's ratio and the shear modulus. Finally, using the relations between stress, strain and displacement and the equation of motion, the elastic wave equation for a homogeneous isotropic medium is derived. Here we also show that in this kind of medium two types of waves, known as P and S waves, propagate.

4.2 The equation of motion under the small-deformation approximation

From the equation of motion (3.5.3) and the symmetry relation (3.5.10) for the stress tensor we obtain

$$\tau_{ij,j} + \rho f_i = \rho \frac{Dv_i}{Dt} \tag{4.2.1}$$

with the following expressions for the acceleration

$$a_i = \frac{Dv_i}{Dt} = \frac{\partial v_i}{\partial t} + \frac{\partial v_i}{\partial x_l} \frac{\partial x_l}{\partial t} \tag{4.2.2}$$

and velocity

$$v_i = \frac{\partial u_i}{\partial t} + \frac{\partial u_i}{\partial x_l} \frac{\partial x_l}{\partial t} \tag{4.2.3}$$

(see §3.2).

To simplify the problem represented by (4.2.1)–(4.2.3) we introduce the small-deformation approximation, which allows us to neglect the spatial derivatives of **u**. Therefore

$$v_i \simeq \frac{\partial u_i}{\partial t}; \qquad a_i \simeq \frac{\partial^2 u_i}{\partial t^2}. \tag{4.2.4}$$

Then, under these approximations (4.2.1) becomes

$$\tau_{ij,j} + \rho f_i = \rho \frac{\partial^2 u_i}{\partial t^2} = \rho \ddot{u}_i, \tag{4.2.5}$$

where the double dots indicate a second derivative with respect to time. Equation (4.2.5) is Cauchy's equation of motion, which in dyadic form is written as

$$\nabla \cdot \mathcal{T} + \rho \mathbf{f} = \rho \frac{\partial^2 \mathbf{u}}{\partial t^2} = \rho \ddot{\mathbf{u}}. \tag{4.2.6}$$

In §4.5 and §4.6 we will derive expressions for τ_{ij} in terms of ε_{ij} which in combination with (4.2.5) will be used to derive the elastic wave equation in §4.8. The reason why (4.2.5) is introduced early in the chapter is that it is needed in §4.4.

4.3 Thermodynamical considerations

By necessity, this section will present a very cursory treatment of important thermodynamics concepts. A very readable presentation can be found in Fermi (1937). Thermodynamics is the branch of physics that deals with the transformation of mechanical work into heat and vice versa. As in continuum mechanics, classical thermodynamics makes an abstraction of the corpuscular nature of matter and deals with the relations between a small number of entities, such as temperature, pressure, energy, heat, work, and entropy, that represent a thermodynamical system. Typical examples of systems are a cylinder with a movable piston that contains gas and a vessel in which a chemical reaction takes place.

At the core of thermodynamics are two basic laws, briefly discussed here. Heat and work are two different manifestations of energy, and the first law is a statement regarding the conservation of energy in a system. In words, the *first law* says that the increase in the energy of a system is equal to the *work* done by external forces plus the *heat* received by the system. The *second law*, which states the impossibility of a transformation whose only effect is to transfer heat from a cooler body to a hotter body, is more difficult to describe. However, its expression (see below) is simple for *reversible processes*, which are defined as processes that can be considered as a sequence of states that differ from a state of equilibrium by an infinitesimal amount. For example, increasing or decreasing the pressure of a gas in a cylinder is a reversible process if it is carried out very slowly. A sudden compression or decompression would be irreversible. It is important to note, however, that a reversible process is an idealized concept unlikely to occur in practice.

To apply the first law to study the deformation of a solid body, the energy is divided into two parts, one corresponding to the total *kinetic energy* K (which depends on mass and velocity) of the body and the other corresponding to the energy U that depends on the temperature and the configuration of the body (Love, 1927). The energy U is known as the *intrinsic energy*. The following discussion is based on Brillouin (1964). Let dw be the work, dQ the heat supplied, and dK and dU the variation in kinetic and internal energies, respectively, during a small deformation. Then,

$$dQ + dw = dK + dU. \tag{4.3.1}$$

For reversible processes the second law states that

$$dQ = T \, dS \tag{4.3.2}$$

where T is the absolute temperature and S is the *entropy* (for irreversible processes the $<$ sign replaces the $=$ sign). Equation (4.3.2) actually defines S.

Introducing (4.3.2) into (4.3.1) gives

$$dU - T\,dS = dw - dK,$$ (4.3.3)

which can be rewritten as

$$d(U - TS) + S\,dT = dw - dK.$$ (4.3.4)

The quantity

$$F = U - TS$$ (4.3.5)

is known as the *free energy* of the system.

Here and in the following it will be assumed that the quantities in (4.3.1) are measured with respect to a reference state of equilibrium that in our applications will be the body before being deformed. It will also be assumed that these quantities and others introduced below are referred to a unit volume in the undeformed body.

The reader may have noted that time does not enter in (4.3.1)–(4.3.5). The reason is the assumption of reversible equilibrium. Even the concept of position within the system is ignored, which means that the values of the thermodynamic variables are assumed to be constant throughout the system. This means that the application of the laws described above requires modification when applied to continuum mechanics problems. In particular, the first law must be expressed in terms of time rates of change (Eringen, 1967)

$$\frac{dQ}{dt} + \frac{dw}{dt} = \frac{dK}{dt} + \frac{dU}{dt}.$$ (4.3.6)

It must be noted, however, that (4.3.6) is a local equation applicable to a volume element around a particle. Thus, it is not an energy balance law for a whole body (Eu, 1992).

It must also be noted that in seismology we do not explicitly consider changes in entropy (dS) or temperature (dT), and the reason for that comes from an inspection of (4.3.3) and (4.3.4). From (4.3.2) we see that $dQ = 0$ implies $dS = 0$ and then, from (4.3.3)

$$dU = dw - dK.$$ (4.3.7)

A process for which $dQ = 0$ is called *adiabatic*. In the laboratory, such a condition is achieved, for example, by insulating the vessel where a chemical reaction takes place in such a way that there is no exchange of heat with the environment. In the case of wave propagation in a solid, adiabatic conditions will be realized if the particle motion is faster than the speed with which heat is transmitted through the solid. This condition is a good approximation for wave propagation in the Earth for the frequencies and wavelengths of interest (Aki and Richards, 1980; Dahlen and Tromp, 1998; Gubbins, 1990; Pilant, 1979).

If we now consider (4.3.4) and (4.3.5) we see that

$$dF = dw - dK \tag{4.3.8}$$

when $dT = 0$.

A process for which $dT = 0$ is called *isothermal*. This condition is achieved when the process takes places so slowly that the temperature within a system is always equal to the temperature of the environment (provided it is constant). In the laboratory this can be accomplished, for example, by slowly mixing chemicals in a vessel immersed in a constant-temperature bath.

4.4 Strain energy

Equations (4.3.7) and (4.3.8) correspond to completely different situations but have the same functional form. An important common feature of these two equations is the presence of the functions dU and dF, which in the case of a deformed solid represent the change in internal energy of the body caused by deformation under two different thermodynamic conditions. These two equations are completely general; to specialize them to the case of elastic deformations we must determine the rate of work done by the surface and external body forces. The following treatment follows Sokolnikoff (1956). When a particle in the body suffers a displacement $\mathbf{u}$ at time t, the displacement in the time interval $(t, t + dt)$ is given by

$$\frac{\partial u_i}{\partial t} \, dt \equiv \dot{u}_i \, dt \tag{4.4.1}$$

(i.e., velocity × time). We also know that work is defined, in general, as the scalar product between the force and displacement. In our case the displacement is given by (4.4.1) and the forces are the body and surface forces. Then, the rate of work is given by

$$\frac{dw}{dt} = \int_V \rho f_i \dot{u}_i \, dV + \int_S T_i \dot{u}_i \, dS. \tag{4.4.2}$$

Here and in the following S again indicates the surface of the volume under consideration.

The surface integral will be modified as follows: (a) replace T_i by its expression in terms of the stress tensor (given by (3.5.11)), (b) use Gauss' theorem to convert the surface integral into a volume integral, (c) perform the required derivatives in the integrand, (d) use the identity

$$\dot{u}_{i,j} \equiv \frac{1}{2}(\dot{u}_{i,j} + \dot{u}_{j,i}) + \frac{1}{2}(\dot{u}_{i,j} - \dot{u}_{j,i}) = \dot{\varepsilon}_{ij} + \dot{\omega}_{ij} \tag{4.4.3}$$

and (e) use the fact that $\tau_{ij}\dot{\omega}_{ij} = 0$ because τ_{ij} and $\dot{\omega}_{ij}$ are symmetric and anti-symmetric, respectively. Thus,

$$\int_S T_i \dot{u}_i \, dS = \int_S \tau_{ij} \dot{u}_i n_j \, dS = \int_V (\tau_{ij} \dot{u}_i)_{,j} \, dV$$

$$= \int_V \tau_{ij,j} \dot{u}_i \, dV + \int_V \tau_{ij} \dot{u}_{i,j} \, dV = \int_V (\tau_{ij,j} \dot{u}_i + \tau_{ij} \dot{\varepsilon}_{ij}) \, dV. \quad (4.4.4)$$

The next step is to write the first term in the rightmost integral in (4.4.4) using the equation of motion (4.2.5). Then (4.4.2) becomes

$$\frac{dw}{dt} = \int_V \rho \ddot{u}_i \dot{u}_i \, dV + \int_V \tau_{ij} \dot{\varepsilon}_{ij} \, dV. \quad (4.4.5)$$

The kinetic energy K of the body is defined as

$$K = \frac{1}{2} \int_V \rho \dot{u}_i \dot{u}_i \, dV \quad (4.4.6)$$

so that the first term on the right-hand side of (4.4.5) is dK/dt. Therefore,

$$\frac{dw}{dt} - \frac{dK}{dt} = \int_V \tau_{ij} \dot{\varepsilon}_{ij} \, dV \quad (4.4.7)$$

(assuming that ρ does not change appreciably with time).

Now assume that there exists a function $W = W(\varepsilon_{ij})$ such that

$$\tau_{ij} = \frac{\partial W(\varepsilon_{ij})}{\partial \varepsilon_{ij}}. \quad (4.4.8)$$

Introducing (4.4.8) in (4.4.7) gives

$$\frac{dw}{dt} - \frac{dK}{dt} = \int_V \frac{\partial W}{\partial \varepsilon_{ij}} \frac{\partial \varepsilon_{ij}}{\partial t} \, dV = \frac{d}{dt} \int_V W \, dV. \quad (4.4.9)$$

Integrating (4.4.9) with respect to t gives

$$dw - dK = \int_V W \, dV. \quad (4.4.10)$$

We have used differentials in (4.4.10) because of our assumption that the quantities involved are measured with respect to a reference state. Comparison of (4.4.10) with (4.3.7) and (4.3.8) shows that the integral in (4.4.10) can be interpreted as dU for an adiabatic process and as dF for an isothermal process. In the first case we write

$$dU = \int_V W \, dV \quad (4.4.11)$$

and call dU the *strain energy* of the body and W the (volume) *density of strain energy* or *elastic potential* (Sokolnikoff, 1956).

4.5 Linear elastic and hyperelastic deformations

A deformation is called elastic if the stress depends on the current value of the strain only, i.e., it does not depend on the rate of strain. A deformation is hyperelastic if there exists a strain energy density that depends on strain only (Hudson, 1980). This density is the function W introduced in (4.4.8). Solids that satisfy these conditions are known as elastic and hyperelastic, respectively. The importance of this distinction will become apparent below.

The usual relation between stress and strain is an extension of Hooke's law of proportionality between forces and deformation. Hooke's experiments were carried out in the late seventeenth century (Timoshenko, 1953) and involved mostly springs. Cauchy generalized Hooke's law to elastic solids by proposing that stress and strain are linearly related. In tensor form the law takes the form

$$\tau_{kl} = c_{klpq}\varepsilon_{pq}. \tag{4.5.1}$$

As Aki and Richards (1980) note, tensors are a relatively recent development. Therefore, equation (4.5.1) represents the modern version of the *generalized Hooke's law*.

Before proceeding two remarks are in order. First, it is implicitly assumed that stress and strain are measured with respect to a reference state in which both are equal to zero. In the Earth this is not true because at a given depth there is stress due to the pressure exerted by the overlying rocks, and for that reason the reference state is assumed to be unstrained but pre-stressed. In this case T° is the stress at zero strain and $T^\circ + T$ is the stress in the strained state (Aki and Richards, 1980). Secondly, c_{ijkl} are independent of ε_{ij}, but they may depend on position, as in the Earth.

The entity c_{klpq} is a fourth-order tensor (Problem 4.1). The number of components of c_{klpq} is 81, but because of the symmetry of τ_{kl} and ε_{pq} the number of independent components is greatly reduced. From

$$\tau_{ij} = \tau_{ji}; \qquad \varepsilon_{kl} = \varepsilon_{lk} \tag{4.5.2}$$

we find that

$$c_{ijkl} = c_{jikl}; \qquad c_{ijkl} = c_{ijlk}. \tag{4.5.3a,b}$$

These relations reduce the number of independent components to 36. A further reduction is possible, but to achieve this it is necessary to stipulate the existence of the function W. This function was introduced by Green (1838, 1839). As shown below, using Green's approach one recovers Hooke's law, which means that a hyperelastic body is also elastic. This distinction, however, is not important in practice (e.g. Hunter, 1976; Pilant, 1979) as it appears that elastic materials are also hyperelastic. On the other hand, Green's work was exceedingly important because his results, unlike those derived by leading early workers in the field of elasticity, did not depend on any molecular theory. This difference in approach resulted in disagreement over the number of parameters required to represent the most general (anisotropic) solid and an isotropic solid. Green's results indicated 21 and two, respectively, while Cauchy and Poisson had found 15 and one, respectively (Timoshenko, 1953). Interestingly, although Cauchy introduced the concept of stress independently of the molecular theories, his later work was based on the same theory used by Navier to derive the results referred to in §3.1. The controversy between the proponents of the two schools of thought lasted several decades and was resolved in favor of Green's results when experimental data of the required quality became available and when it became obvious that the molecular theories were not consistent with the newer atomic theories. We finally note that Green (1839) also derived the equations of motion in a form close to that used today.

To establish the relation between stress, strain, and W we start by expanding W in a Taylor series to second order:

$$W = a + b_{ij}\varepsilon_{ij} + \frac{1}{2}d_{ijkl}\varepsilon_{ij}\varepsilon_{kl}, \qquad (4.5.4)$$

where the coefficients do not depend on strain. Higher-order terms are neglected because of the assumption of small deformations. The coefficient a can be ignored under the assumption that the energy in the reference state is zero. Now we will apply (4.4.8) to (4.5.4), but before proceeding note that in general, if a function F is defined by

$$F = a_i x_i + b_{ij} x_i x_j. \qquad (4.5.5)$$

where x_k indicate the independent variables and the coefficients do not depend on x_k, then

$$\frac{\partial F}{\partial x_k} = a_k + b_{ij}x_{i,k}x_j + b_{ij}x_i x_{j,k} = a_k + b_{kj}x_j + b_{ik}x_i = a_k + (b_{ki} + b_{ik})x_i. \quad (4.5.6)$$

Here we have used $x_{m,n} = \delta_{mn}$ and $b_{kj}x_j = b_{ki}x_i$.

The expression for W is similar to the expression for F, with the single subindices in F playing the role of the pairs of subindices in W. Therefore,

$$\tau_{kl} = \frac{\partial W}{\partial \varepsilon_{kl}} = b_{kl} + \frac{1}{2}(d_{klpq} + d_{pqkl})\varepsilon_{pq} = b_{kl} + c_{klpq}\varepsilon_{pq}, \qquad (4.5.7)$$

where

$$c_{klpq} = \frac{1}{2}(d_{klpq} + d_{pqkl}). \qquad (4.5.8)$$

Because of the assumption of zero initial stress, we must take $b_{kl} = 0$ in (4.5.7).

Therefore,

$$\tau_{kl} = c_{klpq}\varepsilon_{pq} \tag{4.5.9}$$

and

$$W = \frac{1}{2}d_{ijpq}\varepsilon_{ij}\varepsilon_{pq}. \tag{4.5.10}$$

Equation (4.5.9) is similar to (4.5.1), which shows that a hyperelastic material is also elastic. Equation (4.5.8), on the other hand, shows that

$$c_{klpq} = c_{pqkl}. \tag{4.5.11}$$

This symmetry relation is a result of the assumption of the existence of the function W introduced in (4.4.8), together with the small-deformation approximation. Using (4.5.11) it is possible to show that the number of independent components of c_{pqkl} reduces to 21.

Now we will derive the relation between W, τ_{ij}, and ε_{ij}. By a renaming of dummy indices (4.5.10) can be written as

$$W = \frac{1}{2}d_{pqij}\varepsilon_{pq}\varepsilon_{ij} \equiv \frac{1}{2}d_{pqij}\varepsilon_{ij}\varepsilon_{pq}. \tag{4.5.12}$$

Summing (4.5.10) and (4.5.12) together, dividing by two and using (4.5.8) and (4.5.9) gives

$$W = \frac{1}{4}(d_{ijpq} + d_{pqij})\varepsilon_{ij}\varepsilon_{pq} = \frac{1}{2}c_{ijpq}\varepsilon_{ij}\varepsilon_{pq} = \frac{1}{2}\tau_{ij}\varepsilon_{ij}. \tag{4.5.13}$$

4.6 Isotropic elastic solids

Equations (4.5.1) or (4.5.9) apply to the most general elastic solid. When the properties of the medium are the same in any direction, the medium is said to be *isotropic*. Otherwise, it is said to be *anisotropic*. Crystalline substances are typical examples of anisotropic materials. For isotropic media c_{ijkl} is also isotropic and has the following form:

$$c_{ijkl} = \lambda\delta_{ij}\delta_{kl} + \mu(\delta_{ik}\delta_{jl} + \delta_{il}\delta_{jk}) + \nu(\delta_{ik}\delta_{jl} - \delta_{il}\delta_{jk}) \tag{4.6.1}$$

(see (1.4.104)). This is the only fourth-order isotropic tensor, regardless of the nature of c_{ijkl}. However, when the tensor is symmetric in the first two indices, as in our case, it is easy to show that ν has to be equal to zero (Problem 4.2). Therefore, for an isotropic elastic solid c_{ijkl} reduces to

$$c_{ijkl} = \lambda\delta_{ij}\delta_{kl} + \mu(\delta_{ik}\delta_{jl} + \delta_{il}\delta_{jk}). \tag{4.6.2}$$

The two independent variables λ and μ are known as *Lamé's parameters*. Although they are also known as Lamé's constants, it must be recognized that in the Earth they vary with position. Also note that for isotropic media there is no distinction between elasticity and hyperelasticity.

Now using (4.6.2), (4.5.9), the symmetry of ε_{ij} and the relation between the strain tensor and displacement (see (2.4.1)) we obtain

$$\tau_{ij} = \lambda \delta_{ij} \varepsilon_{kk} + 2\mu \varepsilon_{ij} = \lambda \delta_{ij} u_{k,k} + \mu(u_{i,j} + u_{j,i}). \tag{4.6.3}$$

In dyadic form we have

$$\mathcal{T} = \lambda \mathcal{I} \, \nabla \cdot \mathbf{u} + \mu(\mathbf{u}\nabla + \nabla\mathbf{u}). \tag{4.6.4}$$

Now we will write ε_{ij} in terms of τ_{ij}. From the first equality in (4.6.3) we have

$$\varepsilon_{ij} = \frac{1}{2\mu}(\tau_{ij} - \lambda \delta_{ij} \varepsilon_{kk}). \tag{4.6.5}$$

To eliminate ε_{kk} contract the indices i and j (see §1.4.1). This gives

$$\varepsilon_{ii} \equiv \varepsilon_{kk} = \frac{1}{2\mu}(\tau_{ii} - \lambda \delta_{ii} \varepsilon_{kk}) \tag{4.6.6}$$

so that

$$\varepsilon_{kk} = \frac{\tau_{kk}}{(3\lambda + 2\mu)} \tag{4.6.7}$$

and

$$\varepsilon_{ij} = \frac{1}{2\mu} \left(\tau_{ij} - \frac{\lambda \delta_{ij}}{3\lambda + 2\mu} \tau_{kk} \right). \tag{4.6.8}$$

From (4.6.5) or (4.6.8) it is possible to show that ε_{ij} and τ_{ij} have the same principal directions (Problem 4.3).

In dyadic form (4.6.8) becomes

$$\mathcal{E} = \frac{1}{2\mu} \left(\mathcal{T} - \frac{\lambda}{3\lambda + 2\mu} \tau_{kk} \mathcal{I} \right). \tag{4.6.9}$$

Now we will use (4.6.8) to find the strain corresponding to three simple stress tensors. This will help us interpret the physical meaning of the Lamé parameters, and that at the same time will allow the introduction of three important related parameters.

(1) *Uniaxial tension.* Consider a cylindrical bar with axis in the x_1 direction subjected to a tensional force also along x_1 applied to the end of the bar. In this case the only nonzero component of the stress tensor is τ_{11} (> 0). From (4.6.8) we see that the only nonzero strain components are the diagonal components

$$\varepsilon_{11} = \frac{\tau_{11}(\lambda + \mu)}{\mu(3\lambda + 2\mu)} \tag{4.6.10}$$

$$\varepsilon_{22} = \varepsilon_{33} = \frac{-\tau_{11}\lambda}{2\mu(3\lambda + 2\mu)}. \tag{4.6.11}$$

Equations (4.6.10) and (4.6.11) show that the bar suffers a longitudinal extension (i.e., along its axis) and a lateral contraction (i.e., in its cross section). These results are used to introduce two new elastic parameters, *Young's modulus*, indicated by Y, and *Poisson's ratio*, indicated by σ,

$$Y = \frac{\tau_{11}}{\varepsilon_{11}} = \frac{\mu(3\lambda + 2\mu)}{\lambda + \mu} \tag{4.6.12}$$

$$\sigma = -\frac{\varepsilon_{22}}{\varepsilon_{11}} = -\frac{\varepsilon_{33}}{\varepsilon_{11}} = \frac{\lambda}{2(\lambda + \mu)}. \tag{4.6.13}$$

Therefore, Y gives the ratio of the tensional stress to the longitudinal extension and σ gives the ratio of lateral contraction to longitudinal extension.

The Lamé parameters can be obtained from (4.6.12) and (4.6.13),

$$\lambda = \frac{Y\sigma}{(1 + \sigma)(1 - 2\sigma)} \tag{4.6.14}$$

$$\mu = \frac{Y}{2(1 + \sigma)} \tag{4.6.15}$$

(Problem 4.4).

(2) *Simple shear stress.* Consider a bar with rectangular cross section and axis along the x_3 direction subjected to shearing forces of equal magnitude in the (x_1, x_2) plane. The only nonzero components of the stress tensor are $\tau_{12} = \tau_{21}$. The trace of this tensor is zero and the only nonzero components of the corresponding strain tensor are

$$\varepsilon_{12} = \varepsilon_{21} = \frac{\tau_{12}}{2\mu}. \tag{4.6.16}$$

Therefore,

$$\mu = \frac{\tau_{12}}{2\varepsilon_{12}}. \tag{4.6.17}$$

Now recall that $2\varepsilon_{12}$ is equal to the decrease in angle between two line elements in the x_1 and x_2 directions before the deformation (see (2.4.20)). Therefore, μ is the ratio of the shear stress to the decrease in angle, and is known as the *rigidity* or *shear modulus*.

(3) *Hydrostatic pressure.* In this case $\tau_{ij} = -P\delta_{ij}$ with $P > 0$, so that $\tau_{kk} = -3P$. Therefore, from (4.6.8) we obtain

$$\varepsilon_{ij} = \frac{1}{2\mu}\left(-P\delta_{ij} + \frac{3P\lambda}{3\lambda + 2\mu}\delta_{ij}\right) = -\frac{P}{3\lambda + 2\mu}\delta_{ij}. \tag{4.6.18}$$

Consequently, the only nonzero components of ε_{ij} are the diagonal elements, which are equal to each other. Furthermore, the trace of ε_{ij} is

$$\varepsilon_{kk} = -\frac{3P}{3\lambda + 2\mu}. \tag{4.6.19}$$

Now recall that the trace ε_{kk} is the fractional change in volume (see (2.4.22)), which is expected to be negative (i.e., there is a decrease in volume) when a body is compressed. This allows the introduction of an additional elastic parameter, the *bulk modulus* or *modulus of compression* k,

$$k = \frac{P}{-\varepsilon_{kk}} = \lambda + \frac{2}{3}\mu. \tag{4.6.20}$$

Let us discuss some general aspects of these elastic parameters. On physical grounds we postulate that for typical solids

$$\mu \geq 0; \quad k \geq 0. \tag{4.6.21a,b}$$

Relations (4.6.21) will be mathematically justified for hyperelastic materials in the next section.

For materials with no resistance to shear motion, such as gases and inviscid fluids (their viscosity is zero), $\mu = 0$ and, from (4.6.20), $k = \lambda$. For incompressible materials, $\varepsilon_{kk} = 0$ and, from (4.6.20), $k = \infty$. Furthermore, as μ will be, in general, a finite number, from (4.6.20) we find that $\lambda = \infty$, Then, from (4.6.12) and (4.6.13) we see that $Y = 3\mu$ and $\sigma = \frac{1}{2}$ (Problem 4.5). For practical purposes some rubbers may be considered incompressible (e.g., Atkin and Fox, 1980).

From (4.6.20) and (4.6.21b) we find

$$\lambda + \frac{2}{3}\mu \geq 0; \quad \lambda \geq -\frac{2}{3}\mu. \tag{4.6.22a,b}$$

If (4.6.22b) is introduced in (4.6.12) we find $Y \geq 0$, which is again an expected result for a typical solid.

To investigate the range of possible values of σ, k will be rewritten as follows:

$$k = \frac{Y}{3(1 - 2\sigma)} \tag{4.6.23}$$

(Problem 4.6). From (4.6.15), because μ and Y are both positive, $\sigma > -1$. From (4.6.23), because k and Y are positive and k may be equal to infinity, $\sigma \leq \frac{1}{2}$. Therefore,

$$-1 < \sigma \leq \frac{1}{2}. \tag{4.6.24}$$

Some books assume that $\sigma \geq 0$ because it is expected that a longitudinal extension is accompanied by lateral contraction. If this were not the case, the volume of a specimen with negative σ would increase. The assumption of positive σ appears reasonable for most known solids, but in some cases it may be negative. For example, Gregory (1976) measured P- and S-wave velocities in the laboratory and used them to determine σ (see Problem 4.10). The samples analyzed were sandstones and carbonates and the frequency of the waves was 1 MHz. The values of σ had a wide range of variation, depending on the composition of the sample, its porosity, the presence of fluids, the degree of saturation, and the applied pressure. For water- and gas-saturated samples the lowest values of σ were 0.11 and -0.12, respectively. Recent work on artificial foams also shows that negative values of σ are possible, although these materials are not examples of the media considered here (Lakes, 1987a,b; Burns, 1987). We also note that materials with $\sigma = 0$ will not expand laterally when subjected to longitudinal compression. Cork is an example of a material with σ close to zero (Lakes, 1987a).

Finally, we consider the special case of a *Poisson solid*, for which $\sigma = 0.25$. This result, which implies $\lambda = \mu$, was derived by Poisson based on the molecular theory he used in his elasticity studies (Love, 1927). The value of σ for rocks in the Earth's crust and mantle is close to 0.25 and this value is sometimes used in place of the actual ones, but this is not always a good approximation.

4.7 Strain energy density for the isotropic elastic solid

Using (4.5.13) and (4.6.3) we obtain the following expression for the strain energy density:

$$W = \frac{1}{2}\lambda\varepsilon_{kk}\varepsilon_{ii} + \mu\varepsilon_{ij}\varepsilon_{ij} = \frac{1}{2}\lambda\left(\varepsilon_{kk}\right)^2 + \mu\left(\varepsilon_{11}^2 + \varepsilon_{22}^2 + \varepsilon_{33}^2 + 2\varepsilon_{12}^2 + 2\varepsilon_{23}^2 + 2\varepsilon_{13}^2\right)$$

$$(4.7.1)$$

(Problem 4.7). As the reference state (before the deformation) is assumed to be a state of stable equilibrium, W cannot be negative. If W were negative, then the deformation would bring the body to a state of less energy than that which existed before the deformation and the equilibrium would not be stable.[1] Since all the strain terms in (4.7.1) are positive, we conclude that

$$\lambda \geq 0; \qquad \mu \geq 0. \tag{4.7.2}$$

Writing ε_{ij} as a the sum of a deviatoric and an isotropic tensor

$$\varepsilon_{ij} = \bar{\varepsilon}_{ij} + \frac{1}{3}\varepsilon_{kk}\delta_{ij} \tag{4.7.3}$$

---

[1] This qualitative argument is extremely simplified. For a rigorous treatment see Truesdell and Noll (1965). The fact that $W \geq 0$ is essential to proving the uniqueness of the solutions of linear elasticity problems (Sokolnikoff, 1956; Aki and Richards, 1980)

(see §3.7) equation (4.7.1) becomes

$$W = \frac{1}{2}k\,(\varepsilon_{kk})^2 + \mu\bar{\varepsilon}_{ij}\bar{\varepsilon}_{ij} \tag{4.7.4}$$

(Problem 4.8).

Because $W \geq 0$ and the strain terms are all positive, we conclude that

$$k \geq 0; \quad \mu \geq 0 \tag{4.7.5}$$

(Hudson, 1980) which verifies (4.6.21). Alternatively, some books start with (4.7.5) and conclude that W must be positive.

4.8 The elastic wave equation for a homogeneous isotropic medium

Combining the equation of motion, Hooke's law and the relation between strain and displacement (see (4.2.5), (4.5.1), and (2.4.1)) gives an equation for the displacement of the particles of a general anisotropic elastic solid. Solving the resulting equation analytically is extremely difficult, but feasible, if the medium is homogeneous, in which case the Lamé parameters are constant. When the parameters depend on position the equation is generally impossible to solve, and numerical or approximate methods are needed. We will discuss some of these questions in Chapter 8. The case of homogeneous isotropic solids, although complicated, admits exact solutions and for this reason this simplified situation will be discussed here and in most of the book.

Equation (4.2.5) includes the term $\tau_{ij,j}$, which will be written in terms of displacement using (4.6.3), which is valid for isotropic solids:

$$\tau_{ij,j} = \lambda\delta_{ij}u_{k,kj} + \mu(u_{j,ij} + u_{i,jj}) = (\lambda+\mu)u_{j,ji} + \mu u_{i,jj}$$

$$= (\lambda+\mu)[\nabla(\nabla\cdot\mathbf{u})]_i + \mu(\nabla^2\mathbf{u})_i. \tag{4.8.1}$$

The relations $u_{j,ij} = u_{k,ik} = u_{k,ki}$ have been used.

Using (4.8.1) the equation of motion (4.2.5) will be written in vector form:

$$\mu\nabla^2\mathbf{u} + (\lambda+\mu)\nabla(\nabla\cdot\mathbf{u}) + \rho\mathbf{f} = \rho\frac{\partial^2\mathbf{u}}{\partial t^2}, \tag{4.8.2}$$

which can be further rewritten using

$$\nabla^2\mathbf{u} = \nabla(\nabla\cdot\mathbf{u}) - \nabla\times(\nabla\times\mathbf{u}) \tag{4.8.3}$$

(see (1.4.53)) and dividing by ρ:

$$\alpha^2\nabla(\nabla\cdot\mathbf{u}) - \beta^2\,\nabla\times\nabla\times\mathbf{u} + \mathbf{f} = \frac{\partial^2\mathbf{u}}{\partial t^2}, \tag{4.8.4}$$

where

$$\alpha^2 = \frac{\lambda + 2\mu}{\rho}; \qquad \beta^2 = \frac{\mu}{\rho}. \qquad (4.8.5a,b)$$

Either (4.8.2) or (4.8.4) is the elastic wave equation, which some authors call Navier's equation. Note that this is a vector equation, which in view of the comments in §1.1 is valid in any orthogonal coordinate system with the corresponding definitions of gradient, divergence, and curl. Now we will show that (4.8.4) involves the propagation of two types of waves. First, apply the divergence operation to (4.8.4). Using

$$\nabla \cdot [\nabla(\nabla \cdot \mathbf{u})] = \nabla^2(\nabla \cdot \mathbf{u}) \qquad (4.8.6)$$

and

$$\nabla \cdot (\nabla \times \nabla \times \mathbf{u}) = 0 \qquad (4.8.7)$$

(Problem 4.9). Equation (4.8.4) becomes

$$\alpha^2 \nabla^2(\nabla \cdot \mathbf{u}) + \nabla \cdot \mathbf{f} = \frac{\partial^2}{\partial t^2}(\nabla \cdot \mathbf{u}), \qquad (4.8.8)$$

which is a scalar equation because $\nabla \cdot \mathbf{u}$ is a scalar quantity. Equation (4.8.8) represents waves propagating with velocity α (*P waves*). Recall that $\nabla \cdot \mathbf{u}$ is equal to ε_{kk}, which is a fractional volume change (see (2.4.22)).

Now apply the curl operation to (4.8.4). Using

$$\nabla \times [\nabla(\nabla \cdot \mathbf{u})] = \mathbf{0} \qquad (4.8.9)$$

and

$$\nabla \times (\nabla \times \nabla \times \mathbf{u}) = -\nabla^2(\nabla \times \mathbf{u}) \qquad (4.8.10)$$

(Problem 4.9), equation (4.8.4) becomes

$$\beta^2 \nabla^2(\nabla \times \mathbf{u}) + \nabla \times \mathbf{f} = \frac{\partial^2}{\partial t^2}(\nabla \times \mathbf{u}). \qquad (4.8.11)$$

This is a vector equation because $\nabla \times \mathbf{u}$ is a vector quantity. Equation (4.8.11) represents waves propagating with velocity β (*S waves*). Recall that $\nabla \times \mathbf{u}$ corresponds to the small rotation of a line element (see §2.5).

The *P* and *S* waves propagate through the interior of an elastic body and for this reason they are known as *body waves*. The existence of these waves was demonstrated by Poisson, who published his results in 1829 (Timoshenko, 1953; Hudson, 1980).

Equation (4.8.4) with $\mathbf{f} = \mathbf{0}$ will be studied in detail in §5.8. The case of $\mathbf{f} \neq \mathbf{0}$ for a number of special but important forces will be treated in Chapter 9. Results derived in that chapter will then be used to solve the problem of the waves generated by an earthquake source, a question discussed in Chapter 10.

Problems

4.1 Use the quotient theorem (see §1.4.5) to show that the entity c_{klpq} introduced in (4.5.1) is a fourth-order tensor.

4.2 Verify that for an isotropic elastic solid c_{ijkl} reduces to

$$c_{ijkl} = \lambda \delta_{ij} \delta_{kl} + \mu (\delta_{ik} \delta_{jl} + \delta_{il} \delta_{jk}).$$

4.3 Show that ε_{ij} and τ_{ij} have the same principal directions.

4.4 Verify (4.6.14) and (4.6.15).

4.5 Verify that for incompressible materials, $Y = 3\mu$ and $\sigma = \frac{1}{2}$.

4.6 Verify (4.6.23).

4.7 Verify (4.7.1).

4.8 Verify (4.7.4).

4.9 Verify (4.8.6), (4.8.7) and (4.8.9), (4.8.10).

4.10 Show that

$$\sigma = \frac{1}{2} \frac{\alpha^2 - 2\beta^2}{\alpha^2 - \beta^2}.$$

4.11 Show that when body forces are absent the displacement

$$\mathbf{u} = (0, 0, u_3(x_3, t)) = u_3(x_3, t)\mathbf{e}_3$$

satisfies

$$\alpha^2 \frac{\partial^2 u_3}{\partial x_3^2} = \frac{\partial^2 u_3}{\partial t^2},$$

which is a 1-D wave equation for P waves (after White, 1965). An example is provided by P waves at normal incidence (put $c = 0$ in (6.2.4)).

4.12 Show that when body forces are absent the displacement

$$\mathbf{u} = (0, u_2(x_3, t), 0) = u_2(x_3, t)\mathbf{e}_2$$

satisfies

$$\beta^2 \frac{\partial^2 u_2}{\partial x_3^2} = \frac{\partial^2 u_2}{\partial t^2},$$

which is a 1-D wave equation for S waves (after White, 1965). An example is provided by SH waves at normal incidence (put $f = 0$ in (6.2.6)).

5

Scalar and elastic waves in unbounded media

5.1 Introduction

The goal of this chapter is to find three independent vector solutions to the elastic wave equation. The treatment presented here is not the standard one. It began with work by Hansen (1935) and Stratton (1941) on the solution of electromagnetic problems involving the vector wave equation, and was extended to include elastic wave propagation problems by Morse and Feshbach (1953), Eringen and Suhubi (1975), and Ben-Menahem and Singh (1981, and references therein), among a few others. The major advantage of these vector solutions is that they simplify the solution of problems in a number of coordinate systems, including the Cartesian, spherical, and cylindrical systems and a few others. In the particular case of a Cartesian system, the main result will be a system of three vectors that automatically represent the motion corresponding to the P, SV, and SH waves. However, before arriving at that stage it is necessary to present more elementary aspects of wave propagation, beginning with the scalar equation in one and three dimensions. To simplify the presentation and the derivations, the independent variables will be indicated by either x, y, z or x_1, x_2, x_3. Which convention is used will be apparent from the context. Except for §5.5, all the equations are in Cartesian coordinates.

5.2 The 1-D scalar wave equation

Consider the equation

$$\frac{\partial^2 \psi(x, t)}{\partial x^2} = \frac{1}{c^2} \frac{\partial^2 \psi(x, t)}{\partial t^2}, \tag{5.2.1}$$

where c is the velocity of wave propagation and is assumed to be a constant. This equation represents, for example, the propagation of waves in a string.

100

We will solve (5.2.1) using the well-known method of *separation of variables* (e.g., Haberman, 1983), which consists of the decomposition of ψ as the product of a function of x only and a function of t only:

$$\psi(x, t) = X(x)T(t). \tag{5.2.2}$$

Introducing this expression for ψ into (5.2.1) gives:

$$X''(x)T(t) = \frac{1}{c^2}X(x)T''(t), \tag{5.2.3}$$

where the primes indicate second-order derivatives with respect to the argument.

Now (5.2.3) has to be rearranged in a new equation with the quantities depending on x on one side and the quantities depending on t on the other side:

$$c^2\frac{X''(x)}{X(x)} = \frac{T''(t)}{T(t)} = \lambda. \tag{5.2.4}$$

Because the quantities to the left and right of the first equals sign are independent of each other, each of them has to be equal to a constant, indicated by λ and known as the separation constant. Then (5.2.4) can be written as

$$X''(x) = \frac{\lambda}{c^2}X(x) \tag{5.2.5}$$

$$T''(t) = \lambda T(t). \tag{5.2.6}$$

Therefore the original partial differential equation has been transformed into two ordinary second-order differential equations.

When the solution to (5.2.1) is required to satisfy initial and/or boundary conditions, λ cannot take arbitrary values, but since at this point we are not introducing any constraints, λ can be any positive or negative real number or even a complex number. Therefore, the solutions to (5.2.5) and (5.2.6) can be written as

$$T(t) = e^{\pm\sqrt{\lambda}t} \tag{5.2.7}$$

$$X(x) = e^{\pm\sqrt{\lambda}x/c}. \tag{5.2.8}$$

The solution to (5.2.1) is given by the product of $X(x)$ and $T(t)$, as indicated by (5.2.2), but because of the $\pm$ signs in the exponents a general solution is a linear combination of four possible cases. For example,

$$\psi(x, t) = A_\pm e^{\pm\sqrt{\lambda}(t-x/c)} + B_\pm e^{\pm\sqrt{\lambda}(t+x/c)}, \tag{5.2.9}$$

where the factors $A_\pm$ and $B_\pm$ are independent of x and t, but may depend on λ.

Equation (5.2.9) shows that ψ depends on x and t via the combinations $u =$

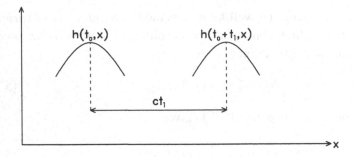

Fig. 5.1. The function $h(t, x) = h(t - x/c)$ for two values of t, t_o and $t_o + t_1$. An observer moving along the x axis with velocity c will not notice any change in the shape of h.

$t - x/c$ and $v = t + x/c$. This suggests that a solution more general than that given by (5.2.9) might be written as

$$\psi(x, t) = h(t - x/c) + g(t + x/c) = h(u) + g(v), \qquad (5.2.10)$$

where h and g are twice-differentiable functions.

   To verify that $\psi(x, t) = h(u)$ satisfies (5.2.1) use the chain rule of differentiation:

$$\frac{\partial h}{\partial x} = \frac{\partial h}{\partial u}\frac{\partial u}{\partial x} = \frac{-1}{c}\frac{\partial h}{\partial u} \qquad (5.2.11)$$

and

$$\frac{\partial^2 h}{\partial x^2} = \frac{\partial}{\partial u}\left(\frac{\partial h}{\partial x}\right)\frac{\partial u}{\partial x} = \frac{1}{c^2}\frac{\partial^2 h}{\partial u^2}. \qquad (5.2.12)$$

Similarly,

$$\frac{\partial^2 h}{\partial t^2} = \frac{\partial^2 h}{\partial u^2} = c^2\frac{\partial^2 h}{\partial x^2}. \qquad (5.2.13)$$

In the last step (5.2.12) was used. Because (5.2.13) and (5.2.1) have the same form, we have shown that $h(u)$ satisfies the one-dimensional wave equation. An analogous argument applies to $g(v)$. Therefore, $h(u) + g(v)$ solves the wave equation. For a more general derivation refer to (5.3.21) with $l = 1$ and $m = n = 0$.

   To give an interpretation to $h(u)$ note that if h_o is the value of $h(u)$ for a given $u_o = t_o - x/c$, then $h(u)$ will remain equal to h_o as long as $u = u_o = t_o + t_1 - (x + ct_1)/c$. Therefore, an observer moving in the positive x direction with speed c will not notice any change in the shape of h (Fig. 5.1). In other words, $h(u)$ can be interpreted as a disturbance (along a string, for example) that propagates with velocity c. The interpretation for $g(v)$ is similar, but in this case motion is in the negative x direction.

5.2.1 Example

Solve the 1-D scalar wave equation under the following initial conditions:

$$\psi(x, 0) = F(x); \qquad \frac{\partial \psi(x, 0)}{\partial t} = 0. \tag{5.2.14}$$

Using (5.2.10) and (5.2.14) we obtain

$$h(-x/c) + g(x/c) = F(x) \tag{5.2.15}$$

$$h'(-x/c) + g'(x/c) = 0. \tag{5.2.16}$$

After integration (5.2.16) becomes

$$-h(-x/c) + g(x/c) = k, \tag{5.2.17}$$

where k is a constant. Subtracting (5.2.17) from (5.2.15) and adding (5.2.15) and (5.2.17) gives

$$h(-x/c) = \frac{1}{2}[F(x) - k] \tag{5.2.18}$$

$$g(x/c) = \frac{1}{2}[F(x) + k]. \tag{5.2.19}$$

Replacing the argument x by $x - ct$ in (5.2.18) and by $x + ct$ in (5.2.19) gives

$$h(t - x/c) = \frac{1}{2}[F(x - ct) - k] \tag{5.2.20}$$

$$g(t + x/c) = \frac{1}{2}[F(x + ct) + k]. \tag{5.2.21}$$

Summing (5.2.20) and (5.2.21) and using (5.2.10) gives

$$\psi(x, t) = \frac{1}{2}[F(x - ct) + F(x + ct)]. \tag{5.2.22}$$

The solution (5.2.22) is plotted in Fig. 5.2 for several values of t assuming that $F(x)$ has a triangular shape.

5.3 The 3-D scalar wave equation

In three dimensions (5.2.1) becomes

$$\nabla^2 \psi(\mathbf{r}, t) = \frac{\partial^2 \psi}{\partial x^2} + \frac{\partial^2 \psi}{\partial y^2} + \frac{\partial^2 \psi}{\partial z^2} = \frac{1}{c^2} \frac{\partial^2 \psi(\mathbf{r}, t)}{\partial t^2}, \tag{5.3.1}$$

where $\mathbf{r} = (x, y, z)$. This equation represents the behavior of a number of important physical systems, such as the vibration of membranes and the propagation of

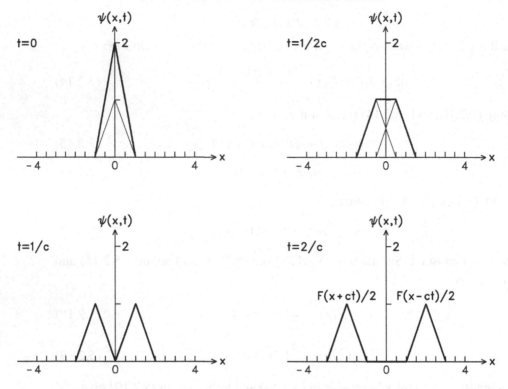

Fig. 5.2. Solution of the 1-D scalar wave equation under the initial conditions (5.2.14) for several times (bold lines). The solution is given by (5.2.22). The velocity of wave propagation is c. The initial disturbance that generates the waves is the triangular function $F(x)$ shown in bold in the plot for $t = 0$. The thin triangles represent $F(x)/2$ when $t = 0$ and the left- and right-shifted versions of $F(x)/2$ when $t < 1/c$.

acoustic waves, and plays a fundamental role in the theory of elastic and electro-magnetic waves.

Equation (5.3.1) will also be solved using the method of separation of variables. Let

$$\psi(\mathbf{r}, t) = F(\mathbf{r})T(t). \tag{5.3.2}$$

Then from (5.3.1) and (5.3.2) and with the same arguments used in §5.2 we find

$$\frac{T''(t)}{T(t)} = \lambda \tag{5.3.3}$$

and

$$c^2 \frac{\nabla^2 F(\mathbf{r})}{F(\mathbf{r})} = \lambda. \tag{5.3.4}$$

The solution to (5.3.3) is given by (5.2.7). To solve (5.3.4) we apply separation of variables again:

$$F(\mathbf{r}) \equiv F(x, y, z) = X(x)Y(y)Z(z), \tag{5.3.5}$$

which gives

$$c^2 \frac{X''YZ + XY''Z + XYZ''}{XYZ} = c^2 \frac{X''}{X} + c^2 \frac{Y''}{Y} + c^2 \frac{Z''}{Z} = \lambda. \tag{5.3.6}$$

In (5.3.6), the terms following the first equals sign are independent of each other, so they can be equated to a different constant under the condition that the sum of the constants is equal to λ. Then

$$X'' = \frac{\lambda_x}{c^2} X; \qquad Y'' = \frac{\lambda_y}{c^2} Y; \qquad Z'' = \frac{\lambda_z}{c^2} Z \tag{5.3.7}$$

with

$$\lambda_x + \lambda_y + \lambda_z = \lambda. \tag{5.3.8}$$

The three equations (5.3.7) are similar to (5.2.5) and have solutions given by

$$X = e^{\pm\sqrt{\lambda_x}x/c}; \qquad Y = e^{\pm\sqrt{\lambda_y}y/c}; \qquad Z = e^{\pm\sqrt{\lambda_z}z/c}. \tag{5.3.9}$$

The solution to (5.3.1) is given by the product

$$\psi(\mathbf{r}, t) = X(x)Y(y)Z(z)T(t) \tag{5.3.10}$$

and from all the possible sign combinations we will choose

$$\psi(\mathbf{r}, t) = A_{\pm}e^{\pm\sqrt{\lambda}[t-(lx+my+nz)/c]} + B_{\pm}e^{\pm\sqrt{\lambda}[t+(lx+my+nz)/c]} \tag{5.3.11}$$

with

$$l = \pm\sqrt{\frac{\lambda_x}{\lambda}}; \qquad m = \pm\sqrt{\frac{\lambda_y}{\lambda}}; \qquad n = \pm\sqrt{\frac{\lambda_z}{\lambda}} \tag{5.3.12}$$

and

$$l^2 + m^2 + n^2 = 1. \tag{5.3.13}$$

The factors $A_{\pm}$ and $B_{\pm}$ are independent of x, y, z, and t, but may depend on l, m, and n.

Equation (5.3.11) shows that the solution to the wave equation depends on the combinations $t \pm (lx + my + nz)/c$, which suggests that a more general solution will be

$$\psi(\mathbf{r}, t) = h\big(t - (lx + my + nz)/c\big) + g\big(t + (lx + my + nz)/c\big). \tag{5.3.14}$$

Equation (5.3.14) is known as the *D'Alembert solution*. To verify that it satisfies (5.3.1) introduce the new variables

$$u = t - \frac{1}{c}(lx + my + nz) \tag{5.3.15}$$

and

$$v = t + \frac{1}{c}(lx + my + nz). \tag{5.3.16}$$

The following intermediate results are needed:

$$\frac{\partial \psi}{\partial t} = \frac{\partial \psi}{\partial u}\frac{\partial u}{\partial t} + \frac{\partial \psi}{\partial v}\frac{\partial v}{\partial t} = \frac{\partial \psi}{\partial u} + \frac{\partial \psi}{\partial v} \tag{5.3.17}$$

$$\frac{\partial^2 \psi}{\partial t^2} = \frac{\partial^2 \psi}{\partial u^2} + \frac{\partial^2 \psi}{\partial v^2} + 2\frac{\partial^2 \psi}{\partial u \partial v} \tag{5.3.18}$$

$$\frac{\partial \psi}{\partial x} = \frac{\partial \psi}{\partial u}\frac{\partial u}{\partial x} + \frac{\partial \psi}{\partial v}\frac{\partial v}{\partial x} = \frac{l}{c}\left(\frac{\partial \psi}{\partial v} - \frac{\partial \psi}{\partial u}\right) \tag{5.3.19}$$

$$\frac{\partial^2 \psi}{\partial x^2} = \frac{l}{c}\left(\frac{\partial^2 \psi}{\partial u \partial v}\frac{\partial u}{\partial x} + \frac{\partial^2 \psi}{\partial v^2}\frac{\partial v}{\partial x} - \frac{\partial^2 \psi}{\partial u^2}\frac{\partial u}{\partial x} - \frac{\partial^2 \psi}{\partial v \partial u}\frac{\partial v}{\partial x}\right)$$

$$= \frac{l^2}{c^2}\left(\frac{\partial^2 \psi}{\partial u^2} + \frac{\partial^2 \psi}{\partial v^2} - \frac{2\partial^2 \psi}{\partial u \partial v}\right). \tag{5.3.20}$$

For the second derivatives with respect to y and z, the corresponding expressions are obtained from (5.3.20) with l replaced by m and n. Introducing the second derivatives into (5.3.1) gives

$$\frac{(l^2 + m^2 + n^2)}{c^2}\left(\frac{\partial^2 \psi}{\partial u^2} + \frac{\partial^2 \psi}{\partial v^2} - 2\frac{\partial^2 \psi}{\partial u \partial v}\right) = \frac{1}{c^2}\left(\frac{\partial^2 \psi}{\partial u^2} + \frac{\partial^2 \psi}{\partial v^2} + 2\frac{\partial^2 \psi}{\partial u \partial v}\right). \tag{5.3.21}$$

Using (5.3.13) we see that (5.3.21) is satisfied if

$$\frac{\partial^2 \psi}{\partial u \partial v} = 0. \tag{5.3.22}$$

Equation (5.3.22) has the solution

$$\psi = h(u) + g(v), \tag{5.3.23}$$

where h and g are arbitrary, twice-differentiable functions. Therefore, equation (5.3.14) satisfies (5.3.1).

Now we will introduce a vector having components l, m, and n,

$$\mathbf{p} = (l, m, n). \tag{5.3.24}$$

From (5.3.13) we see that $\mathbf{p}$ is a unit vector ($|\mathbf{p}| = 1$), and because its components do not depend on x, y, or z, $\mathbf{p}$ is a constant vector.

Using (5.3.24), equation (5.3.14) can be rewritten as

$$\psi(\mathbf{r}, t) = h(t - \mathbf{p} \cdot \mathbf{r}/c) + g(t + \mathbf{p} \cdot \mathbf{r}/c). \tag{5.3.25}$$

Note that for a given value of t, $\psi(\mathbf{r}, t)$ will have the same value for those values of x, y, or z for which $\mathbf{p} \cdot \mathbf{r}$ is a constant, say C. Since $\mathbf{p} \cdot \mathbf{r} = C$ is the equation of a plane, the solutions represented by (5.3.25) are known as *plane waves*. The planes $\mathbf{p} \cdot \mathbf{r} = C$ have $\mathbf{p}$ as the normal vector and are known as *wave fronts*.

5.4 Plane harmonic waves. Superposition principle

In (5.3.3) and (5.3.4) the separation constant λ was arbitrary. If we use $-\omega^2$, instead of λ, we have

$$T''(t) = -\omega^2 T(t) \tag{5.4.1}$$

with the solution

$$T(t) = e^{\pm i\omega t}, \tag{5.4.2}$$

and

$$c^2 \frac{\nabla^2 F}{F} = -\omega^2, \tag{5.4.3}$$

which can be written as

$$\nabla^2 F + k_c^2 F = 0; \qquad k_c = \frac{\omega}{c}. \tag{5.4.4a,b}$$

Equation (5.4.4a) is known as the *Helmholtz equation*. Now write (5.3.6) as follows:

$$\frac{X''}{X} + \frac{Y''}{Y} + \frac{Z''}{Z} = -\frac{\omega^2}{c^2} \tag{5.4.5}$$

and let

$$\frac{\omega^2}{c^2} = k_x^2 + k_y^2 + k_z^2. \tag{5.4.6}$$

Then the equations (5.3.7) become

$$X'' = -k_x^2 X; \qquad Y'' = -k_y^2 Y; \qquad Z'' = -k_z^2 Z \tag{5.4.7}$$

with solutions given by sine or cosine functions (e.g., $\sin k_x x$, $\cos k_x x$) or complex exponentials. We will use the latter for convenience, with the understanding that when solving equations with expected real solutions (such as ground displacement), the real part of the solution must be taken. This approach works well as long as the complex solutions are involved in linear operations (such as integration). For nonlinear operations one must start with real solutions. An example of this will be given in §5.9. With this caveat we write

$$X = e^{\pm ik_x x}; \qquad Y = e^{\pm ik_y y}; \qquad Z = e^{\pm ik_z z}. \tag{5.4.8}$$

Because of (5.4.6), only three out of the four parameters k_x, k_y, k_z, ω can be chosen independently. Let k_z be dependent on the other separation constants:

$$k_z^2 = \frac{\omega^2}{c^2} - k_x^2 - k_y^2 = k_c^2 - k_x^2 - k_y^2. \tag{5.4.9}$$

All the quantities on the right-hand side of (5.4.9) are taken as real, which means that k_z may be real or pure imaginary, depending on the kind of problem being solved. Furthermore, when k_z is imaginary, Z is no longer complex. This will be important in the study of surface waves. Now let

$$\mathbf{k} = (k_x, k_y, k_z) \tag{5.4.10}$$

with

$$|\mathbf{k}| = |k_c| = |\omega|/c. \tag{5.4.11}$$

Equation (5.4.11) follows from (5.4.4b) and (5.4.6). Although ω is usually positive, when working with the Fourier transform it will be allowed to be negative.

Combining (5.4.2) and (5.4.8) we see that the solution to (5.3.1) is of the form $\exp[\pm i(\omega t \pm \mathbf{k} \cdot \mathbf{r})]$. From all the possible sign combinations we will choose

$$\psi(\mathbf{r}, t) \propto e^{i(\omega t \pm \mathbf{k} \cdot \mathbf{r})}. \tag{5.4.12}$$

When ω, k_x, and k_y are positive, the solution (5.4.12) with a minus sign represents a plane harmonic wave propagating away from the origin in the direction of positive x and y. Under these conditions, (5.4.12) with a plus sign represents a wave propagating in the direction of negative x and y.

Using (5.4.10) and (5.4.11) we can define a unit vector

$$\mathbf{p} = \frac{c}{\omega}\mathbf{k}, \tag{5.4.13}$$

which allows us to rewrite (5.4.12) as follows:

$$\psi(\mathbf{r}, t) \propto e^{i\omega(t \pm \mathbf{p} \cdot \mathbf{r}/c)}. \tag{5.4.14}$$

Equations (5.4.12) and (5.4.14) are *harmonic plane waves*, and are a special case of the general plane waves in (5.3.25). The solution (5.4.14) is periodic in both t and $\mathbf{r}$. The corresponding *period T* and *wavelength λ* are derived from

$$e^{i\omega(t - \mathbf{p} \cdot \mathbf{r}/c)} = e^{i\omega[(t+T) - \mathbf{p} \cdot \mathbf{r}/c]} \tag{5.4.15}$$

$$e^{i\omega(t - \mathbf{p} \cdot \mathbf{r}/c)} = e^{i\omega[t - (\mathbf{p} \cdot \mathbf{r} + \lambda)/c]}, \tag{5.4.16}$$

which are satisfied when

$$T = \frac{2\pi}{\omega} = \frac{1}{f}; \qquad \lambda = \frac{2\pi c}{\omega} = \frac{2\pi}{k_c}. \tag{5.4.17a,b}$$

The quantity ω is the *angular frequency* and is equal to $2\pi f$, where f is the *frequency*. If the time t is measured in seconds, then f and ω are measured in cycles per second (or hertz) and radians per second, respectively. Also note that

$$f = \frac{\omega}{2\pi} = \frac{1}{T}; \qquad c = \lambda f; \qquad \frac{k_c}{2\pi} = \frac{1}{\lambda}. \qquad (5.4.18\text{a,b,c})$$

The first and third of these equations show that the equivalent to f in the space domain is $k_c/2\pi$. Some authors call $k_c/2\pi$ the wavenumber, while some others call k_c the *wavenumber*. Here the second usage is followed.

The waves represented by either (5.3.25), (5.4.12), or (5.4.14) are known as *progressive* or *traveling waves*. A somewhat more general solution can be generated by multiplying the two equations (5.4.14) by constant complex numbers $C_1 \exp(i\gamma_1)$ and $C_2 \exp(-i\gamma_2)$, with C_1 and C_2 real, and adding the results:

$$\phi(\mathbf{r}, t, \omega) = C_1 e^{i(\omega t + \mathbf{k} \cdot \mathbf{r} + \gamma_1)} + C_2 e^{i(\omega t - \mathbf{k} \cdot \mathbf{r} - \gamma_2)}. \qquad (5.4.19)$$

In general, $\phi(\mathbf{r}, t, \omega)$ corresponds to a progressive wave, but when $C_1 = C_2 = C$, a different situation arises:

$$\phi(\mathbf{r}, t, \omega) = C \left[e^{i(\omega t + \mathbf{k} \cdot \mathbf{r} + \gamma_1)} + e^{i(\omega t - \mathbf{k} \cdot \mathbf{r} - \gamma_2)} \right]$$

$$= C e^{i[\omega t + (\gamma_1 - \gamma_2)/2]} \left[e^{i[\mathbf{k} \cdot \mathbf{r} + (\gamma_1 + \gamma_2)/2)} + e^{-i[\mathbf{k} \cdot \mathbf{r} + (\gamma_1 + \gamma_2)/2]} \right]$$

$$= 2C e^{i[\omega t + (\gamma_1 - \gamma_2)/2]} \cos[\mathbf{k} \cdot \mathbf{r} + (\gamma_1 + \gamma_2)/2]. \qquad (5.4.20)$$

Equation (5.4.20) does not represent a progressive wave because the combination $\omega t \pm \mathbf{k} \cdot \mathbf{r}$ is no longer present. For this reason, the corresponding waves are known as *standing waves*. This type of wave appears in the study of vibrating strings and plates, and its most important feature is the presence of nodal planes where the motion is zero. These planes satisfy the equation

$$\mathbf{k} \cdot \mathbf{r} + (\gamma_1 + \gamma_2)/2 = \left(n + \frac{1}{2} \right) \pi, \qquad (5.4.21)$$

where n is an integer.

Although harmonic waves are a very special type of waves, they are very important in wave propagation problems because they can be used to generate more general solutions. This is so because in (5.4.14), ω, k_x, and k_y are arbitrary. Therefore, the most general solution to (5.4.1) is obtained by a superposition of the solutions (5.4.14). If ω, k_x, and k_y were discrete (integers), we would write the solution as an infinite sum. However, because these parameters are continuous (real numbers), the solution must be written in integral form. Choosing the minus sign in (5.4.14) we have

$$\psi(\mathbf{r}, t) = \frac{1}{(2\pi)^3} \iiint A(k_x, k_y, z, \omega)e^{i[\omega t - (k_x x + k_y y + \sqrt{k_c^2 - k_x^2 - k_y^2} z)]} \, dk_x \, dk_y \, d\omega.$$

(5.4.22)

Equation (5.4.22) represents the *superposition principle* (Aki and Richards, 1980). Here and in the following, the lower and upper integration limits are $-\infty$ and ∞, respectively. The factor in front of the integral is not required, but is introduced in view of the definitions given below. The function A is, in principle, rather arbitrary (as long as the integral exists). However, when (5.4.22) is used to solve a particular problem, A must be chosen so as to satisfy any appropriate initial or boundary conditions (see Aki and Richards, 1980, for further discussion).

Equation (5.4.22) can be expressed in the following abbreviated form:

$$\psi(\mathbf{r}, t) = \frac{1}{(2\pi)^3} \iiint A(k_x, k_y, z, \omega)e^{i(\omega t - \mathbf{k} \cdot \mathbf{r})} \, dk_x \, dk_y \, d\omega \qquad (5.4.23)$$

(subject to the constraint (5.4.9)).

To give an interpretation of (5.4.23), let us introduce the definition of a Fourier transform in the time and space domains. Given a function $f(\mathbf{r}, t)$, the time and space transform pairs are defined by

$$f(\mathbf{r}, \omega) = \int f(\mathbf{r}, t)e^{-i\omega t} \, dt \qquad (5.4.24)$$

$$f(\mathbf{r}, t) = \frac{1}{2\pi} \int f(\mathbf{r}, \omega)e^{i\omega t} \, d\omega, \qquad (5.4.25)$$

and

$$f(\mathbf{k}, t) = \iiint f(\mathbf{r}, t)e^{i\mathbf{k} \cdot \mathbf{r}} \, dx \, dy \, dz \qquad (5.4.26)$$

$$f(\mathbf{r}, t) = \frac{1}{(2\pi)^3} \iiint f(\mathbf{k}, t)e^{-i\mathbf{k} \cdot \mathbf{r}} \, dk_x \, dk_y \, dk_z. \qquad (5.4.27)$$

The time and space transforms can be combined. For example,

$$f(k_x, k_y, z, \omega) = \iiint f(\mathbf{r}, t)e^{-i(\omega t - k_x x - k_y y)} \, dx \, dy \, dt. \qquad (5.4.28)$$

Comparison of (5.4.22) and (5.4.28) shows that $\psi(\mathbf{r}, t)$ can be considered as the inverse Fourier transform of the function $A(k_x, k_y, z, \omega) \exp[-i(k_c^2 - k_x^2 - k_y^2)^{1/2}z]$.

Note that in the definitions (5.4.24) and (5.4.26) the signs of the exponents are opposite, i.e., negative and positive for the temporal and spatial transforms, respectively. This sign convention is chosen to be consistent with the sign selection in (5.4.23), but is not universal. For example, it is used by Ben-Menahem and Singh (1981), but not by Aki and Richards (1980), who use the opposite convention.

Therefore, care must be taken when comparing results derived by different authors. An example will be given in §7.4.3.

If instead of expressing k_z as a function of k_x, k_y, and ω (see (5.4.9)) we let the three wavenumber components be independent, then from (5.4.10) and (5.4.11),

$$\omega = ck_c \tag{5.4.29}$$

and

$$\psi(\mathbf{r}, t) = \frac{1}{(2\pi)^3} \iiint B(k_x, k_y, k_z) e^{i(\omega t - \mathbf{k} \cdot \mathbf{r})} \, dk_x \, dk_y \, dk_z \tag{5.4.30}$$

for some function B. If the function ψ has a prescribed initial value, say $f(\mathbf{r})$ when $t = 0$, then (5.4.30) gives

$$f(\mathbf{r}) = \frac{1}{(2\pi)^3} \iiint B(k_x, k_y, k_z) e^{-i\mathbf{k} \cdot \mathbf{r}} \, dk_x \, dk_y \, dk_z. \tag{5.4.31}$$

5.5 Spherical waves

Although in this book we will be concerned mostly with plane waves, here we will give a brief introduction to spherical waves to allow a comparison with the plane waves.

Equation (5.3.1) also admits simple solutions under the conditions of spherical symmetry. This can be seen by writing the equation in spherical coordinates with the dependence on the angular variables canceled. Let $\phi = \phi(r, t)$, with $r = |\mathbf{r}|$. Then (5.3.1) becomes

$$\frac{1}{c^2} \frac{\partial^2 \phi}{\partial t^2} = \frac{\partial^2 \phi}{\partial r^2} + \frac{2}{r} \frac{\partial \phi}{\partial r} = \frac{1}{r} \frac{\partial}{\partial r}\left(\phi + r\frac{\partial \phi}{\partial r}\right)$$

$$= \frac{1}{r} \frac{\partial}{\partial r}\left(\frac{\partial}{\partial r}(r\phi)\right) = \frac{1}{r} \frac{\partial^2}{\partial r^2}(r\phi) \tag{5.5.1}$$

(Problem 5.2). Because r does not depend on t, equation (5.5.1) can be rewritten as

$$\frac{1}{c^2} \frac{\partial^2}{\partial t^2}(r\phi) = \frac{\partial^2}{\partial r^2}(r\phi), \tag{5.5.2}$$

which shows that $r\phi(r, t)$ satisfies the one-dimensional wave equation, with the solution given by

$$r\phi(r, t) = h(t - r/c) + g(t + r/c) \tag{5.5.3}$$

so that

$$\phi(r, t) = \frac{1}{r}[h(t - r/c) + g(t + r/c)]. \tag{5.5.4}$$

This solution has several important features. First, note that since r represents the distance from the origin, $h(t_o - r_o/c)$ will be the same for all points on the surface of a sphere with radius r_o. For this reason $h(t - r/c)$ is known as a *spherical wave*. In addition, using an argument similar to that applied to the plane waves it is seen that this wave propagates away from the origin with velocity c. On the other hand, $g(t + r/c)$ is a spherical wave moving towards the origin. However, in unbounded media this wave does not contribute to the solution. This result is based on Sommerfeld's radiation condition (e.g., Stratton, 1941), which can be stated as follows: if the source of waves is confined to a finite region, there are no waves propagating from infinity into the medium (Eringen and Suhubi, 1975).

The factor $1/r$ in (5.5.4) shows that the amplitude of the waves decreases as the distance increases. This is a major difference with plane waves, for which the amplitudes do not depend on the distance. The factor of $1/r$ is known as the *geometric spreading factor*. In spite of the difference in amplitude decay, for sufficiently large values of r a spherical wave front can be approximated locally by a plane wave front (as a tangent to a circle approximates an arc of a circle). This fact is very important because wave propagation problems are generally simpler when plane waves are involved.

Finally, note that $\phi(r, t)$ given by (5.5.4) has a singularity for $r = 0$. This fact is related to the presence of a source for the waves that is located at the origin.

5.6 Vector wave equation. Vector solutions

The vector wave equation

$$\nabla^2 \mathbf{u}(\mathbf{r}, t) = \frac{1}{c^2} \frac{\partial^2 \mathbf{u}(\mathbf{r}, t)}{\partial t^2} \equiv \frac{1}{c^2} \ddot{\mathbf{u}} \tag{5.6.1}$$

where the double dots indicate a second derivative with respect to time, is of importance in both elastic and electromagnetic wave propagation, and will be studied in detail here.

Before proceeding it is important to emphasize the difference between the scalar wave equation (5.3.1) and (5.6.1). In the first case ψ is a scalar variable, which means that to specify ψ at each point in space just one number is required. In the case of (5.6.1), however, three numbers u_1, u_2, u_3 are needed, thus increasing the complexity of the problem of finding the solution of (5.6.1)

If the boundary conditions are such that the use of Cartesian coordinates is appropriate, then (5.6.1) is equivalent to

$$\nabla^2 u_i = u_{i,jj} = \frac{1}{c^2} \ddot{u}_i; \qquad i = 1, 2, 3. \tag{5.6.2}$$

In this particular case the problem reduces to the solution of three scalar wave equations, one for each function u_i.

When the problem requires the use of other coordinate systems, however, the three components of **u** do not separate into three equations. Instead, they become coupled through a system of three simultaneous partial differential equations. This complexity arises because in curvilinear coordinates the Laplacian is expressed as

$$\nabla^2 \mathbf{u} = \nabla(\nabla \cdot \mathbf{u}) - \nabla \times \nabla \times \mathbf{u} \qquad (5.6.3)$$

(see (1.4.53)), which in turn requires that the gradient, divergence, and curl be expressed in curvilinear coordinates. To see the type of systems of equations involved in cylindrical and spherical coordinates refer to, for example, Ben-Menahem and Singh (1981).

Here an alternative approach, based on the references given in §5.1 will be presented, namely, we will look for vector solutions of (5.6.1). As a motivation we start with the following decomposition of **u**:

$$\mathbf{u} = \nabla\psi + \nabla \times \mathbf{v}; \qquad \nabla \cdot \mathbf{v} = 0. \qquad (5.6.4)$$

This is the *Helmholtz decomposition theorem*, which will be proved in §9.3. Although the results to be derived below are quite general, here we will concentrate on Cartesian coordinates. Let

$$\mathbf{L} = \nabla\psi, \qquad (5.6.5)$$

where ψ is a scalar function of **r** and other variables such as t and ω. Then **L** is a solution to (5.6.1) if ψ satisfies the scalar wave equation:

$$\nabla^2\psi - \frac{1}{c^2}\ddot{\psi} = 0. \qquad (5.6.6)$$

To show this, equation (5.6.1) will be written in component form:

$$\nabla^2 u_i - \frac{1}{c^2}\ddot{u}_i = 0; \qquad i = 1, 2, 3. \qquad (5.6.7)$$

Then, using

$$u_i = L_i = (\mathbf{L})_i = \psi_{,i} \qquad (5.6.8)$$

equation (5.6.7) gives

$$L_{i,jj} - \frac{1}{c^2}\ddot{L}_i = \psi_{,ijj} - \frac{1}{c^2}\ddot{\psi}_{,i} = \left(\psi_{,jj} - \frac{1}{c^2}\ddot{\psi}\right)_{,i} = 0. \qquad (5.6.9)$$

From (5.6.9) we see that if

$$\psi_{,jj} - \frac{1}{c^2}\ddot{\psi} = \nabla^2\psi - \frac{1}{c^2}\ddot{\psi} = 0 \qquad (5.6.10)$$

equation (5.6.7) is satisfied and **L** is a solution to (5.6.1).

Now consider the vector

$$\mathbf{M} = \nabla \times \mathbf{a}\phi, \qquad (5.6.11)$$

where $\mathbf{a}$ is either an arbitrary constant vector of unit length or $\mathbf{a} = \mathbf{r}$, and ϕ is a scalar function of $\mathbf{r}$ and other variables such as t and ω. Then $\mathbf{M}$ is a solution to (5.6.1) if ϕ solves the scalar wave equation:

$$\nabla^2 \phi - \frac{1}{c^2}\ddot{\phi} = 0. \tag{5.6.12}$$

The case $\mathbf{a} = $ constant will be treated first,

$$u_i = M_i = (\mathbf{M})_i = \epsilon_{ijk}(\phi a_k)_{,j} = \epsilon_{ijk}\phi a_{k,j} + \epsilon_{ijk}\phi_{,j}a_k = \epsilon_{ijk}\phi_{,j}a_k. \tag{5.6.13}$$

Introducing (5.6.13) into (5.6.7) gives

$$\nabla^2 M_i - \frac{1}{c^2}\ddot{M}_i = M_{i,ll} - \frac{1}{c^2}\ddot{M}_i = \epsilon_{ijk}\phi_{,jll}a_k - \frac{1}{c^2}\epsilon_{ijk}\ddot{\phi}_{,j}a_k$$

$$= \epsilon_{ijk}a_k\left(\phi_{,ll} - \frac{1}{c^2}\ddot{\phi}\right)_{,j} = 0. \tag{5.6.14}$$

Therefore, if

$$\phi_{,ll} - \frac{1}{c^2}\ddot{\phi} = \nabla^2\phi - \frac{1}{c^2}\ddot{\phi} = 0 \tag{5.6.15}$$

equation (5.6.7) is satisfied and $\mathbf{M}$ is a solution to (5.6.1).

Next, consider the case where $\mathbf{a} = \mathbf{r}$. Then,

$$M_i = \epsilon_{ijk}(\phi r_k)_{,j} = \epsilon_{ijk}\phi r_{k,j} + \epsilon_{ijk}\phi_{,j}r_k = \epsilon_{ijk}\phi\delta_{kj} + \epsilon_{ijk}\phi_{,j}r_k = \epsilon_{ijk}\phi_{,j}r_k \tag{5.6.16}$$

and

$$\nabla^2 M_i - \frac{1}{c^2}\ddot{M}_i = \epsilon_{ijk}(\phi_{,j}r_k)_{,ll} - \frac{1}{c^2}\epsilon_{ijk}\ddot{\phi}_{,j}r_k$$

$$= \epsilon_{ijk}\left(\phi_{,jll}r_k + 2\phi_{,jl}r_{k,l} + \phi_{,j}r_{k,ll}\right) - \frac{1}{c^2}\epsilon_{ijk}\ddot{\phi}_{,j}r_k$$

$$= \epsilon_{ijk}r_k\left(\phi_{,ll} - \frac{1}{c^2}\ddot{\phi}\right)_{,j} = 0. \tag{5.6.17}$$

Note that $r_{k,ll} = 0$ and that

$$\phi_{,jl}r_{k,l} = \phi_{,jl}\delta_{kl} = \phi_{,jk}, \tag{5.6.18}$$

which is a symmetric tensor. Also in this case, if

$$\nabla^2\phi - \frac{1}{c^2}\ddot{\phi} = 0 \tag{5.6.19}$$

equation (5.6.7) is satisfied and $\mathbf{M}$ is a solution to (5.6.1).

Finally, let

$$\mathbf{N} = h\nabla \times \mathbf{M}, \tag{5.6.20}$$

where the factor of h is introduced to ensure that $\mathbf{M}$ and $\mathbf{N}$ have the same dimensions. Then,

$$u_i = N_i = (\mathbf{N})_i = h\epsilon_{ijk}M_{k,j}. \tag{5.6.21}$$

Now introduce (5.6.21) into (5.6.7), which gives

$$N_{i,ll} - \frac{1}{c^2}\ddot{N}_i = h\epsilon_{ijk}\left(M_{k,jll} - \frac{1}{c^2}\ddot{M}_{k,j}\right) = h\epsilon_{ijk}\left(M_{k,ll} - \frac{1}{c^2}\ddot{M}_k\right)_{,j} = 0. \tag{5.6.22}$$

Because the expression in parentheses is equal to zero, $\mathbf{N}$ also solves (5.6.1).

The functions ψ and ϕ are known as potentials, and the vectors $\mathbf{L}, \mathbf{M}, \mathbf{N}$ are called *Hansen vectors* by Ben-Menahem and Singh (1981). This usage will be followed here.

5.6.1 Properties of the Hansen vectors

(a) $\mathbf{L}$ is an irrotational vector,

$$\left(\nabla \times \mathbf{L}\right)_i = \epsilon_{ijk}\psi_{,kj} = 0. \tag{5.6.23}$$

(b) $\mathbf{M}$ is a solenoidal vector. For $\mathbf{a} = \mathbf{r}$,

$$\nabla \cdot \mathbf{M} = M_{i,i} = \epsilon_{ijk}(\phi_{,j}r_k)_{,i} = \epsilon_{ijk}(\phi_{,ji}r_k + \phi_{,j}\delta_{ik}) = 0 \tag{5.6.24}$$

on account of the symmetry of $\phi_{,ji}$ and δ_{ik}. This result also holds for $\mathbf{a} =$ constant (Problem 5.3).

(c) $\mathbf{N}$ is a solenoidal vector,

$$\nabla \cdot \mathbf{N} = N_{i,i} = h\epsilon_{ijk}M_{k,ji} = 0, \tag{5.6.25}$$

because $M_{k,ji}$ is symmetric in ji.

(d) Divergence of $\mathbf{L}$,

$$\nabla \cdot \mathbf{L} = L_{i,i} = \psi_{,ii} = \nabla^2\psi = \frac{1}{c^2}\ddot{\psi}. \tag{5.6.26}$$

(e) $\mathbf{M} = \nabla\phi \times \mathbf{a}$, for $\mathbf{a} =$ constant or $\mathbf{a} = \mathbf{r}$.

Let, for example, $\mathbf{a} = \mathbf{r}$. Then, from (5.6.16)

$$M_i = \epsilon_{ijk}\phi_{,j}r_k = (\nabla\phi \times \mathbf{r})_i. \tag{5.6.27}$$

Therefore, if $\phi = \psi$,

$$\mathbf{M} = \mathbf{L} \times \mathbf{a} \tag{5.6.28}$$

in which case $\mathbf{L}$ and $\mathbf{M}$ are perpendicular to each other.

5.6.2 Harmonic potentials

Let **a** be a constant vector, $\phi = \psi$ and

$$\psi(\mathbf{r}, t) = e^{i(\omega t - \mathbf{k} \cdot \mathbf{r})}. \qquad (5.6.29)$$

Then, since $\mathbf{k} \cdot \mathbf{r} = k_i x_i$

$$L_p = \psi_{,p} = -ik_p \psi \qquad (5.6.30)$$

(Problem 5.4) so that

$$\mathbf{L} = -i\mathbf{k}\psi = -i\mathbf{k}e^{i(\omega t - \mathbf{k} \cdot \mathbf{r})} \qquad (5.6.31)$$

and

$$M_p = -i\epsilon_{pjk}k_j a_k \psi = -i(\mathbf{k} \times \mathbf{a})_p \psi \qquad (5.6.32)$$

so that

$$\mathbf{M} = -i(\mathbf{k} \times \mathbf{a})e^{i(\omega t - \mathbf{k} \cdot \mathbf{r})}. \qquad (5.6.33)$$

For the vector **N** we choose $h = 1/k$, where $k = |\mathbf{k}|$. Then, since **a** is a constant vector,

$$N_p = \frac{1}{k}\epsilon_{pjq}M_{q,j} = -i\frac{1}{k}\epsilon_{pjq}[(\mathbf{k} \times \mathbf{a})_q \psi]_{,j} = -\frac{1}{k}\epsilon_{pjq}(\mathbf{k} \times \mathbf{a})_q k_j \psi$$

$$= \frac{1}{k}\epsilon_{pqj}(\mathbf{k} \times \mathbf{a})_q k_j \psi \qquad (5.6.34)$$

(Problem 5.5) so that

$$\mathbf{N} = \frac{1}{k}(\mathbf{k} \times \mathbf{a} \times \mathbf{k})e^{i(\omega t - \mathbf{k} \cdot \mathbf{r})}. \qquad (5.6.35)$$

Equations (5.6.31), (5.6.33), and (5.6.35) show that in this special case the three vectors are perpendicular to each other.

5.7 Vector Helmholtz equation

If the Fourier transform in the time domain is applied to (5.6.1) we obtain the *vector Helmholtz equation*,

$$\nabla^2 \mathbf{u} + k_c^2 \mathbf{u} = 0; \qquad k_c = \frac{\omega}{c} \qquad (5.7.1)$$

(see Problem 9.13) with

$$\mathbf{u}(\mathbf{r}, \omega) = \int_{-\infty}^{\infty} \mathbf{u}(\mathbf{r}, t)e^{-i\omega t} \, dt. \qquad (5.7.2)$$

Note that the same symbol **u** is used regardless of whether it is a function of t or ω.

Equation (5.7.1) is the vector extension of (5.4.4a), and will be used in the solution of the elastic wave equation. To solve (5.7.1) we assume again Cartesian coordinates, so that in component form we have

$$\nabla^2 u_i + k_c^2 u_i = 0. \tag{5.7.3}$$

To solve (5.7.3) we will use Hansen vectors $\mathbf{L}$, $\mathbf{M}$, and $\mathbf{N}$ similar to those introduced in §5.6 with $\phi = \psi$. Let

$$\mathbf{L} = \frac{1}{k_c} \nabla \psi \tag{5.7.4}$$

$$\mathbf{M} = \nabla \times \mathbf{a} \psi \tag{5.7.5}$$

$$\mathbf{N} = \frac{1}{k_c} \nabla \times \mathbf{M}. \tag{5.7.6}$$

Then, equations (5.7.4)–(5.7.6) solve (5.7.3) if ψ solves the scalar Helmholtz equation

$$\nabla^2 \psi + k_c^2 \psi = 0. \tag{5.7.7}$$

In (5.7.5) the vector $\mathbf{a}$ is, as before, either an arbitrary unit vector or it is equal to $\mathbf{r}$. The proofs of these statements are basically those given in §5.6 (Problem 5.6).

Equation (5.7.7) is equivalent to (5.4.4a), the solution of which is proportional to the product of the three functions in (5.4.8), so that

$$\psi \propto e^{\pm i \mathbf{k} \cdot \mathbf{r}} \tag{5.7.8}$$

solves (5.7.7). A particular solution, representing a plane wave, is

$$\psi(\mathbf{r}, \omega) = e^{i(\omega t \pm \mathbf{k} \cdot \mathbf{r})}. \tag{5.7.9}$$

Because $\exp(\pm i \mathbf{k} \cdot \mathbf{r})$ solves (5.7.7), and $\exp(i\omega t)$ is a constant with respect to the spatial variables, their product is also a solution. In §5.8.3, equation (5.7.9) will be used to generate plane wave solutions to the elastic wave equation.

5.8 Elastic wave equation without body forces

In this section we will investigate the vectorial nature of the P- and S-wave motion, and will introduce frequency-domain Hansen vectors for the elastic wave equation. The latter will then be specialized to three vectors representing P, SV, and SH motion for the case of plane waves.

5.8.1 Vector P- and S-wave motion

In §4.8 we saw that the elastic wave equation can be separated into two simpler wave equations, a scalar one and a vector one, involving the divergence and the curl of the displacement, respectively. We found that the corresponding wave velocities were α and β, but we did not make any statements regarding the direction of particle motion. To address this question let us investigate under what conditions a vector plane wave satisfies the elastic vector wave equation in the absence of body forces. For convenience we start with (4.8.2),

$$\mu \nabla^2 \mathbf{u} + (\lambda + \mu) \nabla (\nabla \cdot \mathbf{u}) = \rho \frac{\partial^2 \mathbf{u}}{\partial t^2}. \tag{5.8.1}$$

The case of nonzero body forces will be discussed in Chapter 9. Now let

$$\mathbf{u} = \mathbf{d}h(ct - \mathbf{p} \cdot \mathbf{r}) \tag{5.8.2}$$

where $\mathbf{d}$ is a constant unit vector whose direction must be determined. The function h is the plane wave that appears in (5.3.25) after a minor rewriting of the argument. Recall that $\mathbf{p}$ is also a unit vector. The following intermediate results are needed:

$$\nabla \cdot \mathbf{u} = d_i h_{,i} = -d_i p_i h'(ct - \mathbf{p} \cdot \mathbf{r}) = -(\mathbf{d} \cdot \mathbf{p})h'(ct - \mathbf{p} \cdot \mathbf{r}) \tag{5.8.3}$$

$$\nabla(\nabla \cdot \mathbf{u}) = (p_1, p_2, p_3) (\mathbf{d} \cdot \mathbf{p})h''(ct - \mathbf{p} \cdot \mathbf{r}) = (\mathbf{d} \cdot \mathbf{p})\mathbf{p}h''(ct - \mathbf{p} \cdot \mathbf{r}) \tag{5.8.4}$$

$$\nabla^2 \mathbf{u} = \mathbf{d} \left(p_1^2 + p_2^2 + p_3^2 \right) h''(ct - \mathbf{p} \cdot \mathbf{r}) = \mathbf{d}h''(ct - \mathbf{p} \cdot \mathbf{r}) \tag{5.8.5}$$

$$\ddot{\mathbf{u}} = c^2 \mathbf{d}h''(ct - \mathbf{p} \cdot \mathbf{r}). \tag{5.8.6}$$

Introducing (5.8.4)–(5.8.6) in (5.8.1) gives

$$[\mu \mathbf{d} + (\lambda + \mu)(\mathbf{d} \cdot \mathbf{p})\mathbf{p} - \rho c^2 \mathbf{d}]h''(ct - \mathbf{p} \cdot \mathbf{r}) = 0, \tag{5.8.7}$$

which implies

$$(\mu - \rho c^2)\mathbf{d} + (\lambda + \mu)(\mathbf{d} \cdot \mathbf{p})\mathbf{p} = 0 \tag{5.8.8}$$

(Achenbach, 1973).

To determine $\mathbf{d}$ multiply (5.8.8) scalarly with $\mathbf{p}$ and use $(\mathbf{p} \cdot \mathbf{p}) = 1$. This gives

$$[(\mu - \rho c^2) + (\lambda + \mu)](\mathbf{d} \cdot \mathbf{p}) = 0. \tag{5.8.9}$$

Equation (5.8.9) implies that either the factor in brackets is zero or $(\mathbf{p} \cdot \mathbf{d}) = 0$. In the first case we find

$$c^2 = \frac{\lambda + 2\mu}{\rho} \tag{5.8.10}$$

so that c is the velocity of the P waves. Now, introducing (5.8.10) into (5.8.8) gives

$$(\lambda + \mu)[\mathbf{d} - (\mathbf{d} \cdot \mathbf{p})\mathbf{p}] = 0, \tag{5.8.11}$$

which implies

$$\mathbf{d} = (\mathbf{d} \cdot \mathbf{p})\mathbf{p}. \tag{5.8.12}$$

Multiplying both sides of (5.8.12) scalarly with $\mathbf{d}$ and using the fact that $\mathbf{d}$ is a unit vector gives

$$(\mathbf{d} \cdot \mathbf{p})^2 = 1. \tag{5.8.13}$$

Equation (5.8.13) is satisfied by $\mathbf{d} = \pm\mathbf{p}$. This, in turn, shows that in the case of P waves the direction of particle motion (given by $\mathbf{d}$) is parallel to the direction of wave propagation (given by $\mathbf{p}$). In addition, this type of motion is irrotational, as can be seen by setting $\mathbf{d} = \mathbf{p}$ and taking the curl of $\mathbf{u}$,

$$(\nabla \times \mathbf{u})_i = \epsilon_{ijk} u_{k,j} = \epsilon_{ijk} p_k [h(ct - \mathbf{p} \cdot \mathbf{r})]_{,j} = -\epsilon_{ijk} p_k p_j h'(ct - \mathbf{p} \cdot \mathbf{r}) = 0 \tag{5.8.14}$$

because $p_k p_j$ is symmetric.

Now consider the case where $(\mathbf{d} \cdot \mathbf{p}) = 0$. Under this condition, from (5.8.8) we obtain

$$c^2 = \frac{\mu}{\rho}. \tag{5.8.15}$$

This is the velocity of the S waves. Furthermore, because of the vanishing of the dot product we see that $\mathbf{d}$, and hence $\mathbf{u}$, is perpendicular to $\mathbf{p}$. This motion has the property that its divergence is zero (i.e., it is solenoidal) as can be seen from (5.8.3).

In summary, the elastic wave equation admits two types of solution. One of them corresponds to P waves, with particle motion parallel to the direction of wave propagation. For this reason these waves are known as *longitudinal* waves. The second solution corresponds to S waves, with particle motion in a plane perpendicular to the direction of wave propagation, which motivates calling these waves *transverse*. In addition, because the divergence is zero, these waves are also known as *equivoluminal*. The vectors corresponding to these two types of motion are indicated by $\mathbf{P}$ and $\mathbf{S}$ in Fig. 5.3. In the absence of body forces (and for isotropic media), the direction of $\mathbf{S}$ is unconstrained, and, as shown below, can be decomposed into the so-called SH and SV components. However, we will see in Chapters 9 and 10 that when body forces are considered the direction of the S-wave motion becomes well constrained.

5.8.2   Hansen vectors for the elastic wave equation in the frequency domain

In this case we use the following form of the elastic wave equation:

$$\alpha^2 \nabla(\nabla \cdot \mathbf{u}) - \beta^2 \nabla \times \nabla \times \mathbf{u} = \frac{\partial^2 \mathbf{u}}{\partial t^2} \tag{5.8.16}$$

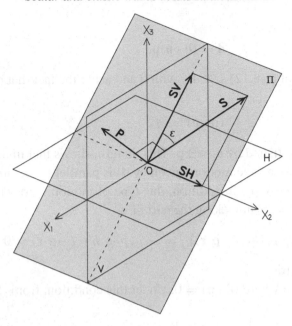

Fig. 5.3. Direction of particle motion corresponding to P and S waves and decomposition of the S-wave motion into SH and SV motion. The vector **P** is perpendicular to the wave front Π and the vector **S** is on the wave front. The vector **SV** is in the intersection of Π with the vertical plane (x_1, x_3). This plane, labeled V, also contains **P**. The vector **SH** is in the intersection of Π with the horizontal plane (x_1, x_2) (labeled H). The coordinate system shown here has been obtained by rotation as discussed in the text. The angle ε between **S** and **SV** is known as the polarization angle of the S wave.

(see (4.8.4)). Equation (5.8.16) will be studied in the frequency domain. After application of the Fourier transform (5.8.16) becomes

$$\alpha^2 \nabla(\nabla \cdot \mathbf{u}) - \beta^2 \nabla \times \nabla \times \mathbf{u} + \omega^2 \mathbf{u} = 0 \qquad (5.8.17)$$

with $\mathbf{u} = \mathbf{u}(\mathbf{r}, \omega)$ given by (5.7.2).

To solve (5.8.17) let

$$\mathbf{u} = \mathbf{u}_\alpha + \mathbf{u}_\beta, \qquad (5.8.18)$$

where $\mathbf{u}_\alpha$ is irrotational and $\mathbf{u}_\beta$ is solenoidal, i.e.

$$\nabla \times \mathbf{u}_\alpha = 0; \qquad \nabla \cdot \mathbf{u}_\beta = 0. \qquad (5.8.19)$$

Introducing (5.8.18) in (5.8.17) and using (5.8.19) gives

$$\alpha^2 \nabla(\nabla \cdot \mathbf{u}_\alpha) - \beta^2 \nabla \times \nabla \times \mathbf{u}_\beta + \omega^2(\mathbf{u}_\alpha + \mathbf{u}_\beta) = 0. \qquad (5.8.20)$$

Furthermore, using (5.6.3) and (5.8.19) we can write

$$\nabla^2 \mathbf{u}_\alpha = \nabla(\nabla \cdot \mathbf{u}_\alpha) \qquad (5.8.21)$$

$$\nabla^2 \mathbf{u}_\beta = -\nabla \times \nabla \times \mathbf{u}_\beta. \tag{5.8.22}$$

Then (5.8.20) can be written as

$$\alpha^2 (\nabla^2 \mathbf{u}_\alpha + k_\alpha^2 \mathbf{u}_\alpha) + \beta^2 (\nabla^2 \mathbf{u}_\beta + k_\beta^2 \mathbf{u}_\beta) = 0, \tag{5.8.23}$$

where

$$k_\alpha = \omega/\alpha; \qquad k_\beta = \omega/\beta. \tag{5.8.24}$$

Equation (5.8.23) is satisfied if

$$\nabla^2 \mathbf{u}_\alpha + k_\alpha^2 \mathbf{u}_\alpha = 0 \tag{5.8.25}$$

and

$$\nabla^2 \mathbf{u}_\beta + k_\beta^2 \mathbf{u}_\beta = 0 \tag{5.8.26}$$

(Ben-Menahem and Singh, 1981). These are vector Helmholtz equations, and because $\mathbf{L}$ is irrotational and $\mathbf{M}$ and $\mathbf{N}$ are solenoidal (see §5.6.1) and satisfy the vector Helmholtz equation (see §5.7), these three vectors constitute solutions to (5.8.17). Following Ben-Menahem and Singh (1981) they will be written as

$$\mathbf{L} = \frac{1}{k_\alpha} \nabla \psi_\alpha \tag{5.8.27}$$

$$\mathbf{M} = \nabla \times (\mathbf{e}_3 \psi_\beta) \tag{5.8.28}$$

$$\mathbf{N} = \frac{1}{k_\beta} \nabla \times \mathbf{M} = \frac{1}{k_\beta} \nabla \times \nabla \times (\mathbf{e}_3 \psi_\beta). \tag{5.8.29}$$

Note that $\mathbf{a} = \mathbf{e}_3$. Equations (5.8.27)–(5.8.29) are also applicable for the case of cylindrical and spherical coordinates with $\mathbf{a}$ equal to $\mathbf{e}_3$ and $\mathbf{r}$, respectively. The potentials ψ_α and ψ_β satisfy the Helmholtz equations:

$$\nabla^2 \psi_\alpha + k_\alpha^2 \psi_\alpha = 0 \tag{5.8.30}$$

$$\nabla^2 \psi_\beta + k_\beta^2 \psi_\beta = 0 \tag{5.8.31}$$

(see (5.7.7)).

5.8.3 *Harmonic elastic plane waves*

In this case the potentials ψ_α and ψ_β are given by

$$\psi_c(\mathbf{r}, \omega) = e^{i(\omega t - \mathbf{k}_c \cdot \mathbf{r})}; \qquad c = \alpha, \beta \tag{5.8.32}$$

(see (5.7.9)), where

$$\mathbf{k}_c = k_c \mathbf{p}; \qquad |k_c| = |\mathbf{k}_c| = \frac{|\omega|}{c} \tag{5.8.33}$$

$$\mathbf{p} = (l, m, n); \qquad l^2 + m^2 + n^2 = 1 \tag{5.8.34a,b}$$

$$\mathbf{k}_c \cdot \mathbf{r} = |\mathbf{k}_c|(lx_1 + mx_2 + nx_3) = k_c\mathbf{p} \cdot \mathbf{r}. \tag{5.8.35}$$

As before, $\mathbf{p}$ is a constant unit vector, $\mathbf{p} \cdot \mathbf{r} = $ constant describes a plane normal to $\mathbf{p}$, and ψ_c is a harmonic plane wave propagating with velocity c and with wave fronts perpendicular to $\mathbf{p}$.

Now replace ψ_c in the expressions (5.8.27)–(5.8.29) for $\mathbf{L}$, $\mathbf{M}$, and $\mathbf{N}$. For $\mathbf{L}$ we have

$$L_j = \frac{1}{k_\alpha}(\nabla\psi_\alpha)_j = \frac{1}{k_\alpha}(\psi_\alpha)_{,j} = \frac{-i}{k_\alpha}(\mathbf{k}_\alpha)_j\psi_\alpha = -i(\mathbf{p})_j\psi_\alpha \tag{5.8.36}$$

so that

$$\mathbf{L} = -i\mathbf{p}\psi_\alpha. \tag{5.8.37}$$

As $\mathbf{L}$ is proportional to $\mathbf{p}$ and is associated with the velocity α, it represents P-wave motion, as discussed in §5.8.1.

For $\mathbf{M}$ we have

$$M_j = \left[\nabla \times (\mathbf{e}_3\psi_\beta)\right]_j = \epsilon_{jlm}(\mathbf{e}_3)_m\psi_{\beta,l} \tag{5.8.38}$$

and, since $\mathbf{e}_3 = (0, 0, 1)$,

$$M_1 = \epsilon_{1l3}\psi_{\beta,l} = \epsilon_{123}\psi_{\beta,2} = -ik_\beta m\psi_\beta \tag{5.8.39}$$

$$M_2 = \epsilon_{2l3}\psi_{\beta,l} = \epsilon_{213}\psi_{\beta,1} = ik_\beta l\psi_\beta \tag{5.8.40}$$

$$M_3 = \epsilon_{3l3}\psi_{\beta,l} = 0 \tag{5.8.41}$$

so that

$$\mathbf{M} = -ik_\beta(m\mathbf{e}_1 - l\mathbf{e}_2)\psi_\beta. \tag{5.8.42}$$

For $\mathbf{N}$ we have

$$\mathbf{N} = \frac{1}{k_\beta}\nabla \times \mathbf{M} = -i\nabla \times [(m\mathbf{e}_1 - l\mathbf{e}_2)\psi_\beta]$$

$$= -i[l\psi_{\beta,3}\mathbf{e}_1 + m\psi_{\beta,3}\mathbf{e}_2 - (l\psi_{\beta,1} + m\psi_{\beta,2})\mathbf{e}_3]$$

$$= [-ln\mathbf{e}_1 - mn\mathbf{e}_2 + (l^2 + m^2)\mathbf{e}_3]k_\beta\psi_\beta. \tag{5.8.43}$$

In summary,

$$\mathbf{L} = -i(l\mathbf{e}_1 + m\mathbf{e}_2 + n\mathbf{e}_3)\psi_\alpha \tag{5.8.44}$$

$$\mathbf{M} = -ik_\beta(m\mathbf{e}_1 - l\mathbf{e}_2)\psi_\beta \tag{5.8.45}$$

$$\mathbf{N} = -k_\beta[ln\mathbf{e}_1 + mn\mathbf{e}_2 - (l^2 + m^2)\mathbf{e}_3]\psi_\beta \tag{5.8.46}$$

(Ben-Menahem and Singh, 1981).

In addition, the three vectors are mutually perpendicular:

$$\mathbf{L} \cdot \mathbf{M} = 0; \qquad \mathbf{L} \cdot \mathbf{N} = 0; \qquad \mathbf{M} \cdot \mathbf{N} = 0. \qquad (5.8.47)$$

Since the vectors $\mathbf{M}$ and $\mathbf{N}$ are perpendicular to $\mathbf{L}$ and are associated with the velocity β, they represent S-wave motion, as discussed in §5.8.1. Moreover, the vector $\mathbf{d}$ discussed there can, in principle, be written as a linear combination of $\mathbf{M}$ and $\mathbf{N}$.

The triplet $(\mathbf{L}, \mathbf{N}, \mathbf{M})$ forms a right-handed orthogonal set. Plane waves propagate away from the origin in the direction of $\mathbf{p}$. Let Π indicate the wave front. Then, $\mathbf{M}$ and $\mathbf{N}$ are vectors on Π, although it must be noted that $\mathbf{M}$ has no $\mathbf{e}_3$ component. Furthermore, because ψ_α and ψ_β are constant on Π for fixed values of t and ω, the values of $\mathbf{L}$, $\mathbf{M}$, and $\mathbf{N}$ are also constant.

5.8.4 *P-, SV-, and SH-wave displacements*

As noted in the previous section, the vector $\mathbf{L}$ represents P-wave motion, while the vectors $\mathbf{M}$ and $\mathbf{N}$ represent S-wave motion. The last two vectors will be used to introduce the SH and SV components of the S-wave motion. This will be done by means of a rotation of coordinates, motivated by the following observation: if m in (5.8.44)–(5.8.46) is set equal to zero, then $\mathbf{M}$ has a component along $\mathbf{e}_2$ only, while $\mathbf{L}$ and $\mathbf{N}$ have components in the (x_1, x_3) plane only. This significant simplification can be achieved by a rotation of coordinates about the x_3 axis such that the new x_1 axis is along the projection of $\mathbf{p}$ onto the (x_1, x_2) plane (see Fig. 5.3). After the rotation $\mathbf{p}$ has no component along the new x_2 axis. At this point we should introduce primes to indicate the new coordinate axes and the new components of $\mathbf{p}$, but this will not be done to simplify the notation. Instead, the unit vectors $\mathbf{e}_1$, $\mathbf{e}_2$, and $\mathbf{e}_3$ will be replaced by unit vectors $\mathbf{a}_1$, $\mathbf{a}_2$, and $\mathbf{a}_3$. In the remainder of this chapter and in Chapters 6 and 7 we will work in the rotated system. The equations below follow Ben-Menahem and Singh (1981).

The expressions for $\mathbf{L}$, $\mathbf{M}$, and $\mathbf{N}$ after the rotation are obtained from (5.8.44)–(5.8.46) and (5.8.32) with $m = 0$:

$$\mathbf{L} = -i(l\mathbf{a}_1 + n\mathbf{a}_3) \exp\left[i\omega \left(t - \frac{lx_1 + nx_3}{\alpha} \right) \right] \qquad (5.8.48)$$

$$\mathbf{M} = ilk_\beta \mathbf{a}_2 \exp\left[i\omega \left(t - \frac{lx_1 + nx_3}{\beta} \right) \right] \qquad (5.8.49)$$

$$\mathbf{N} = -lk_\beta (n\mathbf{a}_1 - l\mathbf{a}_3) \exp\left[i\omega \left(t - \frac{lx_1 + nx_3}{\beta} \right) \right] \qquad (5.8.50)$$

with

$$l^2 + n^2 = 1. \qquad (5.8.51)$$

After rotation, $\mathbf{M}$ is a horizontal vector aligned with the x_2 axis, and the motion represented by this vector is known as SH motion. The vectors $\mathbf{L}$ and $\mathbf{N}$, on the other hand, lie in the vertical plane (x_1, x_3), and, consequently, the motion represented by $\mathbf{N}$ is known as SV motion. Clearly, the H and V in SH and SV denote horizontal and vertical. The plane (x_1, x_3) is known as the *plane of incidence*, and can be visualized as the vertical plane through the seismic source and the receiver that records the seismic waves. As actual three-component seismic data are generally recorded along east–west, north–south, and vertical directions, the S-wave motion observed in the three components is neither SH nor SV. If the location of the seismic source is known, the rotation angle can be determined. Then, with this information it is possible to rotate the two horizontal components, which then will show the so-called SH and SV waves.

Finally, we will write the expressions for the displacements corresponding to P, SV and SH motion in the rotated system, in which

$$\mathbf{p} = l\mathbf{a}_1 + n\mathbf{a}_3. \tag{5.8.52}$$

Introducing this expression in (5.8.48)–(5.8.50) and using $\mathbf{p} \times \mathbf{a}_2 = -(n, 0, -l)$, we obtain

$$\mathbf{u}_P = A\mathbf{p} \exp\left[i\omega\left(t - \frac{\mathbf{p} \cdot \mathbf{r}}{\alpha} \right) \right] \tag{5.8.53}$$

$$\mathbf{u}_{SV} = B(\mathbf{p} \times \mathbf{a}_2) \exp\left[i\omega\left(t - \frac{\mathbf{p} \cdot \mathbf{r}}{\beta} \right) \right] \tag{5.8.54}$$

$$\mathbf{u}_{SH} = C\mathbf{a}_2 \exp\left[i\omega\left(t - \frac{\mathbf{p} \cdot \mathbf{r}}{\beta} \right) \right], \tag{5.8.55}$$

where A, B, and C absorb obvious scalar factors in (5.8.48)–(5.8.50). Equations (5.8.53)–(5.8.55) represent the particle motion corresponding to P, SV, and SH waves and will be used in Chapters 6 and 7. As we will see there, for a given problem, the factors A, B, and C are determined by the boundary conditions appropriate for the problem.

In focal mechanism studies an important parameter is the angle ε between $\mathbf{u}_{SV}$ and $\mathbf{u}_{SV} + \mathbf{u}_{SH}$, which is known as the *polarization angle* (Fig. 5.3, see also §9.9.1 and §10.9).

5.8.4.1 *Polarization of particle motion*

Consider vector displacements of the form

$$\mathbf{u} = (c_1\mathbf{a}_1 + c_3\mathbf{a}_3) \exp[i(\omega t - k_1 x_1)], \tag{5.8.56}$$

where c_1 and c_3 are independent of x_1 and t. Equation (5.8.56) represents, for example, ground motion, for which $x_3 = 0$. Because we are interested in real quantities (5.8.56) must be replaced by

$$\mathbf{u} = (c_1\mathbf{a}_1 + c_3\mathbf{a}_3)\cos(\omega t - k_1 x_1) \equiv u_1\mathbf{a}_1 + u_3\mathbf{a}_3. \tag{5.8.57}$$

The tangent of the angle θ between $\mathbf{u}$ and the x_1 axis is given by

$$\tan\theta = \frac{u_3}{u_1} = \frac{c_3}{c_1}, \tag{5.8.58}$$

which is independent of x_1 and t. Therefore, the vector $\mathbf{u}$ never changes direction, only amplitude. This kind of motion is said to be *linearly polarized*.

A different type of displacement is of the form

$$\mathbf{u} = (c_1\mathbf{a}_1 + ic_3\mathbf{a}_3)\exp[i(\omega t - k_1 x_1)]. \tag{5.8.59}$$

In this case the real part becomes

$$\mathbf{u} = c_1\cos(\omega t - k_1 x_1)\mathbf{a}_1 - c_3\sin(\omega t - k_1 x_1)\mathbf{a}_3 \equiv u_1\mathbf{a}_1 + u_3\mathbf{a}_3 \tag{5.8.60}$$

and

$$\tan\theta = \frac{u_3}{u_1} = -\frac{c_3}{c_1}\tan(\omega t - k_1 x_1) \tag{5.8.61}$$

$$\frac{u_1^2}{c_1^2} + \frac{u_3^2}{c_3^2} = 1 \tag{5.8.62}$$

(Problem 5.7). Equation (5.8.62) represents an ellipse and the corresponding motion is said to be *elliptically polarized*. When $c_1 = c_3$ the polarization is *circular*. An example of elliptical polarization is provided by the Rayleigh waves (see §7.4.1).

The factor i in the x_3 component of $\mathbf{u}$ in (5.8.59) introduces a phase difference of $\pi/2$ between the horizontal and vertical components, but other phase differences may also occur (§6.9.2.3; Haskell, 1962).

5.9 Flux of energy in harmonic waves

Here we will determine the amount of energy in a wave transmitted across a surface element dS, which is part of a surface within an elastic body dividing it into media I and II. This idea is similar to that used to introduce the stress vector (see §3.3) but now the normal $\mathbf{n}$ to dS will be in medium II. The rate of work $\dot{W}$, or power, done by medium I on medium II across dS is given by the scalar product of the surface forces exerted by I on II and the particle velocity at dS. The force is given by the stress vector $\mathbf{T}$ times dS and the velocity is $\dot{\mathbf{u}}$. Then,

$$\dot{W} = -\mathbf{T}\cdot\dot{\mathbf{u}}\,dS. \tag{5.9.1}$$

The minus sign is needed because of the different convention on **n**. Writing the stress tensor in terms of the stress vector in dyadic and component forms (see (3.5.11) and (3.5.13)) we have

$$\dot{W} = -\mathbf{n} \cdot \mathcal{T} \cdot \dot{\mathbf{u}} \, dS = -\tau_{ij} \dot{u}_i n_j \, dS \tag{5.9.2}$$

(Auld, 1990; Hudson, 1980; Ben-Menahem and Singh, 1981). The *energy-flux density vector* **E**, having components E_j, is defined as

$$\mathbf{E} = -\mathcal{T} \cdot \dot{\mathbf{u}}; \qquad E_j = -\tau_{ij} \dot{u}_i. \tag{5.9.3a,b}$$

The power by unit area $\mathcal{P}$, or *power density*, is obtained by dividing $\dot{W}$ by dS. From (5.9.1)–(5.9.3) we have

$$\mathcal{P} = \mathbf{n} \cdot \mathbf{E} = E_j n_j \tag{5.9.4}$$

so that $\mathcal{P}$ is given by the projection of **E** along the direction of **n**.

To find **E** for harmonic P, SV, and SH waves we need the components of τ_{ij} for the corresponding displacements, given by (5.8.53)–5.8.55). However, as (5.9.3a,b) are nonlinear equations, real quantities must be used. Here it will be assumed that the coefficients A, B, and C are real, which means that the real part of the displacements are obtained by replacing the exponential function with the cosine function. In Chapter 6 we will see what happens when some of these coefficients are not real. Using

$$\tau_{ij} = \lambda \delta_{ij} u_{k,k} + \mu(u_{j,i} + u_{i,j}) \tag{5.9.5}$$

(see (4.6.3)) we obtain the following expressions for τ_{ij}. For convenience, they are written in matrix form, which simplifies the operations indicated in (5.9.3b).

(1) *P* waves:

$$\tau_{ij} \to \frac{A}{\alpha} \omega \sin\left[\omega\left(t - \frac{\mathbf{p} \cdot \mathbf{r}}{\alpha}\right)\right] \begin{pmatrix} (\lambda + 2\mu l^2) & 0 & 2\mu l n \\ 0 & \lambda & 0 \\ 2\mu l n & 0 & (\lambda + 2\mu n^2) \end{pmatrix}. \tag{5.9.6}$$

(2) *SV* waves:

$$\tau_{ij} \to \mu\omega\frac{B}{\beta} \sin\left[\omega\left(t - \frac{\mathbf{p} \cdot \mathbf{r}}{\beta}\right)\right] \begin{pmatrix} -2nl & 0 & (l^2 - n^2) \\ 0 & 0 & 0 \\ (l^2 - n^2) & 0 & 2nl \end{pmatrix}. \tag{5.9.7}$$

(3) *SH* waves:

$$\tau_{ij} \to \mu\frac{C}{\beta} \omega \sin\left[\omega\left(t - \frac{\mathbf{p} \cdot \mathbf{r}}{\beta}\right)\right] \begin{pmatrix} 0 & l & 0 \\ l & 0 & n \\ 0 & n & 0 \end{pmatrix} \tag{5.9.8}$$

(Problem 5.8).

Using the real parts of (5.8.53)–(5.8.55) to obtain the expressions for $\dot{u}_i$, equation (5.9.3b) gives

$$\mathbf{E}_P = \rho\alpha\omega^2 A^2 \mathbf{p} \sin^2\left[\omega\left(t - \frac{\mathbf{p}\cdot\mathbf{r}}{\alpha}\right)\right] \tag{5.9.9}$$

$$\mathbf{E}_{SV} = \rho\beta\omega^2 B^2 \mathbf{p} \sin^2\left[\omega\left(t - \frac{\mathbf{p}\cdot\mathbf{r}}{\beta}\right)\right] \tag{5.9.10}$$

$$\mathbf{E}_{SH} = \rho\beta\omega^2 C^2 \mathbf{p} \sin^2\left[\omega\left(t - \frac{\mathbf{p}\cdot\mathbf{r}}{\beta}\right)\right] \tag{5.9.11}$$

(Problem 5.9). These equations show that in every case the energy is transmitted in the direction of $\mathbf{p}$, which is the direction of wave propagation.

In Chapter 6 we will be interested in $\mathcal{P}$ across the plane $x_3 = 0$ in the direction of x_3. In such a case $\mathbf{n} = \mathbf{a}_3$ and from (5.9.4) and (5.9.9)–(5.9.11) we find

$$\mathcal{P}_P = \mathbf{a}_3 \cdot \mathbf{E}_P = \rho\alpha\omega^2 A^2 n \sin^2\left[\omega\left(t - l\frac{x_1}{\alpha}\right)\right] \tag{5.9.12}$$

$$\mathcal{P}_{SV} = \mathbf{a}_3 \cdot \mathbf{E}_{SV} = \rho\beta\omega^2 B^2 n \sin^2\left[\omega\left(t - l\frac{x_1}{\beta}\right)\right] \tag{5.9.13}$$

$$\mathcal{P}_{SH} = \mathbf{a}_3 \cdot \mathbf{E}_{SH} = \rho\beta\omega^2 C^2 n \sin^2\left[\omega\left(t - l\frac{x_1}{\beta}\right)\right] \tag{5.9.14}$$

(Problem 5.10). Furthermore, we will average these results over one cycle. Using $\langle\,\rangle$ to indicate average, from (5.9.12)–(5.9.14) we obtain

$$\langle\mathcal{P}_P\rangle = \frac{1}{2}\rho\alpha n\omega^2 A^2 \tag{5.9.15}$$

$$\langle\mathcal{P}_{SV}\rangle = \frac{1}{2}\rho\beta n\omega^2 B^2 \tag{5.9.16}$$

$$\langle\mathcal{P}_{SH}\rangle = \frac{1}{2}\rho\beta n\omega^2 C^2 \tag{5.9.17}$$

(Problem 5.11). Note that (5.9.15)–(5.9.17) have important implications. Consider for example a P wave with specified values of $\mathbf{p}$, $\langle\mathcal{P}_P\rangle$ and ω. Then, the amplitude A will be inversely proportional to $\sqrt{\rho\alpha}$. This means that given two different elastic media, the amplitude of the wave will be larger in the medium with a smaller value of $\rho\alpha$. Similarly, when A and ω are the only variables, A will be inversely proportional to ω. Analogous results hold for the S waves.

Problems

5.1   Show that the solution to the 1-D scalar wave equation under the following
initial conditions:

$$\psi(x,0) = F(x); \qquad \frac{\partial \psi(x,0)}{\partial t} = G(x)$$

is given by

$$\psi(x,t) = \frac{1}{2}[F(x-ct) + F(x+ct)] + \frac{1}{2c}\int_{x-ct}^{x+ct} G(s)\,ds.$$

5.2   Let $\phi = \phi(r,t)$, where $r = |\mathbf{r}|$. Verify that

$$\nabla^2 \phi = \frac{\partial^2 \phi}{\partial r^2} + \frac{2}{r}\frac{\partial \phi}{\partial r}.$$

5.3   Show that the vector $\mathbf{M}$, defined by (5.6.11) is solenoidal when $\mathbf{a}$ is a constant vector.

5.4   Verify (5.6.30).

5.5   Verify (5.6.34).

5.6   Verify that (5.7.4)–(5.7.6) solve (5.7.3) if ψ satisfies (5.7.7).

5.7   Verify (5.8.62).

5.8   Verify (5.9.6)–(5.9.8).

5.9   Verify (5.9.9)–(5.9.11).

5.10 Verify (5.9.12)–(5.9.14).

5.11 Verify (5.9.15)–(5.9.17). If $f(t)$ is a periodic function with period T, its
average value over one period is given by

$$\langle f \rangle = \frac{1}{T}\int_0^T f(t)\,dt.$$

6

Plane waves in simple models with plane boundaries

6.1 Introduction

After the homogeneous infinite space, the next two simplest configurations are a homogeneous half-space with a free surface and two homogeneous half-spaces (or media, for short) with different elastic properties. The first case can be considered as a special case of the second one with one of the media a vacuum. In either case the boundary between the two media constitutes a surface of discontinuity in elastic properties that has a critical effect on wave propagation. To simplify the problem we will assume plane boundaries and wave fronts. Although in the Earth neither the wave fronts nor the boundaries satisfy these assumptions, they are acceptable approximations as long as the seismic source is sufficiently far from the receiver and/or the wavelength is much shorter than the curvature of the boundary. In addition, the case of spherical wave fronts can be solved in terms of plane wave results (e.g., Aki and Richards, 1980). Therefore, the theory and results described here have a much wider application than could be expected by considering the simplifying assumptions. For example, they are used in teleseismic studies, in the generation of synthetic seismograms using ray theory, and in exploration seismology, particularly in amplitude-versus-offset (AVO) studies.

The interaction of elastic waves with a boundary has a number of similarities with the interaction of acoustic and electromagnetic waves, so that it can be expected that a wave incident on a boundary will generate reflected and transmitted[1] waves (the latter only if the other medium is not a vacuum). In the case of elastic waves, however, an additional process takes place, namely the generation of SV waves by incident P waves and vice versa. This process, which does not affect the SH waves, is known as *mode conversion*, and adds considerable complexity to the problem, which requires a careful set up for its solution. Once that has been done,

---

[1] Transmitted waves are also known as refracted waves, but this term will not be used here to avoid any possible confusion with the refracted waves of exploration seismology.

129

solving the cases described here is fairly straightforward, but before that stage is reached it is necessary to settle one critical matter: how will the P and S waves be represented? Two possibilities exist.

The problem of plane waves incident on a plane boundary was solved initially by Knott and later by Zoeppritz (1919) using a different approach. Knott represented the P and S waves via potentials, from which the P and S displacements can be derived by differentiation (see §9.4). Zoeppritz, instead, worked directly with the displacements. Knott's approach can be found in the majority of seismology books. Displacements are used by Achenbach (1973), Hudson (1980), and Ben-Menahem and Singh (1981), among others, and will constitute the basis of the analysis presented in this chapter and in the next. This choice is motivated by the fact that potentials are nonphysical entities, while displacements represent the actual motion of the medium. Therefore, it seems reasonable to develop the theory in terms of quantities that can be directly compared with observations. The result is a conceptually simpler representation of the physics of wave propagation. Importantly, this simplification does not increase the level of the mathematics involved. In any case, the solutions obtained using the two approaches are related to each other in a simple way (e.g., Aki and Richards, 1980; Miklowitz, 1984). It must also be noted that a seismogram corresponds to a filtered (and amplified) version of the displacement of the ground under the seismometer. For this reason, a direct comparison between theoretical and observed results requires prior removal of the instrument response from the seismogram. This is particularly true when frequency-dependent effects are involved.

Another question to be considered is the representation of the plane wave. In principle, any wave can be used (e.g., Burridge, 1976), but because of their ease of use, it is almost customary to use a harmonic wave. This practice, which will be followed here, is not as restrictive as it may seem, because we know that we can use the Fourier integral to represent any pulse in term of harmonic functions. An example of how that is done is given in §6.5.3.3.

To solve wave propagation problems two main steps are needed. The first one is to write the equation for the displacement at any point in the medium as a sum of the displacements caused by the various waves involved, i.e., SH or P and SV; incident, reflected, and transmitted. The expression for each type of wave is fixed; it is problem-independent. The amplitudes of the individual displacements (or amplitude coefficients), however, are unknown (except for the incident wave) and must be determined under the condition that they are independent of time and position. Regarding the types of wave involved in any particular problem, the equations in §6.4 show that there is a coupling between the P and SV motions, which means that both motions should be taken into account for incident P or SV waves. The SH motion, on the other hand, is completely decoupled from the other two motions, and, consequently, problems involving this type of waves are much easier to solve.

The second step is to apply the *boundary conditions* appropriate to the problem being solved (see §6.3). These conditions are derived from continuity arguments applied to the displacement and stress vectors. The result of applying these conditions is a linear system of equations with the number of unknowns (the amplitude coefficients) equal to the number of equations. Solving the system, either analytically or with a computer, gives the amplitude coefficients in terms of parameters such as velocities, densities, frequency, and incidence angle. At this point the problem has been solved, but this is not the end; usually the solution has a number of important properties that must also be studied.

In the following sections we introduce the expressions for the displacement and stress vectors for the various types of waves referred to above, discuss the boundary conditions for several cases, and then solve three problems with an increasing degree of difficulty: a half-space with a free surface, two elastic half-spaces in welded contact, and a layer over a half-space. The equations involving amplitude coefficients for the first two types of problem are known as *Zoeppritz equations*. The presentation and notation used here follow those of Ben-Menahem and Singh (1981) to facilitate comparison with their work.

6.2 Displacements

The displacements corresponding to incident, reflected, and transmitted P, SV, and SH waves are given in §5.8.4. Each type of wave will be identified by two vectors, $\mathbf{p}$ and $\mathbf{u}$, which indicate the directions of wave propagation and particle displacement, respectively. The relevant geometrical relations are shown in Fig. 6.1. Note that x_3 points downwards. The unit vectors $\mathbf{a}_1$, $\mathbf{a}_2$, and $\mathbf{a}_3$ are along the positive x_1, x_2, and x_3 directions. The vector $\mathbf{p}$ for the incident wave points towards the boundary, while the vector $\mathbf{p}$ for the transmitted and reflected waves point away from the boundary. The last condition is a consequence of the principle of causality, as the incident wave is assumed to originate the other two types of waves (Achenbach, 1973). Although we will use vectors to represent waves, it is important to realize that with each vector $\mathbf{p}$ there is an infinite plane (i.e., a wave front) perpendicular to it. Regarding the displacements, the following convention will be used. For P waves, $\mathbf{u}$ is in the direction of $\mathbf{p}$. For SH waves $\mathbf{u}$ is in the direction of $\mathbf{a}_2$, and for SV waves $\mathbf{u}$ is in the direction of the vector $\mathbf{p} \times \mathbf{a}_2$. Similar conventions, derived exclusively from consideration of the vector nature of the elastic motion and the orthogonality of the P and SV motions, have been used by Achenbach (1973) and Nadeau (1964), and are implicit in all the treatments of the reflection–transmission problem using displacements, including the work

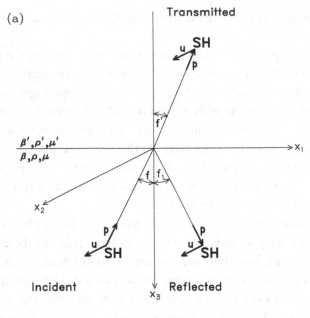

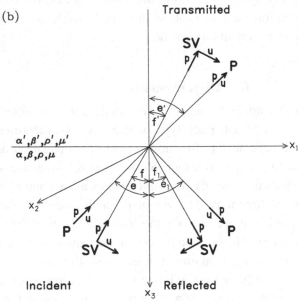

Fig. 6.1. Geometry for the reflection and transmission of plane SH waves (a) and P and SV waves (b) incident on a plane boundary. Note that the x_3 axis points downwards, and that the x_2 axis is drawn so as to make the coordinate system right-handed. The angles e, e', and e_1 refer to P waves and f, f', and f_1 to S waves. The vectors **u** and **p** indicate the positive direction of particle displacement and the normal to the plane wave front, respectively. For the SH wave, **u** is in the direction of the x_2 axis. For the P waves, **u** and **p** are in the same direction. For the SV waves, **u** is in the direction of $\mathbf{p} \times \mathbf{e}_2$. As a mnemonic device, to find the direction of **u** rotate **p** $90°$ clockwise. After Ben-Menahem and Singh (1981).

of Zoeppritz (1919). Finally, the medium that contains the incident and reflected waves will be referred to as the incidence medium, while the other medium will be called the transmission medium.

With these conventions the equations for the displacements of the incident, reflected and transmitted waves are written using (5.8.53)–(5.8.55).

Incident waves. In this case we have

$$\mathbf{p} = \sin \lambda \mathbf{a}_1 - \cos \lambda \mathbf{a}_3 \qquad (6.2.1)$$

$$\mathbf{p} \cdot \mathbf{r} = x_1 \sin \lambda - x_3 \cos \lambda \qquad (6.2.2)$$

$$\mathbf{p} \times \mathbf{a}_2 = \cos \lambda \mathbf{a}_1 + \sin \lambda \mathbf{a}_3, \qquad (6.2.3)$$

where the angle λ is equal to e for P waves and equal to f for S waves (Fig. 6.1), and $\mathbf{r} = (x_1, x_2, x_3)$. Therefore, the displacements are given by

$$\mathbf{u}_P = A(\sin e \mathbf{a}_1 - \cos e \mathbf{a}_3) \exp\left[i\omega\left(t - \frac{x_1 \sin e - x_3 \cos e}{\alpha}\right)\right] \qquad (6.2.4)$$

$$\mathbf{u}_{SV} = B(\cos f \mathbf{a}_1 + \sin f \mathbf{a}_3) \exp\left[i\omega\left(t - \frac{x_1 \sin f - x_3 \cos f}{\beta}\right)\right] \qquad (6.2.5)$$

$$\mathbf{u}_{SH} = C \mathbf{a}_2 \exp\left[i\omega\left(t - \frac{x_1 \sin f - x_3 \cos f}{\beta}\right)\right]. \qquad (6.2.6)$$

Reflected waves. In this case we have

$$\mathbf{p} = \sin \lambda \mathbf{a}_1 + \cos \lambda \mathbf{a}_3 \qquad (6.2.7)$$

$$\mathbf{p} \cdot \mathbf{r} = x_1 \sin \lambda + x_3 \cos \lambda \qquad (6.2.8)$$

$$\mathbf{p} \times \mathbf{a}_2 = -\cos \lambda \mathbf{a}_1 + \sin \lambda \mathbf{a}_3, \qquad (6.2.9)$$

where the angle λ is equal to e_1 for P waves and equal to f_1 for S waves. Therefore, the displacements are given by

$$\mathbf{u}_P = A_1(\sin e_1 \mathbf{a}_1 + \cos e_1 \mathbf{a}_3) \exp\left[i\omega\left(t - \frac{x_1 \sin e_1 + x_3 \cos e_1}{\alpha}\right)\right] \qquad (6.2.10)$$

$$\mathbf{u}_{SV} = B_1(-\cos f_1 \mathbf{a}_1 + \sin f_1 \mathbf{a}_3) \exp\left[i\omega\left(t - \frac{x_1 \sin f_1 + x_3 \cos f_1}{\beta}\right)\right] \qquad (6.2.11)$$

$$\mathbf{u}_{SH} = C_1 \mathbf{a}_2 \exp\left[i\omega\left(t - \frac{x_1 \sin f_1 + x_3 \cos f_1}{\beta}\right)\right]. \qquad (6.2.12)$$

Transmitted waves. This case is similar to the case of the incident wave, with the angles e and f replaced by e' and f'. Therefore, the displacements are given by

$$\mathbf{u}_P = A'(\sin e' \mathbf{a}_1 - \cos e' \mathbf{a}_3) \exp\left[i\omega\left(t - \frac{x_1 \sin e' - x_3 \cos e'}{\alpha'}\right)\right] \qquad (6.2.13)$$

$$\mathbf{u}_{SV} = B'(\cos f' \mathbf{a}_1 + \sin f' \mathbf{a}_3) \exp\left[i\omega\left(t - \frac{x_1 \sin f' - x_3 \cos f'}{\beta'}\right)\right] \qquad (6.2.14)$$

$$\mathbf{u}_{SH} = C' \mathbf{a}_2 \exp\left[i\omega\left(t - \frac{x_1 \sin f' - x_3 \cos f'}{\beta'}\right)\right]. \qquad (6.2.15)$$

6.3 Boundary conditions

Several cases will be considered.

(1) *Free surface.* As a vacuum cannot support stresses, the boundary condition is that the stress vector at the surface should be equal to zero. The surface, however, can move, which means that the surface displacement cannot be specified in advance. To a first approximation, the surface of the earth (including the oceans) is a good example of a free surface because the elastic parameters of the atmosphere are much smaller than those of rocks and water. It must be borne in mind, however, that earthquakes can generate atmospheric waves and that atmospheric explosions can generate surface waves in the Earth (Pilant, 1979; Aki and Richards, 1980; Ben-Menahem and Singh, 1981).

(2) *Two solids in welded contact.* In this case both the displacement and stress vectors must be continuous across the boundary. The continuity of the displacement is required to prevent the interpenetration of mass or the formation of voids at the boundary, which would correspond, for example, to motion in the incidence medium larger and smaller than in the transmission media, respectively.

To show the continuity of the stress vector (or traction) we will use (3.3.5) written as

$$\mathbf{T}(\mathbf{n}) = -\mathbf{T}(-\mathbf{n}). \qquad (6.3.1)$$

Now replace $\mathbf{T}$ by its expression (3.5.11) with $\mathbf{n} = \mathbf{a}_3$. Using the superscripts I and II to indicate the incidence and transmission media we have

$$T_i^{\mathrm{I}}(\mathbf{n}) = \tau_{ij}^{\mathrm{I}}(\mathbf{a}_3)_j = \tau_{i3}^{\mathrm{I}} = \tau_{3i}^{\mathrm{I}} \qquad (6.3.2)$$

and

$$-T_i^{\mathrm{II}}(-\mathbf{n}) = -\tau_{ij}^{\mathrm{II}}(-\mathbf{a}_3)_j = \tau_{3i}^{\mathrm{II}} = T_i^{\mathrm{I}}(\mathbf{n}). \qquad (6.3.3)$$

The last equality follows from (6.3.1). Then (6.3.2) and (6.3.3) give

$$\tau_{3i}^{I} = \tau_{3i}^{II} \qquad i = 1, 2, 3, \tag{6.3.4}$$

which is a boundary condition for the stress tensor. Note that this condition restricts the values that three of the stress tensor components can take, but that it does not place any restriction on the values of the other components.

(3) *Solid–liquid boundary without cavitation.* Two cases should be distinguished. First, the liquid is viscous, which means that it can support shear stresses, as opposed to inviscid fluids (their viscosity is zero), for which the stress tensor is proportional to the Kronecker delta (see (3.7.8)). In the first case, the boundary conditions for the solid–solid case discussed above apply (Pilant, 1979). The term cavitation refers to the formation of a void in a fluid.

For the inviscid case, slip can occur parallel to the boundary. In this case only continuity on the normal components (with respect to the boundary) of the displacement and stress vectors is required. There are no conditions for the tangential components, which means that tangential slip is allowed. For seismological purposes, the oceans and the outer core appear to behave as inviscid fluids (Aki and Richards, 1980).

6.4 Stress vector

To satisfy the boundary conditions discussed in §6.3 the τ_{3i} ($i = 1, 2, 3$) are needed. Using the relation between stress and displacement (see (4.6.3)) and the displacements given in §6.2 we obtain the following equations for the various types of waves:

(1) *P waves:*

$$\tau_{31} = \mu(u_{3,1} + u_{1,3}) = 2\mu u_{1,3} \tag{6.4.1}$$

$$\tau_{32} = \mu(u_{2,3} + u_{3,2}) = 0 \tag{6.4.2}$$

$$\tau_{33} = \lambda(u_{1,1} + u_{3,3}) + 2\mu u_{3,3} = \lambda u_{1,1} + (2\mu + \lambda)u_{3,3}. \tag{6.4.3}$$

(2) *SV waves:*

$$\tau_{31} = \mu(u_{1,3} + u_{3,1}) \tag{6.4.4}$$

$$\tau_{32} = 0 \tag{6.4.5}$$

$$\tau_{33} = \lambda(u_{1,1} + u_{3,3}) + 2\mu u_{3,3} = 2\mu u_{3,3}. \tag{6.4.6}$$

(3) *SH waves:*

$$\tau_{31} = \tau_{33} = 0 \tag{6.4.7}$$

$$\tau_{32} = \mu(u_{3,2} + u_{2,3}) = \mu u_{2,3} \tag{6.4.8}$$

(Problem 6.1). All of these expressions apply to incident, reflected, and transmitted waves. Note that the boundary conditions on the stresses will couple the P and SV displacements, while the SH displacement will be independent of them.

6.5 Waves incident at a free surface

6.5.1 Incident SH waves

The total displacement in the medium is the sum of the displacements of the incident and reflected waves:

$$\mathbf{u} = \mathbf{a}_2 \left\{ C \exp\left[i\omega\left(t - \frac{x_1 \sin f - x_3 \cos f}{\beta} \right) \right] \right.$$

$$\left. + C_1 \exp\left[i\omega\left(t - \frac{x_1 \sin f_1 + x_3 \cos f_1}{\beta} \right) \right] \right\}. \tag{6.5.1}$$

Now use the boundary condition appropriate to SH motion, given by (6.4.8)

$$\tau_{32}\big|_{x_3=0} = \mu u_{2,3}\big|_{x_3=0} = 0. \tag{6.5.2}$$

This gives

$$i\omega\mu \exp(i\omega t) \left[C \cos f \exp\left(-i\omega \frac{x_1 \sin f}{\beta} \right) - C_1 \cos f_1 \exp\left(-i\omega \frac{x_1 \sin f_1}{\beta} \right) \right] = 0. \tag{6.5.3}$$

In (6.5.3) the factor $\omega\mu \exp(i\omega t)$ is different from zero (assuming $\omega \neq 0$), which means that the term in brackets has to be equal to zero, or, equivalently,

$$C = C_1 \frac{\cos f_1}{\cos f} \exp[-i\omega x_1(\sin f_1 - \sin f)/\beta]. \tag{6.5.4}$$

As C and C_1 must be independent of x_1, equation (6.5.4) will be satisfied for all values of x_1 if the difference of the cosine terms in the exponent is equal to zero, which means that

$$f_1 = f \qquad C_1 = C. \tag{6.5.5}$$

A more general argument is given in §6.5.2. Introducing (6.5.5) in (6.5.1) gives

$$\mathbf{u} = \mathbf{a}_2 C \left\{ \exp\left[i\omega\left(t - \frac{x_1 \sin f - x_3 \cos f}{\beta} \right) \right] \right.$$

$$\left. + \exp\left[i\omega\left(t - \frac{x_1 \sin f + x_3 \cos f}{\beta} \right) \right] \right\}. \tag{6.5.6}$$

The surface displacement, indicated by $\mathbf{u}_o$, is obtained from (6.5.6) with $x_3 = 0$:

$$\mathbf{u}_o = 2\mathbf{a}_2 C \exp\left[i\omega\left(t - \frac{x_1 \sin f}{\beta}\right)\right]. \tag{6.5.7}$$

Therefore, the amplitude of the surface displacement is twice the amplitude of the incident wave.

6.5.2 Incident P waves

The total displacement is equal to the sum of the incident P-wave displacement and the displacements of the reflected P and SV waves:

$$\mathbf{u} = A(\sin e \mathbf{a}_1 - \cos e \mathbf{a}_3) \exp\left[i\omega\left(t - \frac{x_1 \sin e - x_3 \cos e}{\alpha}\right)\right]$$

$$+ A_1(\sin e_1 \mathbf{a}_1 + \cos e_1 \mathbf{a}_3) \exp\left[i\omega\left(t - \frac{x_1 \sin e_1 + x_3 \cos e_1}{\alpha}\right)\right]$$

$$+ B_1(-\cos f_1 \mathbf{a}_1 + \sin f_1 \mathbf{a}_3) \exp\left[i\omega\left(t - \frac{x_1 \sin f_1 + x_3 \cos f_1}{\beta}\right)\right]. \tag{6.5.8}$$

Now apply the boundary condition that the components of the stress vector are zero at the surface. Because $\mathbf{u}$ is a combination of P and SV motion, the stress vector associated with $\mathbf{u}$ is obtained from the corresponding combination of stress vectors given in §6.4. After the ensuing equations have been obtained, they have to be evaluated at $x_3 = 0$ and then set equal to zero. From $\tau_{31} = 0$ we obtain

$$\frac{A}{\alpha} \sin 2e \exp\left[i\omega\left(t - \frac{x_1 \sin e}{\alpha}\right)\right] - \frac{A_1}{\alpha} \sin 2e_1 \exp\left[i\omega\left(t - \frac{x_1 \sin e_1}{\alpha}\right)\right]$$

$$+ \frac{B_1}{\beta} \cos 2f_1 \exp\left[i\omega\left(t - \frac{x_1 \sin f_1}{\beta}\right)\right] = 0. \tag{6.5.9}$$

Note that a common factor of $i\omega\mu$ has been canceled out, and that the relations $\cos 2f_1 = \cos^2 f_1 - \sin^2 f_1$ and $\sin \theta \cos \theta = \frac{1}{2} \sin 2\theta$, $\theta = e, e_1$, were used.

From $\tau_{33} = 0$ we obtain

$$\frac{A}{\alpha}(\lambda + 2\mu \cos^2 e) \exp\left[i\omega\left(t - \frac{x_1 \sin e}{\alpha}\right)\right]$$

$$+ \frac{A_1}{\alpha}(\lambda + 2\mu \cos^2 e_1) \exp\left[i\omega\left(t - \frac{x_1 \sin e_1}{\alpha}\right)\right]$$

$$+ B_1 \frac{\mu}{\beta} \sin 2f_1 \exp\left[i\omega\left(t - \frac{x_1 \sin f_1}{\beta}\right)\right] = 0. \tag{6.5.10}$$

After canceling the common factor of $\exp(i\omega t)$, equations (6.5.9) and (6.5.10) are of the form

$$a_1 \exp(ib_1 x_1) + a_2 \exp(ib_2 x_1) + a_3 \exp(ib_3 x_1) = 0 \qquad (6.5.11)$$

with a_i and b_i independent of both x_1 and t. With an argument similar to that used in §6.5.1 it can be concluded that the three phase factors b_1, b_2, and b_3 must be equal to each other. Alternatively, one can use the following more rigorous argument. Assume that the phase factors are all different and multiply (6.5.11) by $\exp(-ib_k x_1)$ and integrate over x_1:

$$\int_{-\infty}^{\infty} \sum_{j=0}^{3} a_j e^{i(b_j - b_k)x_1} \, dx_1 = 2\pi \sum_{j=0}^{3} a_j \delta(b_j - b_k) = 0. \qquad (6.5.12)$$

Then, by letting $k = 1, 2, 3$ we see that $a_1 = a_2 = a_3 = 0$ (Problem 6.2). On the other hand, if a_j are different from zero, equation (6.5.11) is satisfied if

$$b_1 = b_2 = b_3 \qquad (6.5.13)$$

and

$$a_1 + a_2 + a_3 = 0. \qquad (6.5.14)$$

This result will be applied to (6.5.9) and (6.5.10). Except for two special cases ($f_1 = e = e_1 = 0$; $f_1 = \pi/2$) discussed in Problem 6.3, a_1, a_2, a_3 are non-zero, which means that

$$e_1 = e; \qquad \frac{\sin e}{\alpha} = \frac{\sin f_1}{\beta}. \qquad (6.5.15a,b)$$

Equation (6.5.15b) is known as *Snell's law* because of its similarity to the law of the same name in optics. From (6.5.15b) and from the fact that $\alpha > \beta$, we find that $e > f_1$ always.

Now, letting $f_1 = f$, dividing by A, and canceling out the common exponential factor, equations (6.5.9) and (6.5.10) give

$$\frac{A_1}{A} \sin 2e - \frac{B_1}{A} \frac{\alpha}{\beta} \cos 2f = \sin 2e \qquad (6.5.16)$$

and

$$\frac{A_1}{A}(\lambda + 2\mu \cos^2 e) + \frac{B_1}{A} \frac{\alpha}{\beta} \mu \sin 2f = -(\lambda + 2\mu \cos^2 e). \qquad (6.5.17)$$

Before solving for the ratios A_1/A and B_1/A, which are known as *reflection coefficients*, the factors containing λ in (6.5.17) will be rewritten using Snell's law and (4.8.5),

$$\lambda + 2\mu \cos^2 e = \lambda + 2\mu(1 - \sin^2 e) = \lambda + 2\mu - 2\mu\frac{\alpha^2}{\beta^2}\sin^2 f$$

$$= \mu\frac{\alpha^2}{\beta^2}(1 - 2\sin^2 f) = \mu\frac{\alpha^2}{\beta^2}\cos 2f. \tag{6.5.18}$$

Let D be the determinant of the system of equations (6.5.16) and (6.5.17). Then,

$$D = \begin{vmatrix} \sin 2e & -\dfrac{\alpha}{\beta}\cos 2f \\ \mu\left(\dfrac{\alpha}{\beta}\right)^2 \cos 2f & \dfrac{\alpha}{\beta}\mu \sin 2f \end{vmatrix} = \frac{\alpha}{\beta}\mu\left(\sin 2e \sin 2f + \frac{\alpha^2}{\beta^2}\cos^2 2f\right)$$

$$\tag{6.5.19}$$

and

$$\frac{A_1}{A} = \frac{1}{D}\begin{vmatrix} \sin 2e & -\dfrac{\alpha}{\beta}\cos 2f \\ -\mu\left(\dfrac{\alpha}{\beta}\right)^2 \cos 2f & \dfrac{\alpha}{\beta}\mu \sin 2f \end{vmatrix}$$

$$= \frac{\sin 2e \sin 2f - (\alpha/\beta)^2 \cos^2 2f}{\sin 2e \sin 2f + (\alpha/\beta)^2 \cos^2 2f} \tag{6.5.20}$$

$$\frac{B_1}{A} = \frac{1}{D}\begin{vmatrix} \sin 2e & \sin 2e \\ \mu\left(\dfrac{\alpha}{\beta}\right)^2 \cos 2f & -\mu\left(\dfrac{\alpha}{\beta}\right)^2 \cos 2f \end{vmatrix}$$

$$= -\frac{2(\alpha/\beta)\sin 2e \cos 2f}{\sin 2e \sin 2f + (\alpha/\beta)^2 \cos^2 2f}. \tag{6.5.21}$$

Note that the reflection coefficients are independent of ω. However, as shown in §6.5.3.2, there are cases where this does not happen, which introduces significant complexity into the problem.

The dependence of the reflection coefficients on the incidence angle e is not simple, and can be best appreciated by plotting each coefficient as a function of e (Fig. 6.2), but some general comments can be made. The coefficient B_1/A is always negative or zero as long as the Poisson ratio is nonnegative (Problem 6.4). For A_1/A the situation is different. Because the numerator is the difference of two positive quantities, its value can become positive, negative, or zero depending on e and the value of the ratio α/β. For ratios larger than 1.764 the coefficient is always negative, while for smaller ratios there is always a range of angles for which the coefficient is positive (Ben-Menahem and Singh, 1981). This result is important because it means that in some cases the incident and reflected P waves

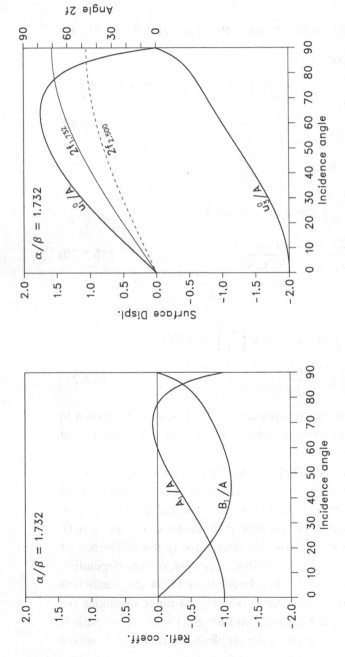

Fig. 6.2. Left: reflection coefficients for the case of a P wave incident on a free surface as functions of the incidence angle e. Note that when the angle is between about $60°$ and $77°$ there is a change in the sign of A_1/A. Right: horizontal (u_1^o/A) and vertical (u_3^o/A) components of particle motion at the free surface and angle $2f$ between the $-x_3$ direction and the direction of motion. The subscript on $2f$ indicates the value of α/β used. Although all the computations were carried out for a value of 1.732, the angle $2f$ for $\alpha/\beta = 2.5$ is shown for comparison.

will have the same polarity (i.e., $A_1/A > 0$), while in some cases the polarity will be reversed, which is equivalent to a phase shift of π.

Because seismic waves are frequently recorded using three-component instruments, it is also useful to write down the expressions for the reflected waves in component form as a fraction of the incident wave. From (6.5.8) and Snell's law (6.5.15b), it is seen that for a point on the surface ($x_3 = 0$), the ratios for the horizontal and vertical components are given as follows.

Reflected P waves:

$$\frac{u_1^{\text{or}}}{u_1^{\text{oi}}} = \frac{A_1}{A}; \qquad \frac{u_3^{\text{or}}}{u_3^{\text{oi}}} = \frac{-A_1}{A}, \tag{6.5.22}$$

where the subscripts 1 and 3 indicate horizontal and vertical components, the superscripts i and r indicate incident and reflected wave, respectively, and the superscript o indicates a point on the surface.

Reflected SV waves:

$$\frac{u_1^{\text{or}}}{u_1^{\text{oi}}} = -\frac{B_1}{A}\frac{\cos f}{\sin e}; \qquad \frac{u_3^{\text{or}}}{u_3^{\text{oi}}} = -\frac{B_1}{A}\frac{\sin f}{\cos e}. \tag{6.5.23}$$

6.5.2.1 Surface displacement

As before, it will be represented by $\mathbf{u}_o$ and is obtained from (6.5.8) with $x_3 = 0$, and using (6.5.15), (6.5.20), and (6.5.21), and simple trigonometric relations:

$$\mathbf{u}_o = A(\sin 2f\,\mathbf{a}_1 - \cos 2f\,\mathbf{a}_3)\frac{2(\alpha/\beta)^2 \cos e}{\sin 2e \sin 2f + (\alpha/\beta)^2 \cos^2 2f}$$

$$\times \exp\left[i\omega\left(t - \frac{x_1 \sin e}{\alpha}\right)\right] \equiv u_1^o\mathbf{a}_1 + u_3^o\mathbf{a}_3 \tag{6.5.24}$$

(Problem 6.5). The identity defines the horizontal components u_1^o and u_3^o. It is worth emphasizing that the three types of waves, incident and reflected P and reflected SV, contribute to the surface motion, not just the incident P wave.

The relative surface displacement $\mathbf{u}_o/A$ for a medium with a velocity ratio α/β equal to $\sqrt{3}$ (corresponding to a Poisson solid) is shown in Fig. 6.2. In general, for A positive, the horizontal component (u_1^o/A) is always positive (or zero), i.e., is in the $+x_1$ direction. The vertical component (u_3^o/A) is always negative (i.e., in the $-x_3$ direction) as long as $\alpha/\beta \geq \sqrt{2}$ (see Problem 6.4). Therefore, the vector $\mathbf{u}_o$ will be in the ($x_1, -x_3$) quadrant. If the plane (x_1, x_2) represents the surface of the Earth, then the angle θ between $\mathbf{u}_o$ and the upward vertical direction (given by

$-\mathbf{a}_3$) is always acute. The tangent of θ is given by

$$\tan \theta = \frac{u_1^o}{-u_3^o} = \tan 2f. \tag{6.5.25}$$

This means that the angle of $\mathbf{u}_o$ with the upward normal is $2f$, while the incidence angle is e. Therefore, if the angle derived from the two components of the displacement measured at the surface is to be used to estimate the incidence angle, it will be necessary to assess the error introduced by this procedure. As Fig. 6.2 shows, for $\alpha/\beta = \sqrt{3}$, $2f$ approximates e quite well for a wide range of angles, with an error of less than than 5° for e less than about 70°. However, when $\alpha/\beta = 2.5$ the error becomes important even at smaller values of e. For example, the error is 10° for e about 40°.

6.5.2.2 *Special cases*

(1) *Normal incidence.* In this case $e = 0$ ($\mathbf{p}$ perpendicular to the free surface), $f = 0$, $A_1 = -A$, and $B_1 = 0$. Therefore, there is no horizontal displacement or reflected SV wave. Furthermore, the fact that $A_1/A = -1$ means that a compressional wave will reflect as a tensional wave and vice versa. This reversal is well known in other elastic problems involving free surfaces and, under appropriate conditions, is capable of causing tensile cracking when the free surface is subjected to a compressive stress. This process is known as *scabbing* or *spalling* (Achenbach, 1973; Graff, 1975).

The surface displacement is given by

$$\mathbf{u}_o = -2A\mathbf{a}_3 \exp\left[i\omega\left(t - \frac{x_1 \sin e}{\alpha}\right)\right]. \tag{6.5.26}$$

This equation shows that the surface displacement has twice the amplitude of the incident wave.

(2) *Grazing incidence.* In this case $e = \pi/2$ ($\mathbf{p}$ parallel to the free surface), $f = \sin^{-1}(\beta/\alpha)$, $A_1 = -A$, and $B_1 = 0$. Note, however, that the total displacement is identically equal to zero everywhere, so this case, as posed, is not physically possible. For a different treatment of grazing incidence see Ewing *et al.* (1957), Miklowitz (1984) and Graff (1975).

(3) *Total mode conversion.* As noted above, A_1/A can be zero for certain values of α/β and e. In those cases, the reflected field will consist of SV waves only, in spite of the fact that the incident wave is a P.

6.5.2.3 *Energy equation*

As shown in §5.9, the average power transmitted by the P wave to the free surface is given by $\frac{1}{2}\rho\alpha n\omega^2 A^2$, with $n = \cos e$. At the surface the power removed by

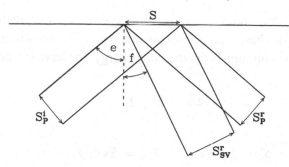

Fig. 6.3. Reflection of a P-wave beam at a free surface. The incident and reflected beams and the corresponding cross-sectional areas are shown. After Achenbach (1973).

the reflected P and SV waves is given by $\frac{1}{2}\rho\alpha\omega^2 \cos e A_1^2$ and $\frac{1}{2}\rho\beta\omega^2 \cos f B_1^2$, respectively. As the surface is free from tractions, there is no dissipation of energy there, and to maintain the energy balance the sum of the reflected energy must be equal to the incident energy:

$$\alpha A^2 \cos e = \alpha A_1^2 \cos e + \beta B_1^2 \cos f. \qquad (6.5.27)$$

After dividing by $\alpha \cos e$ and using Snell's law we obtain the energy equation

$$\left(\frac{A_1}{A}\right)^2 + \frac{\sin 2f}{\sin 2e}\left(\frac{B_1}{A}\right)^2 = 1 \qquad (6.5.28)$$

(Problem 6.6). This equation is extremely useful because it allows one to check the numerical values of the reflection coefficients.

The derivation of (6.5.28) is mathematically straightforward but ignores some of the physical aspects of the problem. As noted earlier, plane waves are infinite in extent, which means that they carry infinite energy. Of course, this is a nonphysical situation, and this is why energy considerations are based on unit areas. An alternative derivation of (6.5.28) based on the consideration of "beams" (Achenbach, 1973; Miklowitz, 1984) of incident and reflected waves having a common surface intersection (Fig. 6.3) will help us understand the physics of the situation. The absolute value of the energy flux in the direction of propagation averaged over one cycle for the beams of incident and two reflected waves (indicated by superscripts i and r) follows from (5.9.9)–(5.9.11):

$$|\Sigma_P^i| = \frac{1}{2}\rho\alpha\omega^2 A^2 S_P^i \qquad (6.5.29)$$

$$|\Sigma_P^r| = \frac{1}{2}\rho\alpha\omega^2 A_1^2 S_P^r \qquad (6.5.30)$$

$$|\Sigma_{SV}^r| = \frac{1}{2}\rho\beta\omega^2 B_1^2 S_{SV}^r \qquad (6.5.31)$$

(Problem 6.7). Here S_P^i, S_P^r, and S_{SV}^r are the cross-sectional areas of the corresponding beams (Fig. 6.3). These equations show that the energy goes to infinity as the areas of the beams go to infinity. From the same energy-conservation arguments used before we have

$$|\Sigma_P^i| = |\Sigma_P^r| + |\Sigma_{SV}^r|. \tag{6.5.32}$$

Also note that

$$S_P^i = S_P^r = S \cos e; \qquad S_{SV}^r = S \cos f, \tag{6.5.33}$$

where S is the area along the boundary (Fig. 6.3).

When (6.5.29)–(6.5.31) and (6.5.33) are used in (6.5.32) there is a common factor of S that can be canceled out. After that is done the resulting equation is the same as (6.5.27).

6.5.3 Incident SV waves

Using Snell's law the displacement can be expressed as

$$\mathbf{u} = \big[B(\cos f \mathbf{a}_1 + \sin f \mathbf{a}_3) \exp(i\omega x_3 \cos f/\beta)$$

$$+ A_1(\sin e \mathbf{a}_1 + \cos e \mathbf{a}_3) \exp(-i\omega x_3 \cos e/\alpha)$$

$$+ B_1(-\cos f \mathbf{a}_1 + \sin f \mathbf{a}_3) \exp(-i\omega x_3 \cos f/\beta)\big] \exp[i\omega(t - x_1 \sin f/\beta)].$$

$$\tag{6.5.34}$$

Proceeding as in the case of the incident P waves gives the following expressions for the reflection coefficients for the reflected P and SV waves:

$$\frac{A_1}{B} = \frac{(\alpha/\beta) \sin 4f}{\sin 2e \sin 2f + (\alpha/\beta)^2 \cos^2 2f} \tag{6.5.35}$$

$$\frac{B_1}{B} = \frac{\sin 2e \sin 2f - (\alpha/\beta)^2 \cos^2 2f}{\sin 2e \sin 2f + (\alpha/\beta)^2 \cos^2 2f}. \tag{6.5.36}$$

Plots of the reflection coefficients are shown in Fig. 6.4.

6.5.3.1 Special cases

(1) *Normal incidence.* In this case $f = 0$, $e = 0$, $A_1 = 0$, and $B_1 = -B$. Therefore, there is no reflected P wave and the displacement is horizontal. At the surface the displacement is given by

$$\mathbf{u}_o = 2B\mathbf{a}_1 \exp[i\omega(t - x_1 \sin f/\beta)]. \tag{6.5.37}$$

As in the case of incident P waves, there is a doubling of the amplitude of the incident wave.

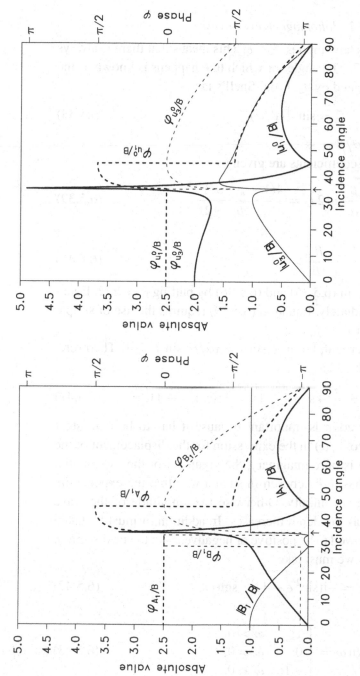

Fig. 6.4. Incident *SV* wave on a free surface. $\alpha/\beta = \sqrt{3}$. Absolute value and the phase of the reflection coefficients (left) and the horizontal and vertical components of the free-surface particle motion (right). The arrow indicates the critical angle, equal to 35.26°.

(2) *Total mode conversion.* There will be no reflected SV waves when the numerator of (6.5.36) is equal to zero, in which case the reflected field will consist of P waves only. For $\alpha/\beta = \sqrt{3}$ this happens for angles of $30°$ and $34.2°$ (Problem 6.8). This can be seen in Fig. 6.4.

6.5.3.2 *Inhomogeneous waves*

It was noted earlier that Snell's law implies $f < e$. This means that there is always a value of f for which $e = \pi/2$. The angle for which that happens is known as the *critical angle* and will be indicated by f_c. From Snell's law

$$f_c = \sin^{-1}(\beta/\alpha). \tag{6.5.38}$$

For example, $f_c = 35.26°$ for $\alpha/\beta = \sqrt{3}$.

When $f = f_c$ the reflection coefficients are given by

$$\frac{A_1}{B} = 2\frac{\beta}{\alpha}\tan 2f_c = \frac{4(1 - \beta^2/\alpha^2)^{1/2}}{(\alpha^2/\beta^2 - 2)} \tag{6.5.39}$$

and

$$\frac{B_1}{B} = -1. \tag{6.5.40}$$

These equations are derived from (6.5.35) and (6.5.36) by putting $e = \pi/2$. Equation (6.5.40) is obtained immediately. Equation (6.5.39) requires the use of simple trigonometric relations and $\sin f_c = \beta/\alpha$.

When $f > f_c$, e is no longer real, because $\sin e = (\alpha/\beta)\sin f > 1$. Therefore, $\cos e$ becomes pure imaginary:

$$\cos e = \pm\sqrt{1 - \sin^2 e} = \pm\sqrt{-(\sin^2 e - 1)} = \pm i(\sin^2 e - 1)^{1/2}. \tag{6.5.41}$$

The selection of the sign for $\cos e$ is important because it has to be consistent with the factor of $\exp(-i\omega x_3 \cos e/\alpha)$ in the expression for the displacement of the reflected P wave. When $\cos e$ is pure imaginary, the argument of the exponential is real, and the sign of $\cos e$ has to be chosen in such a way that the exponential factor goes to zero as x_3 goes to infinity. Otherwise the amplitude of the wave would increase without limit as the depth increases. In addition, it must be taken into account that ω may be allowed to be positive or negative, as in the next section. Based on these considerations we must take

$$\cos e = -i(\sin^2 e - 1)^{1/2}\,\mathrm{sgn}\,\omega, \tag{6.5.42}$$

where

$$\mathrm{sgn}\,\omega = \begin{cases} 1; & \omega > 0, \\ 0; & \omega = 0, \\ -1; & \omega < 0. \end{cases} \tag{6.5.43}$$

Beyond the critical angle the P wave becomes an *inhomogeneous plane wave*, i.e., a wave characterized by an exponential decay with depth. In this sense, inhomogeneous waves behave as surface waves (see Chapter 7), but a major difference is that the former can be assumed to carry no energy. This question is discussed in §6.5.3.4.

Next, we will derive expressions for A_1/B and B_1/B for incidence angles larger than the critical angle. Using (6.5.42), Snell's law, and writing S in place of sgn ω, the denominator in (6.5.35) and (6.5.36) can be written as

$$D = \frac{\alpha^2}{\beta^2}\left[\cos^2 2f - 2i\frac{\beta^2}{\alpha^2}\sin e(\sin^2 e - 1)^{1/2}S\sin 2f\right]$$

$$= \frac{\alpha^2}{\beta^2}\left[\cos^2 2f - 2i\sin f(\sin^2 f - \beta^2/\alpha^2)^{1/2}S\sin 2f\right] = |D|e^{-i\phi S}, \quad (6.5.44)$$

where the vertical bars indicate absolute value and

$$\tan\phi = 2\sin f(\sin^2 f - \beta^2/\alpha^2)^{1/2}\frac{\sin 2f}{\cos^2 2f} = \frac{2\sin f(\sin^2 f - \beta^2/\alpha^2)^{1/2}}{\cos 2f \cot 2f}.$$
$$(6.5.45)$$

Because the numerator of (6.5.36) is minus the complex conjugate of D,

$$\frac{B_1}{B} = -e^{i2\phi S} \equiv e^{-i\pi S}e^{2i\phi S} = e^{i2(\phi - \pi/2)S}. \quad (6.5.46)$$

To obtain an expression for A_1/B rewrite D as follows:

$$D = \frac{\alpha^2}{\beta^2}\cos^2 2f(1 - iS\tan\phi). \quad (6.5.47)$$

Then,

$$\frac{A_1}{B} = 2\frac{\beta}{\alpha}\frac{\tan 2f}{1 - iS\tan\phi} = 2\frac{\beta}{\alpha}\tan 2f\frac{1 + iS\tan\phi}{1 + \tan^2\phi}$$

$$= 2\frac{\beta}{\alpha}\tan 2f\cos\phi(\cos\phi + iS\sin\phi) = 2\frac{\beta}{\alpha}\tan 2f\cos\phi\, e^{i\phi\,\mathrm{sgn}\,\omega}. \quad (6.5.48)$$

Introducing

$$\chi = \frac{\pi}{2} - \phi, \quad (6.5.49)$$

so that

$$\tan\chi = \frac{1}{\tan\phi} = \frac{\cos 2f\cot 2f}{2\sin f(\sin^2 f - \beta^2/\alpha^2)^{1/2}}, \quad (6.5.50)$$

we obtain

$$\frac{B_1}{B} = e^{-2i\chi\,\mathrm{sgn}\,\omega} \quad (6.5.51)$$

and

$$\frac{A_1}{B} = 2\frac{\beta}{\alpha} \tan 2f \sin \chi e^{i(\pi/2 - \chi)\,\text{sgn}\,\omega} \tag{6.5.52}$$

(Nadeau, 1964; Ben-Menahem and Singh, 1981). These results will be used in the next section.

When $f = f_c$, the denominator of (6.5.50) is zero and the phase shift χ becomes $\pi/2$, in which case (6.5.51) and (6.5.52) agree with (6.5.40) and (6.5.39). The phase shift is also $\pi/2$ when $f = \pi/2$, in which case the numerator of (6.5.50) is equal to infinity. For $f = \pi/4$, $\chi = 0$ and A_1 becomes zero, and $B_1 = B$, so that for an incidence angle of $\pi/4$ there is no inhomogeneous wave, only a reflected SV wave without phase change. It must be noted, however, that this result is obtained even when $f_c > \pi/4$, as can be seen from (6.5.35) and (6.5.36) with $f = \pi/4$. However, because critical angles larger than $\pi/4$ require $\alpha/\beta \le \sqrt{2}$, they cannot be reached as long as the Poisson ratio is nonnegative (see Problem 6.4).

The reflection coefficients given by (6.5.51) and (6.5.52) are plotted in Fig. 6.4, and are used with (6.5.34) to compute the horizontal and vertical components of the relative surface displacement $\mathbf{u}_o/B$, which are also shown in Fig. 6.4. Because all the quantities become complex for angles larger than the critical one, they are plotted in terms of amplitudes and phases. An important feature of this figure is the phase difference between the horizontal and vertical components of the displacement. The significance of this difference is that the motion of a given surface particle as a function of time will no longer be linear and will become elliptical (see §5.8.4.1).

6.5.3.3 *Displacement in the time domain*

We know that under fairly general conditions a function $g(t)$ can be expressed as a Fourier integral

$$g(t) = \frac{1}{2\pi} \int_{-\infty}^{\infty} G(\omega) \exp(i\omega t)\, d\omega. \tag{6.5.53}$$

In other words, $g(t)$ can be expressed as an infinite sum of harmonic waves. This fact will be used to analyze plane waves that are not harmonic. In particular, we are interested in expressions of the form $g(t - t_o)$, in which case (6.5.53) is replaced by

$$g(t - t_o) = \frac{1}{2\pi} \int_{-\infty}^{\infty} G(\omega) \exp[i\omega(t - t_o)]\, d\omega. \tag{6.5.54}$$

Equation (6.5.54) is directly related to the reflection coefficients derived in the previous sections. Recall that the P, SV, and SH displacements include factors of the form $\exp[i\omega(t - t_o)]$, where t_o is equal to $\mathbf{p} \cdot \mathbf{r}/c$, with c equal to α or β. Furthermore, $G(\omega)$ can be taken as one of the coefficients A, B, A_1, or B_1. When the reflection coefficients A_1/A, B_1/A, A_1/B, and B_1/B are independent of ω, the reflected pulses will be equal to the incident pulse multiplied by the corresponding reflection coefficient. To see that, let us consider the horizontal components of the displacement for the case of an incident SV wave for incidence angles less than the critical angle.

To analyze the problem it will be assumed that the incident plane wave has a time dependence $b(t)$, with Fourier transform $B(\omega)$. The goal is to find the functions $a_1(t)$ and $b_1(t)$ (corresponding to A_1 and B_1) in terms of $b(t)$. For the horizontal component (indicated with the superscript h) of the incident and the two reflected waves we have from (6.5.34)

$$b^{\mathrm{h}}(t - t_S) = \frac{\cos f}{2\pi} \int_{-\infty}^{\infty} B(\omega) \exp[i\omega(t - t_S)]\, d\omega \qquad (6.5.55)$$

$$a_1^{\mathrm{h}}(t - t_P) = \frac{\sin e}{2\pi} \int_{-\infty}^{\infty} A_1(\omega) \exp[i\omega(t - t_P)]\, d\omega \qquad (6.5.56)$$

$$b_1^{\mathrm{h}}(t - t_{S1}) = -\frac{\cos f}{2\pi} \int_{-\infty}^{\infty} B_1(\omega) \exp[i\omega(t - t_{S1})]\, d\omega. \qquad (6.5.57)$$

If $f \le f_c$, the ratios A_1/B and B_1/B do not depend on ω. Therefore, after multiplying and dividing the integrand of $a_1(t - t_P)$ by B we obtain

$$a_1^{\mathrm{h}}(t - t_P) = \frac{\sin e}{2\pi} \int_{-\infty}^{\infty} \frac{A_1}{B} B(\omega) \exp[i\omega(t - t_P)]\, d\omega$$

$$= \frac{A_1}{B} \frac{\sin e}{2\pi} \int_{-\infty}^{\infty} B(\omega) \exp[i\omega(t - t_P)]\, d\omega. \qquad (6.5.58)$$

The last integral can be written in terms of b^{h} using (6.5.55), which gives

$$a_1^{\mathrm{h}}(t - t_P) = \frac{A_1}{B} \frac{\sin e}{\cos f} b^{\mathrm{h}}(t - t_P). \qquad (6.5.59)$$

Similarly,

$$b_1^{\mathrm{h}}(t - t_{S1}) = -\frac{B_1}{B} b^{\mathrm{h}}(t - t_{S1}). \qquad (6.5.60)$$

Here A_1/B and B_1/B are the ratios given in (6.5.35) and (6.5.36). The implication of (6.5.59) and (6.5.60) is that the reflected P and SV waves have the same shape as the incident wave, while the amplitudes and polarities depend on the angle of incidence. Similar equations can be derived for the vertical components.

The case $f \ge f_c$ is not so simple because the reflection coefficients depend on ω via the combination $\chi \, \mathrm{sgn}\, \omega$ in the phase factor. Starting with the first equality in (6.5.58), using (6.5.52) for A_1/B and multiplying and dividing by $\cos f$ gives

$$a_1^h(t-t_P) = \frac{1}{2\pi} g(f) \sin \chi \int_{-\infty}^{\infty} \cos f \, B(\omega) \exp[i\omega(t-t_P)+i(\pi/2-\chi)\,\text{sgn}\,\omega]\,d\omega,$$

(6.5.61)

where

$$g(f) = \frac{2(\beta/\alpha)\sin e \tan 2f}{\cos f} = 2\tan f \tan 2f.$$

(6.5.62)

For the last equality in (6.5.62) Snell's law was used. Now let $\mathcal{S}$ denote $\text{sgn}\,\omega$ and note that

$$\exp[i(\pi/2 - \chi)\mathcal{S}] = \cos[(\pi/2 - \chi)\mathcal{S}] + i\sin[(\pi/2 - \chi)\mathcal{S}]$$

$$= \cos(\pi/2 - \chi) + i\mathcal{S}\sin(\pi/2 - \chi) = \sin \chi + i\mathcal{S}\cos \chi.$$

(6.5.63)

Therefore,

$$a_1^h(t - t_P) = \frac{1}{2\pi} g(f) \sin \chi \left\{ \sin \chi \int_{-\infty}^{\infty} \cos f \, B(\omega) \exp[i\omega(t - t_P)]\,d\omega \right.$$

$$\left. + \, i\cos \chi \int_{-\infty}^{\infty} \cos f \, B(\omega)\mathcal{S} \exp[i\omega(t - t_P)] \right\}.$$

(6.5.64)

In a similar way, using (6.5.51) we obtain the following expression for b_1^h:

$$b_1(t - t_S) = -\frac{1}{2\pi} \cos f \int \frac{B_1}{B} B \exp[i\omega(t - t_S)]\,d\omega$$

$$= -\frac{1}{2\pi} \int_{-\infty}^{\infty} \cos f \, B(\omega) \exp[i\omega(t - t_S) - i2\chi\mathcal{S}]\,d\omega$$

$$= \frac{1}{2\pi} \left\{ -\cos 2\chi \int_{-\infty}^{\infty} \cos f \, B(\omega) \exp[i\omega(t - t_S)]\,d\omega \right.$$

$$\left. + \, i\sin 2\chi \int_{-\infty}^{\infty} \cos f \, B(\omega)\mathcal{S} \exp[i\omega(t - t_S)]\,d\omega \right\}.$$

(6.5.65)

Note that the first integrals in the final expressions for $a_1^h(t - t_P)$ and $b_1^h(t - t_S)$ are just $2\pi b^h(t - t_P)$ and $2\pi b^h(t - t_S)$, respectively. The corresponding second integrals can be solved using the *convolution theorem*

$$\Im\{r(t) * s(t)\} = \Im\{r(t)\}\,\Im\{s(t)\},$$

(6.5.66)

where $\Im$ indicates the Fourier transform, the asterisk indicates convolution, and $r(t)$ and $s(t)$ represent arbitrary functions of time (Papoulis, 1962). Applying the inverse Fourier transform to both sides of (6.5.66) gives

$$r(t) * s(t) = \Im^{-1}\{\Im\{r(t) * s(t)\}\} = \frac{1}{2\pi}\int_{-\infty}^{\infty}\Im\{r(t)\}\,\Im\{s(t)\}\exp(i\omega t)\,d\omega.$$

(6.5.67)

Another intermediate result we need is

$$\Im\{r(t - t_o)\} = \exp(-i\omega t_o)\Im\{r(t)\},$$

(6.5.68)

which can be proved with a simple change of variables. Now let $\Im\{r(t)\} = \cos f\, B(\omega)$, and $\Im\{s(t)\} = i\,\mathrm{sgn}\,\omega$, which implies that $s(t) = -1/(\pi t)$ (see (A.75)). Therefore,

$$\frac{i}{2\pi}\int_{-\infty}^{\infty}\cos f\, B(\omega)\,\mathrm{sgn}\,\omega\exp[i\omega(t - t_o)]\,d\omega$$

$$= \frac{i}{2\pi}\int_{-\infty}^{\infty}\cos f\, B(\omega)\exp(-i\omega t_o)\,\mathrm{sgn}\,\omega\exp(i\omega t)\,d\omega$$

$$= -\frac{1}{\pi}b^h(t - t_o) * \frac{1}{t}.$$

(6.5.69)

The convolution on the right-hand side is the definition of the Hilbert transform of $b^h(t - t_o)$, which will be indicated with a ˘ symbol

$$\breve{b}^h(t - t_o) = -\frac{1}{\pi}b^h(t - t_o) * \frac{1}{t} = \frac{1}{\pi}\mathcal{P}\int_{-\infty}^{\infty}\frac{b^h(\tau - t_o)}{\tau - t}\,d\tau,$$

(6.5.70)

where $\mathcal{P}$ indicates the Cauchy principal value of the integral, which is computed by excluding any singular value in the integrand (see (A.68)). The Hilbert transform is discussed in Appendix B.

Using (6.5.69) and (6.5.70), from (6.5.64) and (6.5.65) we obtain

$$a_1^h(t - t_P) = g(f)\sin\chi\left[\sin\chi\, b^h(t - t_P) + \cos\chi\,\breve{b}^h(t - t_P)\right]$$

(6.5.71)

$$b_1^h(t - t_S) = -\cos 2\chi\, b^h(t - t_S) + \sin 2\chi\,\breve{b}^h(t - t_S)$$

(6.5.72)

(Ben-Menahem and Singh, 1981). Equations (6.5.71) and (6.5.72) show that the reflected waves are linear combinations of the original wave and its Hilbert transform. Furthermore, because the coefficients in each combination are functions of f, the shapes of the waves will depend on the incidence angle. An example of this change of shape is given in Fig. 6.5, which shows the incident pulse $b^h(t)$, its Hilbert transform $\breve{b}^h(t)$, and the reflection coefficients $a_1^h(t)$ and $b_1^h(t)$ for a number of incidence angles. Note that $\breve{b}^h(t)$ has a value slightly different from zero for $t < 0$, which is not present in $b^h(t)$. This precursor, which also affects the reflection coefficients, is a result of the fact that the $\pi/2$ phase shift that enters

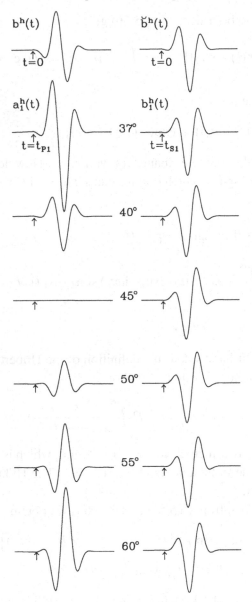

Fig. 6.5. Incident SV wave on a free surface. Plot of the horizontal components of the input waveform $b^h(t)$, of its Hilbert transform $\breve{b}^h(t)$, and of $a_1^h(t)$ and $b_1^h(t)$ (see (6.5.71) and (6.5.72)) for various values of the angle of incidence f (number between waveforms) greater than f_c (equal to $35.26°$ in this case). The horizontal axis is time. All the waveforms have been plotted with the same vertical scale and are shifted to align the arrival times (indicated by arrows). The tips of the arrows coincide with the zero value along the amplitude axis. Note that in some cases the amplitude is different from zero for $t < 0$ (see the text for details).

in the definition of the Hilbert transform (sec (B.13)) is a noncausal operator. See Pilant (1979) for an additional discussion on this matter.

6.5.3.4 Energy equation

When $f < f_c$, an argument similar to that used in §6.5.2.3 leads to

$$\frac{\sin 2e}{\sin 2f}\left(\frac{A_1}{B}\right)^2 + \left(\frac{B_1}{B}\right)^2 = 1. \tag{6.5.73}$$

When $f > f_c$, from (6.5.40) it follows that $|B/B_1| = 1$, so that (6.5.73) is satisfied only if the first term on the left is ignored, and the energy equation becomes

$$\left|\frac{B_1}{B}\right| = 1. \tag{6.5.74}$$

Does (6.5.74) mean that the inhomogeneous P waves carry no energy? The answer is no, but we must consider again that we are dealing with waves of infinite extent. As shown by Miklowitz (1984) the energy in the post-critically reflected P waves is finite when it is integrated between 0 (the free surface) and infinity. This finite energy can be neglected when compared to the infinite amount of energy carried by the incident and reflected SV waves. Ben-Menahem and Singh (1981) discussed in detail a similar problem for SH waves incident on a solid–solid boundary, and found that the energy in the inhomogeneous wave averaged over one cycle is equal to zero. This is another way to show that the inhomogeneous waves can be assumed to carry no energy. See also §6.6.1.2.

6.6 Waves incident on a solid–solid boundary

The two media are assumed to be in welded contact.

6.6.1 Incident SH waves

In this case there is a transmitted wave in addition to the incident and reflected waves. The displacements at any point in the incidence ($x_3 > 0$) and transmission ($x_3 < 0$) media are given by

$$\mathbf{u} = \left\{ C \exp\left[i\omega\left(t - \frac{x_1 \sin f - x_3 \cos f}{\beta}\right)\right] \right.$$
$$\left. + C_1 \exp\left[i\omega\left(t - \frac{x_1 \sin f_1 + x_3 \cos f_1}{\beta}\right)\right] \right\} \mathbf{a}_2 \tag{6.6.1}$$

and

$$\mathbf{u}' = C' \exp\left[i\omega\left(t - \frac{x_1 \sin f' - x_3 \cos f'}{\beta'}\right)\right]\mathbf{a}_2, \qquad (6.6.2)$$

respectively. From the boundary condition $\mathbf{u} = \mathbf{u}'$ at $x_3 = 0$ and with the same argument that lead to (6.5.15) we obtain

$$f_1 = f; \qquad \frac{\sin f}{\beta} = \frac{\sin f'}{\beta'} \qquad (6.6.3\mathrm{a,b})$$

$$C + C_1 = C'. \qquad (6.6.4)$$

Equation (6.6.3b) is Snell's law for this problem. The continuity of the stress vector at $x_3 = 0$ gives

$$\frac{\mu}{\beta} \cos f C - \frac{\mu}{\beta} \cos f C_1 = \frac{\mu'}{\beta'} C' \cos f'. \qquad (6.6.5)$$

From (6.6.4) and (6.6.5) we obtain

$$\frac{C'}{C} - \frac{C_1}{C} = 1 \qquad (6.6.6)$$

and

$$\frac{C'}{C}\frac{\mu'}{\beta'} \cos f' + \frac{C_1}{C}\frac{\mu}{\beta} \cos f = \frac{\mu}{\beta} \cos f. \qquad (6.6.7)$$

Let

$$D = \begin{vmatrix} 1 & -1 \\ (\mu'/\beta') \cos f' & (\mu/\beta) \cos f \end{vmatrix} = \frac{\mu\beta' \cos f + \mu'\beta \cos f'}{\beta\beta'}. \qquad (6.6.8)$$

Then the *transmission coefficient* C'/C is given by

$$\frac{C'}{C} = \frac{1}{D}\begin{vmatrix} 1 & -1 \\ (\mu/\beta) \cos f & (\mu/\beta) \cos f \end{vmatrix} = \frac{2\mu \cos f}{(\mu\beta' \cos f + \mu'\beta \cos f')/\beta'}. \qquad (6.6.9)$$

Now multiply the numerator and denominator of (6.6.9) by $2 \sin f$ and use Snell's law and $\mu'\beta^2/\beta'^2 = \mu\rho'/\rho$. This gives

$$\frac{C'}{C} = \frac{2\mu \sin 2f}{\mu \sin 2f + 2\mu'(\beta/\beta')^2 \sin f' \cos f'} = \frac{2 \sin 2f}{\sin 2f + (\rho'/\rho) \sin 2f'}. \qquad (6.6.10)$$

Similarly, the *reflection coefficient* C_1/C is given by

$$\frac{C_1}{C} = \frac{1}{D} \begin{vmatrix} 1 & 1 \\ (\mu'/\beta') \cos f' & (\mu/\beta) \cos f \end{vmatrix}$$

$$= \frac{\beta'\mu \cos f - \beta\mu' \cos f'}{\beta'\mu \cos f + \beta\mu' \cos f'} = \frac{\sin 2f - (\rho'/\rho) \sin 2f'}{\sin 2f + (\rho'/\rho) \sin 2f'}. \qquad (6.6.11)$$

6.6.1.1 Inhomogeneous waves

From Snell's law (6.6.3) it follows that when $\beta' > \beta$ there will be a critical angle f_c for which $f' = \pi/2$. The critical angle is given by

$$f_c = \sin^{-1} \frac{\beta}{\beta'}. \qquad (6.6.12)$$

When $f \geq f_c$ the transmitted SH wave no longer exists, and is replaced by an inhomogeneous wave. The discussion of this case follows closely that in §6.5.3.2. The definition of $\cos f$ must be such that the transmitted SH wave becomes a wave with amplitude going to zero as x_3 goes to $-\infty$. This requires choosing

$$\cos f' = -\mathrm{i}(\sin f' - 1)^{1/2} \operatorname{sgn} \omega. \qquad (6.6.13)$$

Next, we will derive expressions for C_1/C and C'/C for angles larger than the critical. Let us start by writing (6.6.11) as

$$\frac{C_1}{C} = \frac{1-b}{1+b}, \qquad (6.6.14)$$

where

$$b = \frac{\rho' \sin f' \cos f'}{\rho \sin f \cos f} = -\mathrm{i} \frac{1}{\cos f} \frac{\rho'}{\rho} \frac{\beta'}{\beta} \left(\frac{\beta'^2}{\beta^2} \sin^2 f - 1 \right)^{1/2} S$$

$$= -\mathrm{i} \frac{\mu'}{\mu} \frac{1}{\cos f} \left(\sin^2 f - \beta^2/\beta'^2 \right)^{1/2} S \equiv -\mathrm{i} S a \qquad (6.6.15)$$

(Zoeppritz, 1919), where (6.6.13), Snell's law, and $\mu = \rho\beta^2$ were used, $S = \operatorname{sgn} \omega$, and a is defined by the identity. Then,

$$\frac{C_1}{C} = \frac{1 + \mathrm{i}Sa}{1 - \mathrm{i}Sa} = \mathrm{e}^{2\mathrm{i}\phi S}, \qquad (6.6.16)$$

with

$$\tan \phi = a = \frac{\mu'}{\mu} \frac{1}{\cos f} \left(\sin^2 f - \beta^2/\beta'^2 \right)^{1/2}. \qquad (6.6.17)$$

Equation (6.6.17) will be used in the study of Love waves (§7.3.3).

To determine C'/C use (6.6.4) and (6.6.16):

$$\frac{C'}{C} = 1 + \frac{C_1}{C} = 1 + e^{2i\phi S} = e^{i\phi S}\left(e^{-i\phi S} + e^{i\phi S}\right) = 2\cos\phi\, e^{i\phi S}. \qquad (6.6.18)$$

As done in §6.5.3.2, we will introduce

$$\chi = \frac{\pi}{2} - \phi \qquad (6.6.19)$$

with

$$\tan\chi = \frac{1}{\tan\phi} = \frac{\mu'}{\mu}\frac{\cos f}{\left(\sin^2 f - \beta^2/\beta'^2\right)^{1/2}}. \qquad (6.6.20)$$

Then,

$$\frac{C_1}{C} = e^{i(\pi - 2\chi)\,\mathrm{sgn}\,\omega} = -e^{-2i\chi\,\mathrm{sgn}\,\omega} \qquad (6.6.21)$$

and

$$\frac{C'}{C} = 2\sin\chi\, e^{i(\pi/2 - \chi)\,\mathrm{sgn}\,\omega} \qquad (6.6.22)$$

(Ben-Menahem and Singh, 1981). The displacements for angles larger than the critical are given in Problem 6.11.

Before leaving this subject it is necessary to distinguish between the inhomogeneous waves described here and the head waves, also known as refracted waves in exploration seismology. These waves are generated by curved wave fronts, not by plane wave fronts, and although the inhomogeneous and head waves are directly related to the presence of the critical angle, the theory developed here cannot be used to infer any of the properties of the head waves, which require a much more complicated mathematical analysis (e.g., Aki and Richards, 1980).

6.6.1.2 Energy equation

When $f < f_c$, an argument similar to that used in §6.5.2.3 leads to

$$\rho\beta C^2 \cos f = \rho\beta C_1^2 \cos f + \rho'\beta'C'^2 \cos f'. \qquad (6.6.23)$$

After dividing by $\rho\beta C \cos f$ and using Snell's law we find

$$\left(\frac{C_1}{C}\right)^2 + \frac{\rho'\sin 2f'}{\rho\sin 2f}\left(\frac{C'}{C}\right)^2 = 1. \qquad (6.6.24)$$

When $f > f_c$, we see from (6.6.14) that $|C_1/C| = 1$ which means that (6.6.24) will be satisfied when the second term on the left is ignored. This is consistent with the assumption that the inhomogeneous SH wave carries no energy, as discussed in §6.5.3.4. See also Problem 6.12.

6.6.2 Incident P waves

In this case the displacement at any point in the incidence medium ($x_3 > 0$) is given by

$$\mathbf{u} = A(\sin e\,\mathbf{a}_1 - \cos e\,\mathbf{a}_3) \exp\left[i\omega\left(t - \frac{x_1 \sin e - x_3 \cos e}{\alpha}\right)\right]$$

$$+ A_1(\sin e_1\mathbf{a}_1 + \cos e_1\mathbf{a}_3) \exp\left[i\omega\left(t - \frac{x_1 \sin e_1 + x_3 \cos e_1}{\alpha}\right)\right]$$

$$+ B_1(-\cos f_1\mathbf{a}_1 + \sin f_1\mathbf{a}_3) \exp\left[i\omega\left(t - \frac{x_1 \sin f_1 + x_3 \cos f_1}{\beta}\right)\right], \quad (6.6.25)$$

while in the transmission medium ($x_3 < 0$) the displacement is given by

$$\mathbf{u}' = A'(\sin e'\mathbf{a}_1 - \cos e'\mathbf{a}_3) \exp\left[i\omega\left(t - \frac{x_1 \sin e' - x_3 \cos e'}{\alpha'}\right)\right]$$

$$+ B'(\cos f'\mathbf{a}_1 + \sin f'\mathbf{a}_3) \exp\left[i\omega\left(t - \frac{x_1 \sin f' - x_3 \cos f'}{\beta'}\right)\right]. \quad (6.6.26)$$

The condition that $\mathbf{u} = \mathbf{u}'$ at the boundary provides two constraints: $u_1 = u_1'$ and $u_3 = u_3'$ at $x_3 = 0$, corresponding to the continuity of the horizontal and vertical components of displacement. With the same arguments used before we find that $e_1 = e$, and

$$\frac{\sin e}{\alpha} = \frac{\sin f}{\beta} = \frac{\sin e'}{\alpha'} = \frac{\sin f'}{\beta'} = \frac{1}{c}, \quad (6.6.27)$$

where f_1 was replaced by f. This is the most general form of Snell's law. The rightmost equality introduces the *phase velocity* c. To investigate the meaning of c consider two positions of a wave front separated by a distance Δx_1 along the boundary (Fig. 6.6). The distance between the wave front positions is Δd and the velocity of propagation of the wave front is $\alpha = \Delta d/\Delta t$, while the *apparent velocity* that would be measured along the boundary is

$$V_{\text{app}} = \frac{\Delta x_1}{\Delta t} = \frac{\Delta d/\sin e}{\Delta t} = \frac{\alpha}{\sin e} = c. \quad (6.6.28)$$

Therefore, the *phase velocity* is the apparent velocity of the wave measured along the boundary. If a wave moves in the x_1 direction (grazing incidence), $e = \pi/2$ and $c = \alpha$. If the wave moves in the x_3 direction (normal incidence), $e = 0$ and $c = \infty$; i.e., a collection of receivers along the boundary will detect the wave at the same time. The concept of phase velocity, however, is much broader than that implied by (6.6.28), and is discussed in detail in §7.6.

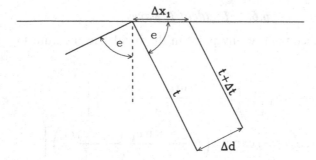

Fig. 6.6. Intersection of a plane wave front with a plane boundary at time positions t and $t + \Delta t$. The incidence angle is e. Δd is the distance between wave fronts and Δx_1 is the distance along the boundary. The wave front moves with a velocity $\alpha = \Delta d/\Delta t$, but along the boundary the apparent velocity of the wave front is $\Delta x_1/\Delta t$.

Now rewrite (6.6.25) and (6.6.26) using (6.6.27):

$$\mathbf{u} = \big[A(\sin e\,\mathbf{a}_1 - \cos e\,\mathbf{a}_3)\exp(i\omega x_3 \cos e/\alpha)$$

$$+ A_1(\sin e\,\mathbf{a}_1 + \cos e\,\mathbf{a}_3)\exp(-i\omega x_3 \cos e/\alpha)$$

$$+ B_1(-\cos f\,\mathbf{a}_1 + \sin f\,\mathbf{a}_3)\exp(-i\omega x_3 \cos f/\beta)\big]\exp[i\omega(t - x_1/c)] \quad (6.6.29)$$

and

$$\mathbf{u}' = \big[A'(\sin e'\,\mathbf{a}_1 - \cos e'\,\mathbf{a}_3)\exp(i\omega x_3 \cos e'/\alpha')$$

$$+ B'(\cos f'\,\mathbf{a}_1 + \sin f'\,\mathbf{a}_3)\exp(i\omega x_3 \cos f'/\beta')\big]\exp[i\omega(t - x_1/c)]. \quad (6.6.30)$$

The condition $u_1 = u_1'$ gives

$$A\sin e + A_1 \sin e - B_1 \cos f = A'\sin e' + B'\cos f', \quad (6.6.31)$$

which can be rewritten as

$$A\sin e = -A_1 \sin e + B_1 \cos f + A'\sin e' + B'\cos f'. \quad (6.6.32)$$

The condition $u_3 = u_3'$ gives

$$-A\cos e + A_1 \cos e + B_1 \sin f = -A'\cos e' + B'\sin f', \quad (6.6.33)$$

which can be rewritten as

$$A\cos e = A_1 \cos e + B_1 \sin f + A'\cos e' - B'\sin f'. \quad (6.6.34)$$

The continuity of the stress vector gives two additional equations, one for the horizontal component and one for the vertical. From the horizontal component we obtain

$$2\mu A\frac{\cos e}{\alpha}\sin e - 2\mu A_1\sin e\frac{\cos e}{\alpha} + \mu B_1\left[\frac{\cos^2 f}{\beta} - \frac{\sin f^2}{\beta}\right]$$

$$= 2A'\frac{\mu'}{\alpha'}\sin e'\cos e' + \mu'B'\left[\frac{\cos f'}{\beta'}\cos f' - \frac{\sin^2 f'}{\beta'}\right], \tag{6.6.35}$$

which can be rewritten as

$$A\sin 2e = A_1\sin 2e - B_1\frac{\alpha}{\beta}\cos 2f + A'\frac{\alpha}{\alpha'}\frac{\mu'}{\mu}\sin 2e' + B'\frac{\alpha}{\beta'}\frac{\mu'}{\mu}\cos 2f'$$

$$= A_1\sin 2e - B_1\frac{\alpha}{\beta}\cos 2f + A'\frac{\rho'}{\rho}\frac{\alpha}{\alpha'}\left(\frac{\beta'}{\beta}\right)^2\sin 2e'$$

$$+ B'\frac{\rho'}{\rho}\frac{\alpha}{\beta'}\left(\frac{\beta'}{\beta}\right)^2\cos 2f'. \tag{6.6.36}$$

In the last step the relation $\mu'/\mu = \rho'\beta'^2/\rho\beta^2$ was used.

From the vertical component of the stress vector in the incidence medium we obtain

$$-A\frac{\lambda}{\alpha}\sin^2 e - A\frac{(\lambda+2\mu)}{\alpha}\cos^2 e$$

$$- A_1\frac{\lambda}{\alpha}\sin^2 e - A_1\frac{(\lambda+2\mu)}{\alpha}\cos^2 e - 2B_1\frac{\mu}{\beta}\sin f\cos f$$

$$= -A\frac{\lambda}{\alpha} - 2A\frac{\mu}{\alpha}\cos^2 e - A_1\frac{\lambda}{\alpha} - 2A_1\frac{\mu}{\alpha}\cos^2 e, \tag{6.6.37}$$

which can be rewritten as

$$-B_1\mu\sin 2f = -\frac{A}{\alpha}(\lambda+2\mu\cos^2 e) - \frac{A_1}{\alpha}(\lambda+2\mu\cos^2 e) - B_1\frac{2\mu}{\beta}\sin 2f$$

$$= -A\mu\frac{\alpha}{\beta^2}\cos 2f - A_1\mu\frac{\alpha}{\beta^2}\cos 2f - B_1\frac{\mu}{\beta}\sin 2f. \tag{6.6.38}$$

In the last step the relation

$$\lambda + 2\mu\cos^2 e = \mu\frac{\alpha^2}{\beta^2}\cos 2f \tag{6.6.39}$$

(see (6.5.18)) was used. For the transmission medium we obtain

$$-A'\frac{\lambda'}{\alpha'}\sin^2 e' - A'\frac{(\lambda'+2\mu')}{\alpha'}\cos^2 e' + 2B'\frac{\mu'}{\beta'}\sin f'\cos f'$$

$$= -A'\mu'\frac{\alpha'}{\beta'^2}\cos 2f' + B'\frac{\mu'}{\beta'}\sin 2f'. \tag{6.6.40}$$

Now multiply (6.6.38) and (6.6.40) by $\beta^2/\mu\alpha$, use

$$\frac{\mu'}{\beta'}\frac{\beta^2}{\mu\alpha} = \frac{\rho'}{\rho}\frac{\beta'}{\alpha}; \qquad \frac{\alpha'\mu'}{\beta'^2}\frac{\beta^2}{\mu\alpha} = \frac{\rho'}{\rho}\frac{\alpha'}{\alpha} \qquad (6.6.41)$$

and equate the two resulting equations. This gives

$$A\cos 2f = -A_1\cos 2f - B_1\frac{\beta}{\alpha}\sin 2f + A'\frac{\rho'}{\rho}\frac{\alpha'}{\alpha}\cos 2f' - B'\frac{\rho'}{\rho}\frac{\beta'}{\alpha}\sin 2f'. \qquad (6.6.42)$$

Equations (6.6.32), (6.6.34), (6.6.36), and (6.6.42) can be written in matrix form as follows:

$$\begin{pmatrix} -\sin e & \cos f & \sin e' & \cos f' \\[2mm] \cos e & \sin f & \cos e' & -\sin f' \\[2mm] \sin 2e & -\dfrac{\alpha}{\beta}\cos 2f & \dfrac{\rho'}{\rho}\dfrac{\alpha}{\alpha'}\left(\dfrac{\beta'}{\beta}\right)^2\sin 2e' & \dfrac{\rho'}{\rho}\dfrac{\alpha}{\beta'}\left(\dfrac{\beta'}{\beta}\right)^2\cos 2f' \\[4mm] -\cos 2f & -\dfrac{\beta}{\alpha}\sin 2f & \dfrac{\rho'}{\rho}\dfrac{\alpha'}{\alpha}\cos 2f' & -\dfrac{\rho'}{\rho}\dfrac{\beta'}{\alpha}\sin 2f' \end{pmatrix} \begin{pmatrix} A_1/A \\[2mm] B_1/A \\[2mm] A'/A \\[2mm] B'/A \end{pmatrix}$$

$$= \begin{pmatrix} \sin e \\[2mm] \cos e \\[2mm] \sin 2e \\[2mm] \cos 2f \end{pmatrix}. \qquad (6.6.43)$$

When A, e, and the densities and velocities in the two media are given, solving the previous equation gives the unknown reflection coefficients A_1/A, B_1/A and transmission coefficients A'/A and B'/A.

The reflection and transmission coefficients have been computed for two cases. In the first case the velocities and densities in the incidence medium are lower than in the transmission medium and the coefficients have simple variations (Fig. 6.7). When the elastic parameters are larger in the transmission medium the amplitudes and phases of the coefficients are more complicated (Fig. 6.8) because of the presence of critical angles (see §6.6.2.2).

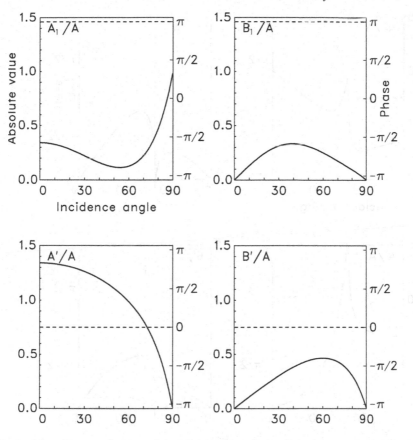

Fig. 6.7. Incident P wave on a solid–solid interface. The velocities and density for the incidence medium (α, β, ρ) are larger than the corresponding values for the transmission medium (α', β', ρ') $(\alpha', \beta', \rho'$: 4.00, 2.31, 2.50; α, β, ρ: 6.45, 3.72, 3.16). Solid and dashed lines indicate absolute values and phases, respectively, of the reflection and transmission coefficients.

6.6.2.1 Normal incidence

This special case is of importance in reflection seismology. Putting $e = f = e' = f' = 0$ in and using a, b, c for obvious ratios of parameters, (6.6.43) gives

$$\begin{pmatrix} 0 & 1 & 0 & 1 \\ 1 & 0 & 1 & 0 \\ 0 & -a & 0 & b \\ -1 & 0 & c & 0 \end{pmatrix} \begin{pmatrix} A_1/A \\ B_1/A \\ A'/A \\ B'/A \end{pmatrix} = \begin{pmatrix} 0 \\ 1 \\ 0 \\ 1 \end{pmatrix}. \tag{6.6.44}$$

Performing the matrix multiplication gives four equations

$$B_1/A + B'/A = 0 \tag{6.6.45}$$

$$A_1/A + A'/A = 1 \tag{6.6.46}$$

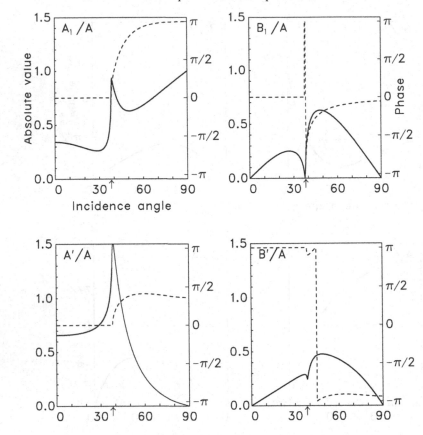

Fig. 6.8. Incident P wave on a solid–solid interface. The velocities and density for the incidence medium (α, β, ρ) are smaller than the corresponding values for the transmission medium (α', β', ρ') (α', β', ρ': 6.45, 3.72, 3.16; α, β, ρ: 4.00, 2.31, 2.50). Solid and dashed lines indicate absolute values and phases, respectively, of the reflection and transmission coefficients. The arrow indicates the critical angle for the transmitted P waves (38.3°). A'/A does not contribute to the energy equation for incidence angles larger than the critical angle.

$$-aB_1/A + bB'/A = 0 \tag{6.6.47}$$

$$-A_1/A + cA'/A = 1. \tag{6.6.48}$$

Equations (6.6.45) and (6.6.47) give

$$B_1 = B' = 0, \tag{6.6.49}$$

which means that the SV waves are absent, and (6.6.46) and (6.6.48) give

$$\frac{A_1}{A} = \frac{c-1}{c+1} = \frac{\rho'\alpha' - \rho\alpha}{\rho'\alpha' + \rho\alpha} \tag{6.6.50}$$

$$\frac{A'}{A} = \frac{2}{c+1} = \frac{2\rho\alpha}{\rho'\alpha' + \rho\alpha}. \tag{6.6.51}$$

The product *density* × *velocity* is known as *acoustic impedance*, and is a special case of the more general definition of *impedance* as the ratio of stress and particle velocity. This definition is the elastic analog of the impedance of electric circuits, and in both cases the underlying concept is the idea of *inertia* (Stratton, 1941), which in the elastic case corresponds to resistance to motion (see also Achenbach, 1973; Aki and Richards, 1980).

Equation (6.6.50) shows that the magnitude of the reflection coefficient depends on the difference of acoustic impedances. In addition, if the impedance in the transmission medium is larger than in the incidence medium, then the coefficient will be positive, but the reflected wave will be affected by a phase shift of π, as can be seen from the expressions for the vertical component of the incident and reflected waves in (6.6.25). For the transmitted wave there is no phase shift. Also note that equation (6.6.46) shows that the sum of the reflection and transmission coefficients is equal to 1.

6.6.2.2 Inhomogeneous waves

From Snell's law we can determine under which conditions critical angles may occur. Because α is always larger than β, there is no critical angle for the reflected SV waves. If $\alpha' > \alpha$, there will be a critical angle e_c^P for the transmitted P waves, corresponding to $e' = \pi/2$. This critical angle is given by

$$e_c^P = \sin^{-1}\frac{\alpha}{\alpha'}. \tag{6.6.52}$$

When $e > e_c^P$, $\cos e'$ has to be replaced by

$$\cos e' = -i\sqrt{\sin^2 e' - 1}. \tag{6.6.53}$$

The minus sign is required for bounded displacements in the transmission medium, where $x_3 < 0$. Under this condition the transmitted P wave becomes an inhomogeneous wave.

If $\beta' > \alpha$, there will also be a critical angle e_c^S for the transmitted SV waves, corresponding to $f' = \pi/2$. This critical angle is given by

$$e_c^S = \sin^{-1}\frac{\alpha}{\beta'}. \tag{6.6.54}$$

When $e > e_c^S$, $\cos f'$ has to be replaced by

$$\cos f' = -i\sqrt{\sin^2 f' - 1} \tag{6.6.55}$$

and the transmitted SV wave becomes inhomogeneous. For the model of Fig. 6.7, there is no e_c^S and e_c^P is equal to 38.3°.

It may be worth emphasizing that the definitions (6.6.53) and (6.6.55) follow from the condition of bounded displacement as x_3 goes to $-\infty$. If $\cos e'$ and $\cos f'$ were defined with the signs changed, the phase relations would be incorrect. This type of error affects several papers dealing with the solid–solid boundary (see Young and Braile (1976) for a summary).

6.6.2.3 *Energy equation*

Using already familiar arguments, for angles $e < e_c^P$,

$$\left(\frac{A_1}{A}\right)^2 + \frac{\sin 2f}{\sin 2e}\left(\frac{B_1}{A}\right)^2 + \frac{\rho'}{\rho}\frac{\sin 2e'}{\sin 2e}\left(\frac{A'}{A}\right)^2 + \frac{\rho'}{\rho}\frac{\sin 2f'}{\sin 2e}\left(\frac{B'}{A}\right)^2 = 1.$$

(6.6.56)

When $e > e_c^P$, some of the reflection coefficients become complex, and (6.6.56) must be modified as follows: replace the reflection and transmission coefficients by their absolute values, and set A'/A equal to zero. If $e > e_c^S$, also set B'/A equal to zero.

6.6.3 *Incident SV waves*

The displacement in the incidence medium is given by

$$\mathbf{u} = \big[B(\cos f\,\mathbf{a}_1 + \sin f\,\mathbf{a}_3)\exp(i\omega x_3\cos f/\beta)$$

$$+ A_1(\sin e\,\mathbf{a}_1 + \cos e\,\mathbf{a}_3)\exp(-i\omega x_3\cos e/\alpha)$$

$$+ B_1(-\cos f\,\mathbf{a}_1 + \sin f\,\mathbf{a}_3)\exp(-i\omega x_3\cos f/\beta)\big]\exp[i\omega(t - x_1/c)].$$

(6.6.57)

The only difference with the case of incident P waves is in the first term. Snell's law also applies, and $f_1 = f$. For the transmission medium the equations are equal to those for the P-wave case (see (6.6.30)), and will not be repeated here.

From continuity of the horizontal and vertical components of the displacements at $x_3 = 0$ we obtain

$$B\cos f = -A_1\sin e + B_1\cos f + A'\sin e' + B'\cos f' \qquad (6.6.58)$$

and

$$B\sin f = -A_1\cos e - B_1\sin f - A'\cos e' + B'\sin f'. \qquad (6.6.59)$$

From the continuity of the horizontal and vertical components of the stress vector at $x_3 = 0$ we obtain

$$B\cos 2f = A_1\frac{\beta}{\alpha}\sin 2e - B_1\cos 2f + A'\frac{\beta}{\alpha'}\frac{\mu'}{\mu}\sin 2e' + B'\frac{\beta}{\beta'}\frac{\mu'}{\mu}\cos 2f'$$

(6.6.60)

and

$$B \sin 2f = A_1 \frac{\alpha}{\beta} \cos 2f + B_1 \sin 2f - A' \frac{\mu'}{\mu} \frac{\beta \alpha'}{\beta'^2} \cos 2f' + B' \frac{\mu'}{\mu} \frac{\beta}{\beta'} \sin 2f'.$$

(6.6.61)

Equations (6.6.58)–(6.6.61) can be written in matrix form as

$$
\begin{pmatrix}
-\sin e & \cos f & \sin e' & \cos f' \\
-\cos e & -\sin f & -\cos e' & \sin f' \\
\dfrac{\beta}{\alpha} \sin 2e & -\cos 2f & \dfrac{\rho' \beta'^2}{\rho \alpha' \beta} \sin 2e' & \dfrac{\rho'}{\rho} \dfrac{\beta'}{\beta} \cos 2f' \\
\dfrac{\alpha}{\beta} \cos 2f & \sin 2f & -\dfrac{\rho'}{\rho} \dfrac{\alpha'}{\beta} \cos 2f' & \dfrac{\rho'}{\rho} \dfrac{\beta'}{\beta} \sin 2f'
\end{pmatrix}
\begin{pmatrix}
A_1/B \\
B_1/B \\
A'/B \\
B'/B
\end{pmatrix}
$$

$$
= \begin{pmatrix}
\cos f \\
\sin f \\
\cos 2f \\
\sin 2f
\end{pmatrix}.
$$

(6.6.62)

Figures 6.9 and 6.10 show the reflection and transmission coefficients for the velocity models used with the P waves (Figs 6.7 and 6.8). In this case there is additional complexity owing to the presence of at least one critical angle regardless of the velocity model, as discussed below.

6.6.3.1 Inhomogeneous waves

These waves will be generated in several cases.

(1) The incidence angle is larger than the critical angle for the reflected P wave, given by

$$f_c^{\mathrm{r}P} = \sin^{-1} \frac{\beta}{\alpha}.$$

(6.6.63)

This angle always exists, and for $f > f_c^{\mathrm{r}P}$ the reflected P wave becomes inhomogeneous and

$$\cos e = -i\sqrt{\sin^2 e - 1}.$$

(6.6.64)

(2) If $\beta < \alpha'$, there will be a critical angle for the transmitted P wave, given by

$$f_c^{\mathrm{t}P} = \sin^{-1} \frac{\beta}{\alpha'}.$$

(6.6.65)

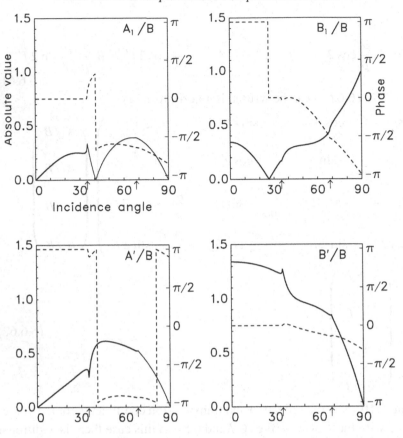

Fig. 6.9. Incident *SV* wave on a solid–solid interface. The velocities and density for the incidence medium (α, β, ρ) are larger than the corresponding values for the transmission medium (α', β', ρ') $(\alpha', \beta', \rho'$: 4.00, 2.31, 2.50; α, β, ρ: 6.45, 3.72, 3.16). Solid and dashed lines indicate absolute values and phases, respectively, of the reflection and transmission coefficients. The arrows indicate the critical angles for the reflected and transmitted *P* waves (35.2° and 68.4°, respectively). A_1/B and A'/B do not contribute to the energy equation for incidence angles larger than their corresponding critical angles.

For $f > f_c^{tP}$ the transmitted *P* wave becomes inhomogeneous and

$$\cos e' = -i\sqrt{\sin^2 e' - 1}. \tag{6.6.66}$$

(3) If $\beta < \beta'$, there will be a critical angle for the transmitted *SV* wave, given by

$$f_c^{tS} = \sin^{-1}\frac{\beta}{\beta'}. \tag{6.6.67}$$

For $f > f_c^{tS}$ the transmitted *SV* wave becomes inhomogeneous and

$$\cos f' = -i\sqrt{\sin^2 f' - 1}. \tag{6.6.68}$$

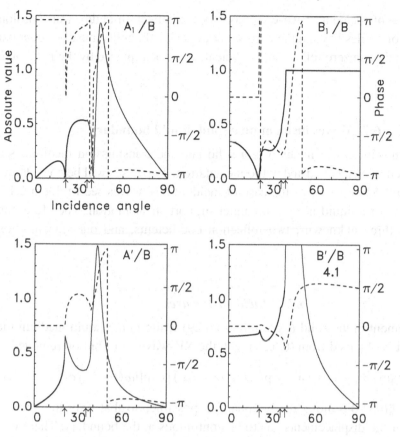

Fig. 6.10. Incident *SV* wave on a solid–solid interface. The velocities and density for the incidence medium (α, β, ρ) are smaller than the corresponding values for the transmission medium (α', β', ρ') (α', β', ρ': 6.45, 3.72, 3.16, α, β, ρ: 4.00, 2.31, 2.50). Solid and dashed lines indicate absolute values and phases, respectively, of the reflection and transmission coefficients. The arrows indicate the critical angles for the reflected and transmitted *P* waves (35.3° and 21°, respectively) and transmitted *SV* wave (38.4°). A_1/B, A'/B, and B'/B do not contribute to the energy equation for incidence angles larger than their corresponding critical angles. The largest value of $|B'/B|$ is 4.1.

For the model shown in Fig. 6.9, f_c^{rP}, f_c^{tP} are equal to 35.2° and 68.4°, respectively. For the model of Fig. 6.10, f_c^{rP}, f_c^{tP}, and f_c^{tS} are equal to 35.3°, 21.0°, and 38.4°, respectively.

6.6.3.2   Energy equation

In this case

$$\frac{\sin 2e}{\sin 2f}\left(\frac{A_1}{B}\right)^2 + \left(\frac{B_1}{B}\right)^2 + \frac{\rho'}{\rho}\frac{\sin 2e'}{\sin 2f}\left(\frac{A'}{B}\right)^2 + \frac{\rho'}{\rho}\frac{\sin 2f'}{\sin 2f}\left(\frac{B'}{B}\right)^2 = 1. \quad (6.6.69)$$

As in the case of incident P waves (§6.6.2.3), the reflection and/or transmission coefficients for angles beyond the corresponding critical angles must be set equal to zero, while the nonzero reflection coefficients must be replaced by their absolute values.

6.7 Waves incident on a solid–liquid boundary

In this case the incidence medium is a solid and the transmission medium is a liquid assumed to be inviscid and without cavitation. Two cases will be considered, incident P and SV waves. For the case of incident SH waves see Problem 6.14. As the rigidity of a liquid is zero, it cannot support shear motion. Therefore, the problem has three unknowns, two reflection coefficients, and one transmission coefficient.

6.7.1 Incident P waves

The displacement in the solid is given by (6.6.29) while in the liquid medium the displacement is obtained from (6.6.30) with the SV-wave contribution removed

$$\mathbf{u}' = \left[A'(\sin e' \mathbf{a}_1 - \cos e' \mathbf{a}_3) \exp(i\omega x_3 \cos e'/\alpha') \right] \exp[i\omega(t - x_1/c)]. \quad (6.7.1)$$

As noted in §6.3, slip along the boundary is possible, so that only the vertical component of the displacements has to be continuous at the boundary. Therefore, the boundary condition is $u_3 = u'_3$ at $x_3 = 0$ (rather than $\mathbf{u} = \mathbf{u}'$). This condition gives

$$A \cos e = A_1 \cos e + B_1 \sin f + A' \cos e'. \quad (6.7.2)$$

The boundary condition that the stress vector is continuous across the boundary gives for the horizontal component

$$A \sin 2e = A_1 \sin 2e - B_1 \frac{\alpha}{\beta} \cos 2f \quad (6.7.3)$$

and for the vertical component

$$A \frac{\mu\alpha}{\beta^2} \cos 2f = -A_1 \frac{\mu\alpha}{\beta^2} \cos 2f - B_1 \frac{\mu}{\beta} \sin 2f + A' \frac{\lambda'}{\alpha'} (\sin^2 e' + \cos^2 e').$$
$$(6.7.4)$$

Multiplication of (6.7.3) by $\beta^2/\mu\alpha$ and use of $\rho'\alpha'^2 = \lambda' + \mu' = \lambda'$ and $\beta^2 = \mu/\rho$ gives

$$A \cos 2f = -A_1 \cos 2f - B_1 \frac{\beta}{\alpha} \sin 2f + A' \frac{\rho'}{\rho} \frac{\alpha'}{\alpha}. \quad (6.7.5)$$

Equations (6.7.2), (6.7.3), and (6.7.5) can be written in matrix form

$$
\begin{pmatrix}
\cos e & \sin f & \cos e' \\
\sin 2e & -\dfrac{\alpha}{\beta}\cos 2f & 0 \\
-\cos 2f & -\dfrac{\beta}{\alpha}\sin 2f & \dfrac{\rho'\,\alpha'}{\rho\,\alpha}
\end{pmatrix}
\begin{pmatrix}
A_1/A \\
B_1/A \\
A'/A
\end{pmatrix}
=
\begin{pmatrix}
\cos e \\
\sin 2e \\
\cos 2f
\end{pmatrix}.
\tag{6.7.6}
$$

6.7.2 Incident SV waves

In this case the displacement in the incident medium is given by (6.6.57), while in the transmission medium the displacement is given by (6.7.1). The continuity of vertical displacements at the boundary give

$$
B\sin f = -A_1\cos e - B_1\sin f - A'\cos e'.
\tag{6.7.7}
$$

The boundary condition that the stress vector is continuous across the boundary gives for the horizontal component

$$
B\cos 2f = A_1\frac{\beta}{\alpha}\sin 2e - B_1\cos 2f
\tag{6.7.8}
$$

and for the vertical component

$$
B\sin 2f = A_1\frac{\alpha}{\beta}\cos 2f + B_1\sin 2f - A'\frac{\lambda'}{\alpha'}\frac{\beta}{\mu}
$$

$$
= A_1\frac{\alpha}{\beta}\cos 2f + B_1\sin 2f - A'\frac{\rho'\,\alpha'}{\rho\,\beta}.
\tag{6.7.9}
$$

In the last step the relations $\alpha^2 = \lambda'/\rho'$ and $\beta^2 = \mu/\rho$ were used.

Equations (6.7.7)–(6.7.9) can be written in matrix form

$$
\begin{pmatrix}
-\cos e & -\sin f & -\cos e' \\
\dfrac{\beta}{\alpha}\sin 2e & -\cos 2f & 0 \\
\dfrac{\alpha}{\beta}\cos 2f & \sin 2f & -\dfrac{\rho'\,\alpha'}{\rho\,\beta}
\end{pmatrix}
\begin{pmatrix}
A_1/B \\
B_1/B \\
A'/B
\end{pmatrix}
=
\begin{pmatrix}
\sin f \\
\cos 2f \\
\sin 2f
\end{pmatrix}.
\tag{6.7.10}
$$

6.8 *P* waves incident on a liquid–solid boundary

In this case the incidence medium is an inviscid liquid without cavitation and the transmission medium is a solid. In the incidence medium there are no reflected SV waves. However, both transmitted *P* and *SV* waves are possible. Therefore, the displacements for the transmission medium are equal to those given for the

solid–solid case (see (6.6.30)), while in the incidence medium the displacement is
given by

$$\mathbf{u} = \big[A(\sin e \mathbf{a}_1 - \cos e \mathbf{a}_3) \exp(i\omega x_3 \cos e/\alpha)$$

$$+ A_1(\sin e \mathbf{a}_1 + \cos e \mathbf{a}_3) \exp(-i\omega x_3 \cos e/\alpha) \big] \exp[i\omega (t - x_1/c)]. \quad (6.8.1)$$

The continuity of horizontal displacements at the boundary gives

$$A \cos e = A_1 \cos e + A' \cos e' - B' \sin f'. \quad (6.8.2)$$

The boundary condition that the stress vector is continuous across the boundary
gives for the horizontal component

$$0 = A_1 \sin 2e' + B' \frac{\alpha'}{\beta'} \cos 2f' \quad (6.8.3)$$

and for the vertical component

$$A = -A_1 + A' \frac{\alpha}{\lambda} \frac{\mu'\alpha'}{\beta'^2} \cos 2f' - B' \frac{\alpha}{\lambda} \frac{\mu'}{\beta'} \sin 2f'$$

$$= -A_1 + A' \frac{\rho'\alpha'}{\rho\alpha} \cos 2f' - B' \frac{\rho'\beta'}{\rho\alpha} \sin 2f'. \quad (6.8.4)$$

In the last step the relations $\rho' = \mu'/\beta^2$ and $\alpha^2 = \lambda/\rho$ (valid for liquids only) have
been used.

Equations (6.8.2)–(6.8.4) can be written in matrix form:

$$\begin{pmatrix} \cos e & \cos e' & -\sin f' \\ 0 & \sin 2e' & \dfrac{\alpha'}{\beta'} \cos 2f' \\ -1 & \dfrac{\rho'}{\rho} \dfrac{\alpha'}{\alpha} \cos 2f' & -\dfrac{\rho'}{\rho} \dfrac{\beta'}{\alpha} \sin 2f' \end{pmatrix} \begin{pmatrix} A_1/A \\ A'/A \\ B'/A \end{pmatrix} = \begin{pmatrix} \cos e \\ 0 \\ 1 \end{pmatrix}. \quad (6.8.5)$$

This case is relevant to marine reflection seismology studies. The P waves transmitted through the water generate P and SV waves within the solid earth. When
these waves reach deeper elastic boundaries they reflect back as P and SV waves.
A sensor at the liquid–solid boundary would record both types of waves, but a
sensor within the liquid records P waves only.

6.9 Solid layer over a solid half-space

The presence of layers in an elastic medium greatly increases the complexity of
the wave propagation problem. For incident P or SV waves and a solid medium,
the effect of each layer boundary is to generate reflected and transmitted P and

SV waves as discussed in the previous sections. As these waves propagate away from the boundary, they interact with the boundaries above and below the previous one, giving rise to new reflected and transmitted waves. When the layer boundaries are curved or nonhorizontal, the problem is not amenable to exact solution and approximate techniques such as ray tracing (see Chapter 8) must be used. Only the case of horizontal layers can be solved exactly, with the method of solution based on a matrix approach introduced by Thomson (1950). The method was later improved by Haskell (1953), who used it to study the dispersion of surface waves and the effect of a layered crust on the propagation of *SH* waves and *P* and *SV* waves (Haskell, 1960, 1962).

In principle, the case of a horizontally layered medium is not essentially different from the single-boundary cases discussed so far. The main difference is that in each layer the displacement is a combination of two pairs of *P* and *SV* waves, one pair propagating in the positive x_3 direction and the other propagating in the negative x_3 direction. The expressions for the corresponding displacements are those given in §6.2 multiplied by unknown coefficients that must be determined. For the case of *SH* waves only two waves per layer are required. Then the customary boundary conditions are applied: continuity of the displacement and stress vectors at each interface, vanishing of the stress vector at the free surface, and, in the case of surface waves, vanishing of the displacements at $x_3 = \infty$. For the *P–SV* case the number of equations and unknowns is of the order of four times the number of boundaries, which makes a direct solution all but impossible except in very simple cases. It was noted by Thomson (1950), however, that the problem can be made tractable by using an approach that reduces the solution of the large system of equations to a product of 4×4 matrices. The Thomson–Haskell method is well described in Ben-Menahem and Singh (1981), and the equations for displacements and stresses given below lead directly to their discussion.

The case to be discussed here, a solid layer over a solid half-space, is the simplest layered model and can be solved exactly without resorting to the matrix approach. In spite of its simplicity, however, this model is exceedingly important because it introduces a number of new results. We will see, in particular, that the effect of a layer is similar to that of a filter, with waves of certain frequencies selectively enhanced in amplitude.

Let *H* be the thickness of the layer. To simplify the notation, the variables x_1, x_2, x_3 will be replaced by *x*, *y*, *z*. The top of the layer is represented by $z = 0$, and the bottom by $z = H$ (Fig. 6.11). Three cases will be discussed, corresponding to incident *SH*, *P*, and *SV* waves impinging at the bottom of the layer. In the layer the *P*- and *S*-wave velocities, rigidity, and density are denoted by α', β', μ', and ρ'. In the half-space the corresponding symbols are α, β, μ, and ρ. The case of incident *SH* waves will also be discussed in the context of ray

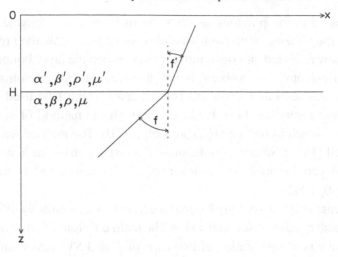

Fig. 6.11. Layer over a half-space and geometry for the case of incident SH waves.

theory (see §8.8.1), which will shed some light on the geometric aspects of the problem.

6.9.1 Incident SH waves

The following analysis is based on Pujol *et al.* (2002). In the layer the displacement, written in component form and indicated by v', is given by

$$v' = C'_u \exp\left[i\omega\left(t - x\frac{\sin f'}{\beta'} + z\frac{\cos f'}{\beta'}\right)\right]$$

$$+ C'_d \exp\left[i\omega\left(t - x\frac{\sin f'}{\beta'} - z\frac{\cos f'}{\beta'}\right)\right]. \qquad (6.9.1)$$

The first and second terms on the right represent waves going up and down, respectively. Their expressions are similar to those written for SH waves in a solid incidence medium (see (6.6.1)).

The equation for the displacement in the half-space is similar to (6.9.1) with all the primed quantities replaced by their unprimed counterparts, but before writing it a few notational changes and simplifications will be introduced. Let

$$c = \frac{\beta}{\sin f} = \frac{\beta'}{\sin f'}; \qquad k = \frac{\omega}{c} \qquad (6.9.2a,b)$$

$$\omega\frac{\sin f'}{\beta'} = \omega\frac{\sin f}{\beta} = k \qquad (6.9.3)$$

and

$$\omega \frac{\cos f'}{\beta'} = \frac{\omega}{\beta'}\sqrt{1 - \sin^2 f'} = \frac{\omega}{\beta'}\sqrt{1 - \frac{\beta'^2}{c^2}} = \frac{\omega}{c}\sqrt{\frac{c^2}{\beta'^2} - 1} = k\sqrt{\frac{c^2}{\beta'^2} - 1} = k\eta'$$

(6.9.4)

with

$$\eta' = \begin{cases} \sqrt{c^2/\beta'^2 - 1}, & \text{if } c > \beta' \\ -i\sqrt{1 - c^2/\beta'^2}, & \text{if } c < \beta'. \end{cases}$$

(6.9.5a,b)

Equation (6.9.2a) is Snell's law (see (6.6.3b)) and c is the phase velocity introduced in (6.6.27). There is an equation similar to (6.9.5) for η, with β' replaced by β. The choice of the minus sign in (6.9.5b) is not critical because z is bounded in the layer and $c \geq \beta$ in the half-space. Moreover, in actual computations using complex arithmetic (6.9.5b) is not necessary.

Before proceeding it is important to make clear the difference between the k in (6.9.2b) and (6.9.3) and the k_c introduced in (5.4.11) (aside from the absolute value in ω). First, note that the subscript c in k_c corresponds to the β' or β considered here. Secondly, k_c is the absolute value of a vector $\mathbf{k}$ perpendicular to the wave front, while from (6.9.3) and (5.4.11) we see that

$$k = k'_\beta \sin f' = k_\beta \sin f,$$

(6.9.6)

which shows that k is the horizontal projection of $\mathbf{k}$. Therefore, k is a *horizontal* wavenumber.

With the definitions introduced above and temporarily ignoring the common factor $\exp[i(\omega t - kx)]$ the displacements in the layer and in the half-space can be written as

$$v' = C'_u e^{izk\eta'} + C'_d e^{-izk\eta'}$$

(6.9.7)

$$v = C_u e^{izk\eta} + C_d e^{-izk\eta}.$$

(6.9.8)

The corresponding stress components are given by

$$\mu' \frac{\partial v'}{\partial z} = i\mu' k\eta'(C'_u e^{izk\eta'} - C'_d e^{-izk\eta'})$$

(6.9.9)

$$\mu \frac{\partial v}{\partial z} = i\mu k\eta(C_u e^{izk\eta} - C_d e^{-izk\eta}).$$

(6.9.10)

In these equation C_u is assumed to be known. For given values of f and ω, the only unknowns are the coefficients C'_u, C'_d, and C_d, which will be derived using the boundary conditions. The continuity of the displacement and stress vectors at

$z = H$, and the vanishing of the stress vector at $z = 0$ give

$$C'_u e^{iHk\eta'} + C'_d e^{-iHk\eta'} = C_u e^{iHk\eta} + C_d e^{-iHk\eta} \tag{6.9.11}$$

$$\frac{\mu'\eta'}{\mu\eta}(C'_u e^{iHk\eta'} - C'_d e^{-iHk\eta'}) = (C_u e^{iHk\eta} - C_d e^{-iHk\eta}) \tag{6.9.12}$$

$$C'_u = C'_d. \tag{6.9.13}$$

Using (6.9.13) equations (6.9.11) and (6.9.12) can be rewritten as

$$2C'_u \cos\theta - C_d e^{-iHk\eta} = C_u e^{iHk\eta} \tag{6.9.14}$$

$$2RiC'_u \sin\theta + C_d e^{-iHk\eta} = C_u e^{iHk\eta}, \tag{6.9.15}$$

where

$$\theta = Hk\eta'; \qquad R = \frac{\mu'\eta'}{\mu\eta}. \tag{6.9.16a,b}$$

To solve for the unknown coefficients let

$$D = \begin{vmatrix} 2\cos\theta & -e^{-iHk\eta} \\ 2iR\sin\theta & e^{-iHk\eta} \end{vmatrix} = 2e^{-iHk\eta}(\cos\theta + iR\sin\theta). \tag{6.9.17}$$

Then,

$$C'_u = \frac{1}{D} \begin{vmatrix} C_u e^{iHk\eta} & -e^{-iHk\eta} \\ C_u e^{iHk\eta} & e^{-iHk\eta} \end{vmatrix} = \frac{C_u e^{iHk\eta}}{\cos\theta + iR\sin\theta} \tag{6.9.18}$$

and

$$C_d = \frac{1}{D} \begin{vmatrix} 2\cos\theta & C_u e^{iHk\eta} \\ 2iR\sin\theta & C_u e^{iHk\eta} \end{vmatrix} = \frac{\cos\theta - iR\sin\theta}{\cos\theta + iR\sin\theta} C_u e^{2iHk\eta}. \tag{6.9.19}$$

Note that the absolute value of the ratio in the right-hand side of equation (6.9.19) is equal to 1. Therefore, $|C_d| = |C_u|$.

6.9.1.1 Surface displacement

Will be indicated with v'_o and is obtained from (6.9.7), (6.9.13), and (6.9.18) with $z = 0$

$$v'_o = 2C'_u e^{i(\omega t - kx)} = 2\frac{C_u e^{iHk\eta}}{\cos\theta + iR\sin\theta} e^{i(\omega t - kx)}. \tag{6.9.20}$$

Equations similar to (6.9.19) and (6.9.20), derived using a layer matrix approach, are given by Haskell (1960) and Ben-Menahem and Singh (1981), although the former does not include the exponential factors involving H.

We will be interested in the amplitude of the ratio v'_o/C_u and its extremal values. If $\beta' < \beta$, then θ and R are real (because $c \geq \beta$). Therefore,

$$\left|\frac{v'_o}{C_u}\right| = \frac{2}{\sqrt{\cos^2\theta + R^2\sin^2\theta}}. \tag{6.9.21}$$

If $\beta' > \beta$, then there will a critical value f_c for the incidence angle f and θ and R will be imaginary. If $f \leq f_c$, $|v_o'/C_u|$ is given by (6.9.21), but if $f > f_c$ the trigonometric functions become hyperbolic functions and (6.9.21) is replaced by

$$\left| \frac{v_o'}{C_u} \right| = \frac{2}{\sqrt{\cosh^2 |\theta| + |R|^2 \sin^2 |\theta|}}. \qquad (6.9.22)$$

To find the extremal values of $|v_o'/C_u|$ given by (6.9.21) set the derivative with respect to θ equal to zero

$$(1 - R^2) \sin 2\theta = 0. \qquad (6.9.23)$$

This, in turn, implies that $\sin 2\theta = 0$ and

$$2\theta = \pi, 2\pi, \ldots m\pi, \ldots, \qquad (6.9.24)$$

where m is an integer. The case $\theta = 0$ is discussed separately below.

For $R < 1$, if θ is an odd multiple of $\pi/2$, the right-hand side of (6.9.21) is equal to $2/R$, which is larger than 2 and is a maximum value. On the other hand, if θ is a multiple of π the right-hand side of (6.9.21) becomes 2, a minimum value. A similar argument shows that when $R > 1$ the maximum value is 2 and the minimum value is $2/R$.

It is useful to express (6.9.24) in terms of the period T_m for which $|v_o'/C_u|$ is maximum:

$$\theta = Hk\eta' = H\frac{\omega}{c}\eta' = \frac{2}{T_m c}\pi H\eta' = (2n + 1)\frac{\pi}{2}; \qquad n = 0, 1, 2, \ldots. \qquad (6.9.25)$$

Therefore,

$$T_m = \frac{4H\eta'}{(2n + 1)c}. \qquad (6.9.26)$$

When $|v_o'/C_u|$ is given by (6.9.22) the points of extremum values are obtained from

$$(1 + R^2) \sinh 2|\theta| = 0, \qquad (6.9.27)$$

which implies $|\theta| = 0$ (see below). For this value of $|\theta|$, $|v_o'/C_u|$ is equal to 2 and corresponds to a maximum value. Also note that as $|\theta|$ goes to infinity, $|v_o'/C_u|$ goes to zero. This happens when k and ω go to infinity, or equivalently, the wavelength λ (equal to $2\pi/k$) and T go to zero.

6.9.1.2 *Special cases*

(1) *Normal incidence.* In this case $f = f' = 0$ and the motion is no longer a function of x (see (6.9.1)). In addition, $c = \infty$ (see (6.9.2a)) and for this

reason η' and η must be defined using a limiting approach. Since c is larger than β and β', from (6.9.5a) and an equivalent expression for η we see that

$$\eta' = \frac{c}{\beta'}; \qquad \eta = \frac{c}{\beta} \tag{6.9.28a,b}$$

in the limit as c goes to infinity. Therefore, the ground displacement becomes

$$v'_o = 2C'_u e^{i\omega t} = 2\frac{C_u e^{iH\omega/\beta}}{\cos\theta + iR\sin\theta}e^{i\omega t} \tag{6.9.29}$$

with

$$\theta = \frac{Hkc}{\beta'} = \frac{H\omega}{\beta'}. \tag{6.9.30}$$

Introducing (6.9.28a) in (6.9.26) gives

$$T_m = \frac{4H}{(2n+1)\beta'}. \tag{6.9.31}$$

Similarly, from (6.9.16b)

$$R = \frac{\mu'\beta}{\mu\beta'} = \frac{\rho'\beta'}{\rho\beta} \tag{6.9.32}$$

so that

$$\max\left|\frac{v'_o}{C_u}\right| = \frac{2}{R} = 2\frac{\rho\beta}{\rho'\beta'}. \tag{6.9.33}$$

When $\beta = \beta'$ and $\rho' = \rho$, the maximum value is 2, as expected from §6.5.1. Therefore, the amplification of the ground motion due to the presence of the layer is given by

$$A = \frac{1}{R} = \frac{\rho\beta}{\rho'\beta'} \tag{6.9.34}$$

(Kanai, 1957).

Equation (6.9.31) can be recast in terms of the wavelength λ'_m, which from (5.4.18) is equal to $T_m\beta'$

$$\lambda'_m = \frac{4H}{2n+1}. \tag{6.9.35}$$

When $n = 0$, equation (6.9.35) gives

$$H = \frac{1}{4}\lambda'_m. \tag{6.9.36}$$

Equation (6.9.36) corresponds to the well-known *quarter-wavelength* rule for the thickness of the layer, which only applies in the case of normal incidence. An early application of this rule was by E. Wiechert, who used it to determine the thickness of the soil in Göttingen, Germany, from the period of a teleseismic recording and the velocity of the S waves (Wiechert and Zoeppritz, 1907).

(2) *Grazing incidence.* In this case $f = \pi/2$, $c = \beta$, $\eta = 0$, $R = \infty$, and the surface displacement becomes zero for all periods except those for which $\sin\theta \equiv 0$. The corresponding values of θ are of the form $m\pi$. These points coincide with the cut-off periods for the Love waves in the layer (see Problem 7.2).

(3) $\theta = 0$. From (6.9.16a) we see that this condition requires that the wavenumber k be equal to zero, or, equivalently, that the wavelength λ be infinity. When $\theta = 0$, $|v'_o|$ is equal to $2C_u$, which is the value obtained for an SH wave incident on a free surface (see §6.5.1). What this means is that when the layer thickness is negligible with respect to the wavelength, the wave is not affected by the presence of the layer. In other words, it can be said that the wave does not see the layer (Savarenskii, 1975).

6.9.1.3 Example: low-velocity layer

This case is discussed in detail because of its relevance in seismic risk studies, as was recognized by, for example, Kanai (1957) and Haskell (1960). Let us consider the case of a thin layer of unconsolidated sediments overlying a high-velocity medium. Let H be 0.1 km, β' and β be 0.25 and 3.5 km s^{-1}, and ρ' and ρ be 1.7 and 2.7 g cm^{-3} (Haskell, 1960). The value of the incidence angle is 10°. Figure 6.12 shows the amplitude and phase of v'_o/C_u for periods between 0.2 and 3 s. Several features of this figure must be noted.

First, note that the plot of $|v'_o/C_u|$ has peaks for the periods predicted by (6.9.26), namely 1.60, 0.53, 0.32, and 0.23 s for $n = 0, 1, 2, 3$. All the peak values are equal to 43.8, and because the incidence angle is close to normal incidence, these values are close to the maximum predicted value of $2/R$ (equal to 44.5) with R given by (6.9.32). The minimum value, on the other hand, is equal to the predicted value of 2.

Figure 6.12 also illustrates two significant effects introduced by the layer. First, the amplitudes for some periods (or frequencies) are selectively amplified. Loosely speaking, it may be said that the layer acts like a filter, and for that reason $|v'_o/C_u|$ can be considered a *crustal transfer function*. The second effect is the large value of A (see (6.9.34)), equal to 22.2. This large amplification is an important factor in seismic risk studies in areas where buildings, bridges, and other constructions are underlain by unconsolidated sediments. Gutenberg (1957) provides an account of early work on these effects. It must be borne in mind however, that the theory developed here applies to plane waves incident on a horizontal boundary, and that (6.9.31) is valid for normal incidence only. The case of incident SH waves (as well

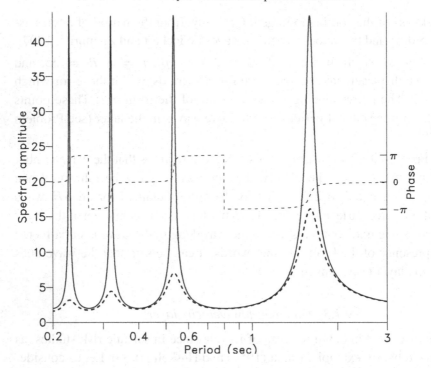

Fig. 6.12. Surface displacement for the case of an SH wave impinging at the bottom of a layer over a half-space. The velocity and density in the layer and in the half-space are β', ρ' (0.25, 1.70) and β, ρ (3.50, 2.70), respectively. The layer thickness and incidence angle are 0.1 km and 10° with $Q = 10$. This example corresponds to low-velocity unconsolidated material overlying much higher-velocity rocks, taken from Haskell (1960). The bold dashed line and thin solid line correspond to the displacements obtained with and without consideration of anelastic attenuation. The thin dashed line corresponds to the phase angle.

as P and SV waves) under more general conditions has been studied by Bard and Bouchon (1985).

These amplitude considerations are not complete, however, unless the effect of attenuation is taken into account. Although attenuation is discussed in Chapter 11, it is introduced here because unconsolidated sediments may strongly attenuate the seismic energy, and for that reason it is important to see how the amplitude peaks are affected. As discussed in §11.4.1, one way to incorporate attenuation is to replace the velocities β' and β by complex velocities obtained as follows. If β_j is the velocity in the layer ($j = 1$) or in the half-space ($j = 2$), then the corresponding complex velocity is given by

$$\beta_j \left(1 + i\frac{1}{2Q_j}\right), \tag{6.9.37}$$

where Q_j is known as the quality factor, whose inverse is a measure of attenuation.

To illustrate the effect of attenuation, $|\tilde{v}_o'(\omega)/C_u|$ was computed using (6.9.37) with $Q_1 = 10$ and $Q_2 = \infty$ (i.e., there is no attenuation in the half-space). This value of Q_1 is widely quoted in the literature, but may not be adequate for all unconsolidated sediments, some of which may have Q values considerably higher (Pujol *et al.*, 2002). As Fig. 6.12 shows, the amplitudes for the higher frequencies (lower periods) are significantly more attenuated than the amplitudes for the lower frequencies (higher periods). Therefore, the effect of attenuation on amplification will depend on the frequency content of the incident wave. This method of introducing attenuation, however, is not complete because it does not account for the dispersion associated with attenuation (see §11.7, §11.8).

6.9.2 *Incident P and SV waves*

In the layer the displacement is a combination of upgoing and downgoing P and SV waves, regardless of whether the incident waves are of P or SV type. The corresponding equations are those corresponding to the transmitted (upgoing) and reflected (downgoing) waves as used for the solid–solid boundary case in §6.6.2. The following convention will be used: the letters A and B represent P- and SV-wave amplitudes, the subscripts u and d indicate upgoing and downgoing waves, and a prime (') will be used to identify quantities related to the layer. Therefore, from (6.6.25) (reflected waves only) and (6.6.26) and using (6.6.27) we obtain

$$
\begin{aligned}
\mathbf{u}' = \big[& A_u'(\sin e' \mathbf{a}_x - \cos e' \mathbf{a}_z) \exp(i\omega z \cos e'/\alpha') \\
& + A_d'(\sin e' \mathbf{a}_x + \cos e' \mathbf{a}_z) \exp(-i\omega z \cos e'/\alpha') \\
& + B_u'(\cos f' \mathbf{a}_x + \sin f' \mathbf{a}_z) \exp(i\omega z \cos f'/\beta') \\
& + B_d'(-\cos f' \mathbf{a}_x + \sin f' \mathbf{a}_z) \exp(-i\omega z \cos f'/\beta') \big] \exp[i\omega(t - x/c)].
\end{aligned}
$$

$$(6.9.38)$$

The equations for the half-space are similar to those above with all the primed quantities replaced by their unprimed counterparts. Furthermore, for incident P waves $B_u = 0$ and A_u is assumed to be known. For incident SV waves $A_u = 0$ and B_u is assumed to be known. This means that there are six unknown quantities: A_u', A_d', B_u', B_d', A_d, and B_d. To solve the problem six equations are needed. They are provided by the usual boundary conditions: the stress vector equal to zero at the surface and displacement and stress vectors continuous at the bottom of the layer. Because we are dealing with vector quantities having two nonzero components, the three conditions give the required six equations.

When a medium with more than one layer is considered, the problem is solved more efficiently using a matrix approach, as introduced by Thomson (1950) and later modified by Haskell (1953). We will not go over that method here, but will write the equations for displacements and stresses in a way similar to that used by Haskell (1953). The final equations presented below follow those in Ben-Menahem and Singh (1981), and lead directly to their general treatment of a layered medium.

As done for the case of the SH waves, the exponential terms in equation (6.9.38) can be rewritten using Snell's law and $k = \omega/c$. For example,

$$\omega \frac{\cos e'}{\alpha'} = \frac{\omega}{\alpha'} \sqrt{1 - \sin^2 e'} = \frac{\omega}{\alpha'} \sqrt{1 - \frac{\alpha'^2}{c^2}} = \frac{\omega}{c} \sqrt{\frac{c^2}{\alpha'^2} - 1} = k \sqrt{\frac{c^2}{\alpha'^2} - 1} = k\eta'_\alpha,$$

(6.9.39)

where

$$\eta'_\alpha = \begin{cases} \sqrt{c^2/\alpha'^2 - 1}, & \text{if } c > \alpha' \\ -i\sqrt{1 - c^2/\alpha'^2}, & \text{if } c < \alpha'. \end{cases}$$

(6.9.40a,b)

Similarly,

$$\omega \frac{\cos f'}{\beta'} = k\eta'_\beta$$

(6.9.41)

with η'_β given by

$$\eta'_\beta = \begin{cases} \sqrt{c^2/\beta'^2 - 1}, & \text{if } c > \beta' \\ -i\sqrt{1 - c^2/\beta'^2}, & \text{if } c < \beta'. \end{cases}$$

(6.9.42a,b)

Before proceeding take the derivative of $\mathbf{u}'$ with respect to t and divide by c:

$$\frac{1}{c} \frac{\partial \mathbf{u}'}{\partial t} \equiv \frac{1}{c} \dot{\mathbf{u}}' = i\frac{\omega}{c} \mathbf{u}' = ik\mathbf{u}'.$$

(6.9.43)

The factors $ik \sin e'$ and $ik \cos e'$ can be rewritten as

$$ik \sin e' = i\frac{k}{\sin e'} \sin^2 e' = ik'_\alpha \left(\frac{\alpha'}{c}\right)^2; \qquad k'_\alpha = \frac{k}{\sin e'}$$

(6.9.44)

and

$$ik \cos e' = i\sqrt{1 - \sin^2 e'}\, k'_\alpha \sin e' = i\sqrt{\frac{1}{\sin^2 e'} - 1}\, k'_\alpha \sin^2 e' = i\eta'_\alpha k'_\alpha \left(\frac{\alpha'}{c}\right)^2.$$

(6.9.45)

The corresponding expressions for $ik \sin f'$ and $ik \cos f'$ are obtained from the previous ones with α replaced by β.

Next decompose $\mathbf{u}'$ into its horizontal (u') and vertical (w') components. This gives

$$\frac{\dot{u}'}{c} = i \left(A'_u k'_\alpha \frac{\alpha'^2}{c^2} e^{ik\eta'_\alpha z} + A'_d k'_\alpha \frac{\alpha'^2}{c^2} e^{-ik\eta'_\alpha z} \right.$$

$$\left. + B'_u \eta'_\beta k'_\beta \frac{\beta'^2}{c^2} e^{ik\eta'_\beta z} - B'_d \eta'_\beta k'_\beta \frac{\beta'^2}{c^2} e^{-ik\eta'_\beta z} \right) e^{i(\omega t - kx)} \qquad (6.9.46)$$

and

$$\frac{\dot{w}'}{c} = i \left(-A'_u \eta'_\alpha k'_\alpha \frac{\alpha'^2}{c^2} e^{ik\eta'_\alpha z} + A'_d \eta_\alpha k'_\alpha \frac{\alpha'^2}{c^2} e^{-ik\eta'_\alpha z} \right.$$

$$\left. + B'_u k'_\beta \frac{\beta'^2}{c^2} e^{ik\eta'_\beta z} + B'_d k'_\beta \frac{\beta'^2}{c^2} e^{-ik\eta'_\beta z} \right) e^{i(\omega t - kx)}. \qquad (6.9.47)$$

Now we need the expressions for the horizontal (τ') and vertical (σ') components of the stress vector. For the P waves they are given by

$$\tau' = 2\mu \frac{\partial u'}{\partial z}; \qquad \sigma' = \lambda \frac{\partial u'}{\partial x} + (\lambda' + 2\mu') \frac{\partial w'}{\partial z} \qquad (6.9.48)$$

and for the SV waves by

$$\tau' = \mu' \left(\frac{\partial u'}{\partial z} + \frac{\partial w'}{\partial x} \right); \qquad \sigma' = 2\mu' \frac{\partial w'}{\partial z}. \qquad (6.9.49)$$

These equations have to be applied to those that give the velocities after the latter are divided by ik. Consider τ' for the upgoing P waves. Aside from the exponential factors we have the following product:

$$i2\mu' k'_\alpha \frac{\alpha'^2}{c^2} \eta'_\alpha A'_u = i2\rho' \beta'^2 k'_\alpha \frac{\alpha'^2}{c^2} \eta'_\alpha A'_u = i\rho' \gamma' k'_\alpha \alpha^2 \eta'_\alpha A'_u, \qquad (6.9.50)$$

where

$$\gamma' = \frac{2\beta'^2}{c^2}. \qquad (6.9.51)$$

The corresponding expression for the downgoing P waves is obtained from the previous one by a change of sign.

For the upgoing SV waves we have

$$i\mu' \left(B'_u \eta'^2_\beta k'_\beta \frac{\beta'^2}{c^2} - B'_d k'_\beta \frac{\beta'^2}{c^2} \right) = i\rho' \beta'^2 k'_\beta \frac{\beta'^2}{c^2} \left(\eta'^2_\beta - 1 \right) B'_u$$

$$= i\rho' \beta'^2 k'_\beta \frac{\beta'^2}{c^2} \left(\frac{c^2}{\beta'^2} - 2 \right) B'_u$$

$$= i\rho'\beta'^2 k'_\beta (1 - \gamma') B'_u = -i\rho' \frac{c^2}{2} k'_\beta \gamma' (\gamma' - 1) B'_u.$$

$$(6.9.52)$$

This expression also applies to the downgoing SV waves.

The vertical component of stress for the upgoing P waves includes the following product:

$$-ik'_\alpha \frac{\alpha'^2}{c^2} \left[\lambda' + (\lambda' + 2\mu') \eta'^2_\alpha \right] A'_u$$

$$= -ik'_\alpha \frac{\alpha'^2}{c^2} \left[(\lambda' + 2\mu') \frac{c^2}{\beta'^2} - 2\mu' \right] A'_u$$

$$= -ik'_\alpha \frac{\alpha'^2}{c^2} \rho' c^2 \left(1 - 2\frac{\beta'^2}{c^2} \right) A'_u = ik'_\alpha \alpha'^2 \rho' (\gamma' - 1) A'_u. \quad (6.9.53)$$

This expression also applies to the downgoing P waves.

For the SV waves we have

$$i2\mu' B'_u k'_\beta \frac{\beta'^2}{c^2} \eta'_\beta = i\rho B'_u k'_\beta 2\frac{\beta'^4}{c^2} \eta'_\beta = i\rho B'_u k'_\beta \gamma'^2 \frac{c^2}{2} \eta'_\beta. \quad (6.9.54)$$

The corresponding expression for the downgoing SV waves is obtained from the previous one by a change of sign.

The expressions for τ' and σ' in the layer are obtained by combining (6.9.50)–(6.9.54):

$$\tau' = \left[i\rho' \gamma' k'_\alpha \alpha'^2 \eta'_\alpha \left(A'_u e^{ik\eta'_\alpha z} - A'_d e^{-ik\eta'_\alpha z} \right) \right.$$

$$\left. - i\rho' \frac{c^2}{2} k'_\beta \gamma' (\gamma' - 1) \left(B'_u e^{ik\eta'_\beta z} + B'_d e^{-ik\eta'_\beta z} \right) \right] e^{i(\omega t - kx)} \quad (6.9.55)$$

$$\sigma' = \left[ik'_\alpha \alpha'^2 \rho' (\gamma' - 1) \left(A'_u e^{ik\eta'_\alpha z} + A'_d e^{-ik\eta'_\alpha z} \right) \right.$$

$$\left. + i\rho k'_\beta \gamma'^2 \frac{c^2}{2} \eta'_\beta \left(B'_u e^{ik\eta'_\beta z} - B'_d e^{-ik\eta'_\beta z} \right) \right] e^{i(\omega t - kx)}. \quad (6.9.56)$$

Finally, the following abbreviations will be introduced:

$$\dot{U}'_\alpha = k'_\alpha \frac{\alpha'^2}{c^2}; \qquad \dot{U}'_\beta = \eta'_\beta k'_\beta \frac{\beta'^2}{c^2} \quad (6.9.57)$$

$$\dot{W}'_\alpha = \eta'_\alpha k'_\alpha \frac{\alpha'^2}{c^2}; \qquad \dot{W}'_\beta = k'_\beta \frac{\beta'^2}{c^2} \quad (6.9.58)$$

$$T'_\alpha = \rho' \gamma' k'_\alpha \alpha'^2 \eta'_\alpha; \qquad T'_\beta = -\rho' \frac{c^2}{2} k'_\beta \gamma' (\gamma' - 1) \quad (6.9.59)$$

$$S'_\alpha = k'_\alpha \alpha'^2 \rho'(\gamma' - 1); \qquad S'_\beta = \rho' k'_\beta \gamma'^2 \frac{c^2}{2} \eta'_\beta \tag{6.9.60}$$

$$E'_{u\alpha} = e^{ik\eta'_\alpha h}; \qquad E'_{u\beta} = e^{ik\eta'_\beta h} \tag{6.9.61}$$

$$E'_{d\alpha} = e^{-ik\eta'_\alpha h}; \qquad E'_{d\beta} = e^{-ik\eta'_\beta h}. \tag{6.9.62}$$

Similar definitions with the primes dropped will be used for the half-space. A further simplification is achieved by temporarily ignoring the factor of $\exp[i(\omega t - kx)]$. Now we are ready to consider P and SV waves independently.

6.9.2.1 P waves

As noted above, $B_u = 0$ and A_u is assumed known. Then the boundary condition that the stresses vanish at the surface ($z = 0$) and the boundary conditions that the displacement and stress vectors are continuous at the bottom of the layer ($z = h$) give the following equations:

$$T'_\alpha A'_u - T'_\alpha A'_d + T'_\beta B'_u + T'_\beta B'_d = 0 \tag{6.9.63}$$

$$S'_\alpha A'_u + S'_\alpha A'_d + S'_\beta B'_u - S'_\beta B'_d = 0 \tag{6.9.64}$$

$$\dot{U}'_\alpha E'_{u\alpha} A'_u + \dot{U}'_\alpha E'_{d\alpha} A'_d + \dot{U}'_\beta E'_{u\beta} B'_u - \dot{U}'_\beta E'_{d\beta} B'_d - \dot{U}_\alpha E_{d\alpha} A_d + \dot{U}_\beta E_{d\beta} B_d = \dot{U}_\alpha E_{u\alpha} A_u \tag{6.9.65}$$

$$- \dot{W}'_\alpha E'_{u\alpha} A'_u + \dot{W}'_\alpha E'_{d\alpha} A'_d + \dot{W}'_\beta E'_{u\beta} B'_u + \dot{W}'_\beta E'_{d\beta} B'_d - \dot{W}_\alpha E_{d\alpha} A_d - \dot{W}_\beta E_{d\beta} B_d$$
$$= -\dot{W}_\alpha E_{u\alpha} A_u \tag{6.9.66}$$

$$T'_\alpha E'_{u\alpha} A'_u - T'_\alpha E'_{d\alpha} A'_d + T'_\beta E'_{u\beta} B'_u + T'_\beta E'_{d\beta} B'_d + T_\alpha E_{d\alpha} A_d - T_\beta E_{d\beta} B_d = T_\alpha E_{u\alpha} A_u \tag{6.9.67}$$

$$S'_\alpha E'_{u\alpha} A'_u + S'_\alpha E'_{d\alpha} A'_d + S'_\beta E'_{u\beta} B'_u - S'_\beta E'_{d\beta} B'_d - S_\alpha E_{d\alpha} A_d + S_\beta E_{d\beta} B_d = S_\alpha E_{u\alpha} A_u. \tag{6.9.68}$$

For given A_u, frequency, velocities, and densities, the amplitude terms depend on the incidence angle e of the P wave. Equations (6.9.63)–(6.9.68) constitute a linear system in six unknown amplitude coefficients. Solving the system basically solves the layer over a half-space problem.

6.9.2.2 SV waves

In this case $A_u = 0$ and B_u is assumed to be known. The only difference with equations (6.9.63)–(6.9.68) is in the four non-zero quantities on the right, which should be replaced by $\dot{U}_\beta$, $\dot{W}_\beta$, T_β, and S_β, each multiplied by $E_{u\beta} B_u$. In this case the amplitude terms depend on the incidence angle f of the SV wave.

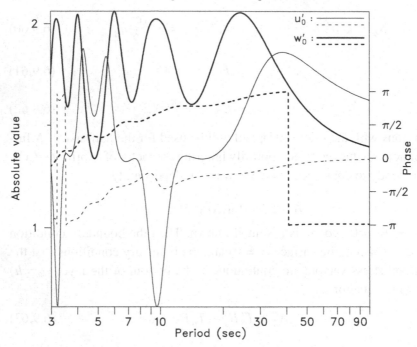

Fig. 6.13. Horizontal (u'_o) and vertical (w'_o) components of surface displacement for the case of a P wave impinging at the bottom of a layer over a half-space. The P- and SV-wave velocities and density in the layer and in the half-space are α', β', ρ' (6.28, 3.63, 2.87) and α, β, ρ (7.96, 4.60, 3.37), respectively. The layer thickness and incidence angle are 37 km and 46°. This example corresponds to a single-layered model for the crust, taken from Haskell (1962). Solid and dashed lines correspond to absolute values and phases.

6.9.2.3 Surface displacement

Once A'_u, A'_d, B'_u, and B'_d have been computed, the horizontal (u'_o) and vertical (w'_o) components of the surface displacement are obtained from (6.9.46) and (6.9.47) after dividing by ik and setting $z = 0$:

$$u'_o = \frac{1}{k}\left[\left(A'_u + A'_d\right)\dot{U}'_\alpha + \left(B'_u - B'_d\right)\dot{U}'_\beta\right]\exp[i(\omega t - kx)] \qquad (6.9.69)$$

and

$$w'_o = \frac{1}{k}\left[\left(-A'_u + A'_d\right)\dot{W}'_\alpha + \left(B'_u + B'_d\right)\dot{W}'_\beta\right]\exp[i(\omega t - kx)]. \qquad (6.9.70)$$

These equations apply to both the P and SV waves, with the amplitude coefficients obtained in §6.9.2.1 and §6.9.2.2, respectively.

The horizontal and vertical components of the displacement for the single-layer crustal model of Haskell (1962) for incident P ($e = 46°$) and SV ($f = 46°$, 25°) waves for periods between 3 and 100 s are shown in Figs 6.13–6.15.

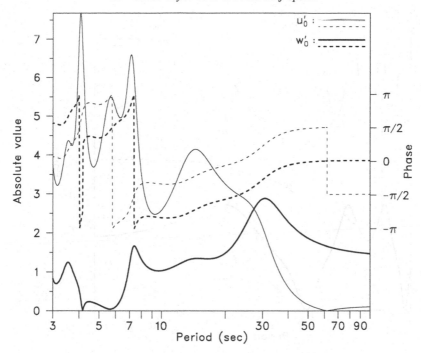

Fig. 6.14. Horizontal (u'_o) and vertical (w'_o) components of surface displacement for the case of an *SV* wave impinging at the bottom of a layer over a half-space. Same model as in Fig. 6.13. The layer thickness and incidence angle are 37 km and 46°. Velocities and densities as in Fig. 6.13.

Two interesting features must be noted. One is the complicated variation of the surface displacement as a function of period, particularly for the *SV* wave in Fig. 6.14. In this case the incidence angle *f* is 46°, which is larger than the critical angle for the transmitted *P* wave (equal to 35.3°). For a comparison, the displacement components when *f* is less than this critical angle are shown in Fig. 6.15.

Another important feature that needs consideration is the difference, for a given period, between the phases for the horizontal and vertical motions. When the phase difference is equal to 0 or π, the motion of a surface particle as a function of time is linear. For any other values of the difference the motion becomes elliptical (see §5.8.4.1), with the degree of ellipticity depending on the phase difference. The nonlinearity of the motion can be particularly severe for *SV* waves, as inferred from Fig. 6.14. In this case the phase difference is close to $\pi/2$ for a wide range of periods. Recall that a similar difference was found for the *SV* waves incident on a free surface for incidence angles larger than the critical value (see §6.5.3.2). These matters are discussed in detail in Haskell (1962).

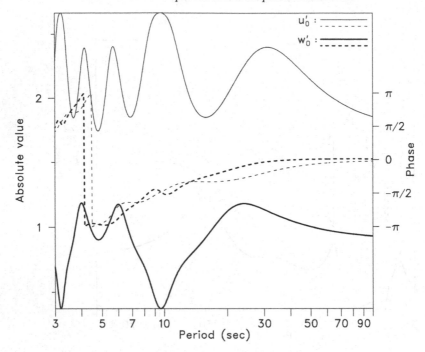

Fig. 6.15. Horizontal (u'_o) and vertical (w'_o) components of surface displacement for the case of an SV wave impinging at the bottom of a layer over a half-space. Same model as in Fig. 6.13. The layer thickness and incidence angle are 37 km and 25°. Velocities and densities as in Fig. 6.13.

Problems

6.1 Verify (6.4.1)–(6.4.8).

6.2 Verify (6.5.12) and the conclusion that $a_1 = a_2 = a_3 = 0$.

6.3 Show that:

(a) when $f_1 = 0$ and $f_1 = \pi/2$, (6.5.10) implies $e_1 = e$ and $A_1 = -A$;

(b) when $f_1 = 0$, (6.5.9) implies that $e = 0$ and $B_1 = 0$;

(c) when $f_1 = \pi/2$, (6.5.9) leads to an equation that does not have a solution.

Verify that these results are special cases of the general results obtained in §6.5.2.

6.4 Verify that the coefficient B_1/A given by (6.5.21) is negative or zero (as long as Poisson's ratio is nonnegative).

6.5 Verify (6.5.24).

6.6 Verify (6.5.28).

6.7 Verify (6.5.29)–(6.5.31).

6.8   Find the angles for which the reflection coefficient B_1/B given by (6.5.36) is equal to zero when $\alpha/\beta = \sqrt{3}$.

6.9   Show that the Hilbert transforms of $\cos at$, $\sin at$, and $\delta(t)$ are given by $-\sin at$, $\cos at$, and $-1/\pi t$, respectively.

6.10   Show that the reflection and transmission coefficients for the case of an SH wave incident on a solid–solid boundary can be written as

$$\frac{C_1}{C} = \frac{\rho\beta \cos f - \rho'\beta' \cos f'}{\rho\beta \cos f + \rho'\beta' \cos f'}$$

$$\frac{C'}{C} = \frac{2\rho\beta \cos f}{\rho\beta \cos f + \rho'\beta' \cos f'}.$$

Then use the definition of impedance (stress/velocity, see §6.6.2.1) to show that $\pm\rho\beta \cos f$ and $\rho'\beta' \cos f'$ are impedances.

6.11   Refer to the problem of SH waves incident on a solid–solid boundary for angles larger than the critical (§6.6.1.1). Show that in this case the displacement is given by

$$\mathbf{u} = 2a_2 C \sin \left(\frac{\omega x_3 \cos f}{\beta} + \chi \right) \exp \left[i\omega(t - x_1/c) + i(\pi/2 - \chi) \right]$$

$$\mathbf{u}' = 2a_2 C \sin \chi \exp \left[\omega x_3 \beta^{-1} \left(\sin^2 f - \beta^2/\beta'^2 \right)^{1/2} \right.$$

$$\left. + i\omega(t - x_1/c) + i(\pi/2 - \chi) \right]$$

(Ben-Menahem and Singh, 1981), where $c = \beta/\sin f$ is the phase velocity introduced in (6.6.27) and $\omega > 0$.

6.12   Refer to §6.6.1.1 and Problem 6.11. Verify that the energy of the inhomogeneous wave, averaged over one period, is equal to zero (Ben-Menahem and Singh, 1981).

6.13   Verify by direct substitution that the reflection and transmission coefficients for the case of normal incidence satisfy (6.6.56).

6.14   Solve the problem of an SH wave incident on a solid–liquid boundary when the incidence medium is solid.

7

Surface waves in simple models – dispersive waves

7.1 Introduction

Surface waves are waves that propagate along a boundary and whose amplitudes go to zero as the distance from the boundary goes to infinity. There are two basic types of surface waves, Love and Rayleigh waves, named after the scientists who studied them first. Love's work was directed to the explanation of waves observed in horizontal seismographs, while Rayleigh predicted the existence of the waves with his name. The main difference between the two types of waves is that the motion is of SH type for Love waves, and of P–SV types for Rayleigh waves. A related type of wave, known as Stoneley waves, consists of P–SV inhomogeneous waves that propagate along the boundary between two half-spaces. In this chapter we will consider these three types of waves. As we shall see, the presence of a layer introduces the phenomenon of *dispersion*, which is characterized by the existence of two velocities, known as the phase and the group velocity, with the property that they are functions of frequency. In §7.6 a detailed analysis of dispersion is presented. The problem of multilayered media will not be considered here, but the groundwork for its analysis has been introduced in §6.9.2.

To solve problems involving surface waves it is necessary to go through the steps described in §6.1, namely, write the equation for the displacement at any point in the medium and then apply appropriate boundary conditions. As the latter are exactly the same as those given in §6.3, here we must consider displacements only.

Because surface waves are inhomogeneous waves, such as those discussed in Chapter 6, some of the expressions derived in §6.9.1 and §6.9.2 can be easily modified to represent surface wave displacements. For example, the quantities introduced in (6.9.5b) and (6.9.40b) and (6.9.42b) were defined to make the corresponding waves inhomogeneous. However, when a model includes layers, the displacements in the layers do not have to be represented by inhomogeneous waves and equations similar to (6.9.5a) and (6.9.40a) and (6.9.42a) should be used. In addition, when layers are present, both upgoing and downgoing waves should be allowed in each of the layers, as in §6.9. Based on these considerations it is relatively simple to write the displacements appropriate for specific problems, but to make this chapter self-contained, in the next section an independent derivation will be given. Except for §7.6, the presentation and notation used here follow those of Ben-Menahem and Singh (1981).

The coordinate system used in this chapter will be that shown in Fig. 6.1, but for most of the discussion the coordinates x_1, x_2, and x_3 will be replaced by x, y, and z. Some of the derivations, however, will be carried out using the subscripted variables.

7.2 Displacements

From (5.8.52)–(5.8.55) we know that the vectors

$$\mathbf{u}_P = A(l\mathbf{a}_x + n\mathbf{a}_z) \exp\left[i\omega\left(t - \frac{lx + nz}{\alpha}\right)\right] \tag{7.2.1}$$

$$\mathbf{u}_{SV} = B(-n\mathbf{a}_x + l\mathbf{a}_z) \exp\left[i\omega\left(t - \frac{lx + nz}{\beta}\right)\right] \tag{7.2.2}$$

$$\mathbf{u}_{SH} = C\mathbf{a}_y \exp\left[i\omega\left(t - \frac{lx + nz}{\beta}\right)\right] \tag{7.2.3}$$

with $n^2 + l^2 = 1$, represent three independent solutions to the wave equation. This means that they can be used to represent the displacements generated by both body waves and surface waves. The difference between the two types of waves is the dependence of $|\mathbf{u}|$ with depth (z), which for surface waves must go to zero as z goes to infinity. Inspection of (7.2.1)–(7.2.3) shows that this kind of behavior is possible if n is pure imaginary. Therefore, the displacements corresponding to surface waves can be derived from (7.2.1)–(7.2.3) by an appropriate modification of the exponents. Before doing that, however, the combination $(lx + nz)/\delta$, where δ is either α or β, will be rewritten for the case of n and l real,

$$\frac{lx + nz}{\delta} = \frac{x + nz/l}{\delta/l}$$

$$= \frac{x \pm \sqrt{(1/l^2) - 1}\, z}{\delta/l} = \frac{x \pm \sqrt{(c^2/\delta^2) - 1)}z}{c}; \qquad \delta = \alpha, \beta, \tag{7.2.4}$$

where

$$n = \pm\sqrt{1 - l^2}; \qquad c = \frac{\delta}{l}, \tag{7.2.5a,b}$$

where c is the phase velocity introduced in (6.6.27). Now introduce another quantity:

$$\eta_\delta = \sqrt{\frac{c^2}{\delta^2} - 1}. \tag{7.2.6}$$

Then, from (7.2.5a,b) and (7.2.6) we obtain

$$\frac{n}{l} = \pm\sqrt{\frac{1}{l^2} - 1} = \pm\eta_\delta. \tag{7.2.7}$$

Equations (7.2.1)–(7.2.3) will now be rewritten in terms of η_α, η_β, and $k = \omega/c$ (the latter being defined in (6.9.2b)):

$$\mathbf{u}_P = A(\mathbf{a}_x \pm \eta_\alpha \mathbf{a}_z) \exp\left[ik\,(ct - x \mp \eta_\alpha z)\right] \tag{7.2.8}$$

$$\mathbf{u}_{SV} = B(\mp\eta_\beta \mathbf{a}_x + \mathbf{a}_z) \exp\left[ik\,(ct - x \mp \eta_\beta z)\right] \tag{7.2.9}$$

$$\mathbf{u}_{SH} = C\mathbf{a}_y \exp\left[ik\,(ct - x \mp \eta_\beta z)\right]. \tag{7.2.10}$$

A factor of $1/l$ has been absorbed in the coefficients A and B; doing that does not restrict the generality of (7.2.8) and (7.2.9) because the coefficients are determined so as to satisfy prescribed boundary conditions. The $-$ and $+$ signs in the exponents indicate two waves moving in the positive and negative z directions, respectively.

Now consider the case of n imaginary. Because the condition

$$n^2 + l^2 = 1 \tag{7.2.11}$$

must be satisfied, l must be larger than 1. Therefore,

$$n = \pm\sqrt{1 - l^2} = \pm i\sqrt{l^2 - 1} \tag{7.2.12}$$

and

$$\frac{n}{l} = \pm i\sqrt{1 - \frac{1}{l^2}} = \pm i\sqrt{1 - \frac{c^2}{\delta^2}}; \qquad \delta = \alpha, \beta. \tag{7.2.13}$$

Now introduce

$$\gamma_\delta = \sqrt{1 - \frac{c^2}{\delta^2}} \tag{7.2.14}$$

and write

$$\frac{n}{l} = \mp i\gamma_\delta. \tag{7.2.15}$$

The sign convention in (7.2.15) is arbitrary, and is not important as long as it is applied consistently.

Equations (7.2.1)–(7.2.3) will now be rewritten in terms of γ_α, γ_β, and k:

$$\mathbf{u}_P = A(\mathbf{a}_x \mp i\gamma_\alpha\mathbf{a}_z)\exp[\mp\gamma_\alpha kz + ik(ct - x)] \qquad (7.2.16)$$

$$\mathbf{u}_{SV} = B(\pm i\gamma_\beta\mathbf{a}_x + \mathbf{a}_z)\exp[\mp\gamma_\beta kz + ik(ct - x)] \qquad (7.2.17)$$

$$\mathbf{u}_{SH} = C\mathbf{a}_y\exp[\mp\gamma_\beta kz + ik(ct - x)]. \qquad (7.2.18)$$

Equations (7.2.16)–(7.2.18) represent surface waves propagating in the positive x direction with their amplitudes decaying exponentially with $\pm z$ (provided that γ_α and γ_β are positive).

7.3 Love waves

The following models will be considered: a half-space, a layer over a half-space, and a vertically heterogeneous medium.

7.3.1 Homogeneous half-space

We will see that in this case Love waves do not exist. In a half-space the displacement must be represented by (7.2.18) with a $-$ sign in the exponent to ensure the expected behavior as z goes to infinity. Therefore,

$$\mathbf{u} = C\mathbf{a}_y e^{-\gamma_\beta kz + ik(ct-x)}. \qquad (7.3.1)$$

The boundary condition that the stress vector be zero at the free surface ($z = 0$) gives

$$\tau_{32} = \mu u_{2,3} = -\mu C\gamma_\beta k = 0; \qquad z = 0 \qquad (7.3.2)$$

(see §6.4), which implies that $C \equiv 0$. In other words, Love waves do not exist in a homogeneous half-space.

7.3.2 Layer over a half-space

Let the thickness of the layer be H (Fig. 7.1). This problem has similarities with the layer over a half-space problem for SH waves discussed in §6.9.1. In the layer we need waves going up and down, and because z is finite, the displacement must be represented by the two equations implied by (7.2.10), corresponding to the $+$ and $-$ signs. In the half-space (7.2.18) should be used. Then

$$\mathbf{u} = \mathbf{a}_y(Ae^{-i\eta_1 kz} + Be^{i\eta_1 kz})e^{ik(ct-x)}; \qquad 0 < z < H \qquad (7.3.3)$$

$$\mathbf{u} = C\mathbf{a}_y e^{-\gamma_2 kz + ik(ct-x)}; \qquad z > H, \qquad (7.3.4)$$

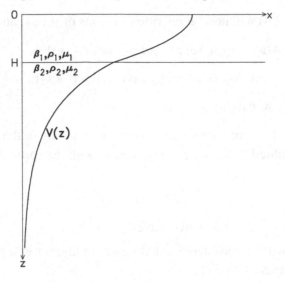

Fig. 7.1. Layer over half-space and plot of the amplitude function $V(z)$ (see (7.3.16)) for the Love wave's fundamental mode.

where

$$\eta_1 = \sqrt{\frac{c^2}{\beta_1^2} - 1}; \qquad \gamma_2 = \sqrt{1 - \frac{c^2}{\beta_2^2}} > 0. \qquad (7.3.5\text{a,b})$$

The condition $\gamma_2 > 0$ is required for $\exp(-\gamma_2 kz)$ to go to zero as z goes to infinity, and implies $\beta_2 > c$. The coefficients A and B in (7.3.3) are unrelated to the coefficients in (7.2.8) and (7.2.9).

The boundary conditions for the problem are: (a) the stress vector at the free surface must be zero; (b) the displacement and stress vectors must be continuous across the interface ($z = H$). These conditions determine the relations between the coefficients A, B, and C.

The expressions for the components of the stress vector are:

$$\tau_{32} = \mu_1 u_{2,3} = \mu_1(-Ai\eta_1 ke^{-i\eta_1 kz} + Bi\eta_1 ke^{i\eta_1 kz})e^{ik(ct-x)}; \qquad 0 < z < H \quad (7.3.6)$$

$$\tau_{32} = \mu_2 u_{2,3} = -\mu_2 C\gamma_2 ke^{-\gamma_2 kz + ik(ct-x)}; \qquad\qquad\qquad z > H. \qquad (7.3.7)$$

From the boundary condition for the stress vector at $z = 0$ and $z = H$ we obtain

$$A = B \qquad (7.3.8)$$

and

$$-i\eta_1(Ae^{-i\eta_1 kH} - Be^{i\eta_1 kH})\mu_1 = -\mu_2 C k\gamma_2 e^{-\gamma_2 kH}. \qquad (7.3.9)$$

Equation (7.3.9) can be written as

$$Ae^{-i\eta_1 kH} - Be^{i\eta_1 kH} + iC\frac{\mu_2\gamma_2}{\mu_1\eta_1}e^{-\gamma_2 kH} = 0.$$ (7.3.10)

The continuity of displacements at $z = H$ gives

$$Ae^{-i\eta_1 kH} + Be^{i\eta_1 kH} - Ce^{-\gamma_2 kH} = 0.$$ (7.3.11)

Using (7.3.8), equation (7.3.11) can be written as

$$C = 2A\cos(\eta_1 kH)e^{\gamma_2 kH}.$$ (7.3.12)

Therefore, equations (7.3.3) and (7.3.4) can be written as

$$\mathbf{u} = \mathbf{a}_y A(e^{-i\eta_1 kz} + e^{i\eta_1 kz})e^{ik(ct-x)}$$

$$= \mathbf{a}_y 2A\cos(\eta_1 kz)e^{ik(ct-x)} \qquad 0 < z < H$$ (7.3.13)

$$\mathbf{u} = \mathbf{a}_y 2A\cos(\eta_1 kH)e^{-\gamma_2 k(z-H)}e^{ik(ct-x)} \qquad z > H.$$ (7.3.14)

Equation (7.3.13) represents a standing wave in the z direction and a propagating wave in the x direction. Equations (7.3.13) and (7.3.14) can be combined as follows:

$$\mathbf{u} = \mathbf{a}_y 2AV(z)e^{ik(ct-x)}$$ (7.3.15)

where

$$V(z) = \begin{cases} \cos(\eta_1 kz) & 0 < z < H \\ \cos(\eta_1 kH)e^{-\gamma_2 k(z-H)} & z > H. \end{cases}$$ (7.3.16)

Note that $V(z)$ is continuous at $z = H$, and that $V(z)$ goes to zero as z goes to infinity. This function is discussed further below.

Finding the equation for the displacement does not solve the problem completely. Aside from the factor A, which can be ignored, the displacement $\mathbf{u}$ includes the quantities η_1, γ_2, and k, which in turn depend on β_1, β_2, μ_1, μ_2, ω, and c. To investigate the relations between these quantities we go back to the equations derived from the boundary conditions. Note that (7.3.8), (7.3.10), and (7.3.11) show that A, B, and C satisfy a homogeneous system of equations. Therefore, for a solution to exist, the determinant of the system must be equal to zero. Let $K = \eta_1 kH$, $L = \gamma_2 kH$, and $M = \mu_2\gamma_2/\mu_1\eta_1$. Then,

$$\begin{vmatrix} 1 & -1 & 0 \\ e^{-iK} & -e^{iK} & iMe^{-L} \\ e^{-iK} & e^{iK} & -e^{-L} \end{vmatrix} = 0.$$ (7.3.17)

After expanding the determinant and operating we obtain

$$iM(e^{iK} + e^{-iK}) = e^{iK} - e^{-iK}.$$ (7.3.18)

Therefore,

$$\tan K = M. \tag{7.3.19}$$

Going back to the original variables we have

$$\tan(\eta_1 k H) = \frac{\mu_2 \gamma_2}{\mu_1 \eta_1}. \tag{7.3.20}$$

Equation (7.3.20) is known as the *period* (or *frequency*) equation, and will be studied in detail.

As noted above, the condition $\gamma_2 > 0$ implies $c < \beta_2$. To find a relation between c and β_1 let us assume that $\beta_1 > c$. Under this condition η_1 is pure imaginary and $\eta_1^2 < 0$, so we can write

$$\eta_1^2 = -\left(1 - \frac{c^2}{\beta_1^2}\right) = -\gamma_1^2, \tag{7.3.21}$$

which implies that $\eta_1 = -i\gamma_1$. Introducing (7.3.21) in (7.3.20) and using the general relation $\tan(-ix) = -i\tanh(x)$ gives

$$-\gamma_1 \tanh(\gamma_1 k H) = \frac{\mu_2}{\mu_1}\gamma_2. \tag{7.3.22}$$

As γ_1, γ_2, μ_1, and μ_2 are positive, equation (7.3.22) has no real solution because the two sides have opposite signs for $k > 0$. Therefore, η_1 is real and $\beta_1 < c$, and

$$\beta_1 < c < \beta_2. \tag{7.3.23}$$

This means that the shear wave velocity in the layer has to be lower than in the half-space.

Equation (7.3.20) must be solved numerically and because the tangent function is periodic, the relation between the values of c and k (or c and ω) that solve (7.3.20) is not simple. The following analysis of this problem follows Hudson (1980). First, let us rewrite η_1 as

$$\eta_1 = \frac{c}{\beta_1}\sqrt{1 - \frac{\beta_1^2}{c^2}} = \frac{c}{\beta_1}\zeta; \qquad \zeta = \sqrt{1 - \frac{\beta_1^2}{c^2}}. \tag{7.3.24a,b}$$

The quantity ζ is always positive because $\beta_1 < c$. Next, rewrite γ_2/η_1 in terms of β_1, β_2, and ζ.

$$\left(\frac{\gamma_2}{\eta_1}\right)^2 = \frac{c^2\left(1/c^2 - 1/\beta_2^2\right)}{c^2\left(1/\beta_1^2 - 1/c^2\right)} = \frac{\beta_1^2/c^2 - \beta_1^2/\beta_2^2}{1 - \beta_1^2/c^2} = \frac{1 - \beta_1^2/\beta_2^2}{\zeta^2} - 1. \tag{7.3.25}$$

In the last step a one was added to, and subtracted from the numerator. Finally,

$$\frac{\gamma_2}{\eta_1} = \sqrt{\frac{1 - \beta_1^2/\beta_2^2}{\zeta^2} - 1}. \tag{7.3.26}$$

Using (7.3.26) and $k = \omega/c$, equation (7.3.20) becomes

$$\tan\left(\frac{\omega H}{\beta_1}\zeta\right) = \frac{\mu_2}{\mu_1}\sqrt{\frac{1 - \beta_1^2/\beta_2^2}{\zeta^2} - 1}. \tag{7.3.27}$$

In (7.3.27) we assume that H, β_1, β_2, and μ_2/μ_1 are fixed, while ω is taken as a parameter. Therefore, the only variable is ζ. Now we will establish certain general properties of the solution of (7.3.27) without actually solving it. First, note that the left-hand side of the equation, corresponding to the tangent function, has an infinite number of branches. The right-hand side represents a curve that approaches infinity as c goes to β_1 and ζ goes to zero, and decreases to zero as c goes to β_2. This is shown in Fig. 7.2(a) for a model with $H = 10$ km, $\beta_1 = 3$ km s^{-1}, $\beta_2 = 4$ km s^{-1}, and $\mu_2/\mu_1 = 2$. The numbered continuous and dashed curves are branches of the tangent function corresponding to $\omega = 5$ and 6 rad s^{-1}.

To simplify the analysis a new variable, $\xi = \omega H/\beta_1$, will be introduced. Then the left-hand side of (7.3.27) becomes $\tan(\xi\zeta)$. This function is equal to zero if $\xi\zeta = (n-1)\pi$ and is equal to infinity if $\xi\zeta = (2m-1)\pi/2$, where n and m are positive integers. On the other hand, the function on the right-hand side of (7.3.27) exists for ζ between 0 and a value $\zeta_{\max}$ given by

$$\zeta_{\max} = \sqrt{1 - \frac{\beta_1^2}{\beta_2^2}}. \tag{7.3.28}$$

Given H, β_1, β_2, and μ_2/μ_1, the number of solutions (roots) of (7.3.27) depends on ω. From Fig. 7.2(a) we see that this equation always has at least one root. Furthermore, if ω is such that $\pi/\xi \le \zeta_{\max} < 2\pi/\xi$, then there is an additional root. This condition can be written as

$$\frac{\pi}{\zeta_{\max}} \le \xi < \frac{2\pi}{\zeta_{\max}}. \tag{7.3.29}$$

In general, if

$$\frac{(N-1)\pi}{\zeta_{\max}} \le \xi < \frac{N\pi}{\zeta_{\max}}; \qquad N = 1, 2, 3, \ldots \tag{7.3.30}$$

there will be N roots. Written in full, equation (7.3.30) becomes

$$\frac{(N-1)\pi}{\sqrt{1 - \beta_1^2/\beta_2^2}} \le \frac{\omega H}{\beta_1} < \frac{N\pi}{\sqrt{1 - \beta_1^2/\beta_2^2}}; \qquad N = 1, 2, 3, \ldots. \tag{7.3.31}$$

Each root corresponds to a *mode*. The first root gives the *fundamental mode*. For a higher mode to exist, ω has to reach a certain value, called the *cut-off frequency*.

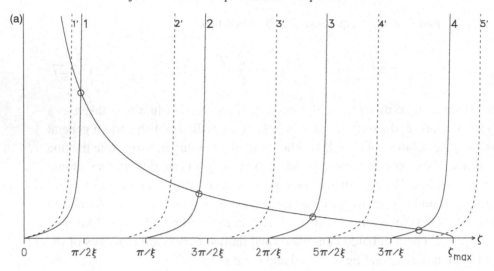

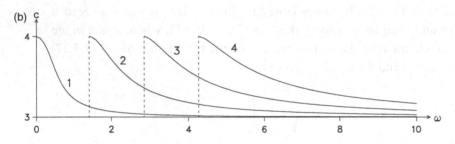

Fig. 7.2. (a) Plot of the functions that appear in the Love wave's period equation (see (7.3.27)) for a model with $H = 10$ km, $\beta_1 = 3$ km s^{-1}, $\beta_2 = 4$ km s^{-1}, and $\mu_2/\mu_1 = 2$. The numbered curves correspond to branches of the tangent function for two values of ω: 5 rad s^{-1} (continuous lines) and 6 rad s^{-1} (dashed lines). The circles indicate the roots of the equation. After Hudson (1980). (b) Love wave phase velocity $c(\omega)$ for the first four modes for the model in (a). The dashed lines indicate the cut-off frequencies.

For the Nth mode this frequency, indicated by ω_{CN}, is obtained from (7.3.31)

$$\omega_{CN} = \frac{(N-1)\pi\beta_1}{H\sqrt{1 - \beta_1^2/\beta_2^2}}. \tag{7.3.32}$$

Once the cut-off frequency has been reached, then the mode exists for all values of ω greater than ω_{CN}.

To complete our study of the period equation let us go back to (7.3.20) and consider two extreme cases. When c goes to β_1, η_1 goes to zero and $\tan(\eta_1 kH)$ goes to infinity (note that η_1 is in the denominator in the right-hand side). This implies that the argument of the tangent function goes to $(2n-1)\pi/2$, so that kH

goes to $(2n - 1)\pi/2\eta_1$, which goes to infinity as η_1 goes to zero. This, in turn, implies that k goes to infinity (as H is finite). Recall that $\omega = kc$ and $\lambda = 2\pi/k$, where λ is wavelength. Therefore, propagation with phase velocity β_1 occurs for all modes at very high frequencies or very short wavelengths.

When $c = \beta_2$, $\gamma_2 = 0$ and $\tan(\eta_1 kH) = 0$, which means that $\eta_1 kH = (n - 1)\pi$. Using $k = \omega/c$ and (7.3.24) we obtain

$$\omega = \frac{(n - 1)\pi c}{H\eta_1} \tag{7.3.33}$$

and

$$\omega = \frac{(n - 1)\pi \beta_1}{H\sqrt{1 - \beta_1^2/c^2}}. \tag{7.3.34}$$

Because we are considering the case $c = \beta_2$, the expression on the right-hand side of (7.3.34) is the cut-off frequency for the nth mode. Therefore, all the modes have velocity β_2 at the cut-off frequency (Fig. 7.2b).

We end this discussion of equation (7.3.20) by emphasizing that the velocity c of the Love waves is a function of frequency. Waves with a frequency-dependent velocity are known as dispersive. The importance of this dependence is that a wave packet will change shape as the wave propagates because of the different velocity of propagation of different wave components. This matter is discussed in §7.6, which also introduces another important concept, namely, that of group velocity.

Finally, we will consider certain properties of the function $V(z)$ defined in (7.3.16). In the layer the cosine factor will become zero if $\eta_1 kz$ is a multiple of $\pi/2$. When that happens the motion will be equal to zero on planes parallel to the free surface, known as *nodal planes*. The number of such planes is not arbitrary; it depends on the mode. Consider first the fundamental mode and investigate whether it is possible that

$$\eta_1 kz = \frac{\omega z \zeta}{\beta_1} = \frac{1}{2}\pi. \tag{7.3.35}$$

In (7.3.35) the variables introduced in (7.3.24) have been used. From Fig. 7.2(a) we see that for the first mode

$$\xi \zeta = \frac{\omega H \zeta}{\beta_1} < \frac{1}{2}\pi. \tag{7.3.36}$$

As $z \leq H$, we see from (7.3.36) that (7.3.35) cannot be satisfied. Therefore, the fundamental mode does not have a nodal plane. A plot of $V(z)$ for this mode is shown in Fig. 7.1. For the second mode equation (7.3.35) is satisfied, but

$$\frac{\omega z \zeta}{\beta_1} = \frac{3}{2}\pi \tag{7.3.37}$$

is not because

$$\frac{\omega H \zeta}{\beta_1} < \frac{3}{2}\pi. \tag{7.3.38}$$

Therefore, the second mode only has one nodal plane. Using the same argument we see that the nth mode has $n - 1$ nodal planes. In the half-space $V(z)$ does not have nodal planes.

7.3.3 Love waves as the result of constructive interference

Here we will show that the Love waves in a layer over a half-space can be interpreted as being the result of constructive interference of multiply reflected SH waves within the layer impinging at the bottom of the layer with angles larger than the critical value. In general, two monochromatic waves with the same frequency and different phases will have the largest *constructive interference* when the phase difference is $2n\pi$, where n is an integer. Fig. 7.3 shows the geometric aspects of the problem. Using ideas developed in Chapter 8, we can say that each line segment with an arrow is a ray representing SH waves impinging either on the free surface or at the bottom of the layer. To set the relevant equation we must consider all the phase changes between the points A and B along the same wave front. From §5.1 we know that there is no phase shift at the free surface while (6.6.16) shows that there is a phase shift of 2ϕ at the solid–solid interface (the sgn ω factor in the exponent can be ignored). To adapt the results derived in §6.6.1.1 to this problem, the layer and the half-space in Fig. 7.3 must be identified with the media having unprimed and primed quantities in Fig. 6.1(a), respectively. Under these conditions, (6.6.17) becomes

$$\tan \phi = \frac{\mu_2}{\mu_1} \frac{(\sin^2 f - \beta_1^2/\beta_2^2)^{1/2}}{(1 - \sin^2 f)^{1/2}}. \tag{7.3.39}$$

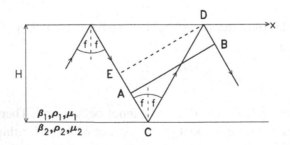

Fig. 7.3. Geometry used to show that Love waves in a layer over a half-space can be explained as the result of constructive interference. The lines AB and ED represent a planar wave front and f is an angle of incidence larger than the critical angle.

To simplify (7.3.39) we use (6.6.27) written as

$$\sin f = \frac{\beta_1}{c}.$$ (7.3.40)

Introducing (7.3.40) into (7.3.39), extracting a common factor of β_1/c in the numerator and denominator and then using (7.3.5a,b) gives

$$\tan \phi = \frac{\mu_2}{\mu_1} \frac{(1 - c^2/\beta_2^2)^{1/2}}{(c^2/\beta_1^2 - 1)^{1/2}} = \frac{\mu_2 \gamma_2}{\mu_1 \eta_1}.$$ (7.3.41)

Another phase shift is introduced when the ray propagates from A to B, which means that the length d of the path $ACDB$ must be determined. From Fig. 7.3,

$$d = \overline{ACDB} = \overline{ECD} = \overline{EC} + \overline{CD} = (\cos 2f + 1)\overline{CD}$$

$$= (\cos 2f + 1)\frac{H}{\cos f} = 2H \cos f.$$ (7.3.42)

To account for the phase shift corresponding to d note that (7.3.3) shows that there is a shift of $-kx$ owing to propagation along the x axis. As k is a horizontal wavenumber (see §6.9.1), for propagation along the ray $ACDB$, k must be replaced by $k/\sin f$. Therefore, using (7.3.40) the phase shift can be written as

$$-2kH\frac{\cos f}{\sin f} = -2kH\frac{c}{\beta_1}(1 - \beta_1^2/c^2)^{1/2} = -2kH\eta_1$$ (7.3.43)

and the condition for constructive interference becomes

$$2\phi - 2kH\eta_1 = 2n\pi.$$ (7.3.44)

Dividing by 2 and applying the tangent to both sides of (7.3.44) and using (7.3.41) gives

$$\tan \phi = \frac{\mu_2 \gamma_2}{\mu_1 \eta_1} = \tan(kH\eta_1),$$ (7.3.45)

which is the period equation (7.3.20).

7.3.4 Vertically heterogeneous medium

In this case the elastic properties of the medium depend on depth (z), so that $\mu = \mu(z)$, $\lambda = \lambda(z)$, and $\rho = \rho(z)$. The corresponding wave equation is

$$\mu\nabla^2\mathbf{u} + (\lambda + \mu)\nabla(\nabla \cdot \mathbf{u}) + \mathbf{a}_z\frac{d\lambda}{dz}\nabla \cdot \mathbf{u} + \frac{d\mu}{dz}\left(2\frac{\partial\mathbf{u}}{\partial z} + \mathbf{a}_z \times \nabla \times \mathbf{u}\right) - \rho\frac{\partial^2\mathbf{u}}{\partial t^2} = 0$$ (7.3.46)

when no external forces are present (Ben-Menahem and Singh, 1981; Problem 7.3).

As Love waves correspond to SH motion, $u_z = 0$ and $\nabla \cdot \mathbf{u} = 0$. Therefore, (7.3.46) reduces to

$$\nabla^2 \mathbf{u} + \frac{1}{\mu} \frac{d\mu}{dz} \left(2 \frac{\partial \mathbf{u}}{\partial z} + \mathbf{a}_z \times \nabla \times \mathbf{u} \right) - \frac{1}{\beta^2(z)} \frac{\partial^2 \mathbf{u}}{\partial t^2} = 0, \tag{7.3.47}$$

where

$$\beta^2(z) = \mu(z)/\rho(z). \tag{7.3.48}$$

Equation (7.3.47) will be solved using a trial solution of the form

$$\mathbf{u} = \mathbf{a}_y V(z) e^{ik(ct-x)}. \tag{7.3.49}$$

This solution is similar to (7.3.15) in form, but here $V(z)$ is a function to be determined under the condition that (7.3.49) satisfies (7.3.47).

For SH motion the Laplacian becomes

$$\nabla^2 \mathbf{u} = \nabla(\nabla \cdot \mathbf{u}) - \nabla \times \nabla \times \mathbf{u} \equiv -\nabla \times \nabla \times \mathbf{u} \tag{7.3.50}$$

(see (1.4.53)). To express $\nabla^2 \mathbf{u}$ in component form we first find the components of $\nabla \times \mathbf{u}$:

$$(\nabla \times \mathbf{u})_i = \epsilon_{ijk} u_{k,j} \tag{7.3.51}$$

$$(\nabla \times \mathbf{u})_1 = \epsilon_{123} u_{3,2} + \epsilon_{132} u_{2,3} = -\frac{dV}{dz} e^{ik(ct-x)} \tag{7.3.52}$$

$$(\nabla \times \mathbf{u})_2 = \epsilon_{213} u_{3,1} + \epsilon_{231} u_{1,3} = 0 \tag{7.3.53}$$

$$(\nabla \times \mathbf{u})_3 = \epsilon_{312} u_{2,1} + \epsilon_{321} u_{1,2} = -ik V e^{ik(ct-x)}. \tag{7.3.54}$$

Let

$$\nabla \times \mathbf{u} = \mathbf{v} = v_1 \mathbf{a}_x + v_3 \mathbf{a}_z, \tag{7.3.55}$$

where v_1 and v_3 are given by (7.3.52) and (7.3.54). Then,

$$\nabla \times \nabla \times \mathbf{u} = \nabla \times \mathbf{v} \tag{7.3.56}$$

and

$$(\nabla \times \mathbf{v})_1 = \epsilon_{123} v_{3,2} + \epsilon_{132} v_{2,3} = 0 \tag{7.3.57}$$

$$(\nabla \times \mathbf{v})_2 = \epsilon_{213} v_{3,1} + \epsilon_{231} v_{1,3} = -V(ik)^2 e^{ik(ct-x)} - \frac{d^2 V e^{ik(ct-x)}}{dz^2}$$

$$= \left(V k^2 - \frac{d^2 V}{dz^2} \right) e^{ik(ct-x)} \tag{7.3.58}$$

$$(\nabla \times \mathbf{v})_3 = \epsilon_{312} v_{2,1} + \epsilon_{321} v_{1,2} = 0. \tag{7.3.59}$$

Now let

$$\mathbf{w} = \mathbf{a}_z \times (\nabla \times \mathbf{u}) = \mathbf{a}_z \times \mathbf{v} \qquad (7.3.60)$$

and take into account that $\mathbf{a}_z = (0, 0, 1)$ and $v_2 = 0$. Then $w_1 = w_3 = 0$, and

$$w_2 = \epsilon_{231} v_1 = -\frac{dV}{dz} e^{ik(ct-x)}. \qquad (7.3.61)$$

Finally,

$$\frac{1}{\beta^2} \frac{\partial^2 \mathbf{u}}{\partial t^2} = \mathbf{a}_y \frac{1}{\beta^2} V(ikc)^2 e^{ik(ct-x)} = -\mathbf{a}_y V \frac{k^2 c^2}{\beta^2} e^{ik(ct-x)} = -\mathbf{a}_y V \frac{\omega^2}{\beta^2} e^{ik(ct-x)}. \qquad (7.3.62)$$

Now introduce the previous results in (7.3.47) and cancel the common factor $e^{ik(ct-x)}$. Note that the relevant vectors have $\mathbf{a}_y$ components only. Therefore we can write

$$-V k^2 \mu + \frac{d^2 V}{dz^2} \mu + \frac{d\mu}{dz} \frac{dV}{dz} = -\frac{\omega^2}{\beta^2} V\mu = -\omega^2 V\rho, \qquad (7.3.63)$$

which can be written in the following compact form:

$$\frac{d}{dz}\left(\mu \frac{dV}{dz}\right) + (\omega^2 \rho - \mu k^2) V = 0. \qquad (7.3.64)$$

Equation (7.3.64) is generally solved numerically, which requires the computation of $d\mu/dz$. However, the computation of this derivative may introduce numerical errors, which will be avoided if (7.3.64) is written as a system of two simultaneous differential equations. Let

$$V = y_1; \qquad \mu(dV/dz) = y_2. \qquad (7.3.65)$$

Note that y_2 is equal to $\mu u_{2,3}$, with $\mathbf{u}$ given by (7.3.49). Therefore, y_2 corresponds to the stress vector. Then, from (7.3.65) and (7.3.64)

$$\frac{dy_1}{dz} = \frac{y_2}{\mu} \qquad (7.3.66)$$

$$\frac{dy_2}{dz} = (\mu k^2 - \rho\omega^2) y_1. \qquad (7.3.67)$$

These two equations can be written in matrix form

$$\frac{d}{dz}\begin{pmatrix} y_1 \\ y_2 \end{pmatrix} = \begin{pmatrix} 0 & 1/\mu \\ \mu k^2 - \rho\omega^2 & 0 \end{pmatrix}\begin{pmatrix} y_1 \\ y_2 \end{pmatrix}. \qquad (7.3.68)$$

The system (7.3.68) must be solved under the the following boundary conditions:

(1) y_2 equal to zero at the surface;
(2) y_1 goes to zero as z goes to infinity;

(3) if the medium has internal surfaces across which the elastic parameters are discontinuous, then the usual conditions that the displacement and stress vectors be continuous across discontinuities must be incorporated into the above formulation.

Techniques to solve (7.3.68) are well known and are described in Aki and Richards (1980) and Ben-Menahem and Singh (1981).

7.4 Rayleigh waves

A half-space and a vertically heterogeneous medium will be considered in detail. The case of a layer over a half-space will be discussed more briefly.

7.4.1 Homogeneous half-space

As noted before, the motion is a combination of P and SV motion. The corresponding displacement is obtained from (7.2.16) and (7.2.17):

$$\mathbf{u} = \left[A(\mathbf{a}_x - i\gamma_\alpha \mathbf{a}_z)e^{-\gamma_\alpha kz} + B(i\gamma_\beta \mathbf{a}_x + \mathbf{a}_z)e^{-\gamma_\beta kz} \right] e^{ik(ct-x)}, \qquad (7.4.1)$$

where

$$\gamma_\delta = \sqrt{1 - \frac{c^2}{\delta^2}}; \qquad \delta = \alpha, \beta. \qquad (7.4.2)$$

These are waves traveling in the positive x direction with phase velocity c and decaying exponentially if γ_δ is real positive, i.e., $c < \delta$, which means that $c < \beta$.

The boundary condition is that the stress vector across the surface should be zero. From §6.4, the components of the stress vector for the P waves are

$$\tau_{31} = 2\mu u_{1,3} \qquad (7.4.3)$$

and

$$\tau_{33} = \lambda u_{1,1} + (\lambda + 2\mu)u_{3,3}, \qquad (7.4.4)$$

while for the S waves they are

$$\tau_{31} = \mu(u_{1,3} + u_{3,1}) \qquad (7.4.5)$$

and

$$\tau_{33} = 2\mu u_{3,3}. \qquad (7.4.6)$$

Consider the horizontal component of the stress vector at the surface. Using (7.4.1) and (7.4.3) and (7.4.5), for the P and SV wave contributions we have:

$$\tau_{31} = -2\mu\gamma_\alpha kA \qquad (7.4.7)$$

and

$$\tau_{31} = \mu(-Bik\gamma_\beta^2 - ikB) = -\mu Bik(1 + \gamma_\beta^2) = -\mu Bik\left(2 - \frac{c^2}{\beta^2}\right), \quad (7.4.8)$$

respectively. For convenience, the exponential term on the right-hand side of (7.4.1) has been dropped. Adding (7.4.7) and (7.4.8) together and setting the result equal to zero gives:

$$-\mu k\left[2\gamma_\alpha A + iB\left(2 - \frac{c^2}{\beta^2}\right)\right] = 0. \quad (7.4.9)$$

For the vertical component the corresponding results, obtained using (7.4.1) and (7.4.4) and (7.4.6), are

$$\tau_{33} = -Aik\lambda + \mu\frac{\alpha^2}{\beta^2}ik\gamma_\alpha{}^2 A = Aik\left(-\lambda + \mu\frac{\alpha^2}{\beta^2}\gamma_\alpha{}^2\right)$$

$$= Aik\left[-\lambda + \mu\frac{\alpha^2}{\beta^2}\left(1 - \frac{c^2}{\alpha^2}\right)\right] = Aik\mu\left(2 - \frac{c^2}{\beta^2}\right) \quad (7.4.10)$$

for the *P* wave (Problem 7.4), and

$$\tau_{33} = -2\mu B\gamma_\beta k \quad (7.4.11)$$

for the *S* wave. Adding (7.4.10) and (7.4.11) together and setting the result equal to zero gives

$$k\mu\left[Ai\left(2 - \frac{c^2}{\beta^2}\right) - 2B\gamma_\beta\right] = 0. \quad (7.4.12)$$

Equations (7.4.9) and (7.4.12) constitute a homogeneous system of equations in *A* and *B*. For nonzero solution the determinant of the system must be zero:

$$\begin{vmatrix} 2\gamma_\alpha & i\left(2 - c^2/\beta^2\right) \\ i\left(2 - c^2/\beta^2\right) & -2\gamma_\beta \end{vmatrix} = 0, \quad (7.4.13)$$

which implies that

$$\left(2 - \frac{c^2}{\beta^2}\right)^2 - 4\gamma_\alpha\gamma_\beta = 0. \quad (7.4.14)$$

Equation (7.4.14) is the *period equation* for Rayleigh waves. Note that since ω does not appear in (7.4.14), c does not depend on ω. Then, in a half-space Rayleigh waves are nondispersive. To study the solution to (7.4.14) we write it as follows:

$$\left(2 - \frac{c^2}{\beta^2}\right)^4 = 16\gamma_\alpha^2\gamma_\beta^2 = 16\left(1 - \frac{c^2}{\alpha^2}\right)\left(1 - \frac{c^2}{\beta^2}\right). \quad (7.4.15)$$

After introducing the new variable $\xi = c^2/\beta^2$ and expanding the left-hand side of (7.4.15) we obtain

$$\xi^3 - 8\xi^2 + 8\xi \left(3 - 2\frac{\beta^2}{\alpha^2}\right) - 16\left(1 - \frac{\beta^2}{\alpha^2}\right) = 0. \qquad (7.4.16)$$

To find out the root(s) of (7.4.16) consider the left-hand side as a function f of ξ, and note that $f(0)$ is a negative number (equal to $-16(1 - \beta^2/\alpha^2)$), and $f(1) = 1$. This means that (7.4.16) has a root, indicated by c_R, between 0 and 1. Therefore, as ξ equal to 0 and 1 imply c equal to 0 and β, respectively, we see that $c_R < \beta$. Additional results regarding the roots of this equation can be found in Achenbach (1973), Eringen and Suhubi (1975), and Hudson (1980).

To get an idea of the kind of values that c_R can take, assume that $\alpha^2/\beta^2 = 3$. This gives

$$3\xi^3 - 24\xi^2 + 56\xi - 32 = 0. \qquad (7.4.17)$$

The three solutions are

$$\xi_1 = 2 - \frac{2}{\sqrt{3}}; \qquad \xi_2 = 4; \qquad \xi_3 = 2 + \frac{2}{\sqrt{3}}. \qquad (7.4.18)$$

The first solution, ξ_1, gives $c_R \simeq 0.92\beta$. The other two solutions give $c_R > \beta$, which violates the condition for the existence of Rayleigh waves.

Now we will derive expressions for the horizontal (u_x) and vertical (u_z) components of the displacement $\mathbf{u}$, given by (7.4.1). First, use (7.4.9) to write A in terms of B:

$$A = \frac{-i}{\gamma_\alpha}\left(1 - \frac{c^2}{2\beta^2}\right)B. \qquad (7.4.19)$$

Using (7.4.19) we obtain the following expression for u_x:

$$u_x = \left[\frac{-i}{\gamma_\alpha}\left(1 - \frac{c^2}{2\beta^2}\right)Be^{-\gamma_\alpha kz} + Bi\gamma_\beta e^{-\gamma_\beta kz}\right]e^{ik(ct-x)}$$

$$= \frac{i}{\gamma_\alpha}\left(1 - \frac{c^2}{2\beta^2}\right)B\left[-e^{-\gamma_\alpha kz} + \frac{\gamma_\alpha \gamma_\beta}{(1 - c^2/2\beta^2)}e^{-\gamma_\beta kz}\right]e^{ik(ct-x)}. \qquad (7.4.20)$$

Now rewrite (7.4.14) as follows:

$$\gamma_\alpha \gamma_\beta = \frac{1}{4}\left(2 - \frac{c^2}{\beta^2}\right)^2 = \left(1 - \frac{c^2}{2\beta^2}\right)^2. \qquad (7.4.21)$$

Then

$$u_x = \frac{i\gamma_\beta}{(1 - c^2/2\beta^2)}B\left[-e^{-\gamma_\alpha kz} + \left(1 - \frac{c^2}{2\beta^2}\right)e^{-\gamma_\beta kz}\right]e^{ik(ct-x)}. \qquad (7.4.22)$$

For u_z we obtain

$$u_z = \left[-i\gamma_\alpha A e^{-\gamma_\alpha kz} + B e^{-\gamma_\beta kz}\right] e^{ik(ct-x)}$$

$$= \left(1 - \frac{c^2}{2\beta^2}\right) B \left[-e^{\gamma_\alpha kz} + \left(1 - \frac{c^2}{2\beta^2}\right)^{-1} e^{-\gamma_\beta kz}\right] e^{ik(ct-x)}$$

$$= \frac{\gamma_\alpha \gamma_\beta}{(1 - c^2/2\beta^2)} B \left[-e^{-\gamma_\alpha kz} + \left(1 - \frac{c^2}{2\beta^2}\right)^{-1} e^{-\gamma_\beta kz}\right] e^{ik(ct-x)}. \qquad (7.4.23)$$

Before proceeding we will take the real parts of u_x (note the i factor) and u_z. After doing that we obtain

$$u_x = Q \left[e^{-\gamma_\alpha kz} - \left(1 - \frac{c^2}{2\beta^2}\right) e^{-\gamma_\beta kz}\right] \sin(\omega t - kx)$$

$$\equiv Q U(z) \sin(\omega t - kx) \qquad (7.4.24)$$

$$u_z = Q\gamma_\alpha \left[-e^{-\gamma_\alpha kz} + \left(1 - \frac{c^2}{2\beta^2}\right)^{-1} e^{-\gamma_\beta kz}\right] \cos(\omega t - kx)$$

$$\equiv Q\gamma_\alpha W(z) \cos(\omega t - kx), \qquad (7.4.25)$$

where the identities define the functions $U(z)$ and $W(z)$, and

$$Q = B\gamma_\beta/(1 - c^2/2\beta^2). \qquad (7.4.26)$$

From (7.4.24) and (7.4.25) we see that

$$\frac{u_x^2}{Q^2 U^2(z)} + \frac{u_z^2}{Q^2 \gamma_\alpha^2 W^2(z)} = 1, \qquad (7.4.27)$$

which is the equation of an ellipse. Therefore the motion of the ground caused by a Rayleigh wave is elliptical in the (x, z) plane (Fig. 7.4). To investigate the ground motion consider a fixed point (i.e., fixed x and z) and study u_x and u_z as functions of t. Let k also be fixed. Then the angle θ (Fig. 7.4) is obtained from

$$\tan\theta = \frac{u_x}{u_z} = \frac{U(z) \sin(\omega t - kx)}{\gamma_\alpha W(z) \cos(\omega t - kx)} = \frac{U(z)}{W(z)\gamma_\alpha} \tan(\omega t - kx). \qquad (7.4.28)$$

As γ_α is always positive, if $U(z)/W(z) > 0$, then θ increases as t grows, so that the motion is counter-clockwise (retrograde). If $U(z)/W(z) < 0$, θ becomes negative as t increases, and the motion is clockwise (prograde or direct). Note that $W(z)$ is always positive because $(1 - c^2/2\beta^2)^{-1} > 1$, and $\gamma_\beta < \gamma_\alpha$. On the other hand, $U(z)$ can be positive, negative or zero. When $z = 0$, U is positive. As z increases, U decreases because $\exp(-\gamma_\alpha kz)$ decays faster than $\exp(-\gamma_\beta kz)$. Therefore, there

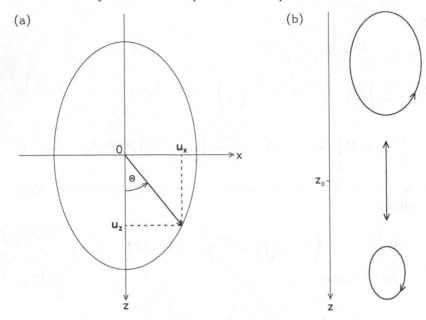

Fig. 7.4. (a) Particle motion generated by Rayleigh waves in a half-space. u_x and u_z are the horizontal and vertical components (see (7.4.24) and (7.4.25)) and θ is the angle introduced in (7.4.28). (b) Schematic variation of particle motion as a function of depth. The motion is elliptical counterclockwise (retrograde) between the surface and a depth z_o at which the motion is purely vertical (see (7.4.29) and Fig. 7.5). Below this depth the motion is clockwise (prograde or direct).

is some value z_o of z at which U becomes zero. From (7.4.24), the value of z_o is obtained by solving

$$e^{(\gamma_\beta - \gamma_\alpha)kz_o} = 1 - \frac{c^2}{2\beta^2}. \tag{7.4.29}$$

At this depth the motion is purely vertical. Below it, U becomes negative. Therefore, the ground motion caused by the Rayleigh wave is retrograde above a certain depth, and prograde below that depth (Fig. 7.4). The functions U and W are shown in Fig. 7.5.

7.4.2 *Layer over a half-space. Dispersive Rayleigh waves*

This case has similarities with the layer over a half-space problem for P and SV waves discussed in §6.9.2 and was first analyzed in general by Love (1911). In the layer we need P and SV waves going up and down, with the displacements represented by the four equations implied by (7.2.8) and (7.2.9). For the half-space

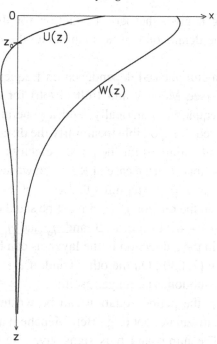

Fig. 7.5. Plot of the functions $U(z)$ and $W(z)$ for Rayleigh waves in a half-space (see (7.4.24) and (7.4.25)).

(7.4.1) must be used. Therefore, using the same convention as in §6.9.2,

$$\mathbf{u}' = \left[A_u'(\mathbf{a}_x - \eta_\alpha'\mathbf{a}_z)e^{ik\eta_\alpha'z} + A_d'(\mathbf{a}_x + \eta_\alpha'\mathbf{a}_z)e^{-ik\eta_\alpha'z} \right.$$

$$\left. + B_d'(\eta_\beta'\mathbf{a}_x + \mathbf{a}_z)e^{ik\eta_\beta'z} + B_u'(-\eta_\beta'\mathbf{a}_x + \mathbf{a}_z)e^{-ik\eta_\beta'z} \right]e^{ik(ct-x)}, \qquad 0 < z < H$$

$$(7.4.30)$$

$$\mathbf{u} = \left[A(\mathbf{a}_x - i\gamma_\alpha\mathbf{a}_z)e^{-k\gamma_\alpha z} + B(i\gamma_\beta\mathbf{a}_x + \mathbf{a}_z)e^{-k\gamma_\beta z} \right]e^{ik(ct-x)}, \qquad z > H,$$

$$(7.4.31)$$

where the primed and unprimed quantities correspond to the layer and the half-space, respectively, and

$$\eta_\delta' = \sqrt{\frac{c^2}{\delta'^2} - 1}; \qquad \gamma_\delta = \sqrt{1 - \frac{c^2}{\delta^2}}; \qquad \delta = \alpha, \beta. \qquad (7.4.32)$$

The boundary conditions are the usual ones, with the stress vector equal to zero at the surface ($z = 0$), and displacement and stress vectors continuous at the bottom of the layer. Writing the corresponding equations down is straightforward and will not be done here. The results of doing that are a homogeneous system of six equations in six unknowns, the coefficients A_u', A_d', B_u', B_d', A, and B. Setting the determinant equal to zero and expanding it gives the period equation for the problem (see Ewing *et al.* (1957), for some details on how to manipulate the determinant and for additional references).

The phase velocity c has a complicated dependence on frequency, layer thickness, and elastic parameters (see Mooney and Bolt (1966) for examples and a method to solve the period equation numerically) and a general analysis of the period equation is quite involved. It is possible to simplify the discussion, however, by analyzing possible ranges of variations for the phase velocity c (Ben-Menahem and Singh, 1981). We first note that as in the case of Rayleigh waves in a half-space, the condition that **u** vanishes at $z = \infty$ requires that $c < \beta$. Secondly, for real solutions of the period equation the relation $\beta' < \beta$ must be satisfied (Mooney and Bolt, 1966). Furthermore, if $c > \alpha'$, then $c > \beta'$ and η'_α and η'_β are real, which means an oscillatory motion in the z direction in the layer, as can be seen from the corresponding exponentials in (7.4.30). On the other hand, if $c < \beta'$, then η'_β and η'_α become imaginary and the motion is no longer oscillatory.

For the case $\alpha' < c < \beta$ the period equation can be written in a way that involves plus or minus a certain square root (e.g., Ben-Menahem and Singh, 1981; Tolstoy and Usdin, 1953). The minus and plus signs give rise to the so-called M_1 and M_2 branches. Furthermore, each branch has an infinite number of modes (M_{1j} and M_{2j}), although the number of real roots for a given frequency and layer thickness is limited. The M_1 and M_2 branches correspond to the symmetrical and anti-symmetrical modes of propagation of a free plate (Tolstoy and Usdin, 1953, and references therein). In a plate, the M_1 and M_2 modes correspond to motion symmetric and anti-symmetric with respect to the median plane of the plate. If u_x and u_z indicate the horizontal and vertical components of the vector displacement and $z = 0$ indicates the median plane, then symmetric and anti-symmetric motion are represented by

$$u_x(-z) = u_x(z); \qquad u_z(-z) = -u_z(z) \qquad (7.4.33)$$

and

$$u_x(-z) = -u_x(z); \qquad u_z(-z) = u_z(z), \qquad (7.4.34)$$

respectively (Eringen and Suhubi, 1975).

A summary of important results follows (Ewing *et al.*, 1957; Eringen and Suhubi, 1975; Ben-Menahem and Singh, 1981).

(1) The particle motion at the surface is retrograde elliptical for the M_1 branch and prograde elliptical for the M_2 branch (see, however, Mooney and Bolt, 1966, p. 64).

(2) There is a cut-off frequency for each mode except for the M_{11} mode (known

as the fundamental mode). The cut-off frequency increases with the mode number. At this frequency the phase velocity is equal to β.

(3) At the short-wave limit (i.e., $kH \to \infty$) the following results apply. The phase velocity of the M_{11} mode approaches that of the Rayleigh waves in the layer. If the conditions for the existence of Stoneley waves are satisfied (see §7.5), there is a mode in the M_2 branch for which the velocity approaches the velocity of the Stoneley waves for two infinite media with the properties of the layer and the half-space. For all other modes the phase velocity approaches β'.

(4) At the long-wave limit (i.e., $kH \to 0$), the phase velocity of the fundamental mode approaches the velocity of Rayleigh waves in the half-space, while for all other modes the phase velocity approaches β.

(5) When $c < \beta'$ and $kH \to \infty$, there is a mode that propagates with phase velocity approaching the Rayleigh wave velocity in the layer, and another mode that propagates with the velocity of the Stoneley waves.

We finally note that Tolstoy and Usdin (1953) showed that Rayleigh waves in a layer over a half-space can be explained as the result of the constructive interference of P and SV waves.

7.4.3 Vertically heterogeneous medium

The equation describing the motion is given by (7.3.46). In analogy with the Rayleigh wave solution obtained in §7.4.1 we will look for a solution of the form

$$\mathbf{u} = [-iU(z)\mathbf{a}_x + W(z)\mathbf{a}_z]e^{ik(ct-x)} \tag{7.4.35}$$

with U and W two functions to be determined. Recall that

$$\nabla^2 \mathbf{u} = \nabla(\nabla \cdot \mathbf{u}) - \nabla \times \nabla \times \mathbf{u}. \tag{7.4.36}$$

The following results are needed:

$$\nabla \cdot \mathbf{u} = u_{i,i} = (-kU + W')e^{ik(ct-x)}, \tag{7.4.37}$$

where the prime represents a derivative with respect to z

$$\nabla \times \mathbf{u} = \mathbf{a}_y(ikW - iU')e^{ik(ct-x)} \tag{7.4.38}$$

$$\nabla(\nabla \cdot \mathbf{u}) = (ik^2U - ikW', 0, -kU' + W'')e^{ik(ct-x)} \tag{7.4.39}$$

$$\nabla \times \nabla \times \mathbf{u} = (-ikW' - iU'', 0, k^2W - kU')e^{ik(ct-x)} \tag{7.4.40}$$

$$\mathbf{a}_z \times \nabla \times \mathbf{u} = (-ikW + iU'), 0, 0)e^{ik(ct-x)} \tag{7.4.41}$$

$$\frac{\partial^2 \mathbf{u}}{\partial t^2} = -k^2 c^2 \mathbf{u} = -\omega^2 \mathbf{u} \tag{7.4.42}$$

(Problem 7.5). After introducing (7.4.37)–(7.4.42) in (7.3.46), combining the terms with $\nabla(\nabla \cdot \mathbf{u})$, canceling the exponential factor, and rearranging we obtain

$$(\lambda + 2\mu)(-k^2 U + kW') + \mu(-kW' + U'') + \mu'(U' + kW) + \omega^2 \rho U$$

$$= U\left[\rho\omega^2 - (\lambda + 2\mu)k^2\right] + \lambda kW' + \mu kW' + \mu U'' + \mu'U' + \mu'kW$$

$$= U\left[\rho\omega^2 - (\lambda + 2\mu)k^2\right] + \lambda kW' + \frac{d}{dz}\left[\mu(U' + kW)\right] = 0 \qquad (7.4.43)$$

for the horizontal component and

$$(\lambda + 2\mu)(-kU' + W'') - \mu(k^2 W - kU') + \lambda'(-kU + W') + 2\mu'W' + \omega^2 W\rho$$

$$= W(\rho\omega^2 - k^2\mu) - \mu kU' + (\lambda + 2\mu)W'' - \lambda kU' - \lambda'kU + \lambda'W' + 2\mu'W'$$

$$= W(\rho\omega^2 - k^2\mu) - \mu kU' + \frac{d}{dz}\left[(\lambda + 2\mu)W' - k\lambda U\right] = 0 \qquad (7.4.44)$$

for the vertical component (Problem 7.6).

Now consider the boundary conditions. The corresponding equations are:

$$\tau_{31} = \mu(u_{3,1} + u_{1,3}) = \mu(-ikW - iU')e^{ik(ct-x)}$$

$$= -i\mu(U' + kW)e^{ik(ct-x)} \qquad (7.4.45)$$

$$\tau_{32} = 0 \qquad (7.4.46)$$

$$\tau_{33} = \lambda u_{1,1} + (\lambda + 2\mu)u_{3,3} = \left[(\lambda + 2\mu)W' - Uk\lambda\right]e^{ik(ct-x)}. \qquad (7.4.47)$$

At $z = 0$ we obtain

$$\mu(U' + kW) = 0 \qquad (7.4.48)$$

and

$$(\lambda + 2\mu)W' - k\lambda U = 0. \qquad (7.4.49)$$

Now introduce the following notation:

$$y_1 = W \qquad (7.4.50)$$

$$y_2 = (\lambda + 2\mu)\frac{dW}{dz} - k\lambda U \qquad (7.4.51)$$

$$y_3 = U \qquad (7.4.52)$$

$$y_4 = \mu\left(\frac{dU}{dz} + kW\right). \qquad (7.4.53)$$

Note the relations between y_2 and τ_{33} and y_4 and τ_{31}. Equations (7.4.50)–(7.4.53) will be rewritten in such a way that does not include spatial derivatives of λ or μ. This will be done in several steps.

From (7.4.50)–(7.4.52)

$$\frac{dy_1}{dz} = \frac{(y_2 + k\lambda U)}{\lambda + 2\mu} = \frac{y_2 + k\lambda y_3}{\lambda + 2\mu}. \tag{7.4.54}$$

From (7.4.44), (7.4.51), and (7.4.53)

$$\frac{dy_2}{dz} = -\rho\omega^2 W + k\mu\left(\frac{dU}{dz} + kW\right) = -\rho\omega^2 y_1 + ky_4. \tag{7.4.55}$$

From (7.4.52), (7.4.53), and (7.4.50)

$$\frac{dy_3}{dz} = \frac{dU}{dz} = -kW + \frac{y_4}{\mu} = -ky_1 + \frac{y_4}{\mu}. \tag{7.4.56}$$

From (7.4.43), (7.4.50), (7.4.52), and (7.4.53)

$$\frac{dy_4}{dz} = -\rho\omega^2 y_3 + (\lambda + 2\mu)k^2 y_3 - \lambda k\frac{dy_1}{dz}. \tag{7.4.57}$$

Now multiply and divide (7.4.51) by $k\lambda$, and use (7.4.50) and (7.4.52)

$$y_2 = \frac{(\lambda + 2\mu)}{k\lambda}\left(k\lambda\frac{dy_1}{dz} - \frac{k^2\lambda^2 y_3}{\lambda + 2\mu}\right) \tag{7.4.58}$$

and add and subtract $k^2\lambda^2 y_3/(\lambda + 2\mu)$ to (7.4.57). This gives

$$\frac{dy_4}{dz} = \frac{-k\lambda}{\lambda + 2\mu}y_2 + \left(-\rho\omega^2 + 4k^2\mu\frac{\lambda + \mu}{\lambda + 2\mu}\right)y_3. \tag{7.4.59}$$

Equations (7.4.54)–(7.4.56) and (7.4.59) can be written in matrix form as

$$\frac{d}{dz}\begin{pmatrix} y_1 \\ y_2 \\ y_3 \\ y_4 \end{pmatrix} = \begin{pmatrix} 0 & \left(\frac{1}{\lambda + 2\mu}\right) & \left(\frac{k\lambda}{\lambda + 2\mu}\right) & 0 \\ (-\rho\omega^2) & 0 & 0 & k \\ -k & 0 & 0 & 1/\mu \\ 0 & \left(\frac{-k\lambda}{\lambda + 2\mu}\right) & K & 0 \end{pmatrix}\begin{pmatrix} y_1 \\ y_2 \\ y_3 \\ y_4 \end{pmatrix}, \tag{7.4.60}$$

where

$$K = -\rho\omega^2 + 4k^2\mu\frac{\lambda + \mu}{\lambda + 2\mu}. \tag{7.4.61}$$

Note that in the notation of Aki and Richards (1980):

$$y_1 = r_2; \qquad y_2 = r_4; \qquad y_3 = r_1; \qquad y_4 = r_3 \tag{7.4.62}$$

and that k is replaced by $-k$ (because of the different definition of **u**).

The system of equations has to be solved under the following boundary conditions:

(1) $y_2 = y_4 = 0$ at the free surface (see (7.4.48), (7.4.49), (7.4.51), and (7.4.53));
(2) y_1 and y_3 go to zero as z goes to infinity;
(3) continuity of the displacement and stress vectors across any discontinuity in the medium.

7.5 Stoneley waves

As mentioned in §7.1, these waves exist along the boundary between two half-spaces. Let $z = 0$ indicate the boundary between them, with the positive z axis directed downwards. Then, the displacement for $z > 0$ is given by (7.4.1):

$$\mathbf{u} = \left[A(\mathbf{a}_x - i\gamma_\alpha \mathbf{a}_z)e^{-\gamma_\alpha kz} + B(i\gamma_\beta \mathbf{a}_x + \mathbf{a}_z)e^{-\gamma_\beta kz} \right] e^{ik(ct-x)}, \tag{7.5.1}$$

while for $z < 0$ we must write

$$\mathbf{u} = \left[C(\mathbf{a}_x + i\gamma_{\alpha'} \mathbf{a}_z)e^{\gamma_{\alpha'} kz} + D(-i\gamma_{\beta'} \mathbf{a}_x + \mathbf{a}_z)e^{\gamma_{\beta'} kz} \right] e^{ik(ct-x)} \tag{7.5.2}$$

in agreement with (7.2.16) and (7.2.17), and the condition that the amplitude goes to zero as z goes to infinity.

The boundary conditions for the problem are the continuity of the displacement and the stress vector at $z = 0$. For $z > 0$ the components of the stress vector are given by (7.4.7), (7.4.8), (7.4.10), and (7.4.11). For $z < 0$ it is straightforward to show that

$$\tau_{31} = 2\mu' \gamma_{\alpha'} kC \tag{7.5.3}$$

$$\tau_{33} = Cik\mu'(1 + \gamma_{\beta'}^2) \tag{7.5.4}$$

and

$$\tau_{31} = -\mu' Dik(1 + \gamma_{\beta'}^2) \tag{7.5.5}$$

$$\tau_{33} = 2\mu' D\gamma_{\beta'} k \tag{7.5.6}$$

for the P and S waves, respectively.

Application of the boundary conditions gives

$$A + i\gamma_\beta B - C + i\gamma_{\beta'} D = 0 \tag{7.5.7}$$

$$-i\gamma_\alpha A + B - i\gamma_{\alpha'} C - D = 0 \tag{7.5.8}$$

$$-2\mu\gamma_\alpha A - \mu i(1 + \gamma_\beta^2)B - 2\mu' \gamma_{\alpha'} C + i\mu'(1 + \gamma_{\beta'}^2)D = 0 \tag{7.5.9}$$

$$i\mu(1 + \gamma_\beta^2)A - 2\mu\gamma_\beta B - i\mu'(1 + \gamma_{\beta'}^2)C - 2\mu' \gamma_{\beta'} D = 0. \tag{7.5.10}$$

In (7.5.9) and (7.5.10) a constant factor of k was canceled out (Problem 7.7).

Equations (7.5.7)–(7.5.10) constitute a homogeneous system of equations in four unknowns (A, B, C, D), which has nonzero solution when its determinant is equal to zero. Let F indicate this determinant. To remove the factor of i multiply the second and fourth rows of F by i first and then multiply the second and fourth columns by $-$i. Finally, divide the third and fourth rows by μ and let $r = \mu'/\mu$. These operations give

$$
F(c) = \begin{vmatrix}
1 & \gamma_\beta & -1 & \gamma_{\beta'} \\
\gamma_\alpha & 1 & \gamma_{\alpha'} & -1 \\
-2\gamma_\alpha & -(1+\gamma_\beta^2) & -2\gamma_{\alpha'}r & (1+\gamma_{\beta'}^2)r \\
-(1+\gamma_\beta^2) & -2\gamma_\beta & (1+\gamma_{\beta'}^2)r & -2\gamma_{\beta'}r
\end{vmatrix} = 0 \qquad (7.5.11)
$$

(Problem 7.8).

The determinant F is a function of c because of (7.4.2). The analysis of the roots of (7.5.11) is exceedingly complicated (e.g., Pilant, 1979), and whether (7.5.11) has a real solution or not depends on the values of the velocities and densities of the two media. In fact, Stoneley waves exist only if ρ and ρ' and β and β' are close to each other. This shows that this type of waves constitutes a rather special occurrence (e.g., Eringen and Suhubi, 1975). If $\beta > \beta'$ there is one positive value of $c < \beta'$ if

$$
F(\beta') < 0 \qquad (7.5.12)
$$

(Hudson, 1980). Moreover, if c_R denotes the velocity of Rayleigh waves in the medium with lower S-wave velocity, then

$$
c_R < c < \beta' < \beta \qquad (7.5.13)
$$

(Eringen and Suhubi, 1975). Finally, because (7.5.11) does not involve the frequency, Stoneley waves are not dispersive.

7.6 Propagation of dispersive waves

Dispersion is a fundamental concept in wave propagation, which can only be understood after a careful discussion. The basic feature of dispersion is the presence of two types of velocities, called the group and the phase velocities, both of which are dependent on frequency. The most obvious consequence of dispersion is a change in the shape of the waves as they propagate away from the source. In general, a wave that originally has a given spatial or temporal extent will become more and more extended as time progresses. Dispersion is a common feature in wave propagation problems, occurring not only in seismology but also in hydrodynamics, acoustics, and electromagnetics, among others. Because the dispersion of elastic waves is directly related to the properties of the medium, it offers a powerful tool for studying the Earth's interior.

7.6.1 Introductory example. The dispersive string

To motivate the following discussion consider the motion of a string subject to a restoring force:

$$\frac{\partial^2 f}{\partial t^2} = v^2 \frac{\partial^2 f}{\partial x^2} - a^2 f, \tag{7.6.1}$$

where $f = f(x,t)$ and a and v are constants that depend on the nature of the problem represented by (7.6.1) (Havelock, 1914; Graff, 1975; Officer, 1974; Whitham, 1974). The restoring force may be the result, for example, of embedding the string in a thin sheet of rubber (Morse and Feshbach, 1953). In quantum mechanics (7.6.1) is known as the Klein–Gordon equation and is discussed in detail by Bleistein (1984) and Morse and Feshbach (1953).

Although equation (7.6.1) is no longer the classical wave equation, we will try a solution that resembles a plane wave propagating with some velocity c (to be determined), not necessarily equal to v:

$$f(x,t) = A e^{i(\omega t - kx)} \tag{7.6.2}$$

where A is a constant and

$$\omega = ck. \tag{7.6.3}$$

We can see by substitution that $f(x,t)$ given by (7.6.2) will solve (7.6.1) if

$$k^2 c^2 = v^2 k^2 + a^2 \tag{7.6.4}$$

(Problem 7.9), which means that

$$c = \frac{\sqrt{v^2 k^2 + a^2}}{k} = \left(v^2 + \frac{a^2}{k^2} \right)^{1/2} \tag{7.6.5}$$

or, using (7.6.3) and (7.6.4)

$$c = v \left(\frac{\omega^2}{\omega^2 - a^2} \right)^{1/2}. \tag{7.6.6}$$

Equations (7.6.5) and (7.6.6) show that the velocity c is not constant, as it depends on frequency (or wavenumber), which is a feature typical of dispersion. In this particular case c is larger than v, with c approaching v as k and ω approach infinity. Furthermore, c increases as k decreases and as $|\omega|$ approaches $|a|$. Also note that for $|\omega| < |a|$, c and k are purely imaginary, which means that (7.6.2) no longer represents a propagating wave. To see this, solve (7.6.4) for k^2 using (7.6.3)

$$k^2 = \frac{\omega^2 - a^2}{v^2} \tag{7.6.7}$$

and let the numerator of (7.6.7) be represented by the real quantity $-\alpha^2$. Then, $k = \pm i\alpha/v$ and (7.6.2) can be written as

$$f(x, t) = Ae^{\pm\alpha x/v}e^{i\omega t}. \tag{7.6.8}$$

The selection of the plus or minus sign in the exponent depends on the sign of x. For $x > 0$ the minus sign should be chosen and in such a case (7.6.8) corresponds to an oscillation that dies out with increasing x.

Equation (7.6.1) is just one example of a dispersive system. In this case the dependence of c on k or ω is relatively simple, but in other cases, such as that of seismic surface waves, it can be much more complicated. However, regardless of the complexity of the system, there are a number of common features of the solutions to dispersive systems that can be derived without actually computing them. Moreover, getting exact solutions may be a difficult or impossible task, but even in these cases it is possible to derive approximate solutions valid for large values of time and distance. These matters are discussed next.

7.6.2 Narrow-band waves. Phase and group velocity

The general analysis of dispersion usually starts with the classical argument (attributed to Stokes) used to introduce group and phase velocities. Consider two plane waves with equal amplitudes and different frequencies and wavenumbers:

$$y_1(x, t) = Ae^{i(\omega_1 t - k_1 x)} \tag{7.6.9}$$

and

$$y_2(x, t) = Ae^{i(\omega_2 t - k_2 x)}, \tag{7.6.10}$$

where $k_1 = k_o + \delta k$, $k_2 = k_o - \delta k$, $\omega_1 = \omega_o + \delta\omega$, and $\omega_2 = \omega_o - \delta\omega$. The superposition of the two waves gives

$$y(x, t) = A\left[e^{i(\delta\omega t - \delta k x)} + e^{-i(\delta\omega t - \delta k x)}\right]e^{i(\omega_o t - k_o x)} = 2A\cos(\delta\omega t - \delta k x)e^{i(\omega_o t - k_o x)}. \tag{7.6.11}$$

If δk and $\delta\omega$ are small, then the exponential factor in the right-hand side of (7.6.11) represents a wave similar to y_1 and y_2 that propagates with a velocity $c = \omega_o/k_o$, known as the *phase velocity*. This is the velocity of propagation of a surface with constant phase, because if

$$\omega_o t - k_o x = \text{constant} \tag{7.6.12}$$

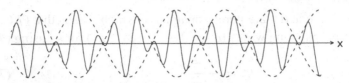

Fig. 7.6. Superposition of two harmonic waves close in frequency and wavenumber. The dashed curve is the envelope of the continuous curve.

then, after differentiation, we find

$$\frac{\mathrm{d}x}{\mathrm{d}t} = \frac{\omega_o}{k_o} = c. \tag{7.6.13}$$

As ω_o and k_o are arbitrary, we can write, in general

$$c = \frac{\omega}{k}. \tag{7.6.14}$$

The cosine factor in the right-hand side of (7.6.11) represents a wave that propagates with velocity U given by

$$U = \frac{\delta\omega}{\delta k} = \frac{\mathrm{d}\omega}{\mathrm{d}k} = c + k\frac{\mathrm{d}c}{\mathrm{d}k}. \tag{7.6.15}$$

In (7.6.15) the second equality is valid in the limit as $\delta\omega$ and δk go to zero. The last equality is obtained using (7.6.14). When $\mathrm{d}c/\mathrm{d}k$ is equal to zero, $U = c$. In other cases U can be larger or smaller than c. Dispersion is generally called *normal* when $U < c$, and *anomalous* when $U > c$. In seismology the term normal dispersion is used when the group velocity increases with period (or wavelength). The opposite situation is termed *inverse dispersion*.

Using U and c, equation (7.6.11) can be written as

$$y = 2A\cos[\delta k(Ut - x)]\,\mathrm{e}^{\mathrm{i}k_o(ct-x)}. \tag{7.6.16}$$

The wave propagating with velocity c is known as the carrier. Because of the presence of the small factor δk, the wave with velocity U varies more slowly than the carrier wave and modulates it. This is seen in Fig. 7.6 (obtained using (7.6.11)), which shows the result of the superposition of two close waves and the well-known phenomenon of "beats". The portion of the carrier wave between two consecutive zeros of the modulation wave, or envelope, constitutes a group (or packet) and, consequently, U is known as *group velocity*.

Equation (7.6.11) represents an extremely simplified situation. To analyze more general cases we will use the Fourier representation discussed in Chapter 5. Two approaches can be taken. One is to let $\omega = \omega(k)$, and the other is to let $k = k(\omega)$. Here the first approach will be taken because it will lead to an equation that can be compared directly with (7.6.16). The basic idea is that an arbitrary wave $f(x, t)$ can be written as an infinite superposition of plane waves:

$$f(x, t) = \frac{1}{2\pi} \int_{-\infty}^{\infty} A(k) e^{i[\omega(k)t - kx]} \, dk. \tag{7.6.17}$$

Strictly speaking, equation (7.6.17) should include waves moving to the left (Whitham, 1974), but ignoring them does not change the following discussion. In addition, we are also ignoring any possible phase terms related to the nature of the source that generates the waves, and to the instrument used to record the waves (Pilant, 1979). This omission, however, is not serious, and is introduced to simplify the derivations (see Problem 7.12).

Now assume that $\omega(k)$ is zero everywhere except in a small interval $(k_o - \delta k, k_o + \delta k)$ in the vicinity of $k = k_o$. Clearly, this case generalizes the previous one by superposing an infinite number of waves with very close wavenumbers and arbitrary amplitudes. The resulting wave is said to be narrow-band (in the k domain, which means that it is broad in the x domain; see Problem 7.10). Under this condition $\omega(k)$ can be approximated using the first two terms of its Taylor expansion:

$$\omega(k) \approx \omega(k_o) + \left.\frac{d\omega}{dk}\right|_{k=k_o} (k - k_o) = \omega_o + \omega'_o(k - k_o), \tag{7.6.18}$$

where the symbols on the right are straightforward abbreviations. With this approximation and the identity $k \equiv k_o + (k - k_o)$ the phase in (7.6.17) becomes

$$\omega t - kx = \omega_o t - k_o x + (\omega'_o t - x)(k - k_o). \tag{7.6.19}$$

Because k_o and ω_o are constants, $f(x, t)$ can be written as

$$f(x, t) = f_o(\omega'_o t - x) e^{i(\omega_o t - k_o x)} \tag{7.6.20}$$

with

$$f_o(\omega'_o t - x) = \frac{1}{2\pi} \int_{k_o - \delta k}^{k_o + \delta k} A(k) e^{i(k - k_o)(\omega'_o t - x)} \, dk. \tag{7.6.21}$$

Comparison of equations (7.6.16) and (7.6.20) shows that in both cases the result of the superposition is equal to the product of a common harmonic wave propagating with velocity c and a modulating factor. In the second case this factor has a constant phase when

$$\omega'_o t - x = \text{constant}, \tag{7.6.22}$$

which means that

$$\frac{dx}{dt} = \omega_o' = \frac{d\omega}{dk}\bigg|_{k=k_o} = U \qquad (7.6.23)$$

as in the first case.

We can obtain additional insight into the nature of $f(x, t)$ by considering the special case of $A(k)$ constant and equal to 1, which corresponds to Dirac's delta in the space domain (see Appendix A). Introducing this value in (7.6.21) gives

$$f_o(\omega_o' t - x) = \frac{\delta k}{\pi} \frac{\sin[(Ut - x)\delta k]}{(Ut - x)\delta k}. \qquad (7.6.24)$$

The second factor on the right is the well-known sinc function, which has a dominant lobe followed by a number of lobes of diminishing amplitudes.

Plots of $f(x, t)$ obtained using (7.6.20) and (7.6.24) for several equispaced values of t and two combinations of c and U values are shown in Fig. 7.7. The plots could be interpreted as snapshots of waves traveling on the surface of a body of water. If we had chosen to plot $f(x, t)$ as a function of time for fixed values of distance, then the plots could be interpreted as seismograms. A number of important properties of the solution can be inferred from this figure.

(1) There is a clear wave packet, which propagates with velocity U. The maximum of the envelope corresponds to the values of x and t that satisfy $Ut - x = 0$, so that $x = Ut$.

(2) The width of the wave group, defined (somewhat arbitrarily) as the distance between the first positive and negative zero crossings of the sinc function is proportional to $1/\delta k$. Therefore, the narrower the width in the wavenumber domain, the larger the width in the space domain (see also Problem 7.10). If this width is indicated by δx, then the product $\delta x \delta k$ is of the order of 1. The famous *Heisenberg uncertainty principle* of quantum mechanics is related to this fact. If we had started with $k = k(\omega)$, we would have ended up with a similar relation for δt and $\delta \omega$.

(3) The plots on the left of the figure were made with $U < c$ and show the propagation of two particular features. One is the maximum value of the envelope, identified with a cross. The other feature is the peak of $f(x, t)$ identified with a circle. The two features propagate with velocities U and c, respectively. As $U < c$, the peak originates at the back of the group, and moves to the front, where it disappears. For the plots on the right, $U > c$, and in this case the peak moves from the front to the back of the packet.

(4) If the time difference Δt between adjacent plots had been chosen larger (say tripled), it would not have been possible to follow the evolution of the individual peak described above. To be able to do that it is necessary that the

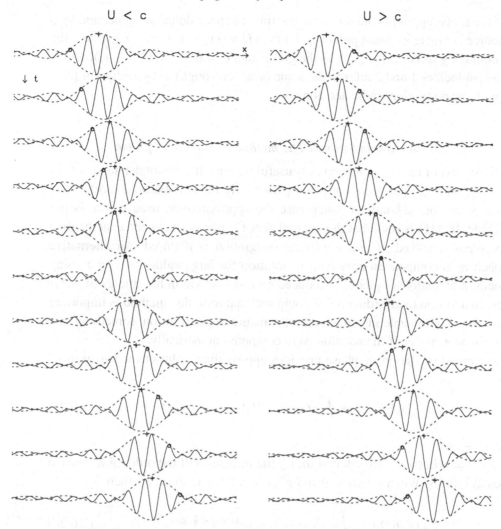

Fig. 7.7. Superposition of an infinite number of harmonic waves close in frequency and wavenumber (see (7.6.20) and (7.6.24)). Two cases are shown: $U < c$ (left) and $U > c$ (right). Each subplot has been generated for a fixed value of t. The different values of t are equispaced. Two special features are shown, the maximum value of the envelope, indicated by the plus, and the same peak in the modulated function, indicated by the circle. The two features move with velocities U and c, respectively.

difference $ct_2 - ct_1 = c\Delta t < \lambda_o$, where $\lambda_o = 2\pi/k_o$ is wavelength. This means $\Delta t < 2\pi/\omega_o = T_o$, where T_o is the period. When $\Delta t > T_o$, it is not possible to tell which peak in one of the plots corresponds to any of the other peaks in the adjacent plot. If the plots were in the time domain for fixed values of x, then the equivalent condition would be $\Delta x < \lambda_o$.

(5) If the plots represent data (either in the time or space domains) generated by a source acting at a certain point (x_o, t_o) then $U = (x_g - x_o)/(t_g - t_o)$, where the subindex g refers to the peak of the group, and $c = (x_2 - x_1)/(t_2 - t_1)$, where the subindices 1 and 2 refer to the same peak (or trough) in two adjacent plots (with the restrictions indicated above).

7.6.3 *Broad-band waves. The method of stationary phase*

The discussion of narrow-band waves is useful because it helps understand certain features of wave propagation, but is not particularly realistic because most waves of interest are broad-band, in which case the approximation used before is not applicable. If $A(k)$ is such that the integral (7.6.17) can be evaluated exactly, then the problem is solved. If not, numerical integration is required. An alternative approach is to compute an approximate solution for large values of x or t using the method of stationary phase, introduced by Lord Kelvin in his studies of water waves. In addition to affording a fast computational tool, this method is important because it gives valuable information on the nature of the solution, something that could not be achieved if the solution were computed numerically.

The method of stationary phase provides approximate solutions to integrals of the form

$$I(a, b, \lambda) = \int_a^b f(x) e^{i\lambda g(x)} \, dx, \qquad \lambda > 0 \qquad (7.6.25)$$

when $\lambda \to \infty$.

If $g'(x) \neq 0$ in the closed interval $[a, b]$, the integral is of the order of λ^{-1} when λ goes to infinity. When $g'(x_o) = 0$ and $g''(x_o) \neq 0$ for x_o in $[a, b]$, then

$$I(a, b, \lambda) \approx \sqrt{\frac{2\pi}{\lambda |g''(x_o)|}} f(x_o) e^{i\lambda g(x_o) + i\frac{\pi}{4} \operatorname{sgn} g''(x_o)} \qquad (7.6.26)$$

(Bleistein, 1984; Segel, 1977). For any function $h(x)$, $\operatorname{sgn} h(x)$ is defined as in (6.5.43) with ω replaced by $h(x)$. When there is more than one stationary point in $[a, b]$, the contributions from all of the points should be added together.

These results are proved using modern analysis techniques, but were originally derived based on the observation that the waves in (7.6.17) interfere constructively only in the neighborhood of the point of stationary phase. Elsewhere the interference will tend to be destructive because of the phase differences (Havelock, 1914). A related argument is that if the exponential factor in the integrand in (7.6.25) oscillates more rapidly than $f(x)$, then the main contributions to the integral will arise from points in the neighborhood of the stationary value of $g(x)$. The following example addresses this point.

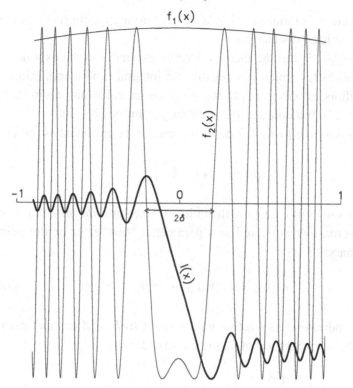

Fig. 7.8. Graphical illustration of the principle of stationary phase. $f_1(x)$ is a slowly varying function with a maximum at $x = 0$. $f_2(x)$ is the function $\cos[\lambda f_1(x)]$, were λ is a parameter. $I(x)$ is the integral of $f_2(x)$ (see (7.6.27)). Because $f_2(x)$ is highly oscillating, the change in the value of $I(x)$ is relatively small except in a neighborhood of $x = 0$.

Consider a function $f_1(x)$ having a stationary value at $x = x_o$, a function $f_2(x) = \cos[\lambda f_1(x)]$ computed for a large value of a parameter λ, and the integral $I(x)$ given by

$$I(x) = \int_{x_1}^{x} f_2(s) \, ds, \tag{7.6.27}$$

which will be computed numerically.

The functions $f_1(x)$ and $f_2(x)$ and $I(x)$ are shown in Fig. 7.8. Let $x_1 = -1$ and $x_o = 0$. The function $f_1(x)$ has a maximum at x_o and has little variation, while $f_2(x)$ is highly oscillating everywhere except in a neighborhood $(-\delta, \delta)$ of x_o. Consequently, its integral $I(x)$ has a small value between -1 and $-\delta$. As the integral is the sum of values of f_2, positive and negative values tend to cancel each other. Between $-\delta$ and δ there is less variation in the values of f_2, and the absolute value of the integral increases significantly. For larger values of x the cancellation takes place again, and the value of the integral does not change much.

If the integrand were $f_2(x)$ multiplied by a slowly varying function of x, the main features of $I(x)$ would be unchanged.

The integral in (7.6.25) has the factor λ written explicitly in the exponent. It is the presence of this factor which ensures that the integral can be approximated as shown. Some authors, however, apply the previous cancellation argument to the phase term in (7.6.17), and consider the stationary point of $\omega t - kx$.

To apply these ideas to equation (7.6.17) we rewrite the exponent as $t\phi(k)$, with

$$\phi(k) = \omega(k) - k\frac{x}{t}, \qquad (7.6.28)$$

and consider large values of t, keeping the ratio x/t fixed at some arbitrary, nonzero value. In other words, x/t is treated as a parameter. Now let k_o be the point for which ϕ is stationary:

$$\phi'(k_o) = \omega'(k_o) - \frac{x}{t} = 0, \qquad (7.6.29)$$

where the prime indicates a derivative with respect to k, and expand $\phi(k)$ in a Taylor series in the neighborhood of k_o to second order:

$$\phi(k) \approx \phi(k_o) + \frac{1}{2}\frac{d^2\phi}{dk^2}\bigg|_{k=k_o} (k - k_o)^2$$

$$= \phi(k_o) + \frac{1}{2}\omega''(k_o) (k - k_o)^2 = \phi_o + \frac{1}{2}\omega''_o (k - k_o)^2. \qquad (7.6.30)$$

As before, the meaning of the symbols on the right should be obvious from the context. Here it is assumed that $\omega''(k) \neq 0$. The case where this condition is not met is treated in the next section. Finally, because the main contributions to the integral come from points in the vicinity of k_o, $A(k)$ can be replaced by $A(k_o)$, which, being a constant, can be placed outside of the integral. Under this conditions (7.6.17) can be written as

$$f(x, t) = \frac{1}{2\pi} \int_{-\infty}^{\infty} A(k)e^{it\phi(k)}\, dk \approx \frac{1}{2\pi} A(k_o) \int_{-\infty}^{\infty} e^{it[\phi_o + \frac{1}{2}\omega''_o (k-k_o)^2]}\, dk$$

$$= \frac{1}{2\pi} A(k_o)e^{it\phi_o} \int_{-\infty}^{\infty} e^{i\frac{1}{2}t\omega''_o (k-k_o)^2}\, dk. \qquad (7.6.31)$$

Strictly speaking, the integral should be limited to a neighborhood of k_o, but because the value of the integral is assumed to be negligible outside of this interval, the integration limits can be extended as in (7.6.31). The integral on the right can be solved using either contour integration methods or a table of integrals (Problem 7.11). The final result is

$$f(x,t) \approx \frac{1}{\sqrt{2\pi t |\omega_o''|}} A(k_o) e^{it\phi_o} e^{i\frac{\pi}{4} \text{sgn}\, \omega_o''} = \frac{1}{\sqrt{2\pi t |\omega_o''|}} A(k_o) e^{i(\omega_o t - k_o x + \frac{\pi}{4} \text{sgn}\, \omega_o'')},$$

(7.6.32)

where $\omega_o = \omega(k_o)$.

The approximation (7.6.32) is valid for values of t that are large compared with the time scales of the problem. Furthermore, the error introduced by this approximation goes to infinity as ω_o'' goes to zero (Pekeris, 1948; Båth, 1968; Whitham, 1974), which means that (7.6.32) is not a good approximation in the vicinity of such a point. This fact can also be seen by the presence of ω_o'' in the denominator. As noted above, if there is more than one point of stationary phase, $f(x,t)$ will be the sum of expressions similar to those in the right-hand side of (7.6.32).

Note that (7.6.29) can be rewritten as

$$\omega'(k_o) = \frac{x}{t}.$$

(7.6.33)

This equation shows that k_o is a function of both x and t. Of course, this statement also applies to ω_o. Therefore, the phase term in (7.6.32) is a function solely of x and t. Let us write the phase as

$$\theta(x,t) = \omega(k)t - k(x,t)x.$$

(7.6.34)

When θ is a slowly varying function, the following definitions of *local* (or *instantaneous*) *frequency* (ω_l) and *local wavenumber* (k_l) are introduced:

$$\omega_l(x,t) = \frac{\partial \theta(x,t)}{\partial t} \equiv \theta_t = \frac{d\omega}{dk}\frac{\partial k}{\partial t}t + \omega - \frac{\partial k}{\partial t}x = \left(\frac{d\omega}{dk}t - x\right)\frac{\partial k}{\partial t} + \omega \quad (7.6.35)$$

and

$$k_l(x,t) = -\frac{\partial \theta(x,t)}{\partial x} \equiv -\theta_x = -\frac{d\omega}{dk}\frac{\partial k}{\partial x}t + \frac{\partial k}{\partial x}x + k = -\left(\frac{d\omega}{dk}t - x\right)\frac{\partial k}{\partial x} + k$$

(7.6.36)

(Segel, 1977; Pilant, 1979). In the case of a pure harmonic wave the phase is $\theta = \omega t - kx$, with ω and k constant. Using the definitions above we see that ω_l and k_l are just ω and k.

If k_o is a stationary point of θ, then (7.6.29) applies, and

$$\omega_l(x,t) = \omega_o$$

(7.6.37)

and

$$k_l(x,t) = k_o.$$

(7.6.38)

The derivative of sgn ω_o'' is Dirac's delta (see Appendix A), which is zero everywhere except at $\omega_o'' = 0$, but since this point has been explicitly excluded, the sgn ω_o'' term in (7.6.32) can be ignored.

Equations (7.6.37) and (7.6.38) can be interpreted in two ways. One is that in the vicinity of a given point (x_1, t_1), with $x_1 = \omega'(k_o)t_1$, the motion represented by (7.6.32) corresponds to a harmonic motion with frequency and wavenumber given by $\omega_o(x_1, t_1)$ and $k_o(x_1, t_1)$. The corresponding period and wavelength are $2\pi/\omega_o$ and $2\pi/k_o$, respectively. The second interpretation is that the wavenumber k_o will be found at points (x, t) that satisfy $x = \omega'(k_o)t$. In other words, $\omega'(k_o)$ is the velocity of propagation of a harmonic component having wavenumber k_o (and frequency $\omega(k_o)$). This interpretation is consistent with our previous definition of ω' as the velocity of a group of waves (see (7.6.15)), although in the present case the concept of group is not as well defined as in the case of narrow-band waves.

The concept of phase velocity introduced before also applies here. When the phase $\theta(x, t)$ is equal to a constant, then after differentiation with respect to t we obtain

$$\theta_x \frac{\mathrm{d}x}{\mathrm{d}t} + \theta_t = 0 \tag{7.6.39}$$

so that

$$\frac{\mathrm{d}x}{\mathrm{d}t} = -\frac{\theta_t}{\theta_x} = \frac{\omega_o}{k_o}. \tag{7.6.40}$$

In summary, equation (7.6.32) represents wave motion characterized by a group velocity $U(x, t)$ and a phase velocity $c(x, t)$ given by

$$U(x, t) = \left.\frac{\mathrm{d}\omega}{\mathrm{d}k}\right|_{k=k_o} = \frac{x}{t} \tag{7.6.41}$$

and

$$c(x, t) = \frac{\mathrm{d}x}{\mathrm{d}t} = \frac{\omega_o}{k_o}. \tag{7.6.42}$$

Because of (7.6.41), the factor ω_o'' in (7.6.32) can be replaced by $\mathrm{d}U/\mathrm{d}k$ computed at $k = k_o$. The existence of two velocities, U and c, and their dependence on x and t, continuously affects the shape of a dispersive wave train. As noted by Whitham (1974), an observer riding on a particular feature of the wave (say a peak) will move with the phase velocity, but will see the local frequency and wavenumber changing. On the other hand, an observer that moves with the group velocity will always see the same local frequency and wavenumber, while peaks and troughs will pass by him.

The group velocity is also important because this is the velocity with which energy propagates. A general analysis of this question in the context of the elastic solid is given by Biot (1957). A more restricted but still useful analysis is provided by Achenbach (1973).

For the surface waves, once the phase velocity as a function of frequency has been obtained (by solving the appropriate period equation) the group velocity can be obtained from (7.6.15) using numerical differentiation, but because of the errors this approach introduces, other methods have been introduced (e.g., Aki and Richards, 1980; Ben-Menahem and Singh, 1981). For the special case of a layer over a half-space there is a closed equation for Love waves (Problem 7.13), while for Rayleigh waves an analytical method has been proposed (Mooney and Bolt, 1966).

7.6.3.1 *Example. The dispersive string*

The method of stationary phase will be applied to the dispersive string (see (7.6.1)). From the preceding discussions we see that the velocity c given by (7.6.5) is the phase velocity. Furthermore, by combining (7.6.3) and (7.6.4) an explicit expression for $\omega(k)$ can be found:

$$\omega(k) = \pm\sqrt{v^2 k^2 + a^2}. \tag{7.6.43}$$

Note that there are two values of ω for each value of k. For the time being we will use the positive solution, but at the end will also account for the negative one.

The group velocity can then be determined using (7.6.41) and (7.6.43):

$$U = \frac{d\omega}{dk} = \frac{v^2 k}{\sqrt{v^2 k^2 + a^2}} = \frac{v^2}{c} = \frac{v^2 k}{\omega}. \tag{7.6.44}$$

For the last two equalities, (7.6.5) and (7.6.3) were used. As noted before, c approaches v as k and ω go to infinity, which means that U also approaches v in the limit. Furthermore, because $c > v$, $U < v < c$.

Using (7.6.33), (7.6.41), and (7.6.44), the point k_o of stationary phase is determined by solving

$$U = \frac{v^2 k_o}{\sqrt{v^2 k_o^2 + a^2}} = \frac{x}{t}, \tag{7.6.45}$$

which gives

$$k_o = \pm\frac{ax}{v\sqrt{v^2 t^2 - x^2}}. \tag{7.6.46}$$

Again, we only use the positive root and consider the negative one at the end.

Equations (7.6.44)–(7.6.46) can be used to determine ω_o:

$$\omega_o = \frac{v^2 t k_o}{x} = \frac{vta}{\sqrt{v^2 t^2 - x^2}}. \tag{7.6.47}$$

Also needed is

$$\omega'' \equiv \frac{d^2\omega}{dk^2} = \frac{dU}{dk} = \frac{d}{dk}\left(\frac{v^2k}{\omega}\right) = \frac{\omega^2 v^2 - v^4 k^2}{\omega^3} = \frac{v^2 a^2}{\omega^3}. \qquad (7.6.48)$$

Here (7.6.44) and (7.6.43) were used. Using (7.6.47) we obtain ω'' at the point of stationary phase:

$$\omega_o'' = \frac{(v^2 t^2 - x^2)^{3/2}}{v t^3 a}. \qquad (7.6.49)$$

Because we will take $vt > x$ (see below), $\omega_o'' > 0$ and sgn $\omega_o'' = 1$. Then, using (7.6.46), and (7.6.47), the phase term in (7.6.32) can be written as

$$\omega_o t - k_o x + \frac{\pi}{4} = \frac{a}{v}\sqrt{v^2 t^2 - x^2} + \frac{\pi}{4}. \qquad (7.6.50)$$

As a further simplification we will assume again that $A(k)$ is equal to 1. Now we must account for the negative root in (7.6.46). When k_o is negative, so is ω_o, as can be seen from the first equality in (7.6.47), and ω_o'', as can be seen from (7.6.48) with $\omega = \omega_o$. Furthermore, sgn $\omega_o'' = -1$. Therefore, the only change in the phase term in (7.6.32) is a sign change. As $A(k) = A(-k)$, the contributions from $+k_o$ and $-k_o$ are two complex conjugate quantities, which combined result in a cosine term. When this fact is taken into account the final solution is given by

$$f(x, t) \approx \sqrt{\frac{2va}{\pi}} \frac{t}{(v^2 t^2 - x^2)^{(3/4)}} \cos\left(\frac{a}{v}\sqrt{v^2 t^2 - x^2} + \frac{\pi}{4}\right) \qquad (7.6.51)$$

(Problem 7.14). An important feature of this solution is that when vt approaches x, the amplitude of $f(x, t)$ goes to infinity. This corresponds to the presence of a wave front (Havelock, 1914) moving with velocity v. When $t < x/v$, $f = 0$. Also note that the dependence of the phase on x and t is nonlinear. This nonlinearity is what causes the changes in the local frequency and wavenumber that characterize dispersion. Using (7.6.35), (7.6.36), and (7.6.51) we find

$$\omega_l = \frac{vta}{\sqrt{v^2 t^2 - x^2}} \qquad (7.6.52)$$

and

$$k_l = \frac{ax}{v\sqrt{v^2 t^2 - x^2}}, \qquad (7.6.53)$$

in agreement with (7.6.47) and (7.6.46). These equations show that when vt is close to x (i.e., near the wave front) both ω_l and k_l are large, so that the period and wavelength are small.

Equation (7.6.51) is plotted in Fig. 7.9 as a function of time for several fixed, equispaced, values of x. Each plot will be referred to as a trace. This figure summarizes everything we have said regarding dispersion. The thin lines across the

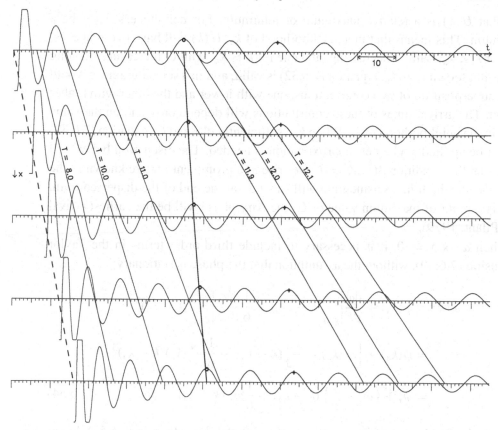

Fig. 7.9. Waves corresponding to a dispersive string (see (7.6.51)). Each subplot has been generated for a fixed value of x. The different values of x are equispaced. The dashed line indicates the arrival time of the wave front. The labeled lines have the same local period T. Circles and pluses identify the same feature. The solid bold line joins the points on the time axes corresponding to the circles. Note the change in local period as a function of x.

traces join the times having the same local periods (indicated by T, and computed using $T = 2\pi/\omega_l$). All the features along each of these lines travel with the corresponding group velocity, given by (7.6.41). As noted above, the concept of a group of waves in the presence of dispersion is not always well defined and this fact is made clear by this example. If we call the cycle in the top trace between about $T = 11.9$ and 12.1 a group, then we see that it expands as the distance increases.

7.6.4 *The Airy phase*

As discussed earlier, equation (7.6.32) is not applicable when $\omega_o'' = 0$, or, equivalently, $dU/dk = 0$. Let us assume first that this situation occurs for a point $k_a \neq 0$, and that $U(k)$ is a relative maximum or minimum. For definiteness, let it be a minimum. This means that in a neighborhood of k_a, $U(k)$ will have two branches, one for $k < k_a$, with U decreasing, and one for $k > k_a$, with U increasing. As long as k is not too close to k_a, equation (7.6.32) is valid, and in a seismogram we would see a superposition of two contributions, one with lower and the other with higher period. The arrival times of these contributions will depend on x and on the value of U in the neighborhood of k_a. When k is sufficiently close to k_a, equation (7.6.32) cannot be applied, and a new approximation is needed. The discussion below will show that the condition $dU/dk = 0$ gives rise to a prominent feature known as the *Airy phase*, which in a seismogram will show up at the end of the dispersed train. If k_a is a point of maximum value of U, the Airy phase will be the earliest arrival (see Pilant, 1979).

When $\omega''(k_a) = 0$, it is necessary to include third-order terms in the Taylor expansion (7.6.30), without the assumption that the phase is stationary:

$$\phi(k) \approx \phi(k_a) + \left.\frac{d\phi}{dk}\right|_{k=k_a} (k - k_a) + \left.\frac{1}{6}\frac{d^3\phi}{dk^3}\right|_{k=k_a} (k - k_a)^3$$

$$= \phi(k_a) + \left[\omega'(k_a) - \frac{x}{t}\right] (k - k_a) + \frac{1}{6}\omega'''(k_a)(k - k_a)^3$$

$$= \phi_a + \left(\omega_a' - \frac{x}{t}\right)(k - k_a) + \frac{1}{6}\omega_a'''(k - k_a)^3. \tag{7.6.54}$$

As k_a is the fixed value of k for which $\omega''(k) = 0$, the derivatives in (7.6.54) do not depend on x or t. Using (7.6.54), (7.6.17) can be written as

$$f(x, t) \approx \frac{1}{2\pi} A(k_a) e^{it\phi_a} \int_{-\infty}^{\infty} e^{i[(t\omega_a' - x)(k - k_a) + t\omega_a'''(k - k_a)^3/6]} \, dk. \tag{7.6.55}$$

The integral above will be rewritten in terms of the Airy function Ai:

$$\mathrm{Ai}(z) = \frac{1}{\pi} \int_0^{\infty} \cos\left(uz + \frac{1}{3}u^3\right) du = \frac{1}{2\pi} \int_{\infty}^{\infty} e^{i(uz + u^3/3)} \, du \tag{7.6.56}$$

(Båth, 1968, Ben-Menahem and Singh, 1981). The second equality arises because the imaginary part of the integral on the right is an odd function of u, so that its contribution is zero. This function is plotted in Fig. 7.10.

Let I indicate the integral on the right-hand side in (7.6.55). To transform I into an Airy function two steps are needed. First, make the change of variable $u = k - k_a$ and let

$$b = t\omega_a' - x; \qquad c = \frac{1}{2}t\omega_a'''. \tag{7.6.57}$$

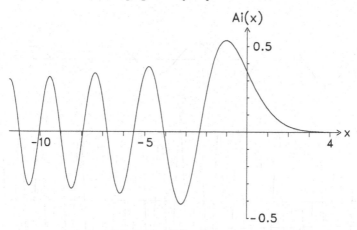

Fig. 7.10. The Airy function (see (7.6.56)).

This gives

$$I = \frac{1}{2\pi} \int_{\infty}^{\infty} e^{i(ub+cu^3/3)} \, du. \tag{7.6.58}$$

Next, make the change of variable $s = |c|^{1/3} u \operatorname{sgn} c$. Now I becomes

$$I = \frac{1}{|c|^{1/3}} \operatorname{Ai} \left(\frac{b \operatorname{sgn} c}{|c|^{1/3}} \right) \tag{7.6.59}$$

(Problem 7.15). When this result is introduced in (7.6.55) we have

$$f(x, t) = A(k_a) \frac{1}{|c|^{1/3}} e^{i(\omega_a t - k_a x)} \operatorname{Ai} \left(\frac{b \operatorname{sgn} c}{|c|^{1/3}} \right). \tag{7.6.60}$$

Writing b and c in full gives

$$f(x, t) = \frac{A(k_a) \sqrt[3]{2}}{\sqrt[3]{t|\omega_a'''|}} e^{i(\omega_a t - k_a x)} \operatorname{Ai} \left(\frac{\sqrt[3]{2} \, (t\omega_a' - x) \operatorname{sgn} \omega_a'''}{\sqrt[3]{t|\omega_a'''|}} \right). \tag{7.6.61}$$

The right-hand side of (7.6.61) represents the Airy phase, which corresponds to a harmonic wave modulated by an Airy function. Note that for points near $x = \omega_a' t$, the amplitude of the Airy phase depends on $t^{-1/3}$, while for the other portions of the dispersive train the amplitude depends on $t^{-1/2}$ (see (7.6.32)). Therefore, for large t (and x) the Airy phase will be the dominant feature (provided that $A(k_a)$ is not too small). Examples of Airy phases in seismograms can be found in Kulhanek (1990).

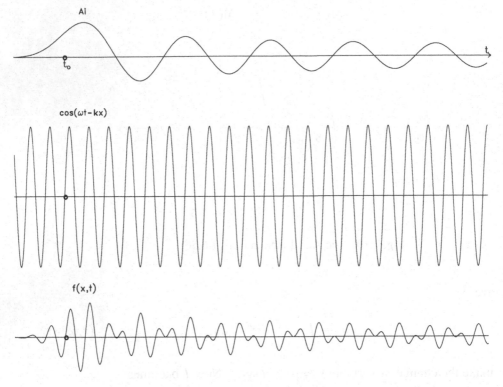

Fig. 7.11. Example of Airy phase. It corresponds to a harmonic wave modulated by an Airy function. The final result is the function $f(x, t)$ (see (7.6.61)). t_o indicates the expected arrival time. After Savage (1969).

Figure 7.11 shows a plot of $f(x, t)$. Here it is assumed that x is fixed and t is variable, and that $U(k_a)$ has a maximum value, which in turn means that ω_a''' is negative. As U is a maximum, the time of the first arrival should be $t_o = x/U(k_a)$, but as the figure shows, the Airy phase has values for times less than t_o. The presence of this precursor was discussed by Tolstoy (1973), who noted that this fact arises from the approximate nature of (7.6.61). If $U(k_a)$ had been a minimum, a tail would have followed the expected last arrival.

Now we will consider two special cases. One is $k_a = 0$. This implies that $\omega_a = 0$. Introducing these values in (7.6.61) gives

$$f(x, t) = \frac{A(0)\sqrt[3]{2}}{\sqrt[3]{t|\omega_a'''|}} \mathrm{Ai}\left(\frac{\sqrt[3]{2}\,(t\omega_a' - x)\,\mathrm{sgn}\,\omega_a'''}{\sqrt[3]{t|\omega_a'''|}} \right). \qquad (7.6.62)$$

This case corresponds to infinite wavelength and period, a situation that occurs near the wave front (Munk, 1949). The significance of (7.6.62) is that when $k_a = 0$, $f(x, t)$ is the Airy function itself, so that the wavelength (or period) of the oscilla-

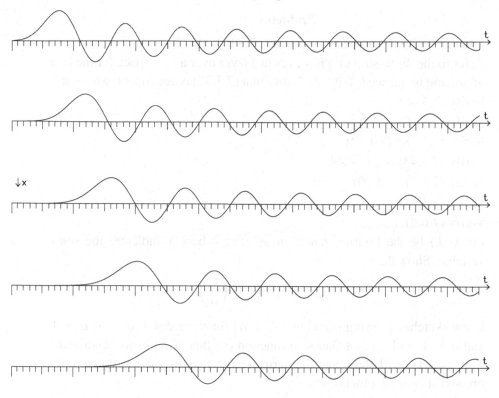

Fig. 7.12. Special case of the Airy phase. The Airy function becomes the function $f(x,t)$ (see (7.6.62)), shown for several equispaced values of x.

tions depends only on the argument of the Airy function. Because of the factor $t^{1/3}$ in the denominator, the period will increase with distance (Fig. 7.12).

The second case is $k_a = k_o$, with k_o the point of stationary phase described above. This situation occurs when $\omega(k)$ has a point of inflection at k_o. Under this condition, the quantity b (see (7.6.57)) is equal to zero. Therefore, from (7.6.61)

$$f(x,t) = \frac{\mathrm{Ai}(0) A(k_o) \sqrt[3]{2}}{\sqrt[3]{t|U_o''|}} \, \mathrm{e}^{\mathrm{i}(\omega_o t - k_o x)}, \qquad (7.6.63)$$

where $\mathrm{Ai}(0) = 3^{-2/3}/\Gamma(2/3) = 0.355$ (Båth, 1968; Ben-Menahem and Singh, 1981), Γ represents the gamma function, and U_o'' is $\mathrm{d}^2 U/\mathrm{d}k^2 = \omega'''(k)$ computed at $k = k_o$. The same result can be obtained by putting $b = 0$ in (7.6.58), separating the integral into real and imaginary parts, recognizing that the imaginary part gives zero, and solving the remaining integral using tabulated results.

Problems

7.1 Show that (7.3.8) can be derived from (7.3.10) and (7.3.11).

7.2 Refer to the discussion of SH waves in a layer over a half-space for the case of grazing incidence (§6.9.1.2). Verify that (7.3.32) is equivalent to $\theta = m\pi$.

7.3 Verify (7.3.46).

7.4 Verify (7.4.10).

7.5 Verify (7.4.38)–(7.4.42).

7.6 Verify (7.4.43) and (7.4.44).

7.7 Verify (7.5.7)–(7.5.10).

7.8 Verify (7.5.11).

7.9 Verify (7.6.4).

7.10 Let $G(k)$ be the Fourier transform of $g(x)$, where x indicates the space variable. Show that

$$\mathcal{F}\{g(ax)\} = \frac{1}{|a|}G\left(\frac{k}{a}\right).$$

Draw sketches showing $g(ax)$ and $G(k/a)$ for three cases: $a = 1$, $a < 1$, and $a > 1$, with $g(x)$ a Gaussian function (so that $G(k)$ is also Gaussian). This exercise will show that the narrower the function in one domain, the broader it is in the other domain.

7.11 Verify (7.6.32).

7.12 Let the phase factor in (7.6.17) include a phase term $\psi(k)$ coming, for example, from the source, the instrument, or propagation effects. Show that in this case (7.6.32) becomes

$$f(x,t) \approx \frac{1}{\sqrt{2\pi t|\phi_o''|}}A(k_o)e^{i[\omega_o t - k_o x + \psi(k_o) + \frac{1}{4}\pi\,\text{sgn}\,\phi_o'']}$$

where

$$\phi_o'' = \omega''(k_o) + \frac{1}{t}\psi''(k_o)$$

(after Pilant, 1979).

7.13 Using (7.3.20) and (7.6.15) show that for the case of Love waves in a layer over a half-space the group velocity is given by

$$U = \frac{\beta_1^2}{c}\left[\frac{(c^2/\beta_1^2) + \Omega}{1 + \Omega}\right],$$

where

$$\Omega = kH\gamma_2\left[\frac{\rho_1}{\rho_2}\left(\frac{c^2 - \beta_1^2}{\beta_2^2 - \beta_1^2}\right) + \frac{\mu_2}{\mu_1}\left(\frac{\beta_2^2 - c^2}{\beta_2^2 - \beta_1^2}\right)\right]$$

(Ben-Menahem and Singh, 1981). Also show that U goes to β_1 and β_2 as c goes to β_1 and β_2, respectively.

7.14   Verify (7.6.51).

7.15   Verify (7.6.59).

8

Ray theory

8.1 Introduction

The elastic wave equation admits exact solutions in relatively few cases involving simple variations in elastic properties. When these properties vary in a two- or three-dimensional way, the equation cannot be solved exactly and either numerical (e.g., finite-difference, finite-element) or approximate solutions must be sought. Ray theory is one of the possible approaches introduced to solve the equation approximately. This theory is traditionally associated with optics, where it originated (for reviews see, e.g., Kline and Kay, 1965; Cornbleet, 1983; Stavroudis, 1972). The extension of the theory to the propagation of electromagnetic waves is due to Luneburg (1964) (work done in the 1940s) while the application to elastic waves is due to Karal and Keller (1959), although simpler ray-theoretic concepts had been used earlier (Cerveny *et al.*, 1977). Russian authors also contributed to the solution of the elastic problem (see, e.g., Cerveny and Ravindra, 1971).

Over the last three decades elastic wave ray theory has grown enormously in scope and complexity and for this reason in this chapter only the most fundamental aspects will be discussed. Two important topics not addressed here are numerical solutions to the ray equations and the considerably more difficult problem of computing amplitudes. The former is well treated by Lee and Stewart (1981), and the latter by Cerveny (2001), who present a thorough discussion of ray theory and includes an extensive reference list. Our starting point will be the 3-D scalar wave equation, which allows the introduction of the main ideas behind ray theory, followed by consideration of the elastic wave equation. Although some books start with the assumption that ray theory applies independently to P and S waves, it is necessary to show that this separation is actually possible. This will be done here and then the general properties of rays, which are curves in space, will be investigated using concepts from differential geometry. The results thus obtained are completely general and apply to any kind of ray, i.e., seismic, acoustic, or electromagnetic. Another topic of importance is Fermat's principle, which will be proved using ray theory and results from variational calculus. The last major topic is an analysis of ray amplitudes for the *P* and *S* waves. The treatment is fairly straightforward for the *P* waves, but is much more complicated for the *S* waves. The most important result here is that there is a coordinate system that allows the separation of the *S*-wave vector into two vectors for which the amplitudes satisfy equations similar to that satisfied by the *P*-wave vector, which results in significant computational savings when amplitudes are computed. This coordinate system also constitutes an essential element in the so-called *dynamic ray tracing*, introduced by Cerveny and Hron (1980). The chapter closes with two examples. One is the effect of a thin low-velocity surface, similar to that treated in §6.9.1.3, although here we concentrate on the normal incidence case. The second example shows synthetic seismograms for a simple 1-D velocity model. Although the actual numerical implementation of the theory is not discussed, this example illustrates certain aspects of the ray method that the reader should be aware of, such as the presence of noncausual arrivals and the effect of distance on wave amplitudes.

8.2 Ray theory for the 3-D scalar wave equation

Let the wave velocity c be a function of position $\mathbf{x}$. We will work in the frequency domain, in which the case the scalar wave equation can be written as

$$u_{,jj} = -\frac{\omega^2}{c^2} u \tag{8.2.1}$$

(see (5.7.3)) with

$$u = u(\mathbf{x}, \omega); \qquad c = c(\mathbf{x}). \tag{8.2.2}$$

Equations (8.2.1) and (8.2.2) follow from (5.7.1) and (5.7.2) with $\mathbf{u}$ replaced by u. We will try to solve (8.2.1) using a solution of the form

$$u(\mathbf{x}, \omega) = A(\mathbf{x}, \omega) e^{i\omega\phi(\mathbf{x})}, \tag{8.2.3}$$

where the amplitude $A(\mathbf{x})$ and phase function $\phi(\mathbf{x})$ are to be determined.

Using (8.2.3), $u_{,j}$ and $u_{,jj}$ are given by

$$u_{,j} = \left(A_{,j} + iA\omega\phi_{,j} \right) e^{i\omega\phi} \tag{8.2.4}$$

and

$$u_{,jj} = \left(A_{,jj} + 2iA_{,j}\omega\phi_{,j} + iA\omega\phi_{,jj} - A\omega^2\phi_{,j}\phi_{,j} \right) e^{i\omega\phi}. \tag{8.2.5}$$

Introducing (8.2.3) and (8.2.5) in (8.2.1), canceling the exponential factor, dividing both sides by $\omega^2 A$ and rearranging gives

$$\left(\phi_{,j}\phi_{,j} - \frac{1}{c^2}\right) - \frac{i}{\omega}\left(\frac{2}{A}A_{,j}\phi_{,j} + \phi_{,jj}\right) - \frac{1}{\omega^2 A}A_{,jj} = 0. \qquad (8.2.6)$$

Because of the presence of the $1/\omega$ and $1/\omega^2$ factors in the last two terms in (8.2.6) it may appear that the first term will be dominant when ω is very large. In such a case the last two terms can be neglected and (8.2.6) can be written as

$$\phi_{,j}\phi_{,j} = |\nabla\phi|^2 = \frac{1}{c^2}. \qquad (8.2.7)$$

This is the well-known *eikonal* equation of optics. However, before asserting that (8.2.7) is generally valid one should ask whether this *high-frequency* approximation is appropriate for the problem represented by (8.2.1). For example, equation (8.2.1) is used to investigate the propagation of acoustic waves in the oceans and in the air. However, the range of frequencies involved in either case may be quite different. Therefore, what is high frequency in one case may not be so in another. In addition, it is conceivable that some of the partial derivatives in the last two terms may become so large as to compensate the effect of large values of ω. For this reason, to study the validity of (8.2.7) it is necessary to go through an order-of-magnitude analysis (Officer, 1974). To do that we will compare the magnitudes of the first and third terms of (8.2.6). To simplify the comparison it is convenient to use ϕ', ϕ'', A', A'', and c' to represent the derivatives ϕ_j, ϕ_{jj}, A_j, A_{jj}, and $c_{,j}$ (in absolute value).

From the first and third terms of (8.2.6) we see that (8.2.7) is valid when

$$\frac{1}{\omega^2}\frac{A''}{A} \ll (\phi')^2 \approx \frac{1}{c^2}. \qquad (8.2.8)$$

This implies

$$\frac{c^2}{\omega^2}\frac{A''}{A} \ll 1 \qquad (8.2.9)$$

or

$$\lambda^2\frac{A''}{A} \ll 1. \qquad (8.2.10)$$

Now approximate A'' by $\Delta A'/\Delta L$, where L indicates length, and use λ in place of ΔL. Then (8.2.10) becomes

$$\frac{\delta A'}{A} \ll \frac{1}{\lambda}, \qquad (8.2.11)$$

where $\delta A'$ is the change in A' over the distance of a wavelength. Equation (8.2.10) will now be used to derive a relation involving c and $\delta c'$, with δ having the same meaning as before. From the second term in (8.2.6) we expect

$$\frac{A'}{A}\phi' \approx \phi'' \tag{8.2.12}$$

or

$$\frac{A'}{A} \approx \frac{\phi''}{\phi'} \approx \frac{(1/c)'}{(1/c)} = -\frac{c'}{c}. \tag{8.2.13}$$

Then, ignoring signs,

$$\left(\frac{A'}{A}\right)' = -\frac{(A')^2}{A^2} + \frac{A''}{A} \approx \left(\frac{c'}{c}\right)' = -\frac{(c')^2}{c^2} + \frac{c''}{c}. \tag{8.2.14}$$

Since from (8.2.13) we see that $(A'/A)^2 \approx (c'/c)^2$, multiplying (8.2.14) by λ^2 and using (8.2.10) gives

$$\lambda^2 \frac{c''}{c} \approx \lambda^2 \frac{A''}{A} \ll 1. \tag{8.2.15}$$

From (8.2.15), using the same argument that led to (8.2.11) gives

$$\delta c' \ll \frac{c}{\lambda} = \frac{1}{2\pi}\omega. \tag{8.2.16}$$

Therefore, the fractional change $\delta c'$ must be much smaller than the frequencies involved.

We finish this discussion of the scalar wave equation by noting that the fact that the second term on the right-hand side of (8.2.6) has been neglected does not mean that this term does not provide useful information. On the contrary, it will be used to derive expressions for the amplitude term $A(\mathbf{x})$ (see §8.7.1).

8.3 Ray theory for the elastic wave equation

We will start with the equation of motion with body forces equal to zero written as follows:

$$\left(c_{ijkl}u_{k,l}\right)_{,j} = \rho \ddot{u}_i, \tag{8.3.1}$$

which is obtained from (4.2.5), (4.5.1), and (2.4.1) and the symmetry relation (4.5.3b). The tensor c_{ijkl} will be allowed to be a function of $\mathbf{x}$. Although here we will be concerned mostly with isotropic media, the analysis is sometimes simpler when the anisotropic case is considered.

We will look for solutions to (8.3.1) of the form

$$\mathbf{u}(\mathbf{x}, t) = \mathbf{U}(\mathbf{x}) f(t - T(\mathbf{x})) \tag{8.3.2}$$

(Cerveny, 1985) where $\mathbf{U}$ and T must be determined under the condition that (8.3.1) is satisfied. As shown below, T can be interpreted as the travel time along rays (to be defined). The surfaces $T(\mathbf{x}) =$ constant are known as wave fronts.

Note that if the Fourier transform is applied to (8.3.2) we obtain

$$\mathbf{u}(\mathbf{x}, \omega) = \mathbf{U}(\mathbf{x})e^{-i\omega T(\mathbf{x})} f(\omega). \tag{8.3.3}$$

Therefore, in the frequency domain $T(\mathbf{x})$ is a phase function and plays a role similar to that of the function $\phi(\mathbf{x})$ in (8.2.3). Furthermore, because (8.3.1) is not dispersive, a pulse in the time domain does not change shape as the wave propagates (Burridge, 1976) and for that reason it is possible to use as trial solutions either (8.2.3) or (8.3.2), which lead to the same set of ray equations (see Hudson (1980) for the equations obtained using (8.2.3)). Also note that (8.3.2) differs from the trial solution given in Aki and Richards (1980) by an interchange of the arguments of $\mathbf{U}$ and f. In either case, these solutions represent the first term of more general solutions written as infinite series. These series solutions are discussed in detail by Karal and Keller (1959), Cerveny *et al.* (1977), Cerveny and Ravindra (1971), Cerveny and Hron (1980), and Cerveny (1985), among others. However, solution (8.3.2) is sufficient to generate most of the ray theoretical results needed in seismological practice and is known as the *zeroth-order solution*, which corresponds to the geometrical optics approximation. Higher-order terms are required to solve problems that cannot be addressed using geometrical optics concepts. For example, they have been used to study diffraction problems (Keller *et al.*, 1956) and head waves (Cerveny and Ravindra, 1971).

To simplify the following equations, c_{ijkl} will be temporarily replaced by c. Then, introducing (8.3.2) in the left-hand side of (8.3.1) gives

$$\left(\mathsf{c}(U_k f)_{,l}\right)_{,j} = \mathsf{c}_{,j}(U_k f)_{,l} + \mathsf{c}(U_k f)_{,lj} = \mathsf{c}_{,j}U_{k,l}f + \mathsf{c}_{,j}U_k f_{,l}$$

$$+ \mathsf{c}U_{k,lj}f + \mathsf{c}U_{k,l}f_{,j} + \mathsf{c}U_{k,j}f_{,l} + \mathsf{c}U_k f_{,lj}. \tag{8.3.4}$$

We need the following relations:

$$f_{,l} = \frac{\partial f(t - T(\mathbf{x}))}{\partial x_l} = \frac{\partial f}{\partial(t - T(\mathbf{x}))}\frac{\partial(t - T(\mathbf{x}))}{\partial x_l} = -\dot{f}\frac{\partial T}{\partial x_l} = -\dot{f}T_{,l} \tag{8.3.5}$$

$$f_{,lj} = \ddot{f}T_{,l}T_{,j} - \dot{f}T_{,lj} \tag{8.3.6}$$

$$\rho\ddot{u}_i = \rho U_i\ddot{f}, \tag{8.3.7}$$

where the dots represent derivatives with respect to the argument. Using (8.3.4)–(8.3.7), (8.3.1) can be rewritten as

$$\left(\mathsf{c}U_k T_{,l}T_{,j} - \rho U_i\right)\ddot{f} - \left(\mathsf{c}_{,j}U_k T_{,l} + \mathsf{c}U_{k,j}T_{,l} + \mathsf{c}U_k T_{,lj} + \mathsf{c}U_{k,l}T_{,j}\right)\dot{f}$$

$$+ \left(\mathsf{c}_{,j}U_{k,l} + \mathsf{c}U_{k,lj}\right)f = 0 \tag{8.3.8}$$

(Problem 8.1).

Now we will introduce the assumption that near the wave front f is much larger than $|U|$ and c_{ijkl} and their derivatives, and $\ddot{f} \gg \dot{f} \gg f$ (Aki and Richards, 1980). This is a high-frequency approximation similar to that used in §8.2 because in the frequency domain $\dot{f}$ and $\ddot{f}$ are proportional to $\omega f(\omega)$ and $\omega^2 f(\omega)$ (see Problem 9.13), respectively. Under these conditions only the first term in (8.3.8) is significant and the other two terms can be neglected. However, at this point one should carry out an analysis similar to that in §8.2 to establish under which conditions it is possible to neglect the last two terms in (8.3.8). According to Cerveny *et al.* (1977) and Cerveny (2001) the ray method is valid when the wavelength λ of interest is much smaller than other characteristic quantities having the dimension of length, such as measures of the inhomogeneity of the medium (e.g. $c/|\nabla c|$) and the radius of curvature of boundaries. The method also fails when the length L of the ray is too large. This condition is expressed as $\lambda \ll l^2/L$, where l is a characteristic length similar to those referred to above. Finally, the method also fails in the vicinity of surfaces such as caustics (see §8.7.1). These conditions are equivalent to those derived for the scalar wave equation (see also Ben-Menahem and Beydoun, 1985). As noted for the scalar wave equation, the second term in (8.3.8) will be used to derive amplitude relations (see §8.7.2).

Under the approximations just discussed, equation (8.3.8) becomes

$$c_{ijkl}T_{,l}T_{,j}U_k - \rho U_i = \left(c_{ijkl}T_{,l}T_{,j} - \rho \delta_{ik}\right)U_k = 0. \qquad (8.3.9)$$

The factor $\ddot{f}$ was canceled because it is generally different from zero.

If we now introduce a matrix $\boldsymbol{\Gamma}$ with elements Γ_{ik} given by

$$\Gamma_{ik} = \frac{1}{\rho}c_{ijkl}T_{,l}T_{,j} \qquad (8.3.10)$$

then from the left-hand side of (8.3.9) equal to zero we obtain

$$\Gamma_{ik}U_k = U_i \qquad (8.3.11)$$

or

$$\boldsymbol{\Gamma}\mathbf{U} = \mathbf{U}, \qquad (8.3.12)$$

which shows that $\mathbf{U}$ is an eigenvector of $\boldsymbol{\Gamma}$ with eigenvalue equal to 1. The symmetric matrix $\boldsymbol{\Gamma}$ (Problem 8.2) is known as the *Christoffel matrix* and is important in the analysis of anisotropic media (Cerveny *et al.*, 1977; Cerveny, 1985; Auld, 1990).

8.3.1 *P and S waves in isotropic media*

Replacing c_{ijkl} by its expression for isotropic media (see (4.6.2)), Γ_{ik} becomes

$$\Gamma_{ik} = \frac{1}{\rho} \left[\lambda \delta_{ij} \delta_{kl} + \mu (\delta_{ik} \delta_{jl} + \delta_{il} \delta_{jk}) \right] T_{,l} T_{,j}$$

$$= \frac{\lambda}{\rho} T_{,i} T_{,k} + \frac{\mu}{\rho} (\delta_{ik} T_{,j} T_{,j} + T_{,i} T_{,k}) = \frac{(\lambda + \mu)}{\rho} T_{,i} T_{,k} + \frac{\mu}{\rho} |\nabla T|^2 \delta_{ik}. \quad (8.3.13)$$

To solve the eigenvalue problem (8.3.12) we start by setting to zero the determinant

$$D = |\Gamma_{ik} - \delta_{ik}| = \left| \left(\frac{\mu}{\rho} |\nabla T|^2 - 1 \right) \delta_{ik} + \frac{(\lambda + \mu)}{\rho} T_{,i} T_{,k} \right|$$

$$= |B \delta_{ik} + C T_{,i} T_{,k}| = 0, \quad (8.3.14)$$

where

$$B = \frac{\mu}{\rho} |\nabla T|^2 - 1; \qquad C = \frac{(\lambda + \mu)}{\rho}. \quad (8.3.15a,b)$$

It is straightforward to show that

$$D = B^2 \left(B + C |\nabla T|^2 \right) = 0 \quad (8.3.16)$$

(Problem 8.3), which means that either $B = 0$ (double root), so that

$$|\nabla T|^2 = \frac{\rho}{\mu} \quad (8.3.17)$$

or the factor in parentheses is equal to zero. From this possibility we obtain

$$\mu |\nabla T|^2 - \rho + (\lambda + \mu) |\nabla T|^2 = 0, \quad (8.3.18)$$

which, in turn, implies that

$$|\nabla T|^2 = \frac{\rho}{\lambda + 2\mu}. \quad (8.3.19)$$

These two equations can be rewritten as

$$|\nabla T|^2 = T_{,i} T_{,i} = \left(\frac{\partial T}{\partial x_1} \right)^2 + \left(\frac{\partial T}{\partial x_2} \right)^2 + \left(\frac{\partial T}{\partial x_3} \right)^2 = \frac{1}{c^2}, \quad (8.3.20)$$

where

$$c = c(\mathbf{x}) = \sqrt{\frac{\lambda + 2\mu}{\rho}} \equiv \alpha(\mathbf{x}) \qquad \text{or} \qquad c(\mathbf{x}) = \sqrt{\frac{\mu}{\rho}} \equiv \beta(\mathbf{x}). \quad (8.3.21a,b)$$

Equation (8.3.20) with c given by (8.3.21) is the eikonal equation for an isotropic elastic medium. Because λ, μ, and ρ depend on $\mathbf{x}$, c, α, and β are functions of $\mathbf{x}$. If the medium is homogeneous, α and β are the *P*- and *S*-wave velocities introduced in (4.8.5). For this reason the waves propagating with velocities $\alpha(\mathbf{x})$ and $\beta(\mathbf{x})$ are known as P and S waves, respectively. Although we will not be concerned with anisotropic media here, it is worth noting that in such media three types of waves are possible, one quasi-compressional and two quasi-shear (the latter two are generally independent of each other) (Cerveny *et al.*, 1977; Cerveny, 2001).

Next, we investigate the direction of motion of the P and S waves. To address this question we start with (8.3.11) rewritten as

$$(\Gamma_{ik} - \delta_{ik}) U_k = 0. \tag{8.3.22}$$

Using (8.3.13), equation (8.3.22) becomes

$$\left[\frac{1}{\rho} (\mu |\nabla T|^2 - \rho) \delta_{ik} + \frac{(\lambda + \mu)}{\rho} T_{,i} T_{,k} \right] U_k = 0. \tag{8.3.23}$$

To find the eigenvectors $\mathbf{U}$ first replace $|\nabla T|^2$ by its expression for the S waves, given by (8.3.17). When this is done the first term of (8.3.23) becomes zero and we find

$$\frac{(\lambda + \mu)}{\rho} T_{,i} T_{,k} U_k = \frac{(\lambda + \mu)}{\rho} (\mathbf{U} \cdot \nabla T) T_{,i} = 0. \tag{8.3.24}$$

Now contract (8.3.24) with $T_{,i}$ (see §1.4.1). This gives

$$\frac{(\lambda + \mu)}{\rho} (\mathbf{U} \cdot \nabla T) |\nabla T|^2 = \frac{(\lambda + \mu)}{\mu} (\mathbf{U} \cdot \nabla T) = 0, \tag{8.3.25}$$

where (8.3.17) was used again. Equation (8.3.25) implies

$$(\mathbf{U} \cdot \nabla T) = 0. \tag{8.3.26}$$

Because (8.3.17) corresponds to a double root, equation (8.3.26) shows that two of the eigenvectors, which indicate the direction of S-wave motion, are perpendicular to ∇T. As noted in §1.4.6, the two eigenvectors can be chosen to be perpendicular to each other. These vectors will be considered further in §8.7.2.2.

To find the eigenvector corresponding to the P wave introduce (8.3.19) into (8.3.23), which gives

$$\left[\frac{(\lambda + \mu)}{\rho} T_{,i} T_{,k} - \frac{(\lambda + \mu)}{\lambda + 2\mu} \delta_{ik} \right] U_k = \frac{(\lambda + \mu)}{\rho} (\mathbf{U} \cdot \nabla T) T_{,i} - \frac{(\lambda + \mu)}{\lambda + 2\mu} U_i = 0 \tag{8.3.27}$$

and then multiply vectorially with $T_{,w}$. The vth component of the resulting vector is

$$\epsilon_{viw} \left[\frac{(\lambda + \mu)}{\rho} (\mathbf{U} \cdot \nabla T) T_{,i} T_{,w} - \frac{(\lambda + \mu)}{\lambda + 2\mu} U_i T_{,w} \right] = -\frac{(\lambda + \mu)}{\lambda + 2\mu} (\mathbf{U} \times \nabla T)_v = 0. \tag{8.3.28}$$

The first term on the left-hand side includes the vector product of ∇T with itself, which is equal to zero. Then from (8.3.28) we see that

$$(\mathbf{U} \times \nabla T) = 0, \tag{8.3.29}$$

which indicates that P-wave motion is parallel to ∇T.

8.4 Wave fronts and rays

A *wave front* moving through space can be represented by

$$t = T(\mathbf{x}) + C, \tag{8.4.1}$$

where C is a constant (Hudson, 1980) Therefore, the same wave front at time $t + dt$ has the equation

$$t + dt = T(\mathbf{x} + d\mathbf{x}) + C, \tag{8.4.2}$$

where $d\mathbf{x}$ is the distance that the wave front moved in time dt. Now expand $T(\mathbf{x} + d\mathbf{x})$ in a Taylor series to first order

$$T(\mathbf{x} + d\mathbf{x}) \simeq T(\mathbf{x}) + \sum_{i=1}^{3} \frac{\partial T}{\partial x_i} dx_i = T(\mathbf{x}) + \nabla T \cdot d\mathbf{x}. \tag{8.4.3}$$

Introducing (8.4.3) in (8.4.2) and subtracting (8.4.1) from the resulting equation gives

$$dt = \nabla T \cdot d\mathbf{x}. \tag{8.4.4}$$

Now let $\mathbf{v}$ be the velocity of the wave front in the direction $d\mathbf{x}$, equal to

$$\mathbf{v} = \frac{d\mathbf{x}}{dt}. \tag{8.4.5}$$

Therefore, dividing (8.4.4) by dt gives

$$1 = \nabla T \cdot \mathbf{v} = |\nabla T| \, |\mathbf{v}| \cos \theta, \tag{8.4.6}$$

where θ is the angle between ∇T and $\mathbf{v}$. If $\mathbf{v}$ is parallel to ∇T, then

$$|\nabla T| \, |\mathbf{v}| = 1 = |\nabla T| c, \tag{8.4.7}$$

where the last equality comes from (8.3.20). Therefore, $|\mathbf{v}| = c$. Furthermore, because ∇T is perpendicular to T (as a general property of the gradient), we see that c is the velocity of the wave front in a direction perpendicular to itself.

In the case of isotropic media, *rays* are defined as curves whose tangents are everywhere perpendicular to a wave front (Hudson, 1980). To describe a ray we will use a parametric representation. Let

$$\mathbf{r}(u) = (x_1(u), x_2(u), x_3(u)) \tag{8.4.8}$$

be the position vector (with respect to a fixed origin) of a point on the ray, with u a parameter that varies along the ray. Because $d\mathbf{r}/du$ is a vector tangent to $\mathbf{r}$ (see, e.g., Spiegel, 1959), it is parallel to ∇T, so that

$$\frac{d\mathbf{r}}{du} = g\nabla T, \tag{8.4.9}$$

where g is a proportionality function that depends on the choice of u and that generally will depend on $\mathbf{r}$. Consider first the case $u = s$, where s is the arc length (or distance) along the ray measured with respect to a fixed point on the ray. To find g first solve (8.4.9) for ∇T

$$\nabla T = \frac{1}{g}\frac{d\mathbf{r}}{ds} = \frac{1}{g}\left(\frac{dx_1}{ds}, \frac{dx_2}{ds}, \frac{dx_3}{ds}\right) \tag{8.4.10}$$

and then introduce this expression in the eikonal equation (8.3.20). This gives

$$|\nabla T|^2 = \nabla T \cdot \nabla T = \frac{1}{g^2}\left[\left(\frac{dx_1}{ds}\right)^2 + \left(\frac{dx_2}{ds}\right)^2 + \left(\frac{dx_3}{ds}\right)^2\right] = \frac{1}{g^2} = \frac{1}{c^2}. \tag{8.4.11}$$

The second equality from the right arises because $ds^2 = dx_1^2 + dx_2^2 + dx_3^2$. Therefore, $g = c$ and

$$\frac{d\mathbf{r}}{ds} = c\nabla T \tag{8.4.12}$$

(Cerveny and Ravindra, 1971). Note that

$$\left|\frac{d\mathbf{r}}{ds}\right| = c|\nabla T| = 1 \tag{8.4.13}$$

as expected because $|d\mathbf{r}| = ds$.

If $u = t$, where t is time, using $dt = ds/c$ and (8.4.12) we obtain

$$\frac{d\mathbf{r}}{dt} = c\frac{d\mathbf{r}}{ds} = c^2\nabla T. \tag{8.4.14}$$

The relation between T and t can be found from

$$\frac{dT}{dt} = \frac{\partial T}{\partial x_1}\frac{dx_1}{dt} + \frac{\partial T}{\partial x_2}\frac{dx_2}{dt} + \frac{\partial T}{\partial x_3}\frac{dx_3}{dt} = \nabla T \cdot \frac{d\mathbf{r}}{dt} = \nabla T \cdot c^2\nabla T = 1, \tag{8.4.15}$$

which means that T and t have the same variation and that T can be interpreted as travel time along the ray (Aki and Richards, 1980).

We will now write (8.4.12) in such a way that T does not appear in the equation. This will produce a differential equation for the space variables only. First, rewrite (8.4.12) in component form:

$$T_{,i} = \frac{1}{c}\frac{dx_i}{ds} \tag{8.4.16}$$

and then take the derivative with respect to s

$$\frac{dT_{,i}}{ds} = \frac{d}{ds}\left(\frac{1}{c}\frac{dx_i}{ds}\right). \tag{8.4.17}$$

From the eikonal equation

$$T_{,i}T_{,i} = \frac{1}{c^2} \tag{8.4.18}$$

(see (8.3.20)) we obtain

$$\frac{d}{ds}(T_{,i}T_{,i}) = 2T_{,i}\frac{dT_{,i}}{ds} = \frac{d}{ds}\left(\frac{1}{c^2}\right) = \frac{-2}{c^3}\frac{\partial c}{\partial x_i}\frac{dx_i}{ds} = \frac{-2}{c^3}\frac{\partial c}{\partial x_i}cT_{,i} = \frac{-2}{c^2}T_{,i}\frac{\partial c}{\partial x_i}, \tag{8.4.19}$$

where (8.4.16) was used. From (8.4.19) we obtain

$$T_{,i}\left(\frac{dT_{,i}}{ds} + \frac{1}{c^2}\frac{\partial c}{\partial x_i}\right) = T_{,i}\left[\frac{d}{ds}\left(\frac{1}{c}\frac{dx_i}{ds}\right) + \frac{1}{c^2}\frac{\partial c}{\partial x_i}\right] = 0, \tag{8.4.20}$$

where (8.4.17) was used.

Contracting (8.4.20) with $T_{,i}$ gives a nonzero factor $1/c^2$ (see (8.4.18)), which means that the expression in parentheses in (8.4.20) must be zero. In addition,

$$\frac{1}{c^2}\frac{\partial c}{\partial x_i} = -\frac{\partial}{\partial x_i}\left(\frac{1}{c}\right) = -\left[\nabla\left(\frac{1}{c}\right)\right]_i. \tag{8.4.21}$$

Combining these results and writing in vector form gives

$$\frac{d}{ds}\left(\frac{1}{c}\frac{d\mathbf{r}}{ds}\right) = \nabla\left(\frac{1}{c}\right). \tag{8.4.22}$$

Equation (8.4.22) will be used with three simple velocity distributions.

8.4.1 Medium with constant velocity

In this case the gradient of $1/c$ is zero and (8.4.22) becomes

$$\frac{d^2\mathbf{r}}{ds^2} = \mathbf{0}. \tag{8.4.23}$$

The solution is

$$\mathbf{r} = \mathbf{a}s + \mathbf{b}, \tag{8.4.24}$$

where $\mathbf{a}$ and $\mathbf{b}$ are constant vectors. Writing the vectors in component form and then solving for s gives

$$s = \frac{x_I - b_I}{a_I}; \qquad I = 1, 2, 3 \tag{8.4.25}$$

(no summation over uppercase indices), or

$$\frac{x_1 - b_1}{a_1} = \frac{x_2 - b_2}{a_2} = \frac{x_3 - b_3}{a_3}. \tag{8.4.26}$$

Equations (8.4.24) or (8.4.26) represent a straight line in 3-D space with direction given by the vector $\mathbf{a}$ passing through a point with coordinates (b_1, b_2, b_3).

To determine the equation of the wave fronts use (8.4.12) and (8.4.24)

$$\frac{d\mathbf{r}}{ds} = \mathbf{a} = c\nabla T, \tag{8.4.27}$$

which means that

$$(\nabla T)_i = \frac{\partial T}{\partial x_i} = \frac{a_i}{c}. \tag{8.4.28}$$

Because of (8.4.13), equation (8.4.27) implies that $|\mathbf{a}| = 1$. A particular solution of (8.4.28) is

$$T = \frac{1}{c}(a_1 x_1 + a_2 x_2 + a_3 x_3) = \frac{1}{c}(\mathbf{r} \cdot \mathbf{a}). \tag{8.4.29}$$

The equation $(\mathbf{r} \cdot \mathbf{a}) = $ constant represents a plane with unit normal $\mathbf{a}$. Therefore, the rays are perpendicular to the wave fronts, as expected, and the corresponding waves are the familiar plane waves.

Another solution can be obtained if $\mathbf{b}$ in (8.4.24) is set equal to zero. In this case the rays are lines through the origin, i.e., they are lines in radial directions. Using $dr = |d\mathbf{r}| = ds$, equation (8.4.27) becomes

$$\frac{d\mathbf{r}}{dr} = \mathbf{e}_r = c\nabla T = c\frac{dT}{dr}\mathbf{e}_r, \tag{8.4.30}$$

where $\mathbf{e}_r$ is the unit vector in the radial direction (vector $\boldsymbol{\Gamma}$ in Fig. 9.10). Equation (8.4.30) implies

$$dT = \frac{dr}{c}, \tag{8.4.31}$$

which, in turn, gives

$$T = \frac{r}{c} \tag{8.4.32}$$

(assuming $T = 0$ at the origin). This solution corresponds to spherical waves. Using

$$r = |\mathbf{r}| = (x_1^2 + x_2^2 + x_3^2)^{1/2} \tag{8.4.33}$$

it is easy to show that r/c satisfies the eikonal equation:

$$\nabla T = \nabla \left(\frac{r}{c}\right) = \frac{1}{rc}(x_1, x_2, x_3) = \frac{1}{rc}\mathbf{r}. \tag{8.4.34}$$

Therefore

$$|\nabla T| = \frac{1}{c}. \tag{8.4.35}$$

8.4.2 Medium with a depth-dependent velocity

In this case $c(\mathbf{r}) = c(x_3)$. We will show that the quantity

$$\mathbf{Q} = \mathbf{e}_3 \times \nabla T = \mathbf{e}_3 \times \frac{1}{c}\frac{d\mathbf{r}}{ds} \tag{8.4.36}$$

(see (8.4.12)) is a constant along rays. To see that, take the derivative with respect to s and use (8.4.22)

$$\frac{d\mathbf{Q}}{ds} = \mathbf{e}_3 \times \frac{d}{ds}\left(\frac{1}{c}\frac{d\mathbf{r}}{ds}\right) = \mathbf{e}_3 \times \nabla\left(\frac{1}{c}\right) = 0 \tag{8.4.37}$$

because $\nabla(1/c)$ is parallel to $\mathbf{e}_3$ (as c is a function of x_3 only). Equation (8.4.37) implies that rays are planar curves parallel to the x_3 axis. To see that expand the vector product in (8.4.36) using $\mathbf{e}_3 = (0, 0, 1)$. This gives

$$\mathbf{e}_3 \times \frac{1}{c}\frac{d\mathbf{r}}{ds} = -\frac{1}{c}\left(\frac{dx_2}{ds}\mathbf{e}_1 - \frac{dx_1}{ds}\mathbf{e}_2\right) = \mathbf{Q}. \tag{8.4.38}$$

As the right-hand side of (8.4.38) is a constant vector, we can write

$$\frac{1}{c}\frac{dx_i}{ds} = \text{constant}; \qquad i = 1, 2. \tag{8.4.39}$$

If the x_1 axis is chosen such that the initial direction of the ray is in the (x_1, x_3) plane, then $dx_2/ds = 0$ and the ray stays in the (x_1, x_3) plane (Hudson, 1980).

If we now compute the absolute value of $\mathbf{Q}$ we obtain

$$|\mathbf{Q}| = |\mathbf{e}_3||\nabla T| \sin i\,(x_3) = \frac{\sin i\,(x_3)}{c\,(x_3)} = \text{constant}, \tag{8.4.40}$$

where $i\,(x_3)$ is the angle (Fig. 8.1) between $\mathbf{e}_3$ and the tangent to the ray (given by ∇T). The angle i at the source is known as the *take-off angle*. It is customary to use p to indicate the constant in (8.4.40) and to call it the *ray parameter*. Then (8.4.40) becomes

$$p = \frac{\sin i\,(x_3)}{c\,(x_3)}, \tag{8.4.41}$$

which is similar to the Snell's law encountered in Chapter 6. It must be noted, however, that here c is a function without discontinuities. When they exist a different treatment is necessary (see §8.7.3).

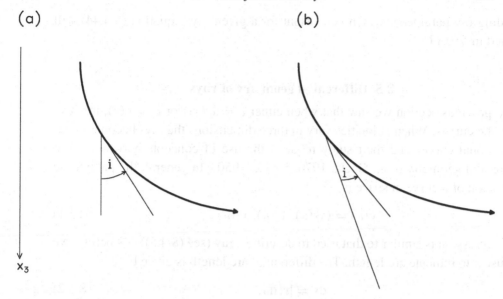

Fig. 8.1. Rays for two particular velocity models. (a) $c = c(x_3)$; (b) $c = c(r)$. In both cases the rays are planar curves. In (a) i is the angle between the tangent to the ray and the x_3 direction and satisfies $\sin i/c = $ constant. In (b) i is the angle between the tangent to the ray and the radial direction and satisfies $r \sin i/c = $ constant.

8.4.3 Medium with spherical symmetry

This kind of medium is also known as radially heterogeneous. To a first approximation, the earth is an example of such a medium. In this case $c = c(r)$ and the quantity

$$\mathbf{Q} = \mathbf{r} \times \nabla T = \mathbf{r} \times \frac{1}{c} \frac{d\mathbf{r}}{ds} \tag{8.4.42}$$

is constant along the ray. To see that, take the derivative of (8.4.42) with respect to s and use (8.4.22):

$$\frac{d\mathbf{Q}}{ds} = \frac{d\mathbf{r}}{ds} \times \frac{1}{c} \frac{d\mathbf{r}}{ds} + \mathbf{r} \times \frac{d}{ds}\left(\frac{1}{c}\frac{d\mathbf{r}}{ds}\right) = \mathbf{r} \times \nabla\left(\frac{1}{c}\right) = \mathbf{0}. \tag{8.4.43}$$

The first and third vector products are equal to zero because they involve pairs of parallel vectors (the gradient of $1/c$ is in the direction of $\mathbf{r}$). The fact that $\mathbf{Q}$ is a constant implies that each ray lies in a vertical plane (see §8.6). Taking the absolute value of $\mathbf{Q}$ gives

$$|\mathbf{Q}| = r|\nabla T|\sin i(r) = \frac{r \sin i(r)}{c(r)} = p, \tag{8.4.44}$$

where $i(r)$ is the angle between the tangent to the ray and $\mathbf{r}$ (Fig. 8.1). Equation (8.4.44) is Snell's law for a medium with spherical symmetry and p is the corre-

sponding ray parameter, which is constant for a given ray. Equation (8.4.44) will
be used in §10.11.

8.5 Differential geometry of rays

In the previous section we saw that when either $c = c(x_3)$ or $c = c(r)$, the rays
are plane curves. When velocities vary in three dimensions the rays become three-
dimensional curves and their study requires the use of concepts borrowed from
differential geometry (e.g., Goetz, 1970; Struik, 1950). In general, the parametric
expression of a curve in space is

$$\mathbf{r}(u) = (x_1(u), x_2(u), x_3(u)). \tag{8.5.1}$$

This expression is similar to that used to describe a ray (see (8.4.8)). As before, we
will use s to indicate arc length. The differential arc length is given by

$$ds = |\dot{\mathbf{r}}| \, du, \tag{8.5.2}$$

where the dot indicates a derivative with respect to the parameter:

$$\dot{\mathbf{r}} = \frac{d\mathbf{r}}{du}. \tag{8.5.3}$$

The unit vector *tangent* to the curve, to be indicated by $\mathbf{t}$, is given by

$$\mathbf{t} = \frac{d\mathbf{r}}{ds} = c\nabla T. \tag{8.5.4}$$

The last equality in (8.5.4) follows from (8.4.12) and is valid for rays only. In terms
of $\mathbf{r}$, $\mathbf{t}$ can be written as

$$\mathbf{t} = \frac{d\mathbf{r}}{ds} = \frac{d\mathbf{r}}{du}\frac{du}{ds} = \frac{\dot{\mathbf{r}}}{|\dot{\mathbf{r}}|}. \tag{8.5.5}$$

Now we will introduce two unit vectors perpendicular to $\mathbf{t}$. Since $\mathbf{t}$ is a unit
vector, $\mathbf{t} \cdot \mathbf{t} = 1$ and

$$\frac{d}{ds}(\mathbf{t} \cdot \mathbf{t}) = 2\mathbf{t} \cdot \frac{d\mathbf{t}}{ds} = 0 \tag{8.5.6}$$

(Problem 8.4), which means that $d\mathbf{t}/ds$ is perpendicular to $\mathbf{t}$. This allows introduction of the so-called *principal normal vector*, given by

$$\mathbf{n} = \frac{1}{\kappa}\frac{d\mathbf{t}}{ds}, \tag{8.5.7}$$

where

$$\kappa(s) = \left|\frac{d\mathbf{t}}{ds}\right| \tag{8.5.8}$$

is known as the *curvature*. Note that the orientation of $d\mathbf{t}/ds$ is determined by the

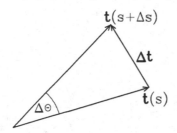

Fig. 8.2. Geometric interpretation of the curvature. The vectors $\mathbf{t}(s + \Delta s)$ and $\mathbf{t}(s)$ are the tangent vectors at two points on a curve a distance Δs apart. The two vectors are referred to a common origin and the angle between them is $\Delta \theta$. The curvature κ is the limit of $\Delta \theta / \Delta s$ as Δs goes to zero.

curve, while $\mathbf{n}$ has two possible orientations (Struik, 1950). By choosing κ positive, $\mathbf{n}$ and $d\mathbf{t}/ds$ have the same orientation, which points in the direction where the curve is concave (Goetz, 1970). These considerations have more than an academic interest; near the end of this section they will be used to show that rays bend toward regions of lower velocity.

To give a geometric interpretation to the curvature consider two points on the curve represented by the vectors $\mathbf{r}(s)$ and $\mathbf{r}(s+\Delta s)$. The corresponding unit tangent vectors are $\mathbf{t}(s)$ and $\mathbf{t}(s + \Delta s)$. Let $\Delta \theta$ be the angle between these two vectors referred to a common origin (Fig. 8.2). As Δs goes to zero $\Delta \theta$ approaches $|\mathbf{t}(s + \Delta s) - \mathbf{t}(s)|$ (Problem 8.5). Therefore,

$$\lim_{|\Delta s| \to 0} \frac{\Delta \theta}{|\Delta s|} = \lim_{|\Delta s| \to 0} \frac{|\mathbf{t}(s + \Delta s) - \mathbf{t}(s)|}{|\Delta s|} = \left| \frac{d\mathbf{t}}{ds} \right| = \kappa. \tag{8.5.9}$$

From this definition we see that the particular case of $\kappa = 0$ corresponds to a straight line (see also Problem 8.6). Another useful way to look at the curvature is to consider the *osculating circle*, which is the circle with radius $1/\kappa$ and center along $\mathbf{n}$ a distance $1/\kappa$ from the curve. Therefore, for any given point on the curve there is a circle with the same curvature as the curve. Clearly, the curve and the osculating circle have at least one point in common. The quantity $1/\kappa$ is known as the *radius of curvature*.

The second normal vector perpendicular to $\mathbf{t}$, known as the *binormal vector* and indicated by $\mathbf{b}$, is defined as the vector product

$$\mathbf{b} = \mathbf{t} \times \mathbf{n}. \tag{8.5.10}$$

The vector $\mathbf{b}$ has unit length and the vectors $\mathbf{t}$, $\mathbf{n}$, and $\mathbf{b}$ form a right-handed system known as the *Frenet* or *moving trihedral*.

Next, we investigate the variation of $\mathbf{b}$ and $\mathbf{n}$ with respect to s. Since $(\mathbf{b} \cdot \mathbf{t}) = 0$,

(a)   (b)

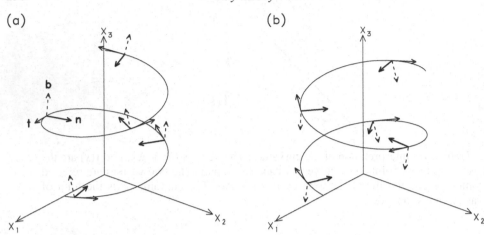

Fig. 8.3. Graph of (a) the right-handed helix $(a \cos u, a \sin u, b\, u)$ and (b) the left-handed helix $(a \cos u, -a \sin u, b\, u)$, where a and b are positive constants. The vectors **t**, **n**, and **b** are shown for several points. The vector **n**, in particular, is parallel to the (x_1, x_2) plane. A cylindrical standard screw corresponds to a right-handed helix. The difference between the two helices is in the sign of the torsion, which is positive when the helix is right-handed and negative otherwise. See also Problem 8.8.

taking the derivative with respect to s gives

$$\frac{d}{ds}(\mathbf{b} \cdot \mathbf{t}) = \frac{d\mathbf{b}}{ds} \cdot \mathbf{t} + \frac{d\mathbf{t}}{ds} \cdot \mathbf{b} = \frac{d\mathbf{b}}{ds} \cdot \mathbf{t} + \kappa\mathbf{n} \cdot \mathbf{b} = \frac{d\mathbf{b}}{ds} \cdot \mathbf{t} = 0. \qquad (8.5.11)$$

This means that $d\mathbf{b}/ds$ is perpendicular to **t**. In addition, from an argument similar to that used to obtain (8.5.6) we find that $d\mathbf{b}/ds$ is also perpendicular to **b**. Therefore, $d\mathbf{b}/ds$ must be parallel to **n** and we can write

$$\frac{d\mathbf{b}}{ds} = -\tau(s)\mathbf{n}, \qquad (8.5.12)$$

where τ is known as the *torsion* of the curve and

$$|\tau(s)| = \left|\frac{d\mathbf{b}}{ds}\right|. \qquad (8.5.13)$$

Note that unlike the curvature, which is positive, the torsion can be positive or negative. Curves for which the torsion is positive or negative are said to be *right-handed* or *left-handed*. Figure 8.3 shows these two possibilities. The geometric interpretation of the torsion is similar to that of the curvature, with $\Delta\theta$ in (8.5.9) the angle between binormal vectors and the limit equal to τ (Goetz, 1970).

To find $d\mathbf{n}/ds$ use $\mathbf{n} = \mathbf{b} \times \mathbf{t}$ and (8.5.7), (8.5.10), and (8.5.12)

$$\frac{d\mathbf{n}}{ds} = \mathbf{b} \times \frac{d\mathbf{t}}{ds} + \frac{d\mathbf{b}}{ds} \times \mathbf{t} = -\kappa\mathbf{t} + \tau\mathbf{b}. \qquad (8.5.14)$$

Equations (8.5.7), (8.5.12), and (8.5.14) are known as Frenet–Serret formulas. An application of these formulas is given in §8.7.2.2.

An important vector linking $\mathbf{t}$ and $\mathbf{b}$ is the *Darboux vector*, defined as

$$\mathbf{d} = \tau\mathbf{t} + \kappa\mathbf{b} \tag{8.5.15}$$

and having the property that

$$\mathbf{d} \times \mathbf{t} = \frac{d\mathbf{t}}{ds}; \qquad \mathbf{d} \times \mathbf{n} = \frac{d\mathbf{n}}{ds}; \qquad \mathbf{d} \times \mathbf{b} = \frac{d\mathbf{b}}{ds} \tag{8.5.16}$$

(Problem 8.9).

The Darboux vector will be used to investigate the rotation of the Frenet trihedral $(\mathbf{t}, \mathbf{n}, \mathbf{b})$ around the ray. Let $\mathbf{r}$ be an arbitrary vector referred to this trihedral. Then

$$\mathbf{r} = r_1\mathbf{t} + r_2\mathbf{n} + r_3\mathbf{b}, \tag{8.5.17}$$

where r_i are independent of s. Multiplying the three equations (8.5.16) by r_1, r_2, and r_3, respectively, and then adding the corresponding results gives

$$\mathbf{d} \times \mathbf{r} = r_1\frac{d\mathbf{t}}{ds} + r_2\frac{d\mathbf{n}}{ds} + r_3\frac{d\mathbf{b}}{ds} = \frac{d\mathbf{r}}{ds}. \tag{8.5.18}$$

When $s = t$, $d\mathbf{r}/ds$ is the velocity vector $\mathbf{v}$ and (8.5.18) becomes

$$\mathbf{v} = \frac{d\mathbf{r}}{ds} = \mathbf{d} \times \mathbf{r} \tag{8.5.19}$$

(Rey Pastor *et al.*, 1957). This equation is similar to

$$\mathbf{v} = \boldsymbol{\omega} \times \mathbf{r}, \tag{8.5.20}$$

which gives the velocity of a point of a rigid body with position vector $\mathbf{r}$ that rotates about an axis with angular velocity ω. The rotation is characterized by the *angular velocity vector* $\boldsymbol{\omega}$ and has absolute value equal to ω (e.g., Arya, 1990; Davis and Snider, 1991). The corresponding geometry is shown in Fig. 8.4. Then, comparison of (8.5.19) and (8.5.20) shows that the Darboux vector represents the instantaneous rotation of the Frenet trihedral. From (8.5.15) we see that when τ is equal to zero,

$$\mathbf{d} \cdot \mathbf{t} = \kappa\mathbf{b} \cdot \mathbf{t} = 0, \tag{8.5.21}$$

which means that $\mathbf{d}$ is perpendicular to $\mathbf{t}$, so that $\mathbf{d}$ has no component in the direction of $\mathbf{t}$, which in turn means that there is no rotation about the ray (Lewis, 1966). We will use this fact in §8.7.2.2.

So far we have considered general properties of curves. Now we will derive a result of importance in seismology. To simplify the derivation we will introduce the *slowness*, which is defined as

$$S = \frac{1}{c}. \tag{8.5.22}$$

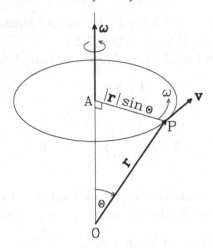

Fig. 8.4. Rigid-body rotation and angular velocity vector. A point P rotates about an axis with constant angular velocity ω. The tangential velocity is $v = |AP|\omega = |\mathbf{r}|\omega \sin\theta$. The angular velocity is $\mathbf{v} = \boldsymbol{\omega} \times \mathbf{r}$, where $\boldsymbol{\omega}$ is the angular velocity vector, with $|\boldsymbol{\omega}| = \omega$, and $|\mathbf{v}| = v$. After Davis and Snieder (1991).

Then, using (8.4.22) and (8.5.4) and (8.5.7) we obtain

$$\frac{d}{ds}(S\mathbf{t}) = \nabla S = \mathbf{t}\frac{dS}{ds} + S\frac{d\mathbf{t}}{ds} = \mathbf{t}\frac{dS}{ds} + \kappa S\mathbf{n} \qquad (8.5.23)$$

so that

$$\mathbf{n} = \frac{1}{\kappa S}\left(\nabla S - \mathbf{t}\frac{dS}{ds}\right). \qquad (8.5.24)$$

Contracting (8.5.24) with $\mathbf{n}$ gives

$$\mathbf{n} \cdot \mathbf{n} = 1 = \frac{1}{\kappa S}\nabla S \cdot \mathbf{n} \qquad (8.5.25)$$

because $\mathbf{t}$ is perpendicular to $\mathbf{n}$. Therefore,

$$\kappa = \frac{1}{S}\nabla S \cdot \mathbf{n}. \qquad (8.5.26)$$

Writing the gradient in component form and using velocity again we have

$$\frac{1}{S}(\nabla S)_i = c\frac{\partial}{\partial x_i}\left(\frac{1}{c}\right) = -\frac{1}{c}\frac{\partial c}{\partial x_i} = -\frac{1}{c}(\nabla c)_i = -[\nabla(\ln c)]_i. \qquad (8.5.27)$$

Then, from (8.5.26) and (8.5.27) we obtain

$$\kappa = -\frac{1}{c}\nabla c \cdot \mathbf{n} = -\nabla(\ln c) \cdot \mathbf{n}. \qquad (8.5.28)$$

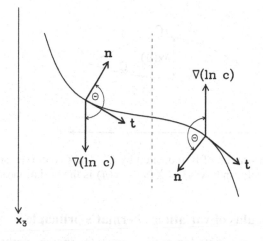

Fig. 8.5. Ray in a medium in which $c = c(x_3)$. The velocity increases in the direction of x_3 to the left of the dashed line and decreases to the right. The angle θ must satisfy (8.5.29). Consequently, rays bend in the direction of decreasing velocity.

Because k is generally positive (i.e., is not zero), this equation implies that the rays bend toward regions of lower velocity. To see that write (8.5.28) as follows:

$$-|\mathbf{n}| \, |\nabla(\ln c)| \cos \theta = \kappa > 0, \tag{8.5.29}$$

where θ is the angle between $\mathbf{n}$ and $\nabla(\ln c)$. To satisfy (8.5.29) θ must be between $90°$ and $270°$. Fig. 8.5 shows this situation for the case of depth-dependent velocity.

Finally we will derive an expression for $\mathbf{b}$ in terms of $\mathbf{t}$ and κ. First, use (8.5.4) and (8.5.22) to write

$$S\mathbf{t} = \nabla T \tag{8.5.30}$$

and then apply the curl operation to both sides of (8.5.30). Using (1.4.61) we obtain

$$\nabla \times (S\mathbf{t}) = S\nabla \times \mathbf{t} + \nabla S \times \mathbf{t} = \mathbf{0}. \tag{8.5.31}$$

Introducing in (8.5.31) the expression for ∇S given in (8.5.23) we obtain

$$S\nabla \times \mathbf{t} + S\kappa \mathbf{n} \times \mathbf{t} = \mathbf{0} \tag{8.5.32}$$

so that

$$\nabla \times \mathbf{t} = -\kappa \mathbf{n} \times \mathbf{t} = \kappa \mathbf{t} \times \mathbf{n} = \kappa \mathbf{b} \tag{8.5.33}$$

(Ben-Menahem and Singh, 1981).

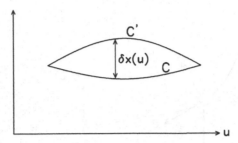

Fig. 8.6. A curve C and a variation of it, indicated by C'. The two curves are represented by $x(u)$ and $X(u)$, respectively, and $\delta x(u) = X(u) - x(u)$ is the variation of x.

8.6 Calculus of variations. Fermat's principle

This well-known principle originated in optics and states that raypaths are paths of stationary travel time. The principle can be proved using concepts from variational calculus, which was devised to find stationary values of definite integrals when the integrand includes an unknown function. The final product is the Euler equations, which are differential equations involving the unknown function (e.g., Båth, 1968; Lanczos, 1970). The derivation given below is less general but proceeds from first principles.

Consider a curve C between two points and let C' be a curve obtained by an arbitrary infinitesimal modification of C keeping the end points fixed (Fig. 8.6). The curves C and C' will be represented by $x(u)$ and $X(u)$, respectively. Examples of x and u are the x_i and u introduced in (8.5.1). The variation δx of x is defined as

$$\delta x(u) = X(u) - x(u) \qquad (8.6.1)$$

so that

$$X(u) = x(u) + \delta x(u). \qquad (8.6.2)$$

Note that δx is an arbitrary function of u. Next, we will consider the variation of expressions involving functions of $x(u)$.

(1) *Variation of* dx/du.

$$\delta \frac{dx}{du} = \frac{dX}{du} - \frac{dx}{du} = \frac{d}{du}(\delta x). \qquad (8.6.3)$$

Equation (8.6.3) shows that δ and the derivative operator d/du commute.
(2) *Variation of* dx.

$$\delta(dx) = dX - dx = d(x + \delta x) - dx = d(\delta x). \qquad (8.6.4)$$

Therefore, dx and δx also commute. It is important to note, however, a major difference between the two: dx is the change in x due to a change du in the independent variable, while δx is a change in x that generates a new function X (Lanczos, 1970).

(3) *Variation of a definite integral of $x(u)$.* Let

$$I[x] = \int_a^b x(u)\, du. \tag{8.6.5}$$

Note that the argument of I is a function but that I is a number. The variation of I is given by

$$\delta I = I[X] - I[x] = \int_a^b X(u)\, du - \int_a^b x(u)\, du$$

$$= \int_a^b [X(u) - x(u)]\, du = \int_a^b \delta x\, du. \tag{8.6.6}$$

Therefore,

$$\delta I[x] = I[\delta x]. \tag{8.6.7}$$

(4) *Variation of a definite integral of $f(x(u))$.* Let

$$I[f(x)] = \int_a^b f(x(u))\, du. \tag{8.6.8}$$

Then, proceeding as before,

$$\delta I = I[f(X)] - I[f(x)] = \int_a^b [f(X) - f(x)]\, du = \int_a^b \delta f(x)\, du. \tag{8.6.9}$$

Therefore,

$$\delta I[f(x)] = I[\delta f(x)]. \tag{8.6.10}$$

(5) *Variation of $f(x_1, x_2, x_3)$.* Here each of the x_i is a function of u having a variation given by (8.6.1) with x replaced by x_i,

$$\delta f = f(X_1, X_2, X_3) - f(x_1, x_2, x_3)$$

$$= f(x_1 + \delta x_1, x_2 + \delta x_2, x_3 + \delta x_3) - f(x_1, x_2, x_3). \tag{8.6.11}$$

After expanding the first term on the right-hand side in a Taylor series to first order, equation (8.6.11) becomes

$$\delta f = \frac{\partial f}{\partial x_i} \delta x_i = \nabla f \cdot \delta \mathbf{r}. \tag{8.6.12}$$

where

$$\delta \mathbf{r} = (\delta x_1, \delta x_2, \delta x_3). \tag{8.6.13}$$

(6) *Variation of* $g(dx_1, dx_2, dx_3)$. An argument similar to that used to derive (8.6.12) shows that

$$\delta g = \frac{\partial g}{\partial (dx_i)} \delta(dx_i). \tag{8.6.14}$$

For example, if

$$g = ds = \sqrt{dx_1^2 + dx_2^2 + dx_3^2} \tag{8.6.15}$$

then

$$\delta(ds) = \frac{1}{ds} dx_i \delta(dx_i) = \frac{1}{ds} d\mathbf{r} \cdot \delta(d\mathbf{r}) = \mathbf{t} \cdot d(\delta\mathbf{r}), \tag{8.6.16}$$

where $\mathbf{t}$ is the tangent vector introduced in (8.5.4) and a generalization of (8.6.4) was used.

(7) *Variation of the product* $f(x_1, x_2, x_3)g(x_1, x_2, x_3)$

$$\delta\left(f(x_1, x_2, x_3)g(x_1, x_2, x_3)\right) = f(X_1, X_2, X_3)g(X_1, X_2, X_3)$$

$$- f(x_1, x_2, x_3)g(x_1, x_2, x_3). \tag{8.6.17}$$

Replacing X_i by $x_i + \delta x_i$, expanding in Taylor series as before, performing the corresponding multiplications, neglecting second-order terms, and using (8.6.12) we obtain

$$\delta\left(f(x_1, x_2, x_3)g(x_1, x_2, x_3)\right) = f(x_1, x_2, x_3)\nabla g \cdot \delta\mathbf{r} + (\nabla f \cdot \delta\mathbf{r})g(x_1, x_2, x_3)$$

$$= f(x_1, x_2, x_3)\delta g + g(x_1, x_2, x_3)\delta f. \tag{8.6.18}$$

Now we are ready to prove Fermat's principle, which will be done following Ben-Menahem and Singh (1981). The time between two points P and Q along a ray is given by

$$t_{PQ} = \int_P^Q \frac{ds}{c} = \int_P^Q S\, ds, \tag{8.6.19}$$

where $S = S(x_1, x_2, x_3)$ is the slowness introduced in (8.5.22) and the integration is along the ray. Using (8.6.10) (valid for functions of more than one variable) and (8.6.18), the variation of t_{PQ} is given by

$$\delta t_{PQ} = \delta \int_P^Q S\, ds = \int_P^Q [(\delta S)\, ds + S\delta(ds)] = \int_P^Q [(\nabla S \cdot \delta\mathbf{r})\, ds + S\mathbf{t} \cdot d(\delta\mathbf{r})]. \tag{8.6.20}$$

The last equality was obtained using (8.6.12) with $f = S$ and (8.6.16). Now apply integration by parts to the second term on the right-hand side of (8.6.20). Considering the contribution from the x_i component we obtain

$$\int_P^Q St_i\, d(\delta x_i) = St_i \delta x_i \Big|_P^Q - \int_P^Q \delta x_i \frac{d}{ds}(St_i)\, ds = -\int_P^Q \delta x_i \frac{d}{ds}(St_i)\, ds \tag{8.6.21}$$

because δx_i is zero at the end points. In vector form (8.6.21) can be written as

$$\int_P^Q S\mathbf{t} \cdot d(\delta \mathbf{r}) = -\int_P^Q \frac{d}{ds}(S\mathbf{t}) \cdot \delta \mathbf{r}\, ds. \qquad (8.6.22)$$

Therefore, from (8.6.20) and (8.6.22) we obtain

$$\delta t_{PQ} = \int_P^Q \delta \mathbf{r} \cdot \left[\nabla S - \frac{d}{ds}(S\mathbf{t}) \right] ds = 0 \qquad (8.6.23)$$

because the term in brackets is equal to zero, as can be seen from the first equality in (8.5.23).

Equation (8.6.23) is valid for any path adjacent to the ray, and, consequently, δt_{PQ} is stationary along the ray (its variation is zero). Note that (8.6.23) does not say that the time is a maximum or a minimum, although usually it is a minimum. These questions are discussed in detail in Luneburg (1964), Born and Wolf (1975), and Hanyga (1985). Important examples of nonminimum traveltime paths are discussed by Choy and Richards (1975) in the context of caustics (see §8.7.1). It must also be mentioned that the above derivation is valid for an isotropic medium only. However, Fermat's principle also applies to anisotropic media, as discussed in Hanyga (1985).

Fermat's principle can be used to show that in a medium with spherical symmetry the rays are plane curves. The following proof is based on Hanyga (1985). Consider a ray in a spherical earth model. The arc element in spherical coordinates r, θ, and ϕ (see Fig. 9.10) is given by

$$ds = \sqrt{(dr)^2 + r^2(d\theta)^2 + r^2 \sin^2 \theta\, (d\phi)^2} \qquad (8.6.24)$$

(Spiegel, 1959). If r, θ, and ϕ are written in terms of a parameter u, the expression for ds becomes

$$ds = \frac{ds}{du}\, du = \sqrt{\dot{r}^2 + r^2\dot{\theta}^2 + r^2 \sin^2 \theta \dot{\phi}^2}\, du, \qquad (8.6.25)$$

where the dot represents a derivative with respect to the argument, and the time between two points A and B on the ray is given by the integral

$$\int_{u_1}^{u_2} \frac{1}{c(r)} \sqrt{\dot{r}^2 + r^2\dot{\theta}^2 + r^2 \sin^2 \theta\, \dot{\phi}^2}\, du, \qquad (8.6.26)$$

where u_1 and u_2 are the values of u at points A and B. For convenience the coordinate system will be chosen such that $\phi = 0$ at u_1. Equation (8.6.26) shows that the time will be minimum as long as $\dot{\phi}$ is zero (because all the quantities in the integrand are positive). Therefore, the ray remains in the plane $\phi = 0$.

8.7 Ray amplitudes

In our discussion of (8.2.6) and (8.3.8) we stated that under certain conditions the last two terms of each equation could be neglected. Each second term contains information concerning ray amplitudes and is known as the *transport equation*, which can be considered equal to zero. In the following the transport equation for the scalar and elastic waves will be considered in detail. Strictly speaking, we will derive ray-theory amplitude functions for the different types of waves (Cerveny, 2001), but will use expressions such as *P-* or *S-wave amplitudes* for simplicity.

8.7.1 Scalar wave equation

We start with the second term of (8.2.6) set equal to zero

$$\frac{2}{A}A_{,j}T_{,j} + T_{,jj} = 0 \tag{8.7.1}$$

and the following relation applicable to any function of position g:

$$g_{,j}T_{,j} = \frac{1}{c}\frac{\partial g}{\partial x_j}\frac{dx_j}{ds} = \frac{1}{c}\frac{dg}{ds}, \tag{8.7.2}$$

where in the first equality (8.4.16) was used.

Applying (8.7.2) to (8.7.1) with $g = A$ gives

$$\frac{2}{A}\frac{1}{c}\frac{dA}{ds} + \nabla^2 T = 0. \tag{8.7.3}$$

Integrating (8.7.3) we obtain

$$\int_{s_o}^{s}\frac{1}{A}\frac{dA}{ds}\,ds = -\int_{s_o}^{s}\frac{c}{2}\nabla^2 T\,ds = 0. \tag{8.7.4}$$

The integral on the left-hand side gives $\ln(A/A_o)$, with $A_o = A(s_o)$. Therefore,

$$A = A_o\exp\left(-\int_{s_o}^{s}\frac{c}{2}\nabla^2 T\,ds\right). \tag{8.7.5}$$

Equation (8.7.5) shows that once a ray has been found, the amplitude variations along the ray can be determined in a rather simple way if the amplitude is known for some value s_o of the parameter s.

Using a standard approach an even simpler relation for A can be derived. To see that multiply (8.7.1) by A^2. This gives

$$2AA_{,j}T_{,j} + A^2T_{,jj} = \left(A^2T_{,j}\right)_{,j} = \nabla\cdot(A^2\nabla T) = 0. \tag{8.7.6}$$

Now integrate (8.7.6) over the volume inside a narrow tube of rays with the two end surfaces (denoted by S_o and S_1) corresponding to the wave front at two different

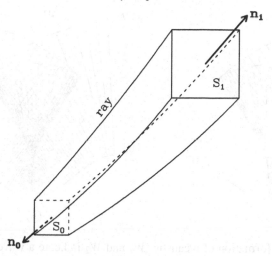

Fig. 8.7. A narrow tube of rays used to derive ray amplitudes. The surfaces S_o and S_1 are portions of the wave front at times t_o and t_1. After Cerveny and Ravindra (1971).

times t_o and t_1 (Fig. 8.7). Let S_2 be the surface formed by the rays. Furthermore, because the right-hand side of (8.7.6) involves a divergence we can use Gauss' theorem (see §1.4.9). If S represents the three surfaces S_o, S_1, S_2, we have

$$\int_S A^2 \nabla T \cdot \mathbf{n} = 0, \tag{8.7.7}$$

where $\mathbf{n}$ is the outer unit vector normal to S. As S is made up of three surfaces, there will be three expressions for $\mathbf{n}$. The normals to S_o and S_1 are the rays. This means that the outer unit normals corresponding to S_o and S_1 are given by $(-c\nabla T)$ and $c\nabla T$. For the surface S_2 the normal is also normal to the rays, which means that $\nabla T \cdot \mathbf{n} = 0$. Under these conditions (8.7.7) becomes

$$\int_{S_1} c A^2 \nabla T \cdot \nabla T \, dS = \int_{S_o} c A^2 \nabla T \cdot \nabla T \, dS. \tag{8.7.8}$$

Using the eikonal equation (see (8.3.20)), equation (8.7.8) can be further simplified

$$\int_{S_1} \frac{1}{c} A^2 \, dS = \int_{S_o} \frac{1}{c} A^2 \, dS. \tag{8.7.9}$$

Under the assumption that the ray tube is very narrow, the integrands in (8.7.9) can be approximated, for example, by their values at the centers of the respective surfaces and (8.7.9) becomes

$$\left(\frac{1}{c} A^2 \right)_1 \delta S_1 = \left(\frac{1}{c} A^2 \right)_o \delta S_o, \tag{8.7.10}$$

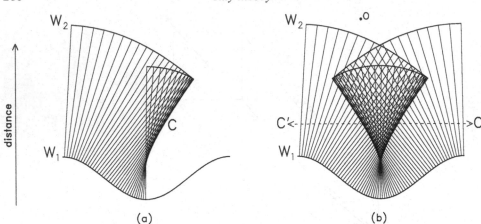

(a)   (b)

Fig. 8.8. Example of the formation of a caustic. W_1 and W_2 indicate a wave front at two different times. (a) Rays associated with the left half of the wave front. (b) All the rays. The caustic, indicated by CC', corresponds to a triplication of the wave front. After Whitham (1974).

where δS_0 and δS_1 indicate cross-sectional areas of the ray tube. Therefore, if A is known at some point on a ray, its value at any other point can be determined using (8.7.10). As a simple example, consider the case of constant velocity. Then for plane waves the surfaces δS_o and δS_1 are equal to each other and the corresponding amplitudes are also equal, as we already know.

Equation (8.7.10) can be written as

$$\frac{1}{c}A^2\delta S = \text{constant along a ray.} \qquad (8.7.11)$$

It must be noted, however, that (8.7.11) fails when δS goes to zero, which means that A must go to infinity and ray theory is no longer applicable. The regions where A becomes infinite are known as *caustics*. The quantitative analysis of the problem is quite difficult (Aki and Richards, 1980; Ben-Menahem and Singh, 1981), although simplified treatments are available (e.g., Boyles, 1984). Here we will give an example of the formation of a caustic. Figures 8.8(a) and (b) show part of a hypothetical wave front W_1 (corresponding to time t_1) with a concave portion. This concavity may be interpreted as being a result of the effect of a low-velocity zone on a wave front that was initially circular. Figure 8(a) shows the rays associated with the left half of the wave front W_2 (corresponding to time t_2) determined under the assumption that the velocity of the medium ahead of the wave front W_1 is a constant. Two important features should be noticed. One is the folding of the wave front W_2. The other is the absence of rays to the right of the curve C and the concentration of rays in the vicinity of C, which is the envelope of the rays.

This envelope is known as a caustic. When all the rays are considered (Fig. 8.8b), we see a triplication of the wave front, which means that a receiver located at a point such as O will receive three different arrivals, and a region bounded by the caustic CC' where each point is covered by three rays. When the rays touch either C or C' the area of a ray tube goes to zero and the amplitude goes to infinity in the ray theory approximation. In general, ray theory is applicable away from caustics, but when a ray crosses (or touches) a caustic there is a phase shift of $\pi/2$. Rather loosely, the presence of this phase shift can be explained by a change in the sign of the surface element δS after passing the caustic (the area is different from zero on both sides of the caustic). This introduces a factor of $\sqrt{-1}$ when taking the square root of A^2 (Cerveny and Ravindra, 1971; Hill, 1974; Choy and Richards, 1975). Therefore, the waveform associated with a ray will be distorted (see §6.5.3.3) after crossing (or touching) a caustic. If a ray crosses more than one caustic, every crossing contributes an additional phase shift of $\pi/2$. An important property of rays that touch a caustic is that they do not have the property of minimal travel paths (Choy and Richards, 1975).

8.7.2 Elastic wave equation

Here we will proceed in two steps. The first one will produce an expression similar to (8.7.11), which will then be specialized to the amplitude of P waves. The case of S waves will require a more elaborate analysis.

The transport equations for the P and S waves are obtained from the second term of (8.3.8) set equal to zero:

$$c_{ijkl,j}U_kT_{,l} + c_{ijkl}U_{k,j}T_{,l} + c_{ijkl}U_kT_{,lj} + c_{ijkl}U_{k,l}T_{,j}$$

$$= \left(c_{ijkl}U_kT_{,l}\right)_{,j} + c_{ijkl}U_{k,l}T_{,j} = 0. \tag{8.7.12}$$

The following derivation is based on Burridge (1976). Contract (8.7.12) with U_i, and in the second term on the right-hand side use $c_{ijkl} = c_{klij}$ (see (4.5.11)) and interchange the dummy indices i and k and j and l. With these operations (8.7.12) becomes

$$U_i\left(c_{ijkl}U_kT_{,l}\right)_{,j} + U_{i,j}c_{ijkl}U_kT_{,l} = \left(c_{ijkl}U_iU_kT_{,l}\right)_{,j} = 0. \tag{8.7.13}$$

Now rewrite the left-hand side of (8.3.9), which is equal to zero, as

$$c_{ijkl}T_{,l}T_{,j}U_k = (c^2T_{,k}T_{,k})\rho U_i, \tag{8.7.14}$$

where c is α or β. The factor in parentheses is equal to one (eikonal equation, (8.3.20)) and is introduced to allow the following derivation. When not needed, the factor will be removed. Next, take the derivative of (8.7.14) with respect to $T_{,p}$:

$$c_{ijkl}\left(\delta_{lp}T_{,j}+\delta_{jp}T_{,l}\right)U_k+c_{ijkl}T_{,l}T_{,j}\frac{\partial U_k}{\partial T_{,p}}=2c^2\delta_{kp}T_{,k}\rho U_i+\rho\frac{\partial U_i}{\partial T_{,p}}. \quad (8.7.15)$$

This equation can be rewritten as

$$\left(c_{ijkp}T_{,j}+c_{ipkl}T_{,l}\right)U_k=2c^2T_{,p}\rho U_i-\left(c_{ijkl}T_{,j}T_{,l}-\rho\delta_{ik}\right)\frac{\partial U_k}{\partial T_{,p}}. \quad (8.7.16)$$

Next contract (8.7.16) with U_i. The second term on the right-hand side gives

$$\left(c_{ijkl}T_{,j}T_{,l}-\rho\delta_{ik}\right)U_i\frac{\partial U_k}{\partial T_{,p}}=\left(c_{klij}T_{,l}T_{,j}-\rho\delta_{ik}\right)U_k\frac{\partial U_i}{\partial T_{,p}}$$

$$=\left(c_{ijkl}T_{,l}T_{,j}-\rho\delta_{ik}\right)U_k\frac{\partial U_i}{\partial T_{,p}}=0. \quad (8.7.17)$$

To obtain (8.7.17) the following facts were used. The four indices are dummy ones, the indices i and k and j and l were interchanged, the symmetry of c_{ijkl} was used, and the right-hand side of (8.7.17) involves (8.3.9), which gives zero. Therefore, after contraction equation (8.7.16) becomes

$$\left(c_{ijkp}T_{,j}+c_{ipkl}T_{,l}\right)U_kU_i=2c^2T_{,p}\rho U_iU_i=2c^2T_{,p}\rho|\mathbf{U}|^2. \quad (8.7.18)$$

After a simple index manipulation we can also show that the two terms on the left-hand side of (8.7.18) are equal to each other (Problem 8.11), so that (8.7.18) becomes

$$c_{ipkl}T_{,l}U_kU_i=c^2T_{,p}\rho|\mathbf{U}|^2. \quad (8.7.19)$$

Replacing p by j in (8.7.19) and using (8.7.13) gives

$$\left(c_{ijkl}T_{,l}U_kU_i\right)_{,j}=\left(c^2\rho T_{,j}|\mathbf{U}|^2\right)_{,j}=0. \quad (8.7.20)$$

The right-hand side of (8.7.20) is similar to (8.7.6) for the scalar wave equation with A replaced by $c\sqrt{\rho}|\mathbf{U}|$. This means that we can use the ray tube argument of §8.7.1, which gives

$$c\rho|\mathbf{U}|^2\delta S=\text{constant}, \quad (8.7.21)$$

where δS is the cross-sectional area of the ray tube.

Equation (8.7.21) can be rewritten as

$$c_1\rho_1|\mathbf{U}_1|^2\delta S_1=c_2\rho_2|\mathbf{U}_2|^2\delta S_2, \quad (8.7.22)$$

where the subscripts 1 and 2 indicate two positions along the ray tube. This relation shows that if the amplitude $|\mathbf{U}|$ is known at a given point, then it can be determined at any other point after computation of the corresponding cross-sectional areas. Although (8.7.22) can, in principle, be used in the computation of amplitudes, it is numerically more convenient to use equations derived using other approaches, one of which is discussed in the following. It must also be noted that (8.7.22) is based on the implicit assumption that the elastic energy is confined to the tube of rays and that there is no flow of energy through the walls of the tube. This assumption is valid only for the zeroth-order ray theory (Cerveny and Ravindra, 1971). An important application of (8.7.22) is given in §10.11.

8.7.2.1 P-wave amplitudes

Equation (8.7.20) will be used to derive a differential equation for the amplitude of the P-wave motion. Let the corresponding vector $\mathbf{U}$ be equal to $A\mathbf{t}$, with A a real constant that represents the ray amplitude of the P wave. Replacing c by α and $|\mathbf{U}|^2$ by A^2 in (8.7.20) gives

$$\left(A^2\alpha^2\rho T_{,j}\right)_{,j} = 2AA_{,j}\alpha^2\rho T_{,j} + A^2\alpha^2\rho T_{,jj} + A^2(\alpha^2\rho)_{,j}T_{,j} = 0. \qquad (8.7.23)$$

After using (8.7.2) and dividing by $2A\alpha\rho$, equation (8.7.23) becomes

$$\frac{dA}{ds} + \frac{1}{2}A\left[\alpha\nabla^2 T + \frac{1}{\alpha^2\rho}\frac{d}{ds}(\alpha^2\rho)\right] = \frac{dA}{ds} + \frac{1}{2}A\left[\alpha\nabla^2 T + \frac{d}{ds}\ln(\alpha^2\rho)\right] = 0. \qquad (8.7.24)$$

Once a ray has been determined, equation (8.7.24) can be used to compute the amplitude variations along the ray. Note that (8.7.24) is valid for isotropic and anisotropic media.

Equation (8.7.24) can also be written in terms of t. To do that use

$$\frac{d}{ds} = \frac{dt}{ds}\frac{d}{dt} = \frac{1}{\alpha}\frac{d}{dt} \qquad (8.7.25)$$

in (8.7.24) and multiply by α. This gives

$$\frac{dA}{dt} + \frac{1}{2}A\left[\alpha^2\nabla^2 T + \frac{d}{dt}\ln(\alpha^2\rho)\right] = 0. \qquad (8.7.26)$$

8.7.2.2 S-wave amplitudes. Ray-centered coordinate system

As we saw before, the vector representing the S waves is perpendicular to ∇T and because the vectors $\mathbf{n}$ and $\mathbf{b}$ are perpendicular to ∇T, in the earlier literature they were chosen as the basis vectors for the decomposition of the S-wave vector. This selection is inconvenient because the equations for the amplitudes of the two S-wave components form a coupled system (Cerveny and Ravindra, 1971), which complicates the numerical computations (Cerveny and Hron, 1980). Only when the torsion is zero are the equations decoupled. This happens, for example, in the case of radial velocity variations. However, it is possible to introduce a pair of vectors $\mathbf{e}^I$ and $\mathbf{e}^{II}$ perpendicular to ∇T which allow decoupling of the S-wave components. These vectors were introduced by Popov and Psencik (Cerveny and Hron, 1980;

Psencik, 1979). The following derivation showing the conditions under which the decoupling is possible follows Burridge (1976). The starting point is the transport equation (8.7.12):

$$\left(c_{ijkl}U_k T_{,l}\right)_{,j} + c_{ijkl}U_{k,l}T_{,j} = 0. \tag{8.7.27}$$

Replacing c_{ijkl} by its expression for an isotropic medium, given by (4.6.2), equation (8.7.27) becomes

$$\left(\lambda U_l T_{,l}\right)_{,i} + \left(\mu U_i T_{,j}\right)_{,j} + \left(\mu U_j T_{,i}\right)_{,j} + \lambda U_{l,l}T_{,i} + \mu U_{i,j}T_{,j} + \mu U_{j,i}T_{,j} = 0. \tag{8.7.28}$$

The first term is zero because it involves the scalar product of **U** and ∇T, which are perpendicular to each other. Now introduce two unit vectors

$$\mathbf{e}^{\mathrm{I}} = (e_1^{\mathrm{I}}, e_2^{\mathrm{I}}, e_3^{\mathrm{I}}); \qquad \mathbf{e}^{\mathrm{II}} = (e_1^{\mathrm{II}}, e_2^{\mathrm{II}}, e_3^{\mathrm{II}}) \tag{8.7.29}$$

perpendicular to each other and to ∇T:

$$\mathbf{e}^{\mathrm{I}} \cdot \mathbf{e}^{\mathrm{II}} = 0; \qquad \mathbf{e}^{\mathrm{I}} \cdot \nabla T = \mathbf{e}^{\mathrm{II}} \cdot \nabla T = 0. \tag{8.7.30a,b}$$

The vector **U** will be written as a linear combination of $\mathbf{e}^{\mathrm{I}}$ and $\mathbf{e}^{\mathrm{II}}$:

$$\mathbf{U} = A_1 \mathbf{e}^{\mathrm{I}} + A_2 \mathbf{e}^{\mathrm{II}} = A_m \mathbf{e}^m; \qquad m = \mathrm{I, II}, \tag{8.7.31}$$

where A_1 and A_2 are functions of position to be determined together with the two unit vectors. In component form we have

$$U_i = A_1 e_i^{\mathrm{I}} + A_2 e_i^{\mathrm{II}} = A_m e_i^m; \qquad i = 1, 2, 3, \quad m = \mathrm{I, II}. \tag{8.7.32}$$

Upon contraction with e_i^l ($l = \mathrm{I, II}$) the fourth term of (8.7.28) gives zero because of the orthogonality relations (8.7.30b). After expanding the derivatives the four other terms of (8.7.28) give

$$e_i^l \mu U_{i,j} T_{,j} + e_i^l U_i \left(\mu T_{,j}\right)_{,j} + \left(\mu U_j\right)_{,j} e_i^l T_{,i} + e_i^l U_j \mu T_{,ij} + e_i^l \mu U_{i,j} T_{,j} + e_i^l \mu U_{j,i} T_{,j}. \tag{8.7.33}$$

The third term of (8.7.33) gives zero because of (8.7.30b). The fourth and sixth terms can be combined to give

$$e_i^l \mu U_j T_{,ij} + e_i^l \mu U_{j,i} T_{,j} = e_i^l \mu \left(T_{,j} U_j\right)_{,i} = 0 \tag{8.7.34}$$

because ∇T and **U** are perpendicular to each other.

Finally, after noting that the first and fifth terms are equal (8.7.33) becomes

$$2 e_i^l \mu T_{,j} U_{i,j} + e_i^l \left(\mu T_{,j}\right)_{,j} U_i = 0. \tag{8.7.35}$$

Using (8.7.2) and (8.7.32) the first term of (8.7.35) will be written as

$$2 e_i^l \mu T_{,j} U_{i,j} = 2 e_i^l \frac{\mu}{\beta} \frac{\mathrm{d}U_i}{\mathrm{d}s} = 2 e_i^l \frac{\mu}{\beta} \frac{\mathrm{d}}{\mathrm{d}s}\left(A_m e_i^m\right). \tag{8.7.36}$$

Using (8.7.32) and the orthogonality of $\mathbf{e}^{\mathrm{I}}$ and $\mathbf{e}^{\mathrm{II}}$ the second term of (8.7.35) gives

$$\left(\mu T_{,j}\right)_{,j} e_i^l U_i = \left(\mu T_{,j}\right)_{,j} e_i^l A_m e_i^m = \left(\mu T_{,j}\right)_{,j} A_m \delta_{lm} = \left(\mu T_{,j}\right)_{,j} A_l. \quad (8.7.37)$$

Introducing (8.7.36) and (8.7.37) in (8.7.35) and expanding the derivative with respect to s gives

$$2\frac{\mu}{\beta}\left(\delta_{lm}\frac{\mathrm{d}A_m}{\mathrm{d}s} + e_i^l A_m \frac{\mathrm{d}e_i^m}{\mathrm{d}s}\right) + \left(\mu T_{,j}\right)_{,j} A_l$$

$$= 2\frac{\mu}{\beta}\frac{\mathrm{d}A_l}{\mathrm{d}s} + 2\frac{\mu}{\beta}A_m e_i^l \frac{\mathrm{d}e_i^m}{\mathrm{d}s} + \left(\mu T_{,j}\right)_{,j} A_l = 0. \quad (8.7.38)$$

Note the coupling between A_1 and A_2 introduced by the sum over m in the second term of (8.7.38). However, this coupling disappears if $\mathbf{e}^l$ and $\mathrm{d}\mathbf{e}^m/\mathrm{d}s$ are perpendicular to each other, which in turn requires $\mathrm{d}\mathbf{e}^m/\mathrm{d}s$ parallel to ∇T. Under these conditions and replacing μ by $\beta^2 \rho$, equation (8.7.38) becomes

$$\frac{\mathrm{d}A_l}{\mathrm{d}s} + \frac{1}{2\beta\rho}\left(\beta^2 \rho T_{,j}\right)_{,j} A_l = 0; \qquad l = 1, 2. \quad (8.7.39)$$

Equation (8.7.39) can be rewritten as

$$\frac{\mathrm{d}A_l}{\mathrm{d}s} + \frac{1}{2}A_l \beta \nabla^2 T + \frac{1}{2\beta\rho}(\beta^2 \rho)_{,j} T_{,j} A_l$$

$$= \frac{\mathrm{d}A_l}{\mathrm{d}s} + \frac{1}{2}A_l\left[\beta\nabla^2 T + \frac{\mathrm{d}}{\mathrm{d}s}\ln(\beta^2\rho)\right] = 0; \qquad l = 1, 2. \quad (8.7.40)$$

To obtain the last term we used the same arguments as for the P waves in §8.7.2.1. Note that the amplitudes of the P waves and the two components of the S waves satisfy equations of the same form.

To complete the analysis of the S-wave amplitudes it is necessary to derive the expressions for the vectors $\mathbf{e}^{\mathrm{I}}$ and $\mathbf{e}^{\mathrm{II}}$. To do that first use the fact that $\mathbf{e}^m$ ($m = \mathrm{I}, \mathrm{II}$) is perpendicular to $\mathbf{t}$, so that

$$(\mathbf{e}^m \cdot \mathbf{t}) = 0; \qquad m = \mathrm{I}, \mathrm{II}. \quad (8.7.41)$$

Recall that $\mathbf{t}$ is parallel to ∇T (see (8.5.4)).

Differentiation of (8.7.41) with respect to s gives

$$\frac{\mathrm{d}\mathbf{e}^m}{\mathrm{d}s}\cdot \mathbf{t} = -\mathbf{e}^m \cdot \frac{\mathrm{d}\mathbf{t}}{\mathrm{d}s}; \qquad m = \mathrm{I}, \mathrm{II}. \quad (8.7.42)$$

Now, since $\mathrm{d}\mathbf{e}^m/\mathrm{d}s$ is parallel to $\mathbf{t}$, we can write

$$\frac{\mathrm{d}\mathbf{e}^m}{\mathrm{d}s} = -\left(\mathbf{e}^m \cdot \frac{\mathrm{d}\mathbf{t}}{\mathrm{d}s}\right)\mathbf{t} = -\kappa(\mathbf{e}^m \cdot \mathbf{n})\mathbf{t}; \qquad m = \mathrm{I}, \mathrm{II} \quad (8.7.43)$$

(Problem 8.13) where (8.5.7) has been used (Burridge, 2000, personal communication).

Because $\mathbf{e}^m$ and $\mathbf{n}$ are unit vectors and $\mathbf{e}^I$ and $\mathbf{e}^{II}$ are perpendicular to each other we can write

$$\mathbf{e}^I \cdot \mathbf{n} = \cos\theta; \qquad \mathbf{e}^{II} \cdot \mathbf{n} = \sin\theta, \tag{8.7.44}$$

where θ is to be determined. Introducing (8.7.44) into (8.7.43) gives

$$\frac{d\mathbf{e}^I}{ds} = -\kappa \cos\theta\, \mathbf{t} \tag{8.7.45}$$

and

$$\frac{d\mathbf{e}^{II}}{ds} = -\kappa \sin\theta\, \mathbf{t}. \tag{8.7.46}$$

To find the angle θ note that the vectors $\mathbf{n}$ and $\mathbf{b}$ can be obtained by a rotation of $\mathbf{e}^I$ and $\mathbf{e}^{II}$ through θ:

$$\mathbf{n} = \mathbf{e}^I \cos\theta + \mathbf{e}^{II} \sin\theta \tag{8.7.47}$$

$$\mathbf{b} = -\mathbf{e}^I \sin\theta + \mathbf{e}^{II} \cos\theta. \tag{8.7.48}$$

Taking the derivatives of (8.7.47) and (8.7.48) with respect to s and using (8.7.45)–(8.7.48) gives

$$\frac{d\mathbf{n}}{ds} = \frac{d\mathbf{e}^I}{ds}\cos\theta - \mathbf{e}^I \sin\theta \frac{d\theta}{ds} + \frac{d\mathbf{e}^{II}}{ds}\sin\theta + \mathbf{e}^{II} \cos\theta \frac{d\theta}{ds} = -\kappa\mathbf{t} + \frac{d\theta}{ds}\mathbf{b} \tag{8.7.49}$$

$$\frac{d\mathbf{b}}{ds} = -\frac{d\mathbf{e}^I}{ds}\sin\theta - \mathbf{e}^I \cos\theta \frac{d\theta}{ds} + \frac{d\mathbf{e}^{II}}{ds}\cos\theta - \mathbf{e}^{II} \sin\theta \frac{d\theta}{ds} = -\frac{d\theta}{ds}\mathbf{n}. \tag{8.7.50}$$

Comparison of (8.7.49) and (8.7.50) with the Frenet–Serret equations (8.5.14) and (8.5.12) shows that

$$\tau = \frac{d\theta}{ds} \qquad \text{or} \qquad d\theta = \tau\, ds. \tag{8.7.51a,b}$$

Integrating (8.7.51b) along a ray between some initial value s_o and s gives

$$\theta(s) = \theta(s_o) + \int_{s_o}^{s} \tau\, ds. \tag{8.7.52}$$

The vectors $\mathbf{e}^I$ and $\mathbf{e}^{II}$ can be obtained from (8.7.45) and (8.7.46) by numerical integration, but because one of the vectors can be obtained from the other, only one equation must be integrated. In addition, the computations can be simplified using

$$-\kappa\cos\theta = \frac{1}{c}\nabla c \cdot \mathbf{e}^I; \qquad -\kappa\sin\theta = \frac{1}{c}\nabla c \cdot \mathbf{e}^{II} \tag{8.7.53a,b}$$

(Psencik, 1979; Cerveny, 1985; Problem 8.14).

The vectors $\mathbf{t}$, $\mathbf{e}^{\mathrm{I}}$ and $\mathbf{e}^{\mathrm{II}}$ constitute the *ray-centered coordinate system* or *polarization trihedral* and the vectors $\mathbf{e}^{\mathrm{I}}$ and $\mathbf{e}^{\mathrm{II}}$ are the *polarization vectors*. Now we will investigate the rotation of the polarization trihedral. The following derivation is due to Lewis (1966). The first step is to write the polarization vectors in terms of $\mathbf{n}$ and $\mathbf{b}$. From (8.7.47) and (8.7.48) we obtain

$$\mathbf{e}^{\mathrm{I}} = \mathbf{n}\cos\theta - \mathbf{b}\sin\theta \qquad (8.7.54)$$

$$\mathbf{e}^{\mathrm{II}} = \mathbf{n}\sin\theta + \mathbf{b}\cos\theta \qquad (8.7.55)$$

(Problem 8.15). To simplify the notation let

$$\mathbf{z}_1 = \mathbf{t}; \qquad \mathbf{z}_2 = \mathbf{e}^{\mathrm{I}}; \qquad \mathbf{z}_3 = \mathbf{e}^{\mathrm{II}}. \qquad (8.7.56)$$

We want to find the angular velocity vector $\boldsymbol{\omega}$ that satisfies

$$\frac{d\mathbf{z}_i}{ds} = \boldsymbol{\omega} \times \mathbf{z}_i; \qquad i = 1, 2, 3 \qquad (8.7.57)$$

(see (8.5.20)). Because the $\mathbf{z}_i$ are orthogonal unit vectors, $\boldsymbol{\omega}$ can be written as

$$\boldsymbol{\omega} = \omega_1 \mathbf{z}_1 + \omega_2 \mathbf{z}_2 + \omega_3 \mathbf{z}_3 = \omega_i \mathbf{z}_i, \qquad (8.7.58)$$

where ω_i must be determined. Introducing (8.7.58) in (8.7.57) and using

$$\mathbf{z}_J \times \mathbf{z}_J = \mathbf{0}; \qquad J = 1, 2, 3 \qquad (8.7.59)$$

(no summation over uppercase indices) and

$$\mathbf{z}_1 \times \mathbf{z}_2 = \mathbf{z}_3; \qquad \mathbf{z}_2 \times \mathbf{z}_3 = \mathbf{z}_1; \qquad \mathbf{z}_3 \times \mathbf{z}_1 = \mathbf{z}_2 \qquad (8.7.60)$$

we obtain

$$\frac{d\mathbf{z}_1}{ds} = \omega_i \mathbf{z}_i \times \mathbf{z}_1 = 0\mathbf{z}_1 + \omega_3 \mathbf{z}_2 - \omega_2 \mathbf{z}_3 \qquad (8.7.61\mathrm{a})$$

$$\frac{d\mathbf{z}_2}{ds} = \omega_i \mathbf{z}_i \times \mathbf{z}_2 = -\omega_3 \mathbf{z}_1 + 0\mathbf{z}_2 + \omega_1 \mathbf{z}_3 \qquad (8.7.61\mathrm{b})$$

$$\frac{d\mathbf{z}_3}{ds} = \omega_i \mathbf{z}_i \times \mathbf{z}_3 = \omega_2 \mathbf{z}_1 - \omega_1 \mathbf{z}_2 + 0\mathbf{z}_3. \qquad (8.7.61\mathrm{c})$$

Now we will write the expressions for $d\mathbf{z}_i/ds$ and by comparison with (8.7.61) determine ω_i. From (8.5.7) and (8.7.47) we obtain

$$\frac{d\mathbf{z}_1}{ds} = \frac{d\mathbf{t}}{ds} = \kappa\cos\theta\, \mathbf{z}_2 + \kappa\sin\theta\, \mathbf{z}_3. \qquad (8.7.62)$$

Using (8.7.54), (8.7.55), (8.7.51a), and (8.5.12) and (8.5.14) we obtain

$$\frac{d\mathbf{z}_2}{ds} = \frac{d\mathbf{e}^{\mathrm{I}}}{ds} = \cos\theta\frac{d\mathbf{n}}{ds} - \tau\sin\theta\,\mathbf{n} - \sin\theta\frac{d\mathbf{b}}{ds} - \tau\cos\theta\,\mathbf{b}$$

$$= -\tau\mathbf{e}^{\mathrm{II}} + \cos\theta(-\kappa\mathbf{t} + \tau\mathbf{b}) + \tau\sin\theta\,\mathbf{n}$$

$$= -\kappa\cos\theta\,\mathbf{t} = -\kappa\cos\theta\,\mathbf{z}_1 \tag{8.7.63}$$

and

$$\frac{d\mathbf{z}_3}{ds} = \frac{d\mathbf{e}^{\mathrm{II}}}{ds} = \sin\theta\frac{d\mathbf{n}}{ds} + \tau\cos\theta\,\mathbf{n} + \cos\theta\frac{d\mathbf{b}}{ds} - \tau\sin\theta\,\mathbf{b}$$

$$= \tau\mathbf{e}^{\mathrm{I}} + \sin\theta(-\kappa\mathbf{t} + \tau\mathbf{b}) - \tau\cos\theta\,\mathbf{n} = -\kappa\sin\theta\,\mathbf{t} = -\kappa\sin\theta\,\mathbf{z}_1. \tag{8.7.64}$$

Comparison of (8.7.61) with (8.7.62)–(8.7.64) shows that

$$\omega_1 = 0; \qquad \omega_2 = -\kappa\sin\theta; \qquad \omega_3 = \kappa\cos\theta \tag{8.7.65}$$

so that from (8.7.58) and (8.7.48) we obtain

$$\boldsymbol{\omega} = -\kappa\sin\theta\,\mathbf{z}_2 + \kappa\cos\theta\,\mathbf{z}_3 = k(-\sin\theta\,\mathbf{e}^{\mathrm{I}} + \cos\theta\,\mathbf{e}^{\mathrm{II}}) = \kappa\mathbf{b}. \tag{8.7.66}$$

Equation (8.7.66) has important implications, because

$$\boldsymbol{\omega}\cdot\mathbf{t} = \kappa\mathbf{b}\cdot\mathbf{t} = 0 \tag{8.7.67}$$

so that the polarization trihedral does not rotate around the ray (see the discussion following (8.5.21)). To get a better understanding of this fact note that (8.7.51a) gives the rate at which the polarization trihedral rotates about the Frenet trihedral. The latter, in turn, rotates about the ray at the same rate but in an opposite direction, as can be seen from (8.5.12) (i.e., $d\mathbf{b}/ds = -\tau\mathbf{n}$), and the geometric interpretation of the torsion. In conclusion, the rate of rotation of the polarization trihedral about the ray is zero.

8.7.3 Effect of discontinuities in the elastic parameters

In §8.3 we saw that in isotropic media the P and S waves satisfy independent eikonal equations, which means that the two types of waves are decoupled and can be treated independently. This result is valid as long as the variation in velocities (or ρ, λ, and μ) is smooth. However, if the medium includes first-order interfaces (i.e., surfaces across which these parameters are discontinuous) ray theory is no longer applicable. Moreover, from our discussion in Chapter 6 it can be expected that the interaction of the P or S waves with a surface of discontinuity will result, in general, in the generation of reflected and transmitted P and S waves. Considering that the surfaces involved may be curved and that the waves involved are not necessarily plane, the question is how to treat this problem. The answer is that under the zeroth-order approximation to ray theory, at the point where the ray intersects the surface the waves can be considered to be plane and the surface can be replaced by its tangent plane (Cerveny and Ravindra, 1971; Cerveny *et al.*, 1977). The analysis of the problem is similar to the analysis of the P, SV, and SH waves carried out in Chapters 5 and 6, and has been described in detail by Cerveny and Ravindra (1971). The main points of their analysis are summarized below.

A plane of incidence (see §5.8.4) and a local Cartesian coordinate system are defined as follows. The former is the plane determined by the normal to the surface and the tangent to the ray at the point of intersection. The coordinate system has its origin at the intersection point, the z axis is normal to the surface, the x axis is in the incidence plane and perpendicular to the z axis, and the y axis is such that the three axes form a right-handed system. The y axis is equivalent to the x_2 axis in Fig. 5.3, although a major difference is that the y axis is not necessarily horizontal because the surface of discontinuity need not be so. Also note that since the ray may not be a plane curve, only a portion of the ray may lie in the incidence plane. To solve the problem it is necessary to write the ray solutions for the incident, reflected, and transmitted P and S waves, and to apply the boundary conditions discussed in §6.3. Three unit vectors are defined as: $\mathbf{n}_P$, equal to $\mathbf{t}$ (see (8.5.4)) and equivalent to the vector $\mathbf{p}$ introduced in (5.8.52), $\mathbf{n}_{SV}$, perpendicular to $\mathbf{n}_P$ and in the incidence plane, and $\mathbf{n}_{SH}$, equivalent to $\mathbf{a}_2$ in (5.8.55). Finally, the S-wave vector is decomposed into components SV and SH similar to those defined in §5.8.4, although the SV and SH components are not necessarily in a vertical plane and along a horizontal axis, respectively. Taking these definitions and conventions into account, the result of Cerveny and Ravindra's (1971) analysis is that for the zeroth-order approximation the equations for the reflection and transmission coefficients coincide with the Zoeppritz equations derived in Chapter 6. In addition, Snell's law (see (6.6.27)) is also valid. Since during the ray tracing the S waves are written in terms of the vectors $\mathbf{e}^I$ and $\mathbf{e}^{II}$ (or $\mathbf{b}$ and $\mathbf{n}$), which are in the same plane as $\mathbf{n}_{SV}$ and $\mathbf{n}_{SH}$, a rotation of coordinates is needed before applying the Zoeppritz equations.

8.8 Examples

Here we will consider two examples. The first one is the case of SH waves at normal incidence in a layer over a half-space. This problem was treated in §6.9.1.2 using a wave theory approach, but here we will show that a simpler ray theory argument produces similar results. The second example uses ray theory synthetic seismograms generated for a simple 1-D model to illustrate certain features of the method and of wave propagation.

8.8.1 SH waves in a layer over a half-space at normal incidence

The following discussion is based on Savarenskii (1975). Let H be the thickness of the layer and β' and ρ' the corresponding velocity and density, and consider a plane wave impinging at the bottom of the layer at normal incidence. Let $f(t)$ indicate the time dependence of the wave and $|f(\omega)|$ the amplitude of one of its harmonic components (obtained from the Fourier transform of $f(t)$). The amplitude of the ray transmitted into the layer is given by $|f(\omega)|c_t$, where c_t is the transmission coefficient, which can be obtained from (6.6.9) with $C = 1$ and $f = 0$. Using β and ρ for the velocity and density in the half-space we obtain

$$c_t = \frac{2\mu\beta'}{\mu\beta' + \mu'\beta} = \frac{2\rho\beta^2\beta'}{\rho\beta^2\beta' + \rho'\beta'^2\beta} = \frac{2\rho\beta}{\rho\beta + \rho'\beta'}. \tag{8.8.1}$$

When the ray reaches the surface it reflects back into the layer without modification (see §6.5.1) and travels down until it encounters the bottom of the layer, where it reflects again, this time in the upward direction. Upon reflection the amplitude of the ray becomes $|f(\omega)|c_t c_r'$, where c_r' is the reflection coefficient, which can be determined using (6.6.6), but because the incidence and transmission media now are the layer and the half-space, respectively, we must interchange ρ and ρ' and β and β'. After doing that we find

$$c_t' = \frac{2\rho'\beta'}{\rho\beta + \rho'\beta'} \tag{8.8.2}$$

and, upon setting $C = 1$,

$$c_r' = C_1 = c_t' - 1 = \frac{\rho'\beta' - \rho\beta}{\rho\beta + \rho'\beta'}. \tag{8.8.3}$$

Note that $|c_r'| < 1$ and that $c_r' < 0$ when $\rho'\beta' < \rho\beta$.

This pattern of rays reflected at the top and bottom of the layer keeps repeating, with the amplitude of each ray multiplied by c_r' at every reflection from the bottom. Because $|c_r'| < 1$, the amplitude of the rays eventually become negligible.

So far we have considered ray amplitudes only. To account for their phases we will take as the origin time the time at which the incident wave arrives at the bottom of the layer. With this convention the transmitted ray and the successive reflected rays reach the top of the layer at times H/β', $3H/\beta'$, $5H/\beta'$, and so on. Since a time shift t_o corresponds to a phase shift of $\exp(-\omega t_o)$ in the frequency domain, the surface displacement is given by the infinite sum

$$v_o(\omega) = 2|f(\omega)|c_{\mathrm{t}}\left[e^{-i\theta} + c'_{\mathrm{r}}e^{-i3\theta} + c_{\mathrm{r}}'^2 e^{-i5\theta} + \cdots + c_{\mathrm{r}}'^n e^{-i(2n+1)\theta} + \cdots\right]$$

$$= |f(\omega)|2c_{\mathrm{t}}e^{-i\theta}\sum_{n=0}^{\infty}\left(c'_{\mathrm{r}}e^{-i2\theta}\right)^n, \tag{8.8.4}$$

where

$$\theta = \frac{H\omega}{\beta'}. \tag{8.8.5}$$

The factor of 2 in (8.8.4) comes from the doubling of the surface displacement (see §6.5.1). Note that the expression for θ is equal to that obtained from (6.9.16a) and (6.9.28a) for the case of normal incidence. The rightmost factor in (8.8.4) is a geometric series, with sum equal to

$$\frac{1}{1 - c'_{\mathrm{r}}e^{-i2\theta}}. \tag{8.8.6}$$

Introducing (8.8.1), (8.8.3), and (8.8.6) in (8.8.4) gives

$$v_o(\omega) = \frac{4\rho\beta|f(\omega)|}{e^{i\theta}\left[\rho\beta + \rho'\beta' - (\rho'\beta' - \rho\beta)e^{-i2\theta}\right]}$$

$$= \frac{4\rho\beta|f(\omega)|}{2\rho\beta\cos\theta + 2i\rho'\beta'\sin\theta} = \frac{2|f(\omega)|}{\cos\theta + iR\sin\theta}, \tag{8.8.7}$$

where

$$R = \frac{\rho'\beta'}{\rho\beta}. \tag{8.8.8}$$

It is convenient to consider the normalized surface displacement, given by

$$\tilde{v}_o(\omega) = \frac{v_o(\omega)}{|f(\omega)|} = \frac{2}{\cos\theta + iR\sin\theta}. \tag{8.8.9}$$

Aside from a phase factor, equation (8.8.9) is similar to (6.9.29), so that we can use the results derived in §6.9.1.1 and §6.9.1.2. Thus, $\tilde{v}_o(\omega)$ has a maximum absolute value given by

$$\max|\tilde{v}_o(\omega)| = \frac{2}{R} = 2\frac{\rho\beta}{\rho'\beta'} \tag{8.8.10}$$

for periods T_m equal to

$$T_m = \frac{4H}{(2n+1)\beta'}; \qquad n = 0, 1, 2, \ldots, \tag{8.8.11}$$

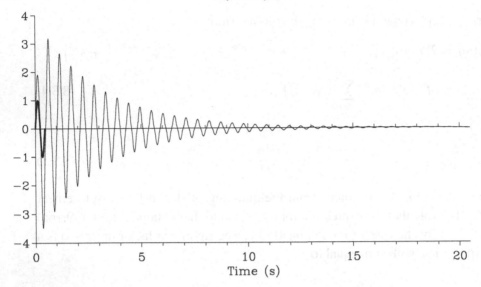

Fig. 8.9. A function $f(t)$ (bold curve) and the function $g(t)$ obtained using (8.8.13) (thin curve).

while the amplification of the ground motion due to the presence of the layer is given by

$$A = \frac{\rho\beta}{\rho'\beta'}. \tag{8.8.12}$$

Now we will look at the same problem from a different point of view. As before, we consider a function $f(t)$ and its amplitude spectrum $|f(\omega)|$. Let ω_M be the value of ω for which $|f(\omega)|$ is a maximum and let T_M be the corresponding period. Now consider the function $g(t)$ defined by

$$g(t) = c_t \sum_{n=0}^{N} c_r'^n f(t - nT) \tag{8.8.13}$$

with c_t and c_r' as before, N large enough to make the last element in the sum negligible, and T the time shift that maximizes the amplitude $|g(\omega)|$ of the Fourier transform of $g(t)$. If T is equal to $2H/\beta'$, equation (8.8.13) is the truncated time domain representation of (8.8.4) (aside from a constant factor and a phase shift). In addition, aside from constant factors, equation (8.8.13) can also be obtained using a wave theory approach (Murphy *et al.*, 1971).

Equation (8.8.13) was used with the functions $f(t)$ and $|f(\omega)|$ shown in Figs 8.9 and 8.10 and with a model having $\beta' = 0.25$ km s^{-1}, $\rho' = 1.7$ g cm^{-3}, $\beta = 3.5$ km s^{-1}, and $\rho = 2.7$ g cm^{-3} (also used to produce Fig. 6.12). At this point we are not concerned with the layer thickness. The time shift T, determined using a

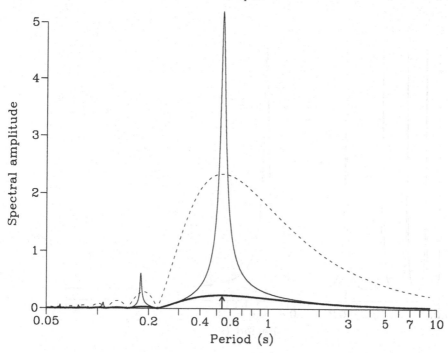

Fig. 8.10. Amplitude spectra of the functions $f(t)$ and $g(t)$ shown in Fig. 8.9 (bold and thin curves, respectively). For display purposes, the spectrum of $f(t)$ has been multiplied by 10 (dashed curve). The arrow corresponds to the period for which the spectrum of $g(t)$ is maximum.

computer search, was found to be equal to $T_M/2$. This shift maximizes the interference of the shifted functions of alternating polarity that contribute to $g(t)$ in (8.8.13). This interference process is qualitatively easy to understand because the function $f(t)$ is very simple, having a time width close to T_M (equal to 0.535 s). The function $g(t)$ obtained with $T = T_M/2$ and its amplitude spectrum are shown in Figs 8.9 and 8.10. The long duration of $g(t)$ results from the low velocity in the layer. Also note that the largest amplitude of $g(t)$ is about 3.5 times larger than the largest amplitude of $f(t)$. This large magnification is the result of large values of c'_r and c_t (equal to -0.91 and 1.91, respectively). In the frequency domain the largest amplitude of $|g(\omega)|$ is 22.2 times larger than the largest amplitude of $|f(\omega)|$. This value is equal to the amplification factor computed using (8.8.12). In addition, a plot of the ratio $|g(\omega)|/|f(\omega)|$ (Fig. 8.11) shows a function with peaks at periods given by (8.8.11) with $4H/\beta' = T_M$ and peak values also equal to 22.2.

The peak values in Fig. 8.11 agree with the expected amplification, given by (8.8.12). To understand the position of the peak values, note that (8.8.11) gives the periods that will be preferentially amplified when H and β' are given. In our

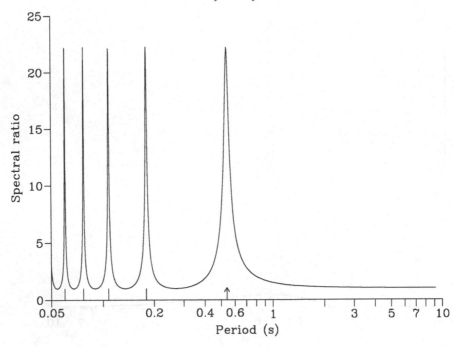

Fig. 8.11. Ratio of the amplitude spectra of $g(t)$ and $f(t)$ (see Fig. 8.10). The tick marks above the period axis were determined as follows. The period indicated by the arrow (corresponding to a peak value of the ratio) was taken as the T_m that would be obtained using (8.8.11) with $n = 0$. The position of the tick marks were determined using $T_m/(2n + 1)$, with $n = 1, \ldots, 4$, and agree with the other peaks of the ratio.

example we assume that β' is given and want to find the value of H for which $T_m = T_M$ when $n = 0$. This gives $H = 0.0335$ m. When this H is used in (8.8.11), the values of T_m for $n = 1, \ldots, 4$ coincide with the periods corresponding to the peak values in Fig. 8.11.

8.8.2 *Ray theory synthetic seismograms*

The synthetic seismograms were computed using the program SEIS81, written by V. Cerveny and I. Psencik with some modifications by Herrmann (1998). In this program rays are traced using a *shooting method*, in which the take-off angle at the source is varied until the ray reaches a desired receiver location. The software was used with a simple model, corresponding to a constant-velocity layer over a half-space (Fig. 8.12). The model parameters are $H = 3$ km, $\alpha' = 4$ km s^{-1}, $\beta' = 2.3$ km s^{-1}, $\rho' = 2.2$ g cm^{-3}, $\alpha = 5.5$ km s^{-1}, $\beta = 3.2$ km s^{-1}, and $\rho = 2.6$ g cm^{-3}. The source and the receivers are on the surface. The source generates P and S waves, and the boundaries of the layer generate multiply reflected waves

Distance (km)

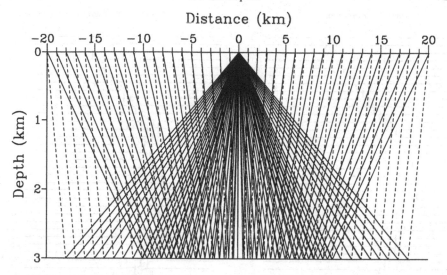

Fig. 8.12. Raypaths for reflected *P* wave (solid lines) and *S* waves (dashed lines) for a flat homogeneous layer. The *P*- and *S*-wave velocities are 4.0 and 2.3 km s^{-1} in the layer, and 5.5 and 3.2 km s^{-1} in the half-space.

with and without conversion, but to simplify the example here we only consider source-generated *P* waves reflected as *P* and *S* waves. The ray trajectories are shown in Fig. 8.12 and the vertical and horizontal components of the ground displacement in Fig. 8.13. For a comparison, the expected arrival times of the *P* reflections (Problem 8.16) are shown in Fig. 8.13. Several features should be noted. One of them is the change in the shape of the wave pulses as a function of distance. For distances between −6 and 6 km for the *P* waves and between −4 and 4 km for the *S* waves, the pulse is similar in shape to the pulse generated at the source. For other distances the pulse is a combination of the source pulse and its Hilbert transform and is the result of an incidence angle larger than the critical value. This can be seen using arguments similar to those in §6.5.3.3. In our case there are two possible critical angles: one for the *P* waves incident at the bottom of the layer (equal to 46.7°) and one for the *S* waves incident at the free surface (equal to 35.1°). The corresponding critical distances are 6.4 and 4.2 km, respectively, in agreement with the observations in the synthetic seismograms. A second observation is that for distances larger than the critical the *P* waves arrive before they are expected to. This noncausal behavior is inherent to ray theory because the Hilbert transform is generally noncausal even for causal functions (Cerveny, 2001). Finally, note the variation in amplitude as a function of distance, with the largest amplitudes for distances close to, but larger than, the critical distances, and the overall larger amplitudes of the *P* waves in the horizontal components.

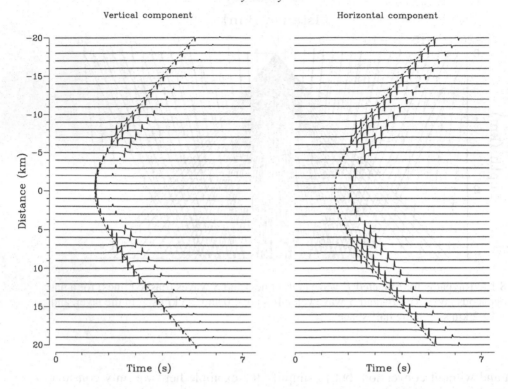

Fig. 8.13. Ray-theoretic synthetic seismogram for the rays shown in Fig. 8.12. The dashed curve shows the expected *P*-wave arrival times (see Problem 8.16). The ray tracing package Seis81, written by Cerveny and Psencik and modified by Herrmann (1998), was used.

Problems

8.1   Verify (8.3.8).

8.2   Show that the matrix $\boldsymbol{\Gamma}$ in (8.3.12) is symmetric.

8.3   Verify the first equality in (8.3.16).

8.4   Verify (8.5.6).

8.5   Refer to Fig. 8.2. Show that $\Delta\theta$ approaches $|\Delta\mathbf{t}|$ as Δs goes to zero.

8.6   Using (8.5.7) show that a curve with zero curvature is a straight line.

8.7   Show that a curve with zero torsion is a plane curve.

8.8   (a) Show that the curvature and torsion of the helix defined by $(a\cos u, a\sin u, b\,u)$ with a and b nonzero constants and $a > 0$ are given by

$$\kappa = \frac{a}{a^2 + b^2}; \qquad \tau = \frac{b}{a^2 + b^2}.$$

Therefore, the helix will be right-handed or left-handed depending on whether b is positive or negative.

(b) Show that the helix defined by $(a \cos u, -a \sin u, b\, u)$ with a and b positive constants is left-handed.

8.9 Verify (8.5.16).

8.10 Consider a spherical wave in a medium with constant velocity. Using (8.7.10) show that the ratio of amplitudes A_1/A_o is equal to r_o/r_1, where r indicates distance from the origin. (After Whitham, 1974.)

8.11 Show that the two terms on the left of (8.7.18) are equal to each other.

8.12 Refer to §8.7.1. Show that

$$\frac{dA}{dt} + \frac{1}{2}Ac^2\nabla^2 T = 0.$$

Compare with (8.7.26).

8.13 Verify (8.7.43).

8.14 Verify (8.7.53).

8.15 Verify (8.7.54) and (8.7.55).

8.16 Consider Fig. 8.12. Let x denote horizontal distance (measured from the source location), H the thickness of the layer, and α the P-wave velocity. Show that the travel times for the reflected P waves satisfy the equation

$$t^2 = \frac{x^2}{\alpha^2} + t_o^2,$$

where $t_o = 2H/\alpha$.

9

Seismic point sources in unbounded homogeneous media

9.1 Introduction

In the previous chapters we studied the propagation of plane waves without consideration of the source of the waves. Although this approach is very fruitful, it does not allow investigation of the waves generated by seismic sources, either natural or artificial. Earthquakes are the most important natural sources, and the study of the waves they generate has played a major role in our understanding of the inner structure of the Earth and the nature of the earthquake source, which will be the subject of the next chapter. However, before reaching the point where it can be analyzed it is necessary to start with simpler problems, which will be done in this chapter.

The simplest problem corresponds to a spatially concentrated force (or *point source*) directed along one of the coordinate axes. Even in this case, however, solving the elastic wave equation is a rather complicated task that requires considerable mathematical background, which is provided below. The starting point is the scalar wave equation with a source term, which is first solved for an impulsive source, in which case the solution is known as Green's function for the problem. Then the Helmholtz decomposition theorem, which we have already encountered in §5.6, is used to reduce the solution of the elastic wave equation to the solution of two simpler ones. After this series of steps, and considerable additional work, the problem of the concentrated force can be solved. Then it is relatively easy to investigate the problem involving pairs of parallel forces of equal magnitude and opposite directions a small distance apart. This combination of forces is extremely important because it allows the introduction of the concept of a moment tensor, which plays a fundamental role in the theory and is treated in considerable detail.

Solving these elastic problems results in a vector solution describing the displacement of the medium, but this is not the end of the analysis. In each case it is necessary to investigate the nature of the solution, which requires consideration

278

of the type of motion (e.g., *P*- and *S*-wave motion) and the dependence of the solution on distance (e.g., near and far fields) and on orientation (i.e., radiation patterns). These questions are also discussed in detail here.

9.2 The scalar wave equation with a source term

It is represented by the equation

$$\frac{\partial^2 \psi(\mathbf{x}, t)}{\partial t^2} = c^2 \nabla^2 \psi(\mathbf{x}, t) + F(\mathbf{x}, t), \tag{9.2.1}$$

where the velocity c is a constant and $F(\mathbf{x}, t)$ is the source term. This equation can be solved in two steps, as follows.

(1) Impulsive source. In this case

$$F(\mathbf{x}, t) = \delta(\mathbf{x} - \boldsymbol{\xi})\delta(t - t_o), \tag{9.2.2}$$

where, in Cartesian coordinates,

$$\delta(\mathbf{x} - \boldsymbol{\xi}) = \delta(x_1 - \xi_1)\delta(x_2 - \xi_2)\delta(x_3 - \xi_3). \tag{9.2.3}$$

Equation (9.2.2) represents a concentrated source acting at point $\boldsymbol{\xi}$ (Fig. 9.1) and time t_o. Under these conditions the function ψ is known as *Green's function* for the wave equation and is indicated by $G(\mathbf{x}, t; \boldsymbol{\xi}, t_o)$. Equation (9.2.1) now becomes

$$\frac{\partial^2 G}{\partial t^2} = c^2 \nabla^2 G + \delta(\mathbf{x} - \boldsymbol{\xi})\delta(t - t_o). \tag{9.2.4}$$

To satisfy causality (i.e., waves cannot exist before the source acts) the additional condition $G = 0$ for $t < t_o$ must be imposed. As shown in Appendix C,

$$G(\mathbf{x}, t; \boldsymbol{\xi}, t_o) = \frac{1}{4\pi c r}\delta[r - c(t - t_o)], \tag{9.2.5}$$

where

$$r = |\mathbf{x} - \boldsymbol{\xi}| = \left[(x_1 - \xi_1)^2 + (x_2 - \xi_2)^2 + (x_3 - \xi_3)^2\right]^{1/2}. \tag{9.2.6}$$

Equation (9.2.5) represents a spherical surface centered at $\boldsymbol{\xi}$ that expands with velocity c and has an amplitude inversely proportional to the distance r. Note that since the delta is different from zero only when its argument is equal to zero, a point $\mathbf{x}$ will be affected by the source only when $r = c(t - t_o)$. Also note that G depends on t and t_o via the combination $t - t_o$, which is sometimes called the elapsed time. This means that a translation of the time axis will not affect Green's function.

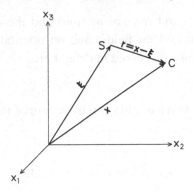

Fig. 9.1. Geometry for the solution of the scalar wave equation with a concentrated source. S and C indicate source and receiver locations, respectively.

Green's function can be rewritten using the general properties of Dirac's delta

$$\delta(\mathbf{x} - \boldsymbol{\xi}) = \delta(\boldsymbol{\xi} - \mathbf{x}); \qquad \delta(c\mathbf{x}) = \frac{1}{c}\delta(\mathbf{x}). \qquad (9.2.7\text{a,b})$$

Equation (9.2.7a) follows from the fact that the delta is even (see Appendix A). For (9.2.7b) see Problem 9.1. Then,

$$G(\mathbf{x}, t; \boldsymbol{\xi}, t_o) = \frac{1}{4\pi c^2 r}\delta\left(t - t_o - \frac{r}{c}\right). \qquad (9.2.8)$$

(2) Arbitrary sources. Let

$$F(\mathbf{x}, t) = \frac{1}{\rho}\Phi(\mathbf{x}, t), \qquad (9.2.9)$$

where ρ is a constant introduced for later use. $\Phi(\mathbf{x}, t)$ is a function that may be extended in space and time. The wave equation now becomes

$$\frac{\partial^2 \psi(\mathbf{x}, t)}{\partial t^2} = \ddot{\psi}(\mathbf{x}, t) = c^2\nabla^2\psi(\mathbf{x}, t) + \frac{1}{\rho}\Phi(\mathbf{x}, t). \qquad (9.2.10)$$

The solution ψ can be written in terms of Green's function given in (9.2.5). In an unbounded medium and under the assumption of zero initial conditions, which means that ψ and its time derivative are zero for $t = 0$, ψ is given by

$$\psi(\mathbf{x}, t) = \frac{1}{\rho}\int_0^{t^+} dt_o \int_V G(\mathbf{x}, t; \boldsymbol{\xi}, t_o)\Phi(\boldsymbol{\xi}, t_o)\,dV_\xi$$

$$= \frac{1}{4\pi c\rho}\int_0^{t^+} dt_o \int_V \frac{1}{r}\delta[r - c(t - t_o)]\Phi(\boldsymbol{\xi}, t_o)\,dV_\xi \qquad (9.2.11)$$

(Morse and Feshbach, 1953; Haberman, 1983). The time t^+ is equal to $t + \delta t$, with δt arbitrarily small. The spatial integration must be carried out over the volume V inside which the integrand is nonzero. Here and in the following, the subscript attached to the volume element indicates the integration variable. Thus, dV_ξ is equal to $d\xi_1 \, d\xi_2 \, d\xi_3$. Note that the first equality in (9.2.11) can be interpreted as the superposition of a distribution of point sources, each having an amplitude given by $\Phi(\mathbf{x}, t)/\rho$.

Because of the presence of Dirac's delta in (9.2.11), the integration over time can be performed easily by noting that, in general,

$$\int_{-\infty}^{\infty} \delta(s - (t - \tau)) f(s) \, ds = f(t - \tau) \tag{9.2.12}$$

(Problem 9.2). In addition, the delta appearing in (9.2.11) can be written as

$$\delta[r - c(t - t_o)] = \frac{1}{c}\delta[t_o - (t - r/c)] \tag{9.2.13}$$

(Problem 9.3).

Then, using (9.2.12) and (9.2.13), equation (9.2.11) becomes

$$\psi(\mathbf{x}, t) = \frac{1}{4\pi c^2 \rho} \int_V \frac{\Phi(\boldsymbol{\xi}, t - r/c)}{r} \, dV_\xi = \frac{1}{4\pi c^2 \rho} \int_V \frac{\Phi(\boldsymbol{\xi}, t - |\mathbf{x} - \boldsymbol{\xi}|/c)}{|\mathbf{x} - \boldsymbol{\xi}|} \, dV_\xi.$$
$$\tag{9.2.14}$$

Two aspects of this solution are worth noting. First, because $\Phi(\mathbf{x}, t)$ is assumed to be zero for $t < 0$, $\Phi(\boldsymbol{\xi}, t - r/c)$ is nonzero when $t - r/c \geq 0$, or $r = |\mathbf{x} - \boldsymbol{\xi}| \leq ct$. This is the equation of a sphere with center at $\mathbf{x}$, and defines the volume V. Secondly, the value of ψ at a particular time t involves source contributions corresponding to the earlier times $t - r/c$. For this reason $\psi(\mathbf{x}, t)$ is known as a *retarded solution* (or *potential*).

9.3 Helmholtz decomposition of a vector field

Let $\mathbf{Z}(\mathbf{x})$ be a vector field. Then there exist a scalar $V(\mathbf{x})$ and a vector $\mathbf{Y}(\mathbf{x})$, known as scalar and vector potentials, such that

$$\mathbf{Z} = \nabla V + \nabla \times \mathbf{Y}; \qquad \nabla \cdot \mathbf{Y} = 0. \tag{9.3.1}$$

To prove this consider the vector Poisson equation

$$\nabla^2 \mathbf{W}(\mathbf{x}) = \mathbf{Z}(\mathbf{x}) \tag{9.3.2}$$

and use the definition of the Laplacian of a vector

$$\nabla^2 \mathbf{W} = \nabla(\nabla \cdot \mathbf{W}) - \nabla \times (\nabla \times \mathbf{W}) \equiv \nabla V + \nabla \times \mathbf{Y} \tag{9.3.3}$$

(see (1.4.53)) with

$$V = \nabla \cdot \mathbf{W}; \qquad \mathbf{Y} = -\nabla \times \mathbf{W} \qquad (9.3.4\text{a,b})$$

and

$$\nabla \cdot \mathbf{Y} = -\nabla \cdot (\nabla \times \mathbf{W}) = 0. \qquad (9.3.5)$$

Equation (9.3.5) corresponds to a general property of vector fields (see §1.4.5).

Equations (9.3.4a,b) show how to find V and $\mathbf{Y}$ when $\mathbf{W}$ is given. To complete the decomposition it is necessary to solve (9.3.2), which in Cartesian coordinates can be written as three scalar Poisson equations. The latter can be solved either from first principles (Haberman, 1983), or by removing the time dependence in (9.2.14), which gives

$$\mathbf{W}(\mathbf{x}) = -\frac{1}{4\pi} \int_V \frac{\mathbf{Z}(\boldsymbol{\xi})}{r} \, dV_\xi = -\frac{1}{4\pi} \int_V \frac{\mathbf{Z}(\boldsymbol{\xi})}{|\mathbf{x} - \boldsymbol{\xi}|} \, dV_\xi. \qquad (9.3.6)$$

The minus sign arises because in Poisson's equation the source function $\mathbf{Z}$ and $\nabla^2 \mathbf{W}$ are on different sides of the equal sign. Also, $c = \rho = 1$.

This presentation of the Helmholtz decomposition has ignored two important facts. One is that the derivation given here is valid for finite volumes. For infinite space the following condition must be satisfied. If s indicates the distance from the origin, then $|\mathbf{Z}|$ must go to zero at least as fast as k/s^2, where k is a constant. The other is that $\mathbf{W}$ must satisfy certain continuity and differentiability conditions. More details are given in Achenbach (1973) and Miklowitz (1984) and references therein.

9.4 Lamé's solution of the elastic wave equation

Here we will show that a displacement of the form

$$\mathbf{u}(\mathbf{x}, t) = \nabla \phi(\mathbf{x}, t) + \nabla \times \boldsymbol{\psi}(\mathbf{x}, t) \qquad (9.4.1)$$

solves the elastic wave equation with $\mathbf{f}$ now a force per unit volume

$$\rho \ddot{\mathbf{u}} = (\lambda + 2\mu)\nabla(\nabla \cdot \mathbf{u}) - \mu \nabla \times (\nabla \times \mathbf{u}) + \mathbf{f} \qquad (9.4.2)$$

(see (4.8.2) and (4.8.3)) provided that ϕ and $\boldsymbol{\psi}$ satisfy the following equations:

$$\ddot{\phi} = \alpha^2 \nabla^2 \phi + \frac{1}{\rho} \Phi; \qquad \alpha^2 = \frac{\lambda + 2\mu}{\rho} \qquad (9.4.3\text{a,b})$$

$$\ddot{\boldsymbol{\psi}} = \beta^2 \nabla^2 \boldsymbol{\psi} + \frac{1}{\rho} \boldsymbol{\Psi}; \qquad \beta^2 = \frac{\mu}{\rho} \qquad (9.4.4\text{a,b})$$

$$\nabla \cdot \boldsymbol{\psi} = 0. \qquad (9.4.5)$$

The velocities α and β are the P- and S-wave velocities introduced in (4.8.5a,b). This solution for $\mathbf{f} = \mathbf{0}$ was presented by Lamé in 1852 (Miklowitz, 1984). The following proof follows Achenbach (1973). Let $\dot{\mathbf{u}}(\mathbf{x}, 0)$ and $\mathbf{u}(\mathbf{x}, 0)$ be the initial displacement and velocity. These initial conditions and the force $\mathbf{f}$ will be rewritten using the Helmholtz decomposition:

$$\dot{\mathbf{u}}(\mathbf{x}, 0) = \nabla A + \nabla \times \mathbf{B} \tag{9.4.6}$$

$$\mathbf{u}(\mathbf{x}, 0) = \nabla C + \nabla \times \mathbf{D} \tag{9.4.7}$$

$$\mathbf{f} = \nabla \Phi + \nabla \times \mathbf{\Psi} \tag{9.4.8}$$

with

$$\nabla \cdot \mathbf{B} = 0; \qquad \nabla \cdot \mathbf{D} = 0; \qquad \nabla \cdot \mathbf{\Psi} = 0. \tag{9.4.9}$$

After rewriting (9.4.2) as

$$\ddot{\mathbf{u}} = \alpha^2 \nabla (\nabla \cdot \mathbf{u}) - \beta^2 \nabla \times (\nabla \times \mathbf{u}) + \frac{1}{\rho} \mathbf{f} \tag{9.4.10}$$

and integrating twice with respect to time we obtain

$$\mathbf{u} = \alpha^2 \nabla \int_0^t d\tau \int_0^\tau (\nabla \cdot \mathbf{u}) \, ds - \beta^2 \nabla \times \int_0^t d\tau \int_0^\tau (\nabla \times \mathbf{u}) \, ds$$

$$+ \int_0^t d\tau \int_0^\tau \frac{1}{\rho} \mathbf{f} \, ds + t \dot{\mathbf{u}}(\mathbf{x}, 0) + \mathbf{u}(\mathbf{x}, 0). \tag{9.4.11}$$

Now introduce (9.4.6)–(9.4.9) into (9.4.11) and define

$$\phi = \alpha^2 \int_0^t \int_0^\tau \left(\nabla \cdot \mathbf{u} + \frac{1}{\rho \alpha^2} \Phi \right) ds \, d\tau + At + C \tag{9.4.12}$$

$$\mathbf{\psi} = -\beta^2 \int_0^t \int_0^\tau \left(\nabla \times \mathbf{u} - \frac{1}{\rho \beta^2} \mathbf{\Psi} \right) ds \, d\tau + \mathbf{B}t + \mathbf{D}. \tag{9.4.13}$$

With these definitions $\mathbf{u}$ becomes

$$\mathbf{u}(\mathbf{x}, t) = \nabla \phi + \nabla \times \mathbf{\psi} \tag{9.4.14}$$

which verifies (9.4.1). Also note that from (9.4.13), (9.4.9) and $\nabla \cdot (\nabla \times \mathbf{u}) = 0$ (see (9.3.5)) it follows that

$$\nabla \cdot \mathbf{\psi} = 0, \tag{9.4.15}$$

which proves (9.4.5). Next differentiate (9.4.12) and (9.4.13) twice with respect to time, which gives

$$\ddot{\phi} = \alpha^2 \nabla \cdot \mathbf{u} + \frac{1}{\rho} \Phi \tag{9.4.16}$$

$$\ddot{\boldsymbol{\psi}} = -\beta^2 \nabla \times \mathbf{u} + \frac{1}{\rho}\boldsymbol{\Psi}. \tag{9.4.17}$$

Finally, application to (9.4.1) of the divergence and curl operations give

$$\nabla \cdot \mathbf{u} = \nabla^2 \phi + \nabla \cdot (\nabla \times \boldsymbol{\psi}) \equiv \nabla^2 \phi \tag{9.4.18}$$

and

$$\nabla \times \mathbf{u} = \nabla \times (\nabla \phi) + \nabla \times (\nabla \times \boldsymbol{\psi}) \equiv \nabla \times (\nabla \times \boldsymbol{\psi}) = -\nabla^2 \boldsymbol{\psi} + \nabla(\nabla \cdot \boldsymbol{\psi}) \equiv -\nabla^2 \boldsymbol{\psi} \tag{9.4.19}$$

(Problem 9.4). Then, equations (9.4.16)–(9.4.19) give

$$\ddot{\phi} = \alpha^2 \nabla^2 \phi + \frac{1}{\rho}\Phi \tag{9.4.20}$$

$$\ddot{\boldsymbol{\psi}} = \beta^2 \nabla^2 \boldsymbol{\psi} + \frac{1}{\rho}\boldsymbol{\Psi} \tag{9.4.21}$$

in agreement with (9.4.3) and (9.4.4).

Lamé's solution is important because it reduces the solution of the complicated elastic wave equation to the solution of two simpler equations and provides an alternative approach to that introduced in §5.8. The displacements $\nabla \phi$ and $\nabla \times \boldsymbol{\psi}$ are the P- and S-wave components of $\mathbf{u}$, respectively. If the coordinate system is chosen as in §5.8.4, the potentials ϕ and ψ will be functions of x_1, x_3, and t, but not of x_2. In addition, if the body force is equal to zero and $\boldsymbol{\psi}$ is of the form

$$\boldsymbol{\psi}(x_1, x_3, t) = (0, \psi(x_1, x_3, t), 0), \tag{9.4.22}$$

then (9.4.15) is satisfied, equation (9.4.21) becomes a scalar wave equation, and

$$u_1 = \frac{\partial \phi}{\partial x_1} - \frac{\partial \psi}{\partial x_3} \tag{9.4.23a}$$

$$u_2 = 0 \tag{9.4.23b}$$

$$u_3 = \frac{\partial \phi}{\partial x_3} + \frac{\partial \psi}{\partial x_1} \tag{9.4.23c}$$

(Problem 9.5). These are the equations used to solve P–SV problems in terms of potentials. For the SH waves, characterized by

$$\mathbf{u} = (0, u_2(x_1, x_3, t), 0), \tag{9.4.24}$$

it is possible to find an appropriate potential (Miklowitz, 1984), but this is not necessary because in this case u_2 itself satisfies a scalar wave equation (Problem 9.6).

9.5 The elastic wave equation with a concentrated force in the x_j direction

The force $\mathbf{f}$ in (9.4.10) will be assumed to be spatially concentrated and applied in the x_j direction at the point $\boldsymbol{\xi}$. To simplify the treatment we start with the case $j = 1$. The force will be represented by

$$\mathbf{f}(\mathbf{x}, t; \boldsymbol{\xi}) = T(t)\delta(\mathbf{x} - \boldsymbol{\xi})\mathbf{e}_1 = T(t)\delta(\mathbf{x} - \boldsymbol{\xi})(1, 0, 0). \tag{9.5.1}$$

The solution to this problem was presented by Stokes in 1849, and was later confirmed by Love in 1904 using a different approach. Here we follow Miklowitz (1984), who in turn followed the work of Love. To obtain Stokes' solution the first step is to apply the Helmholtz decomposition to $\mathbf{f}$, which involves finding the vector $\mathbf{W}$, which is obtained from (9.3.6) and (9.5.1):

$$\mathbf{W}(\mathbf{x}, t; \boldsymbol{\xi}) = -\frac{T(t)}{4\pi}\mathbf{e}_1 \int_V \frac{\delta(\boldsymbol{\chi} - \boldsymbol{\xi})}{|\mathbf{x} - \boldsymbol{\chi}|} \, dV_\chi$$

$$= -\frac{T(t)}{4\pi}\frac{1}{|\mathbf{x} - \boldsymbol{\xi}|}(1, 0, 0) = -\frac{T(t)}{4\pi}\frac{1}{r}(1, 0, 0). \tag{9.5.2}$$

Note that in (9.3.6) $\boldsymbol{\xi}$ denotes the integration variable, where now $\boldsymbol{\xi}$ is used to indicate the source position. Therefore, the integration variable must be given a different name. Also note that $\mathbf{W}$ has nonzero components in the x_1 direction only.

The potentials Φ and $\boldsymbol{\Psi}$ are obtained from (9.3.4a,b)

$$\Phi = \nabla \cdot \mathbf{W} = -\frac{T(t)}{4\pi}\frac{\partial}{\partial x_1}\frac{1}{r} \tag{9.5.3}$$

$$\boldsymbol{\Psi} = -\nabla \times \mathbf{W} = \frac{T(t)}{4\pi}\left(0, \frac{\partial}{\partial x_3}\frac{1}{r}, -\frac{\partial}{\partial x_2}\frac{1}{r}\right) \tag{9.5.4}$$

(Problem 9.7). Now introduce (9.5.3) and (9.5.4) in (9.4.20) and (9.4.21), respectively:

$$\ddot{\phi} = \alpha^2 \nabla^2 \phi - \frac{T(t)}{4\pi\rho}\frac{\partial}{\partial x_1}\frac{1}{r} \tag{9.5.5}$$

$$\ddot{\boldsymbol{\psi}} = \beta^2 \nabla^2 \boldsymbol{\psi} + \frac{T(t)}{4\pi\rho}\left(0, \frac{\partial}{\partial x_3}\frac{1}{r}, -\frac{\partial}{\partial x_2}\frac{1}{r}\right). \tag{9.5.6}$$

The solution of (9.5.5) is given by (9.2.14),

$$\phi(\mathbf{x}, t; \boldsymbol{\xi}) = \frac{-1}{(4\pi\alpha)^2 \rho} \int_V \frac{T(t - |\mathbf{x} - \boldsymbol{\chi}|/\alpha)}{|\mathbf{x} - \boldsymbol{\chi}|}\frac{\partial}{\partial \chi_1}\frac{1}{|\mathbf{x} - \boldsymbol{\xi}|} \, dV_\chi$$

$$= \frac{-1}{(4\pi\alpha)^2 \rho} \int_V \frac{T(t - h/\alpha)}{h}\frac{\partial}{\partial \chi_1}\frac{1}{R} \, dV_\chi, \tag{9.5.7}$$

where $h = |\mathbf{x} - \boldsymbol{\chi}|$ and $R = |\boldsymbol{\chi} - \boldsymbol{\xi}|$.

The integral on the right-hand side of (9.5.7) can be solved by dividing the

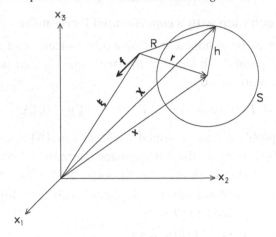

Fig. 9.2. Geometry for the integration indicated in (9.5.7). Vectors $\boldsymbol{\xi}$, $\mathbf{x}$, and $\boldsymbol{\chi}$ correspond to the source and receiver locations and to the integration variable, respectively. σ indicates a sphere centered at the receiver location having radius h. Vector $\mathbf{f}$ indicates the force, which acts in the direction of x_1. (After Miklowitz, 1984.)

volume into thin spherical shells centered at the observation point $\mathbf{x}$ and having radius equal to h (Fig. 9.2). On the shell the function $T(t - h/\alpha)$ is a constant, so that (9.5.7) becomes

$$\phi(\mathbf{x}, t; \boldsymbol{\xi}) = \frac{-1}{(4\pi\alpha)^2 \rho} \int_0^\infty \frac{T(t - h/\alpha)}{h} \, \mathrm{d}h \int_\sigma \frac{\partial}{\partial \chi_1} \frac{1}{R} \, \mathrm{d}\sigma, \qquad (9.5.8)$$

where σ is the surface of the shell at a distance h and $\mathrm{d}\sigma$ is the corresponding surface element. The surface integral appears in potential theory and is equal to

$$\int_\sigma \frac{\partial}{\partial \chi_1} \frac{1}{R} \, \mathrm{d}\sigma = \begin{cases} 0 & \text{for } h > r \\ 4\pi h^2 \dfrac{\partial}{\partial x_1}(1/r) & \text{for } h < r \end{cases} \qquad (9.5.9)$$

(see Aki and Richards (1980) for a discussion of this integral). Equation (9.5.9) indicates that the upper limit for the integral over h in (9.5.8) is r. With this modification and introducing the change of variable $h = \alpha\tau$, equation (9.5.8) becomes

$$\phi(\mathbf{x}, t; \boldsymbol{\xi}) = \frac{-1}{4\pi\rho} \left(\frac{\partial}{\partial x_1} \frac{1}{r} \right) \int_0^{r/\alpha} \tau T(t - \tau) \, \mathrm{d}\tau. \qquad (9.5.10)$$

In a similar way, the vector $\boldsymbol{\Psi}$ is found to be

$$\boldsymbol{\Psi}(\mathbf{x}, t; \boldsymbol{\xi}) = \frac{1}{4\pi\rho} \left(0, \frac{\partial}{\partial x_3} \frac{1}{r}, -\frac{\partial}{\partial x_2} \frac{1}{r} \right) \int_0^{r/\beta} \tau T(t - \tau) \, \mathrm{d}\tau \qquad (9.5.11)$$

(Problem 9.8).

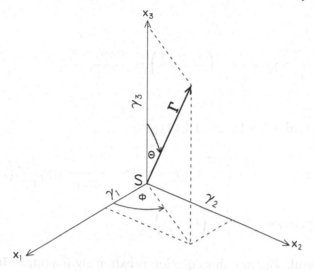

Fig. 9.3. Definition of the direction cosines γ_1, γ_2, and γ_3 and vector $\boldsymbol{\Gamma}$. S indicates the source location. The receiver is in the direction of $\boldsymbol{\Gamma}$.

To determine the displacement $\mathbf{u}$ we use (9.4.1). The ith component is given by

$$u_{i1}(\mathbf{x}, t; \boldsymbol{\xi}) = (\nabla\phi)_i + (\nabla \times \boldsymbol{\Psi})_i = \frac{1}{4\pi\rho} \left(\frac{\partial^2}{\partial x_i \partial x_1} \frac{1}{r} \right) \int_{r/\beta}^{r/\alpha} \tau T(t - \tau)\, d\tau$$

$$+ \frac{1}{4\pi\rho\alpha^2 r} \left(\frac{\partial r}{\partial x_i} \frac{\partial r}{\partial x_1} \right) T(t - r/\alpha)$$

$$+ \frac{1}{4\pi\rho\beta^2 r} \left(\delta_{i1} - \frac{\partial r}{\partial x_i} \frac{\partial r}{\partial x_1} \right) T(t - r/\beta). \tag{9.5.12}$$

The derivation of (9.5.12) is given in Appendix D. The subscript 1 in u_{i1} is needed to indicate the direction of the force.

Equation (9.5.12) will be rewritten using the direction cosines of the vector $\mathbf{x} - \boldsymbol{\xi}$, given by

$$\gamma_i = \frac{x_i - \xi_i}{r} = \frac{\partial r}{\partial x_i}. \tag{9.5.13}$$

The last equality in (9.5.13) follows from the definition of r (see (9.2.6)). γ_i constitute the elements of a unit vector

$$\boldsymbol{\Gamma} = (\gamma_1, \gamma_2, \gamma_3) \tag{9.5.14}$$

along the source–receiver direction (Fig. 9.3).

Now the direction x_1 will be replaced by any of the three coordinate directions, labeled x_j. Then,

$$\frac{\partial^2}{\partial x_i x_j}\frac{1}{r} = \frac{2}{r^3}\gamma_i\gamma_j - \frac{1}{r^2}\frac{\partial}{\partial x_i}\left(\frac{x_j - \xi_j}{r}\right) = \frac{3\gamma_i\gamma_j - \delta_{ij}}{r^3} \qquad (9.5.15)$$

(Problem 9.9).

Introducing (9.5.13) and (9.5.15) in (9.5.12) gives

$$u_{ij}(\mathbf{x}, t) = \frac{1}{4\pi\rho}(3\gamma_i\gamma_j - \delta_{ij})\frac{1}{r^3}\int_{r/\alpha}^{r/\beta}\tau T(t - \tau)\,\mathrm{d}\tau + \frac{1}{4\pi\rho\alpha^2}\gamma_i\gamma_j\frac{1}{r}T\left(t - \frac{r}{\alpha}\right)$$

$$- \frac{1}{4\pi\rho\beta^2}(\gamma_i\gamma_j - \delta_{ij})\frac{1}{r}T\left(t - \frac{r}{\beta}\right). \qquad (9.5.16)$$

This is our final result. Because this equation is extremely important, it will be discussed in detail. Two questions will be addressed. First, what type of motion (i.e. *P* wave, *S* wave, other?) does u_{ij} represent? Secondly, what is the relative importance of the three terms in (9.5.16) as a function of the source–receiver distance *r* (i.e., near and far fields)?

9.5.1 Type of motion

For given values of ρ, α, and β, each of the last two terms on the right-hand side of (9.5.16) is the product of three factors: a vectorial factor which depends on γ_i and γ_j, and which determines the type of motion, and two scalar factors, r^{-1} and $T(t - r/c)$, with $c = \alpha, \beta$. The factor $T(t - r/c)$ indicates wave propagation with *P*- and *S*-wave velocities. The factor r^{-1} is discussed below. For the second term in (9.5.16) the vectorial factor $\gamma_i\gamma_j$ is a multiple of γ_i because γ_j is fixed. Therefore, this term is in the direction of Γ, which corresponds to *P*-wave motion (see §5.8.1). For the third term in (9.5.16), the factor $\gamma_i\gamma_j - \delta_{ij}$ corresponds to a vector perpendicular to Γ, as can be seen by computing their scalar product:

$$\gamma_i(\gamma_i\gamma_j - \delta_{ij}) = \gamma_i\gamma_i\gamma_j - \gamma_i\delta_{ij} = 0. \qquad (9.5.17)$$

Here the properties $\gamma_i\gamma_i = 1$ and $\gamma_i\delta_{ij} = \gamma_j$ were used. Therefore, the third term corresponds to *S*-wave motion (see §5.8.1).

The first term in (9.5.16) is more complicated than the last two, as can be seen by writing the vectorial factor as $2\gamma_i\gamma_j + (\gamma_i\gamma_j - \delta_{ij})$, which shows that this factor represents a combination of *P* and *S* motion (see Problems 9.10 and 9.11). The contribution of the integral to the displacement can be better understood if it is written as

$$I \equiv \int_{r/\alpha}^{r/\beta} \tau T(t-\tau)\,d\tau = \int_{-\infty}^{\infty} \tau \left[H\left(\tau - \frac{r}{\alpha}\right) - H\left(\tau - \frac{r}{\beta}\right) \right] T(t-\tau)\,d\tau$$

$$= t \left[H\left(t - \frac{r}{\alpha}\right) - H\left(t - \frac{r}{\beta}\right) \right] * T(t). \tag{9.5.18}$$

The function $H(t)$ represents the Heaviside unit step function, and the difference of the two Heaviside functions gives the "box car" function, with a value of one within the interval $[r/\alpha, r/\beta]$ and zero elsewhere. Therefore, $t[H(t-r/\alpha)-H(t-r/\beta)]$ is a discontinuous function with steps of r/α and r/β. Within the box this function is a straight line with slope equal to one. Therefore, the contribution of I to u_{ij} will be a roughly linear trend, depending on the nature of $T(t)$, over the time interval $[r/\alpha, (r/\beta) + w]$, where w is the width (or duration) of $T(t)$. The importance of this contribution is discussed below.

9.5.2 Near and far fields

Here we will discuss the dependence of u_{ij} on r, and the relative importance of the three terms on the right-hand side of (9.5.16). The last two terms depend on r^{-1}, which means that they will dominate for sufficiently large values of r. Therefore, they are known as *far-field* terms. The first term, on the other hand, has a factor proportional to r^{-3} times an integral whose limits depend on r. When $T(t)$ is close to a delta this factor is proportional to r^{-2} (see §9.6), which dominates over r^{-1} as r goes to zero. For this reason the first term is known as the *near-field* term. However, as we will see below, when discussing the contributions of the near and far fields to the displacement, in addition to distance we should consider the wavelengths involved and the nature of function $T(t)$. To examine the importance of these other variables, (9.5.16) will be written first in a way that does not involve the integral. This will be done by introducing a new function $J(t)$ such that $T(t) = J''(t)$, and then integrating by parts (Haskell, 1963, following Keilis-Borok). When that is done (9.5.16) becomes

$$4\pi\rho u_{ij}(\mathbf{x}, t) = (3\gamma_i\gamma_j - \delta_{ij})\frac{1}{r^3}J(t - r/\alpha) + (3\gamma_i\gamma_j - \delta_{ij})\frac{1}{\alpha r^2}J'(t - r/\alpha)$$

$$+ \gamma_i\gamma_j\frac{1}{\alpha^2 r}J''(t - r/\alpha) - (3\gamma_i\gamma_j - \delta_{ij})\frac{1}{r^3}J(t - r/\beta)$$

$$- (3\gamma_i\gamma_j - \delta_{ij})\frac{1}{\beta r^2}J'(t - r/\beta) - (\gamma_i\gamma_j - \delta_{ij})\frac{1}{\beta^2 r}J''(t - r/\beta)$$

$$\tag{9.5.19}$$

(Problem 9.12). This expression makes it easier to compute u_{ij} when $J''(t)$ is given, and shows that the integral in (9.5.16) contributes terms with r^{-2} and r^{-3} dependences, but because the derivative of a function has a different frequency content than the original function (Problem 9.13), we cannot use (9.5.19) to assess the importance of the various terms. A comparison is possible, however, when working in the frequency domain (Aki and Richards, 1980). Thus, applying the Fourier transform to both sides of (9.5.19) we obtain

$$4\pi\rho u_{ij}(\mathbf{x}, \omega) = T(\omega)\left\{\frac{1}{\alpha^2 r}\exp(-i\omega r/\alpha)\left[-(3\gamma_i\gamma_j - \delta_{ij})\left(\frac{\alpha}{\omega r}\right)^2\right.\right.$$

$$-\ i(3\gamma_i\gamma_j - \delta_{ij})\left(\frac{\alpha}{\omega r}\right) + \gamma_i\gamma_j\Bigg]$$

$$-\frac{1}{\beta^2 r}\exp(-i\omega r/\beta)\left[-(3\gamma_i\gamma_j - \delta_{ij})\left(\frac{\beta}{\omega r}\right)^2\right.$$

$$\left.\left.-\ i(3\gamma_i\gamma_j - \delta_{ij})\left(\frac{\beta}{\omega r}\right) + (\gamma_i\gamma_j - \delta_{ij})\right]\right\}, \qquad (9.5.20)$$

where $T(\omega)$ represents the Fourier transform of $T(t)$ (Problem 9.14).

Equation (9.5.20) shows that both r and ω must be considered when discussing the relative importance of the different terms in (9.5.16) and (9.5.19) and (9.5.20). Moreover, from (9.5.20) we see that the importance of the near- and far-field terms actually depends on the adimensional factor $c/\omega r$ ($c = \alpha, \beta$) or $\lambda/2\pi r$, where $\lambda = 2\pi c/\omega$ is wavelength (see (5.4.17)). If λ/r is on the order of one, none of the terms dominate. If $\lambda/r \gg 1$, or $\lambda \gg r$ the near-field terms dominate, and if $\lambda \ll r$ the far-field terms dominate. The following example shows that the width of the source time function (or source pulse) can also be used to estimate the importance of the terms in (9.5.16).

Equation (9.5.19) will be used to generate synthetic seismograms with the functions $J(t)$, $J'(t)$, and $J''(t)$ given by

$$J(t) = \int_0^t J'(\tau)\,d\tau = H(t)\left[t + \frac{2}{a}\left(e^{-at} - 1\right) + te^{-at}\right] \qquad (9.5.21)$$

$$J'(t) = H(t)\left(1 - e^{-at} - at\,e^{-at}\right) \qquad (9.5.22)$$

$$J''(t) = H(t)\,a^2t\,e^{-at} \qquad (9.5.23)$$

(Problem 9.15).

The function $J'(t)$ was introduced by Ohnaka and used by Harkrider (1976). Here a is a parameter with the dimension of inverse time, which controls the width (or frequency content) and amplitude of the source time function $J''(t)$. As *a* goes to infinity, the width goes to zero and the amplitude goes to infinity. Thus, in the limit, $J''(t)$ tends to Dirac's delta. Also in the limit, J and J' become $tH(t)$ and $H(t)$, respectively. Ohnaka's pulse is also used in §10.10, where $J(t)$ and its derivatives are plotted.

To examine the effect of distance and pulse width on the amplitude of u_{ij}, let $\mathbf{\Gamma} = (0.750, 0.433, 0.500)$, $\alpha = 6$ km s^{-1}, $\beta = 3.5$ km s^{-1}, $\rho = 2.8$ g cm^{-3}, $r = 100, 200$ km, and $a = 1, 3$. Also, to see the effect of the source direction on the three components of the displacement we consider all the ij combinations. However, because u_{ij} is symmetric in ij, only six combinations are required. Each of the individual plots in Fig. 9.4 shows the total field, computed using all the terms in (9.5.19), and the far field, corresponding to the terms in (9.5.19) that have a $1/r$ dependence. When $r = 100$ km and $a = 1$, the difference between the total and the far fields can be very large. Also note the linear trend between the *P*- and *S*-wave arrivals (indicated by the arrows), which comes from (9.5.18). This difference can be reduced by increasing *a* or *r* (or both), as shown in Fig. 9.4. The effect of larger *r* when the source pulse is unchanged is clear from (9.5.19): it will result in a relative reduction of the near-field terms. The effect of a change in the value of *a* when *r* is unchanged can be justified by noting that the convolution in (9.5.18) will give larger values for a wider pulse than for a narrower pulse. This fact explains qualitatively why the near field is less important for $a = 3$ than for $a = 1$. If the pulse were even narrower, the effect of the near field would be even smaller (see §9.6).

9.5.3 Example. The far field of a point force at the origin in the x_3 direction

The far-field displacement caused by this force will be discussed in detail. We will be interested, in particular, in the analysis of the *radiation pattern*, which gives the dependence of the displacement on the source–receiver direction. Therefore we will assume that $r = 1$ and will ignore all constant terms. Then, from the last two terms of (9.5.16)

$$u_i^P = \gamma_3 \gamma_i \qquad (9.5.24)$$

and

$$u_i^S = -\gamma_3 \gamma_i + \delta_{i3}, \qquad (9.5.25)$$

where the equal signs have to be interpreted to within a multiplicative constant.

In spherical coordinates (see Fig. 9.3) the direction cosines are given by

$$\gamma_1 = \sin\theta\cos\phi \qquad \gamma_2 = \sin\theta\sin\phi \qquad \gamma_3 = \cos\theta \qquad (9.5.26)$$

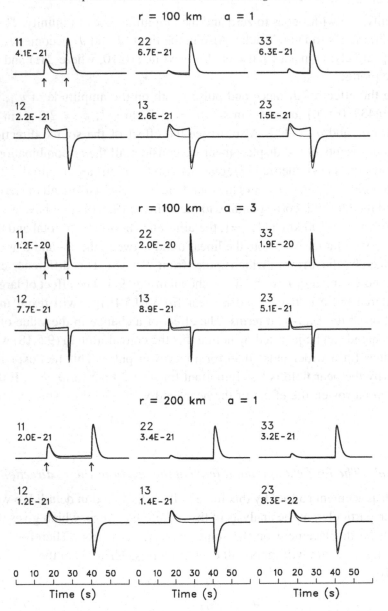

Fig. 9.4. Synthetic seismograms computed using (9.5.19) with the functions J, J', and J'' defined in (9.5.21)–(9.5.23) and three combinations of r and a. The P- and S-wave velocities and the density are given by $\alpha = 6$ km s^{-1}, $\beta = 3.5$ km s^{-1}, and $\rho = 2.8$ g cm^{-3}. The large two-digit number on the upper left-hand corner of each plot corresponds to the subindices ij. The bold curves indicate the total field (i.e., all the terms in (9.5.19) included). The thin curves indicate the far field, corresponding to terms with $1/r$ dependence. The number in exponential notation is the amplitude of u_{ij} (total field); its dimension is cm when the force is given in dynes. The arrows indicate the times t/α and t/β. (After Pujol and Herrmann, 1990.)

Introducing these expressions into (9.5.24) and (9.5.25) gives:

$$u_1^P = \frac{1}{2} \sin 2\theta \cos \phi \qquad u_2^P = \frac{1}{2} \sin 2\theta \sin \phi \qquad u_3^P = \cos^2 \theta \qquad (9.5.27)$$

and

$$u_1^S = -\frac{1}{2} \sin 2\theta \cos \phi \qquad u_2^S = -\frac{1}{2} \sin 2\theta \sin \phi \qquad u_3^S = \sin^2 \theta. \qquad (9.5.28)$$

We will consider amplitudes first. They are given by

$$|\mathbf{u}^P| = |\gamma_3| = |\cos \theta| \qquad (9.5.29)$$

and

$$|\mathbf{u}^S| = \left| \sqrt{1 - \gamma_3^2} \right| = |\sin \theta|. \qquad (9.5.30)$$

Since $|\mathbf{u}^P|$ and $|\mathbf{u}^S|$ do not depend on ϕ, they are symmetric about the x_3 axis. The plot of $f(\theta) = |\cos \theta|$ corresponds to two circles with centers on the x_3 axis (Fig. 9.5a, Problem 9.17). Therefore, because of the axial symmetry already noted, the radiation pattern for the P wave has the shape of two spheres (Fig. 9.5c). In the (x_1, x_3) plane ($\theta = 90°$) the motion is zero. The plot of $f(\theta) = |\sin \theta|$ corresponds to two circles with centers on the x_2 axis (Fig. 9.5b) so that the radiation pattern for the S wave resembles a doughnut without the central hole (Fig. 9.5d). Note that along the x_3 axis ($\theta = 0°$) there is no S-wave motion.

The radiation patterns just described give information concerning amplitudes. The direction of motion for the P and S waves has already been described (in the direction of, and perpendicular to Γ, respectively). To determine the sense of motion (positive or negative direction) it is necessary to analyze the sign of the vector components. Consider, for example, the P-wave displacement in the (x_1, x_3) plane. For positive x_1, $\phi = 0°$, and

$$u_1^P = \frac{1}{2} \sin 2\theta \qquad u_2^P = 0 \qquad u_3^P = \cos^2 \theta, \qquad (9.5.31)$$

while for negative x_1, $\phi = 180°$, and

$$u_1^P = -\frac{1}{2} \sin 2\theta \qquad u_2^P = 0 \qquad u_3^P = \cos^2 \theta. \qquad (9.5.32)$$

In both cases u_3^P is always positive, so the sense of motion depends on the sign of u_1^P. For points with positive x_3 ($\theta < 90°$), motion is away from the source, while for points with negative x_3 ($\theta > 90°$), motion is towards the source (Fig. 9.6a). This means that a seismometer located in the half-space $x_3 > 0$ will experience a push or compression, and a pull or dilatation if it is located in the half-space $x_3 < 0$.

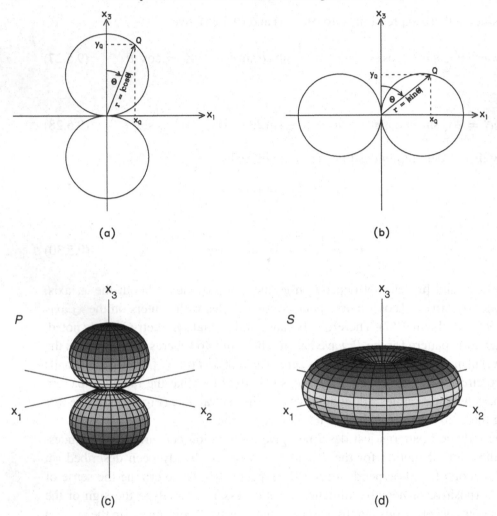

Fig. 9.5. Plot of (a) $|\cos\theta|$ and (b) $|\sin\theta|$ in polar coordinates (r, θ). The argument θ varies between $0°$ and $180°$ for x_1 positive or negative. (a) and (b) are two-dimensional representations of the far-field P- and S-wave radiation patterns generated by a point source in the x_3 direction ((see (9.5.29) and (9.5.30)). (c) and (d) are the corresponding three-dimensional representations. (2-D plots after Pujol and Herrmann, 1990.)

For the S waves, for $\phi = 0°$ and for $\phi = 180°$ we have

$$u_1^S = -\frac{1}{2}\sin 2\theta \qquad u_2^S = 0 \qquad u_3^S = \sin^2\theta \qquad (9.5.33)$$

and

$$u_1^S = \frac{1}{2}\sin 2\theta \qquad u_2^S = 0 \qquad u_3^S = \sin^2\theta \qquad (9.5.34)$$

respectively. The corresponding plots are shown in Fig. 9.6(b).

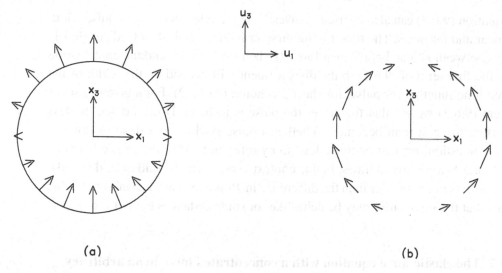

Fig. 9.6. Plot in the (x_1, x_3) plane of (a) the P and (b) S wave far-field motion generated by a point source in the x_3 direction. To draw the plots, assume that there is a pair of axes u_1 and u_3 on each point of the circle and plot (9.5.31) and (9.5.32) for (a), and (9.5.33) and (9.5.34) for (b). (After Pujol and Herrmann, 1990.)

9.6 Green's function for the elastic wave equation

Green's function is obtained from (9.5.16) with $T(t)$ replaced by $\delta(t)$ and is given by

$$G_{ij}(\mathbf{x}, t; \boldsymbol{\xi}, 0) = \frac{1}{4\pi\rho}(3\gamma_i\gamma_j - \delta_{ij})\frac{1}{r^3}\left[H\left(t - \frac{r}{\alpha}\right) - H\left(t - \frac{r}{\beta}\right)\right]t$$

$$+ \frac{1}{4\pi\rho\alpha^2}\gamma_i\gamma_j\frac{1}{r}\delta\left(t - \frac{r}{\alpha}\right) - \frac{1}{4\pi\rho\beta^2}(\gamma_i\gamma_j - \delta_{ij})\frac{1}{r}\delta\left(t - \frac{r}{\beta}\right)$$

$$\tag{9.6.1}$$

(Burridge, 1976; Hudson, 1980). Note that G_{ij} is a tensor-valued function (or tensor, for short) (Problem 9.18). The arguments $\boldsymbol{\xi}$ and 0 in G_{ij} indicate the location of the source and the time it acts. The first term on the right-hand side of (9.6.1) can be obtained directly from (9.5.18) because $\delta(t)$ is the unit element with respect to the operation of convolution (see Appendix A).

From (9.5.16) and (9.6.1) and using the relation

$$T(t) * \delta(t - t_o) = T(t - t_o) \tag{9.6.2}$$

it follows that

$$u_{ij}(\mathbf{x}, t) = T(t) * G_{ij}(\mathbf{x}, t; \boldsymbol{\xi}, 0). \tag{9.6.3}$$

Equation (9.6.3) can also be used to investigate the relation between pulse width and near and far fields. The time t in the first term on the right-hand side of (9.6.1) ranges between r/α and r/β. Therefore t/r^3 behaves as $1/r^2$ and the contribution from the first term of (9.6.1) to the displacement will depend on the width of the source time function (or pulse, for short), as noted in (9.5.2). For a given value of r, from (9.6.3) we see that the closer the pulse is to being Dirac's delta, the less important the first term becomes. Whether a pulse is close to a delta or not is a relative question, but can be made less so by reference to the difference between the P- and S-wave arrival times. In this context, a pulse can be said to be delta-like if its width is much smaller than the difference in P- and S-wave arrival times. This means that the same pulse may be delta-like for some distances but not for others.

9.7 The elastic wave equation with a concentrated force in an arbitrary direction

If the body force $\mathbf{f}(\mathbf{x}, t)$ is still applied at a point $\mathbf{x} = \boldsymbol{\xi}$ but has an arbitrary orientation represented by the vector $\mathbf{F}(t)$, the latter can be decomposed in terms of three forces in the directions of the coordinate axes:

$$\mathbf{f}(\mathbf{x}, t) = \mathbf{F}(t)\delta(\mathbf{x} - \boldsymbol{\xi}) = (F_1(t), F_2(t), F_3(t))\delta(\mathbf{x} - \boldsymbol{\xi}) \qquad (9.7.1)$$

and the total displacement becomes the sum of the displacements generated by the forces F_1, F_2, and F_3 directed along the x_1, x_2, and x_3 directions. Therefore, using (9.6.3) we obtain

$$u_i(\mathbf{x}, t) = F_1 * G_{i1} + F_2 * G_{i2} + F_3 * G_{i3} = F_j(t) * G_{ij}(\mathbf{x}, t; \boldsymbol{\xi}, 0), \qquad (9.7.2)$$

where the summation convention over repeated indices has been used in the last expression. Equation (9.7.2) can also be used to show that G_{ij} is a tensor (Problem 9.19).

Introducing the last two terms of (9.6.1) in (9.7.2) we obtain the following expression for the far field:

$$u_i = \frac{1}{4\pi\rho\alpha^2}\gamma_i\gamma_j\frac{1}{r}F_j\left(t - \frac{r}{\alpha}\right) - \frac{1}{4\pi\rho\beta^2}(\gamma_i\gamma_j - \delta_{ij})\frac{1}{r}F_j\left(t - \frac{r}{\beta}\right). \qquad (9.7.3)$$

As in the case of (9.5.16), the first term on the right-hand side of (9.7.3) represents motion in the direction of $\boldsymbol{\Gamma}$, which corresponds to P-wave motion, while the second term represents motion perpendicular to the direction of $\boldsymbol{\Gamma}$, which corresponds to S-wave motion. Also note that the maximum amplitude of the S wave is $(\alpha/\beta)^2$ times larger than the maximum amplitude of the P wave. This occurs because $|\gamma_i\gamma_j - \delta_{ij}|$ is equal to either $|\gamma_i\gamma_j|$ $(i \neq j)$ or $|\gamma_i^2 - 1|$, so that it is always ≤ 1, which in turn implies that

$$\left|\frac{1}{\alpha^2}\gamma_i\gamma_j\right| \le \frac{1}{\alpha^2}; \qquad \left|\frac{1}{\beta^2}(\gamma_i\gamma_j - \delta_{ij})\right| \le \frac{1}{\beta^2}. \qquad (9.7.4)$$

9.8 Concentrated couples and dipoles

The study of couples constitutes the first step in the development of the theory of waves generated by an earthquake source and to introduce them we will consider pairs of parallel forces of equal magnitude and opposite directions some distance apart. Here the forces will be parallel to the coordinate axes separated by very small distances. Following Love (1927), we will refer to these pairs of forces as *double forces*. If the two forces have different lines of action they constitute a *couple*; otherwise they constitute a *vector dipole* (or dipole). Let us consider now the couple with forces in the positive and negative x_3 directions separated by a distance D along the x_2 direction (Fig. 9.7). The two forces are $\mathbf{F}_3(t) = (0, 0, F_3(t))$ and $-\mathbf{F}_3(t) = (0, 0, -F_3(t))$ acting at points $(\boldsymbol{\xi} + \mathbf{d}/2)$ and $(\boldsymbol{\xi} - \mathbf{d}/2)$, respectively, where $\mathbf{d} = D\mathbf{e}_2$ is a vector of length D in the x_2 direction. The displacement caused by this couple is equal to the sum of the displacements caused by each force, so that from (9.7.2)

$$u_k(\mathbf{x}, t) = DF_3(t) * \left[\frac{G_{k3}(\mathbf{x}, t; \boldsymbol{\xi} + \frac{1}{2}D\mathbf{e}_2, 0) - G_{k3}(\mathbf{x}, t; \boldsymbol{\xi} - \frac{1}{2}D\mathbf{e}_2, 0)}{D}\right].$$

$$(9.8.1)$$

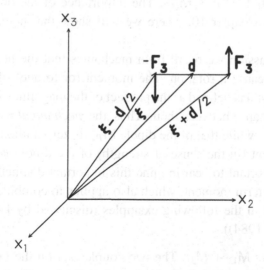

Fig. 9.7. Pair of forces ($\mathbf{F}_3$ and $-\mathbf{F}_3$) of equal magnitude in the $+x_3$ and $-x_3$ directions a small distance $D = |\mathbf{d}|$ apart in the x_2 direction. (After Ben-Menahem and Singh, 1981; Pujol and Herrmann, 1990.)

In (9.8.1) the factor D has been included in the numerator and denominator to allow for the incremental quotient in brackets, which is an approximation to the partial derivative of G_{k3} with respect to ξ_2. Taking the limit of u_k as F_3 goes to infinity and D goes to zero in such a way that the product $DF_3(t)$ remains finite gives

$$u_k(\mathbf{x}, t) = M_{32}(t) * \frac{\partial G_{k3}(\mathbf{x}, t; \boldsymbol{\xi}, 0)}{\partial \xi_2}, \qquad (9.8.2)$$

where the quantity $M_{32}(t) = DF_3(t)$ is the *moment* of the couple and is a measure of the magnitude or strength of the source.

If the couple has forces in the x_P direction separated along the x_Q direction (the direction of the arm of the couple), equation (9.8.2) becomes

$$u_k = M_{PQ} * \frac{\partial G_{kP}}{\partial \xi_Q}, \qquad (9.8.3)$$

where the arguments have been deleted for simplicity and there is no summation over uppercase subindices. The sign of F_P is important in the definition of M_{PQ}. For $\boldsymbol{\xi} = \mathbf{0}$, F_P is in the positive x_P direction when located on the positive x_Q axis. Reversing this convention gives $-M_{PQ}$. Although equation (9.8.3) was derived for couples, it is also applicable to dipoles, for which $P = Q$, although in this case it is necessary to qualify the meaning of M_{PP}, as discussed below. To identify the double force having moment M_{PQ} we will use the symbol M_{PQ}, which in turn represents the nine possible combinations shown in Fig. 9.8. In addition to couples we will also be concerned with the superposition of pairs of couples M_{PQ} and M_{QP} having the same strength. This combination, known as a *double couple*, will be indicated by the symbol $\mathsf{M}_{PQ} + \mathsf{M}_{QP}$. The importance of the double couple will become apparent in Chapter 10, where we will show that it represents the earthquake source.

To complete this discussion we recall from mechanics that the net effect of a couple is a tendency to cause a rotation. The moment (or torque) of the couple quantifies this tendency, and is defined as the product of the magnitude of the forces and the perpendicular distance between them. Thus, the *mechanical* moment of the couple in Fig. 9.7 is DF_3, while that of the dipole M_{PP} is zero, although this does not mean that the moment (in the sense of strength) of the dipole is necessarily zero. Therefore, it is important to bear in mind this important distinction between the two meanings of the term moment, which also applies to combinations of the double forces, as shown in the following examples (discussed by Love in 1904 (Love, 1927; Miklowitz, 1984)).

(1) *The double couple* $\mathsf{M}_{13} + \mathsf{M}_{31}$. The two couples act on the (x_1, x_3) plane (Fig. 9.9a) and their mechanical moment is zero. Note that the individual effects of the couples M_{13} and M_{31} are counterclockwise and clockwise

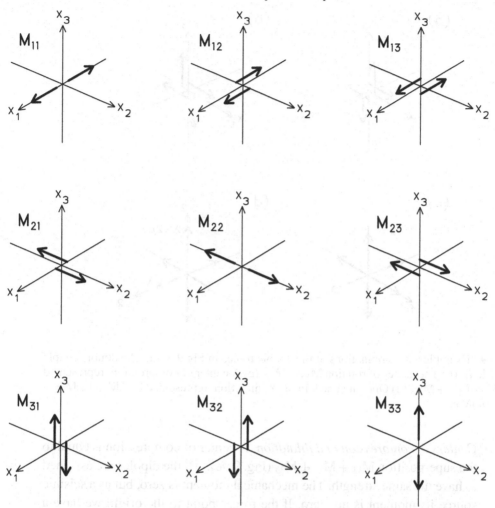

Fig. 9.8. Representation of the nine double forces M_{PQ}. Three of them are dipoles ($P = Q$) and the rest are couples. For the latter the subscripts P and Q denote the directions of the forces and arm of the couple, respectively. (After Aki and Richards, 1980.)

rotations, respectively, about the x_2 axis, so that the two rotations cancel each other. These results apply to all the double couples $M_{PQ} + M_{QP}$.

(2) *Center of rotation about the x_2 axis.* This can be represented by the super-position of the couples M_{13} and $-M_{31}$ (Fig. 9.9b) and will be represented by $M_{13} - M_{31}$. In this case the mechanical moment is not zero, and the effect of this source is a rotation about the x_2 axis. We can also say that the center of rotation has a net moment. These results apply to all the centers of rotation $M_{PQ} - M_{QP}$ ($P \neq Q$).

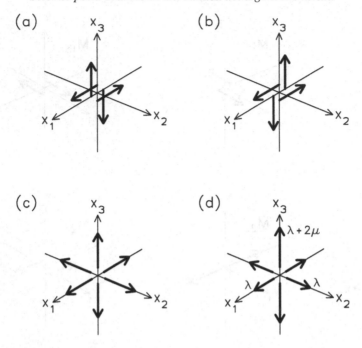

Fig. 9.9. Examples of combinations of the double forces in Fig.9.8. (a) The double couple $M_{13} + M_{31}$. (b) The center of rotation $M_{13} - M_{31}$. (c) A center of compression, represented by $M_{11} + M_{22} + M_{33}$. (d) Opening crack in the x_3 direction, represented by $\lambda M_{11} + \lambda M_{22} + (\lambda + 2\mu)M_{33}$

(3) *Centers of compression and dilatation.* A center of compression is equal to the superposition $M_{11} + M_{22} + M_{33}$ (Fig. 9.9c). All the dipoles are assumed to have the same strength. The mechanical moment is zero, but as a seismic source its moment is not zero. If the forces point to the origin we have a *center of dilatation*, which can be represented by $-(M_{11} + M_{22} + M_{33})$.

(4) *Opening crack in the x_3 direction.* This is represented by the combination $\lambda M_{11} + \lambda M_{22} + (\lambda + 2\mu)M_{33}$ (Fig. 9.9d), where λ and μ are the usual Lamé parameters (see Problem 10.6). This source does not have net moment.

9.9 Moment tensor sources. The far field

Here we will consider the seismic sources that can be represented by a combination of the double forces (dipoles and couples) shown in Fig. 9.8. Let $M_{ij}(t)$ be the moment of either the dipole in the x_i direction ($i = j$) or the couple with forces and arm in the x_i and x_j directions, respectively ($i \neq j$). Depending on the source, a particular double force M_{pq} may not contribute to the combination, in which case the corresponding moment M_{pq} will be assigned a value of zero. With this convention, the displacement generated by the source will be the sum of the displacements owing to the nine individual double forces and because each of them contributes a term similar to that in (9.8.3), the total displacement is given by

$$u_k = M_{ij} * \frac{\partial G_{ki}}{\partial \xi_j} = M_{ij} * G_{ki,j}, \tag{9.9.1}$$

where the usual summation convention applies. The collection of nine terms M_{ij} constitutes a tensor (Problem 9.20) known as the *source moment tensor* (Ben-Menahem and Singh, 1981) and will be represented by a matrix $\mathbf{M}$ with elements M_{ij}. The components of the moment tensor are functions of time, but it is usually assumed that they have the same time dependence, in which case $\mathbf{M}$ can be written as the product of two factors:

$$\mathbf{M}(t) = s(t)\bar{\mathbf{M}}, \tag{9.9.2}$$

where $s(t)$ is a scalar function of time and source strength, and $\bar{\mathbf{M}}$ is a tensor that describes the spatial nature of the source. Several examples of $\bar{\mathbf{M}}$ are given below. Because the first two, and particularly the second one, are important in the following developments, they will be identified with special symbols.

(1) Single couple M_{31}*:*

$$\bar{\mathbf{M}}^{sc} = \begin{pmatrix} 0 & 0 & 0 \\ 0 & 0 & 0 \\ 1 & 0 & 0 \end{pmatrix}. \tag{9.9.3}$$

(2) Double couple $M_{13} + M_{31}$*:*

$$\bar{\mathbf{M}}^{dc} = \begin{pmatrix} 0 & 0 & 1 \\ 0 & 0 & 0 \\ 1 & 0 & 0 \end{pmatrix}. \tag{9.9.4}$$

(3) Center of rotation $M_{13} - M_{31}$*:*

$$\bar{\mathbf{M}} = \begin{pmatrix} 0 & 0 & 1 \\ 0 & 0 & 0 \\ -1 & 0 & 0 \end{pmatrix}. \tag{9.9.5}$$

This source generates S waves but not P waves (Problem 9.21).

(4) Center of compression:

$$\bar{\mathbf{M}} = \begin{pmatrix} 1 & 0 & 0 \\ 0 & 1 & 0 \\ 0 & 0 & 1 \end{pmatrix} = \mathbf{I}, \tag{9.9.6}$$

where $\mathbf{I}$ is the identity matrix. This source generates P waves but not S waves (Problem 9.22) and can be used to represent an explosion.

(5) Opening crack in the $\mathbf{x}_3$ *direction:*

$$\bar{\mathbf{M}} = \begin{pmatrix} \lambda & 0 & 0 \\ 0 & \lambda & 0 \\ 0 & 0 & \lambda + 2\mu \end{pmatrix}. \tag{9.9.7}$$

Although the decomposition (9.9.2) is convenient, it is not necessary, and will not be used, to derive the displacement indicated by (9.9.1). Because the final expression is rather complicated, its treatment is deferred to §9.13. Here we will concentrate on the far-field displacement, which corresponds to the waves commonly investigated in earthquake seismology. The derivation of the far-field expressions requires consideration of the last two terms of (9.6.1) only. Furthermore, after the derivative of Green's function is computed, only the terms with factors $1/r$ will be retained. The following relations are needed:

$$\frac{\partial r}{\partial \xi_j} = -\gamma_j \tag{9.9.8}$$

and

$$\frac{\partial \gamma_i}{\partial \xi_j} = \frac{\gamma_i \gamma_j - \delta_{ij}}{r} \tag{9.9.9}$$

(Problem 9.23). Then, in the far field

$$\frac{\partial G_{ki}}{\partial \xi_j} = \frac{1}{4\pi\rho\alpha^3} \gamma_k \gamma_i \gamma_j \frac{1}{r} \frac{\partial \delta(t - r/\alpha)}{\partial(t - r/\alpha)} - \frac{1}{4\pi\rho\beta^3} (\gamma_k \gamma_i - \delta_{ki}) \gamma_j \frac{1}{r} \frac{\partial \delta(t - r/\beta)}{\partial(t - r/\beta)} \tag{9.9.10}$$

(see terms J and L in Appendix E). To compute the convolution indicated in (9.9.1) note that, in general,

$$f(t) * \frac{\mathrm{d}g(t)}{\mathrm{d}t} = \frac{\mathrm{d}f(t)}{\mathrm{d}t} * g(t) \tag{9.9.11}$$

and assume that this relation can be applied to Dirac's delta. Then we obtain

$$u_k = \frac{1}{4\pi\rho\alpha^3} \gamma_k \gamma_i \gamma_j \frac{1}{r} \dot{M}_{ij} \left(t - \frac{r}{\alpha} \right) - \frac{1}{4\pi\rho\beta^3} (\gamma_k \gamma_i - \delta_{ki}) \gamma_j \frac{1}{r} \dot{M}_{ij} \left(t - \frac{r}{\beta} \right). \tag{9.9.12}$$

The dot over M_{ij} indicates a derivative with respect to the argument.

Equation (9.9.12) can be written in matrix form as

$$\mathbf{u} = \frac{1}{4\pi\rho\alpha^3} \frac{1}{r} (\mathbf{\Gamma}^\mathrm{T} \dot{\mathbf{M}} \mathbf{\Gamma}) \mathbf{\Gamma} - \frac{1}{4\pi\rho\beta^3} \frac{1}{r} [(\mathbf{\Gamma}^\mathrm{T} \dot{\mathbf{M}} \mathbf{\Gamma}) \mathbf{\Gamma} - \dot{\mathbf{M}} \mathbf{\Gamma}]. \tag{9.9.13}$$

The scalar function (it has no free indices) $\boldsymbol{\Gamma}^T \mathbf{M} \boldsymbol{\Gamma}$ corresponds to the scalar product $\boldsymbol{\Gamma} \cdot \mathbf{M} \boldsymbol{\Gamma}$ (see §1.2). In (9.9.13), $\boldsymbol{\Gamma}$ is a column vector.

Equation (9.9.13) contains two terms, the first one is in the direction of $\boldsymbol{\Gamma}$, so it corresponds to P-wave motion, and the second in a direction perpendicular to $\boldsymbol{\Gamma}$, as can be seen by computing the scalar product with $\boldsymbol{\Gamma}$, which gives

$$(\boldsymbol{\Gamma}^T \mathbf{M} \boldsymbol{\Gamma}) \boldsymbol{\Gamma}^T \boldsymbol{\Gamma} - \boldsymbol{\Gamma}^T \mathbf{M} \boldsymbol{\Gamma} = 0, \tag{9.9.14}$$

where $\boldsymbol{\Gamma} \cdot \boldsymbol{\Gamma} = \boldsymbol{\Gamma}^T \boldsymbol{\Gamma} = 1$ was used. Therefore, the second term corresponds to S-wave motion and the vector $\mathbf{u}$ can be written as

$$\mathbf{u} = \mathbf{u}^P + \mathbf{u}^S, \tag{9.9.15}$$

where $\mathbf{u}^P$ and $\mathbf{u}^S$ are equal to the first and second terms on the right of (9.9.13), respectively.

9.9.1 Radiation patterns. SV and SH waves

Here we will extend the ideas introduced in §5.8.4. The vector $\mathbf{u}^P$ in (9.9.15) is in the direction of $\boldsymbol{\Gamma}$, while $\mathbf{u}^S$ lies in a plane perpendicular to $\boldsymbol{\Gamma}$. To introduce the SV and SH waves the vector $\mathbf{u}^S$ will be decomposed into two vectors, one in a vertical plane that contains the source and the receiver, and the other in a horizontal plane. This decomposition is easily achieved by introducing unit vectors $\boldsymbol{\Gamma}$, $\boldsymbol{\Phi}$, and $\boldsymbol{\Theta}$ in spherical coordinates (Fig. 9.10). Recall that $\boldsymbol{\Gamma}$ is in the radial (or source–receiver) direction. The vector $\boldsymbol{\Theta}$ is tangent to the great circle, while $\boldsymbol{\Phi}$ is tangent to the small circle parallel to the (x_1, x_2) plane, so that it has no component on the x_3 direction. In terms of angles θ and ϕ these vectors are written as

$$\boldsymbol{\Gamma} = \begin{pmatrix} \sin\theta \cos\phi \\ \sin\theta \sin\phi \\ \cos\theta \end{pmatrix}; \qquad \boldsymbol{\Theta} = \begin{pmatrix} \cos\theta \cos\phi \\ \cos\theta \sin\phi \\ -\sin\theta \end{pmatrix}; \qquad \boldsymbol{\Phi} = \begin{pmatrix} -\sin\phi \\ \cos\phi \\ 0 \end{pmatrix}. \tag{9.9.16}$$

It is easy to verify that each of the vectors is perpendicular to the other two, that the absolute value of each of them is equal to 1, and that the three vectors form a right-handed system ($\boldsymbol{\Phi} = \boldsymbol{\Gamma} \times \boldsymbol{\Theta}$) (Problem 9.24).

To find the expressions for $\mathbf{u}^{SH}$ and $\mathbf{u}^{SV}$ note that, in general, if $\mathbf{a}$ and $\hat{\mathbf{b}}$ are a vector and a unit vector, respectively, then the vector obtained by projecting $\mathbf{a}$ along the direction $\hat{\mathbf{b}}$ is given by $(\hat{\mathbf{b}} \cdot \mathbf{a})\hat{\mathbf{b}}$ or $(\hat{\mathbf{b}}^T \mathbf{a})\hat{\mathbf{b}}$. Then, using (9.9.13) and the orthogonality relations $\boldsymbol{\Theta}^T \boldsymbol{\Gamma} = \boldsymbol{\Phi}^T \boldsymbol{\Gamma} = 0$, the P, SH, and SV displacements are given by

$$\mathbf{u}^P = \frac{1}{4\pi\rho\alpha^3} \frac{1}{r} (\boldsymbol{\Gamma}^T \mathbf{M} \boldsymbol{\Gamma}) \boldsymbol{\Gamma} \tag{9.9.17a}$$

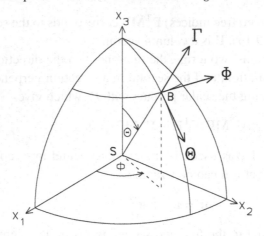

Fig. 9.10. Unit vectors $\boldsymbol{\Gamma}$, $\boldsymbol{\Phi}$, and $\boldsymbol{\Theta}$ in spherical coordinates. The source and receiver locations are indicated by S and B, respectively.

$$\mathbf{u}^{SV} = \frac{1}{4\pi\rho\beta^3}\frac{1}{r}(\boldsymbol{\Theta}^{\mathrm{T}}\dot{\mathbf{M}}\boldsymbol{\Gamma})\boldsymbol{\Theta} \qquad (9.9.17b)$$

$$\mathbf{u}^{SH} = \frac{1}{4\pi\rho\beta^3}\frac{1}{r}(\boldsymbol{\Phi}^{\mathrm{T}}\dot{\mathbf{M}}\boldsymbol{\Gamma})\boldsymbol{\Phi}. \qquad (9.9.17c)$$

In (9.9.17) the terms in parentheses are the corresponding radiation patterns, to be indicated by $\mathcal{R}$, which can be written in detail as follows:

$$\mathcal{R}^{P} = \boldsymbol{\Gamma}^{\mathrm{T}}\dot{\mathbf{M}}\boldsymbol{\Gamma} = \gamma_i \dot{M}_{ij}\gamma_j = \gamma_1 v_1 + \gamma_2 v_2 + \gamma_3 v_3 \qquad (9.9.18a)$$

$$\mathcal{R}^{SV} = \boldsymbol{\Theta}^{\mathrm{T}}\dot{\mathbf{M}}\boldsymbol{\Gamma} = \Theta_i \dot{M}_{ij}\gamma_j = \Theta_1 v_1 + \Theta_2 v_2 + \Theta_3 v_3 \qquad (9.9.18b)$$

$$\mathcal{R}^{SH} = \boldsymbol{\Phi}^{\mathrm{T}}\dot{\mathbf{M}}\boldsymbol{\Gamma} = \Phi_i \dot{M}_{ij}\gamma_j = \Phi_1 v_1 + \Phi_2 v_2 + \Phi_3 v_3, \qquad (9.9.18c)$$

where Θ_i and Φ_i indicate the components of $\boldsymbol{\Theta}$ and $\boldsymbol{\Phi}$, respectively, and

$$v_i = \dot{M}_{i1}\gamma_1 + \dot{M}_{i2}\gamma_2 + \dot{M}_{i3}\gamma_3. \qquad (9.9.19)$$

If $\mathbf{M}$ can be decomposed as in (9.9.2), then

$$\dot{\mathbf{M}} = \dot{s}\,\bar{\mathbf{M}} \qquad (9.9.20)$$

and (9.9.18) can be used with $\dot{\mathbf{M}}$ replaced with $\bar{\mathbf{M}}$.

The *angle of polarization* ε of the S wave was defined in §5.8.4 as the angle between $\mathbf{u}^{S}$ and $\mathbf{u}^{SV}$, and is obtained from

$$\tan\varepsilon = \mathcal{R}^{SH}/\mathcal{R}^{SV}. \qquad (9.9.21)$$

The angle ε is used in focal mechanism studies based on S waves (Stauder, 1960, 1962; Herrmann, 1975).

Finally, we will find out the directions that produce extremal values of $\mathcal{R}^P$ when M_{ij} is symmetric. To do that we will determine the vector $\boldsymbol{\Gamma}$ that makes (9.9.18a) an extremum under the constraint that $|\boldsymbol{\Gamma}|^2 = \gamma_i \gamma_i = 1$. As in §3.9, we will use the method of Lagrange multipliers. In our case the function of interest is

$$F = \gamma_i \dot{M}_{ij} \gamma_j + \lambda(1 - \gamma_i \gamma_i). \tag{9.9.22}$$

Taking the derivative of F with respect to γ_k and setting it equal to zero gives

$$\frac{\partial F}{\partial \gamma_k} = \dot{M}_{kj} \gamma_j + \gamma_i \dot{M}_{ik} - 2\lambda \gamma_k = 2(\dot{M}_{kj} \gamma_j - \lambda \gamma_k) = 0. \tag{9.9.23}$$

Here the symmetry of $\dot{M}_{ij}$ was used. From (9.9.23) we obtain

$$\dot{M}_{kj} \gamma_j = \lambda \gamma_k. \tag{9.9.24}$$

Therefore, the extremal values of $\mathcal{R}^P$ are in the directions of the eigenvectors of $\dot{M}_{ij}$ (see (1.4.87)). Furthermore, if $\boldsymbol{\Gamma}^e$ indicates an extremal direction, then from (9.9.18a) and (9.9.24) the corresponding extremal value of $\mathcal{R}^P$ is given by

$$\gamma_i^e \dot{M}_{ij} \gamma_j^e = \gamma_i^e \lambda \gamma_i^e = \lambda. \tag{9.9.25}$$

In summary, by solving the eigenvalue problem (9.9.24) we immediately obtain the extremal directions and values of $\mathcal{R}^P$ (provided that the moment tensor is symmetric). An example of these results is given in §9.12.

9.10 Equivalence of a double couple and a pair of compressional and tensional dipoles

Here we will show that a pair of tensional and compressional dipoles at right angles with respect to each other and at $45°$ with respect to the x_1 and x_3 axes (Fig. 9.11) generates the same radiation pattern as the double couple $M_{13}+M_{31}$ (see §9.8). This equivalence was shown by Nakano in 1923 (Ben-Menahem, 1995) and to prove it is enough to show that the two sources have the same moment tensor. This will be done using the transformation laws of tensor components derived in Chapter 1. In the (x_1, x_2, x_3) system the double couple has a moment tensor given by (9.9.4)

$$\bar{\mathbf{M}}^{dc} = \begin{pmatrix} 0 & 0 & 1 \\ 0 & 0 & 0 \\ 1 & 0 & 0 \end{pmatrix}, \tag{9.10.1}$$

while the pair of dipoles has a moment tensor $\bar{\mathbf{M}}$ to be determined. To find out $\bar{\mathbf{M}}$ first note that in the rotated system (x_1', x_2', x_3') the dipole has a moment tensor $\bar{\mathbf{M}}'$

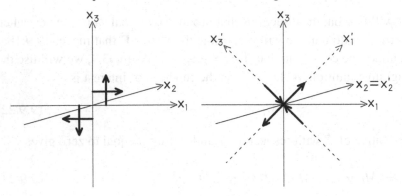

Fig. 9.11. The double couple $M_{13} + M_{31}$ on the left and the pair of dipoles in the (x_1, x_3) plane on the right have the same moment tensor representation and thus generate the same radiation pattern. (After Ben-Menahem and Singh, 1980; Pujol and Herrmann, 1990.)

given by

$$\bar{\mathbf{M}}' = \begin{pmatrix} 1 & 0 & 0 \\ 0 & 0 & 0 \\ 0 & 0 & -1 \end{pmatrix}. \tag{9.10.2}$$

Now recall from (1.4.20) that $\bar{\mathbf{M}}$ and $\bar{\mathbf{M}}'$ are related by

$$\bar{\mathbf{M}} = \mathbf{R}^{\mathrm{T}} \bar{\mathbf{M}}' \mathbf{R}, \tag{9.10.3}$$

where in this case $\mathbf{R}$ is a counterclockwise rotation of $45°$ about the x_2 axis, given by

$$\mathbf{R} = \begin{pmatrix} \cos 45° & 0 & \sin 45° \\ 0 & 1 & 0 \\ -\sin 45° & 0 & \cos 45° \end{pmatrix}. \tag{9.10.4}$$

After performing the product indicated in (9.10.3) it is found that

$$\bar{\mathbf{M}} = \bar{\mathbf{M}}^{\mathrm{dc}} \tag{9.10.5}$$

(Problem 9.26) which shows that the two sets of forces are equivalent. We also see that these two tensors have zero trace and determinant.

9.11 The tension and compression axes

The equivalence of the double couple and the pair of tensional and compressional dipoles discussed above is extremely important because it allows the introduction of two unit vectors

$$\mathbf{t} = (1/\sqrt{2}) \begin{pmatrix} 1 \\ 0 \\ 1 \end{pmatrix}; \qquad \mathbf{p} = (1/\sqrt{2}) \begin{pmatrix} 1 \\ 0 \\ -1 \end{pmatrix}, \qquad (9.11.1)$$

which are directed along the axes that contain the dipoles. The axis corresponding to $\mathbf{t}$, known as the T or *tension axis*, lies in one of the two quadrants in the (x_1, x_3) plane that generate compressions *at the receiver*, while the P or *compression axis*, which corresponds to $\mathbf{p}$, is in a dilatational quadrant.

Next we show that $\mathbf{t}$ and $\mathbf{p}$ are eigenvectors of $\bar{\mathbf{M}}^{dc}$. To see that solve

$$\bar{\mathbf{M}}^{dc}\mathbf{v} = \lambda \mathbf{v} \qquad (9.11.2)$$

(see (1.4.88)) or

$$\begin{pmatrix} 0 & 0 & 1 \\ 0 & 0 & 0 \\ 1 & 0 & 0 \end{pmatrix} \begin{pmatrix} v_1 \\ v_2 \\ v_3 \end{pmatrix} = \lambda \begin{pmatrix} v_1 \\ v_2 \\ v_3 \end{pmatrix}. \qquad (9.11.3)$$

Because $\bar{\mathbf{M}}^{dc}$ is symmetric its eigenvalues must be real (see §1.4.6). By equating components (9.11.3) gives

$$v_3 = \lambda v_1; \qquad 0 = \lambda v_2; \qquad v_1 = \lambda v_3. \qquad (9.11.4a,b,c)$$

From (9.11.4a,c) it is seen that $\lambda^2 = 1$, or $\lambda = \pm 1$. Therefore, equation (9.11.3) is satisfied by $\mathbf{t}$ and $\mathbf{p}$, with eigenvalues $\lambda = 1$ and -1, respectively. Equation (9.11.4b) shows that $\lambda = 0$ is also an eigenvalue, with unit eigenvector $\mathbf{b}$ given by

$$\mathbf{b} = \begin{pmatrix} 0 \\ 1 \\ 0 \end{pmatrix}, \qquad (9.11.5)$$

which defines the so-called B or *null axis*. Note that

$$\mathbf{b} = \mathbf{t} \times \mathbf{p}, \qquad (9.11.6)$$

so that the vectors $\mathbf{p}$, $\mathbf{b}$ and $\mathbf{t}$ form a right-handed coordinate system.

As discussed in §1.4.6, the matrix formed with the eigenvectors of a given tensor can be used to reduce it to a diagonal form, with the diagonal elements equal to the eigenvalues of the tensor. In fact, if $\mathbf{R}$ is the matrix having as rows the vectors $\mathbf{t}$, $\mathbf{b}$, and $\mathbf{p}$, then

$$\bar{\mathbf{M}}' = \mathbf{R}^T \bar{\mathbf{M}}^{dc} \mathbf{R}, \qquad (9.11.7)$$

where $\bar{\mathbf{M}}'$ is given by (9.10.2) (Problem 9.27).

9.12 Radiation patterns for the single couple M_{31} and the double couple $M_{13} + M_{31}$

The moment tensor corresponding to the single couple M_{31} (Fig. 9.8) is given by (9.9.3)

$$\bar{\mathbf{M}}^{\text{sc}} = \begin{pmatrix} 0 & 0 & 0 \\ 0 & 0 & 0 \\ 1 & 0 & 0 \end{pmatrix}, \tag{9.12.1}$$

which is not symmetric.

The P, SV, and SH radiation patterns are obtained using (9.9.18),

$$\mathcal{R}^P = \gamma_1 \gamma_3 = \tfrac{1}{2} \sin 2\theta \cos \phi \tag{9.12.2a}$$

$$\mathcal{R}^{SV} = \gamma_1 \Theta_3 = - \sin^2 \theta \cos \phi \tag{9.12.2b}$$

$$\mathcal{R}^{SH} = \gamma_1 \Phi_3 = 0. \tag{9.12.2c}$$

The last equation shows that this source does not generate SH waves.

The radiation patterns $\mathcal{R}^P$ and $\mathcal{R}^{SV}$ depend on θ and ϕ, so the axial symmetry that characterized the single force discussed in §9.5.3 is lost here. It is possible, however, to have an idea of their shape by looking at their maxima and minima. $\mathcal{R}^P$ has four lobes, is zero on the (x_1, x_2) and (x_2, x_3) planes (known as *nodal planes*) and reaches maximum absolute values along lines at $45°$ with respect to the x_1 and x_3 axes (Fig. 9.12a,c). Dilatations and compressions form an alternating pattern. $\mathcal{R}^{SV}$ has only two lobes, is zero on the (x_2, x_3) plane, and has maximum absolute values along the x_1 axis (Fig. 9.12b,d). These radiation patterns will be presented using an equal-area projection in §10.9.

The double couple $M_{13} + M_{31}$, discussed in §9.10 and §9.11, has the moment tensor $\bar{\mathbf{M}}^{\text{dc}}$ given by

$$\bar{\mathbf{M}}^{\text{dc}} = \begin{pmatrix} 0 & 0 & 1 \\ 0 & 0 & 0 \\ 1 & 0 & 0 \end{pmatrix}. \tag{9.12.3}$$

The radiation patterns are obtained from (9.9.18)

$$\mathcal{R}^P = 2\gamma_1 \gamma_3 = \sin 2\theta \cos \phi \tag{9.12.4a}$$

$$\mathcal{R}^{SV} = \gamma_1 \Theta_3 + \gamma_3 \Theta_1 = \cos 2\theta \cos \phi \tag{9.12.4b}$$

$$\mathcal{R}^{SH} = \gamma_1 \Phi_3 + \gamma_3 \Phi_1 = - \cos \theta \sin \phi. \tag{9.12.4c}$$

The extremal values of $\mathcal{R}^P$ (Fig. 9.13) are attained for directions that coincide with the directions of the T and P axes, as expected from the discussion in §9.9.1

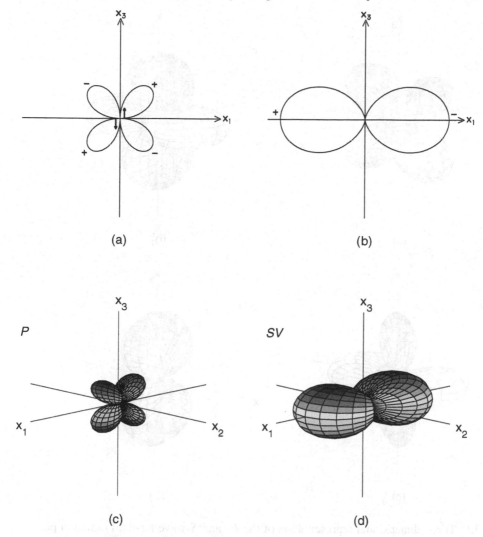

Fig. 9.12. Two- and three-dimensional representations of the *P*- and *S*-wave far-field radiation patterns generated by the single couple M_{31} (obtained using (9.12.2)). (a), (c) $|\mathcal{R}^P|$; (b), (d) $|\mathcal{R}^{SV}|$.

(Problem 9.28). The *B* axis, on the other hand, is not an extremal direction; it lies in the intersection of the two nodal planes.

The plots of $\mathcal{R}^{SH}$ and $\mathcal{R}^{SV}$ are quite complicated, as can be seen from their absolute values (Fig. 9.13), and can be better appreciated when displayed as equal-area projection plots (see §10.9). In absolute value the radiation pattern of $\mathbf{u}^S$ is given by

$$|\mathcal{R}^S| = [(\mathcal{R}^{SH})^2 + (\mathcal{R}^{SV})^2]^{1/2} = (\cos^2 2\theta \cos^2 \phi + \cos^2 \theta \sin^2 \phi)^{1/2}. \quad (9.12.5)$$

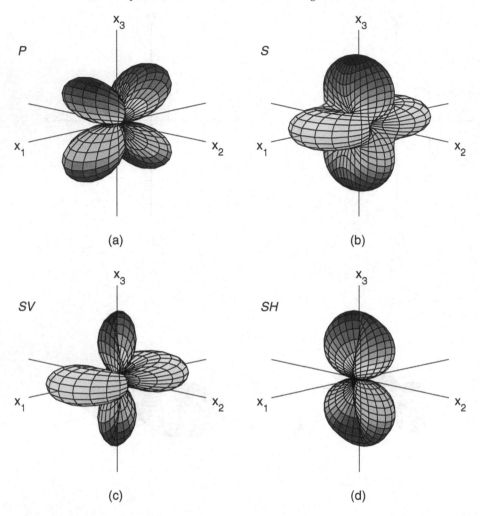

Fig. 9.13. Three-dimensional representation of the P- and S-wave far-field radiation patterns generated by the double couple $M_{13} + M_{31}$. (a) $|\mathcal{R}^P|$, (b) $|\mathcal{R}^S|$, (c) $|\mathcal{R}^{SV}|$, (d) $|\mathcal{R}^{SH}|$. See (9.12.4a) for (a), (9.12.5) for (b), and (9.12.4b,c) for (c) and (d).

$|\mathcal{R}^S|$ has four lobes with maximum values along the x_1 and x_3 axes (Fig. 9.13). The motion is zero along the lines for which $\mathcal{R}^P$ is maximum.

It is instructive to compare the radiation patterns for the single and double couples. From (9.12.2a) and (9.12.4a) we see that for the P wave they are the same (aside from a factor of 2), so that the two sources are indistinguishable from each other when only P-wave information is available. For the S waves, however, the patterns are different, and this allows one to distinguish one type of source from the other. These properties of the radiation patterns of the single and double couples played a critical role in the early controversy over which system of forces better represented the earthquake source (Stauder, 1962).

9.13   Moment tensor sources. The total field

Although the far field accounts for most of the seismological observations, a thorough understanding of the earthquake rupture process, still the subject of research, requires observations close to the source, i.e., within the near field. Therefore, it is necessary to consider all the terms in (9.9.1), which is done in Appendix E. The final result is

$$
4\pi\rho u_k(\mathbf{x}, t) = (15\gamma_k\gamma_i\gamma_j - 3\gamma_k\delta_{ij} - 3\gamma_i\delta_{kj} - 3\gamma_j\delta_{ki})\frac{1}{r^4}\int_{r/\alpha}^{r/\beta} \tau M_{ij}(t - \tau)\, d\tau
$$

$$
+ (6\gamma_k\gamma_i\gamma_j - \gamma_k\delta_{ij} - \gamma_i\delta_{kj} - \gamma_j\delta_{ki})\frac{1}{\alpha^2 r^2} M_{ij}\left(t - \frac{r}{\alpha}\right)
$$

$$
- (6\gamma_k\gamma_i\gamma_j - \gamma_k\delta_{ij} - \gamma_i\delta_{kj} - 2\gamma_j\delta_{ki})\frac{1}{\beta^2 r^2} M_{ij}\left(t - \frac{r}{\beta}\right)
$$

$$
+ \gamma_k\gamma_i\gamma_j\frac{1}{\alpha^3 r}\dot{M}_{ij}\left(t - \frac{r}{\alpha}\right) - (\gamma_k\gamma_i - \delta_{ki})\gamma_j\frac{1}{\beta^3 r}\dot{M}_{ij}\left(t - \frac{r}{\beta}\right)
$$

$$
\tag{9.13.1}
$$

(Aki and Richards, 1980; see also Haskell, 1964).

To simplify (9.13.1) we will assume that $M_{ij}(t)$ can be decomposed as in (9.9.2)

$$
M_{ij}(t) = D(t)\bar{M}_{ij}. \tag{9.13.2}
$$

Writing in vector form (9.13.1) becomes

$$
4\pi\rho\mathbf{u}(\mathbf{x}, t) = \mathbf{A}^{\mathrm{N}}\frac{1}{r^4}\int_{r/\alpha}^{r/\beta} \tau D(t - \tau)\, d\tau + \mathbf{A}^{\mathrm{I}\alpha}\frac{1}{\alpha^2 r^2}D\left(t - \frac{r}{\alpha}\right)
$$

$$
- \mathbf{A}^{\mathrm{I}\beta}\frac{1}{\beta^2 r^2}D\left(t - \frac{r}{\beta}\right) + \mathbf{A}^{\mathrm{FP}}\frac{1}{\alpha^3 r}\dot{D}\left(t - \frac{r}{\alpha}\right)
$$

$$
- \mathbf{A}^{\mathrm{FS}}\frac{1}{\beta^3 r}\dot{D}\left(t - \frac{r}{\beta}\right) \tag{9.13.3}
$$

where

$$
\mathbf{A}^{\mathrm{N}} = 3[5(\mathbf{\Gamma}^{\mathrm{T}}\bar{\mathbf{M}}\mathbf{\Gamma})\mathbf{\Gamma} - \mathrm{tr}(\bar{\mathbf{M}})\mathbf{\Gamma} - \bar{\mathbf{M}}^{\mathrm{T}}\mathbf{\Gamma} - \bar{\mathbf{M}}\mathbf{\Gamma}] \tag{9.13.4}
$$

$$
\mathbf{A}^{\mathrm{I}\alpha} = 6(\mathbf{\Gamma}^{\mathrm{T}}\bar{\mathbf{M}}\mathbf{\Gamma})\mathbf{\Gamma} - \mathrm{tr}(\bar{\mathbf{M}})\mathbf{\Gamma} - \bar{\mathbf{M}}^{\mathrm{T}}\mathbf{\Gamma} - \bar{\mathbf{M}}\mathbf{\Gamma} \tag{9.13.5}
$$

$$
\mathbf{A}^{\mathrm{FP}} = (\mathbf{\Gamma}^{\mathrm{T}}\bar{\mathbf{M}}\mathbf{\Gamma})\mathbf{\Gamma} \tag{9.13.6}
$$

$$\mathbf{A}^{I\beta} = 6(\mathbf{\Gamma}^T\bar{\mathbf{M}}\mathbf{\Gamma})\mathbf{\Gamma} - \mathrm{tr}(\bar{\mathbf{M}})\mathbf{\Gamma} - \bar{\mathbf{M}}^T\mathbf{\Gamma} - 2\bar{\mathbf{M}}\mathbf{\Gamma} \qquad (9.13.7)$$

$$\mathbf{A}^{FS} = (\mathbf{\Gamma}^T\bar{\mathbf{M}}\mathbf{\Gamma})\mathbf{\Gamma} - \bar{\mathbf{M}}\mathbf{\Gamma}. \qquad (9.13.8)$$

$\bar{\mathbf{M}}$ represents the tensor with components $\bar{M}_{ij}$ and tr denotes trace (Problem 9.29). The superscripts N, I, and F indicate near, intermediate and far fields, respectively, and P and S indicate the type of wave. The reason for using superscripts α and β for the intermediate field, instead of P and S as used for the far field, is that $\mathbf{A}^{I\alpha}$ and $\mathbf{A}^{I\beta}$ are not in the direction of $\mathbf{\Gamma}$ and perpendicular to $\mathbf{\Gamma}$, respectively, which are the criteria we used to define P- and S-wave motions (see §5.8.1 and §9.5.1). For symmetric moment tensors some simplifications are possible because $\bar{\mathbf{M}}^T\mathbf{\Gamma} = \bar{\mathbf{M}}\mathbf{\Gamma}$.

For computational purposes it is convenient to introduce a function $J(t)$ such that

$$J'(t) = D(t) \qquad (9.13.9)$$

(compare with §9.5.2). After integrating by parts and letting

$$G(t) = \int_0^t J(\tau)\,d\tau; \qquad E(t) = J(t); \qquad D'(t) = J''(t) \qquad (9.13.10)$$

(Harkrider, 1976), equation (9.13.3) becomes

$$4\pi\rho\mathbf{u}(\mathbf{x}, t) = \mathbf{A}^N\left[\frac{1}{r^4}G\left(t - \frac{r}{\alpha}\right) + \frac{1}{\alpha r^3}E\left(t - \frac{r}{\alpha}\right)\right]$$

$$+ \mathbf{A}^{I\alpha}\frac{1}{\alpha^2 r^2}D\left(t - \frac{r}{\alpha}\right) + \mathbf{A}^{FP}\frac{1}{\alpha^3 r}\dot{D}\left(t - \frac{r}{\alpha}\right)$$

$$- \mathbf{A}^N\left[\frac{1}{r^4}G\left(t - \frac{r}{\beta}\right) + \frac{1}{\beta r^3}E\left(t - \frac{r}{\beta}\right)\right] - \mathbf{A}^{I\beta}\frac{1}{\beta^2 r^2}D\left(t - \frac{r}{\beta}\right)$$

$$- \mathbf{A}^{FS}\frac{1}{\beta^3 r}\dot{D}\left(t - \frac{r}{\beta}\right), \qquad (9.13.11)$$

(Problem 9.30). This equation will be used in §10.10, where we will investigate the effect of the distance and the frequency content of the source time function on the displacement caused by an earthquake. We note here, however, that terms with an r^{-2} dependence are never dominant, and for this reason the expression "intermediate field" associated with this factor is somewhat misleading (Aki and Richards, 1980).

Equation (9.13.11) gives the components of $\mathbf{u}(\mathbf{x})$ in Cartesian coordinates. To obtain the components in spherical coordinates the radiation patterns given below can be used, but it is simpler to apply the rotation of coordinates described in §10.10.

9.13.1 Radiation patterns

As done in §9.9.1, the displacement **u** given in (9.13.11) will be decomposed along the vectors $\mathbf{\Gamma}$, $\mathbf{\Theta}$, and $\mathbf{\Phi}$. In addition, we will concentrate on the radiation patterns, in which case we obtain

$$\mathcal{R}^{N\Gamma} = \pm\mathbf{\Gamma}^T\mathbf{A}^N \tag{9.13.12a}$$

$$\mathcal{R}^{N\Theta} = \pm\mathbf{\Theta}^T\mathbf{A}^N \tag{9.13.12b}$$

$$\mathcal{R}^{N\Phi} = \pm\mathbf{\Phi}^T\mathbf{A}^N, \tag{9.13.12c}$$

where the $+$ or $-$ signs are used with waves that propagate with velocities α or β, respectively,

$$\mathcal{R}^{I\alpha\Gamma} = \mathbf{\Gamma}^T\mathbf{A}^{I\alpha} \tag{9.13.13a}$$

$$\mathcal{R}^{I\alpha\Theta} = \mathbf{\Theta}^T\mathbf{A}^{I\alpha} \tag{9.13.13b}$$

$$\mathcal{R}^{I\alpha\Phi} = \mathbf{\Phi}^T\mathbf{A}^{I\alpha} \tag{9.13.13c}$$

$$\mathcal{R}^{I\beta\Gamma} = -\mathbf{\Gamma}^T\mathbf{A}^{I\beta} \tag{9.13.14a}$$

$$\mathcal{R}^{I\beta\Theta} = -\mathbf{\Theta}^T\mathbf{A}^{I\beta} \tag{9.13.14b}$$

$$\mathcal{R}^{I\beta\Phi} = -\mathbf{\Phi}^T\mathbf{A}^{I\beta}. \tag{9.13.14c}$$

The expressions for the far field are given in (9.9.18) and will not be repeated here. Using (9.13.4)–(9.13.8) and assuming that the moment tensor is symmetric, (9.13.12)–(9.13.14) become

$$\mathcal{R}^{N\Gamma} = \pm[9\mathbf{\Gamma}^T\bar{\mathbf{M}}\mathbf{\Gamma} - 3\,\mathrm{tr}(\bar{\mathbf{M}})] \tag{9.13.15a}$$

$$\mathcal{R}^{N\Theta} = \pm(-6\mathbf{\Theta}^T\bar{\mathbf{M}}\mathbf{\Gamma}) \tag{9.13.15b}$$

$$\mathcal{R}^{N\Phi} = \pm(-6\mathbf{\Phi}^T\bar{\mathbf{M}}\mathbf{\Gamma}) \tag{9.13.15c}$$

where the $+$ and $-$ signs have the same meaning as in (9.13.12),

$$\mathcal{R}^{I\alpha\Gamma} = 4\mathbf{\Gamma}^T\bar{\mathbf{M}}\mathbf{\Gamma} - \mathrm{tr}(\bar{\mathbf{M}}) \tag{9.13.16a}$$

$$\mathcal{R}^{I\alpha\Theta} = -2\mathbf{\Theta}^T\bar{\mathbf{M}}\mathbf{\Gamma} \tag{9.13.16b}$$

$$\mathcal{R}^{I\alpha\Phi} = -2\mathbf{\Phi}^T\bar{\mathbf{M}}\mathbf{\Gamma} \tag{9.13.16c}$$

$$\mathcal{R}^{I\beta\Gamma} = -3\mathbf{\Gamma}^T\bar{\mathbf{M}}\mathbf{\Gamma} + \mathrm{tr}(\bar{\mathbf{M}}) \tag{9.13.17a}$$

$$\mathcal{R}^{I\beta\Theta} = 3\mathbf{\Theta}^T\bar{\mathbf{M}}\mathbf{\Gamma} \tag{9.13.17b}$$

$$\mathcal{R}^{l\beta\Phi} = 3\Phi^{\mathrm{T}}\bar{\mathbf{M}}\Gamma \tag{9.13.17c}$$

(Problem 9.31).

Problems

9.1   Prove that

$$\int_{-\infty}^{\infty} f(ax)\varphi(x)\,\mathrm{d}x = \frac{1}{a}\int_{-\infty}^{\infty} f(x)\varphi(x/a)\,\mathrm{d}x; \qquad a > 0.$$

In the language of Appendix A, the left-hand side of this equation is a regular distribution. The extension to the delta can be written in a symbolic way as in (9.2.7b) (Zemanian 1965).

9.2   Verify (9.2.12).

9.3   Verify (9.2.13).

9.4   Verify (9.4.18) and (9.4.19).

9.5   Verify (9.4.23).

9.6   Show that the u_2 component of the displacement given in (9.4.24) satisfies

$$\frac{\partial^2 u_2}{\partial t^2} = \beta^2 \nabla^2 u_2; \qquad \frac{\partial^2 u_2}{\partial x_2^2} \equiv 0.$$

9.7   Verify (9.5.4).

9.8   Verify (9.5.11).

9.9   Verify (9.5.15).

9.10 Why is the following statement incorrect. The dot product of $(3\gamma_i\gamma_j - \delta_{ij})$ (see (9.5.16)) and γ_i is equal to $2\gamma_j$. Therefore, $(3\gamma_i\gamma_j - \delta_{ij})$ represents motion in the direction of Γ. Give two reasons.

9.11 Show that the vector $3\gamma_i\gamma_j - \delta_{ij}$ is neither perpendicular to nor in the direction of Γ.

9.12 Verify (9.5.19).

9.13 Show that if $F(\omega)$ is the Fourier transform of $f(t)$, then

$$\mathcal{F}\left\{\frac{\mathrm{d}f}{\mathrm{d}t}\right\} = i\omega F(\omega)$$

and

$$\mathcal{F}\left\{\frac{\mathrm{d}^2 f}{\mathrm{d}t^2}\right\} = -\omega^2 F(\omega).$$

Briefly comment on the relation between the energy (see §11.6.2) of $f(t)$ and its derivatives and the frequency content of $f(t)$.

9.14 Verify (9.5.20).

9.15 Given (9.5.23), verify (9.5.21) and (9.5.22).

9.16 Find the amplitude spectrum of the function $J''(t)$ given in (9.5.23) and plot it for $a = 1, 2$.

9.17 Verify that the plots of $|\cos\theta|$ and $|\sin\theta|$ are pairs of circles with centers at $(0, \pm\frac{1}{2})$ and $(\pm\frac{1}{2}, 0)$, respectively, and radii equal to $\frac{1}{2}$.

9.18 Show that Green's function introduced in §9.6 is a tensor.

9.19 Use (9.7.2) to give an alternative proof that Green's function is a tensor.

9.20 Verify that the collection of nine terms M_{ij} introduced in §9.9 constitutes a tensor.

9.21 Verify that the center of rotation M_{13}–M_{31} generates S waves but not P waves in the near or far fields.

9.22 Verify that a center of compression generates P waves but not S waves in the near or far fields.

9.23 Verify (9.9.8) and (9.9.9).

9.24 Verify (9.9.16) and that the three vectors form a right-handed system.

9.25 Refer to §9.9.1. Show that when M_{ij} is symmetric the extremal values of $\mathcal{R}^P$ are attained for values of θ and ϕ that make $\mathcal{R}^{SV}$ and $\sin\theta\, \mathcal{R}^{SH}$ equal to zero simultaneously.

9.26 Verify that the product indicated in (9.10.3) gives (9.10.5).

9.27 Verify (9.11.7).

9.28 Verify that the extremal values of (9.12.4a) are attained for directions that coincide with the directions of the T and P axes. Also verify that for these directions $\mathcal{R}^{SV}$ and $\mathcal{R}^{SH}$ are equal to zero.

9.29 Verify (9.13.3).

9.30 Verify (9.13.11).

9.31 Verify (9.13.15)–(9.13.17) and apply them to $\bar{\mathbf{M}}^{dc}$.

10

The earthquake source in unbounded media

10.1 Introduction

Most earthquakes can be represented by slip on a fault, which for simplicity will be modeled as a planar feature. When an earthquake occurs, the two sides of the fault suffer a sudden relative displacement with respect to each other, which in turn is the source of seismic waves. Assuming that the Earth is initially at rest and that there are no external forces, the occurrence of an earthquake is the result of a localized and temporary breakdown of the stress–strain relationship, which is the only basic equation of elasticity that is not a fundamental law of physics (Backus and Mulcahy, 1976). Let us compare this situation with that encountered in Chapter 9. There we used (9.4.2) to study the displacement field caused by a given body force, while here the displacement at points away from the fault is caused by a *discontinuity in displacement* across the fault, and because of the breakdown of Hooke's law, we cannot use (9.4.2) directly. Therefore, we are faced with the following problem: what is the *equivalent body force* that in the absence of the fault will cause exactly the same displacement field as slip on the fault. Getting an answer to this question took a considerable amount of time and effort (for a review, see Stauder, 1962). As a result of work done during the 1920–50s, two possible models were introduced, based on single and double couples similar to those discussed in Chapter 9. The single-couple model made sense from a physical point of view but a major drawback was the fact that it had a net mechanical moment, which in turn implied a nonequilibrium condition inside the Earth. The double-couple model did not have this problem, but was more difficult to justify. As noted in §9.12 (see also §10.9), the two models generate different S-wave radiation patterns, but the quality of the data available was not sufficient to discriminate between them. However, even if the data had allowed the identification of the right model, the fact would have remained that there was no theoretical justification for either of them. This critical problem was solved by Burridge and Knopoff (1964),

316

who developed the theory of equivalent body forces for heterogeneous anisotropic media. In the particular case of slip in the (x_1, x_2) plane in the x_1 direction, Burridge and Knopoff (1964) showed that under the assumption of isotropy the equivalent body force is proportional to the double couple $M_{13} + M_{31}$, introduced in §9.8. This result closed the long-standing debate regarding the nature of the earthquake source. In this context, two other authors that made important contributions are Maruyama (1963) and Haskell (1964), who found the double-couple equivalent for the case of homogeneous isotropic media. In the seismic literature the earthquakes generated by slip on a fault are known as dislocations.[1] A dislocation can be visualized through the following thought experiment, based on Steketee (1958). Consider a cut made over a surface Σ within an elastic body. After the cut has been made there are two surfaces, indicated by Σ^+ and Σ^-, which will be deformed differently by application of some force distribution. If the combined system of forces is in static equilibrium, then the body will remain in the original equilibrium state. The result of this operation is a discontinuity in the displacement across Σ, known as a *dislocation*, which is accomodated by deformation within the body. A good discussion of the relation between earthquakes and dislocations is provided by Steketee (1958), although his analysis was restricted to the static case. The contributions made by other authors is described by Stauder (1962). Burridge and Knopoff (1964) extended the analysis to the dynamic case, and as we will see in §10.4, the displacement field anywhere in the medium depends on the discontinuities in both the displacement and stress vectors across the fault. Because of equilibrium considerations, in the case of an earthquake the stress discontinuity must be zero, so that an earthquake is also known as a displacement dislocation.

In this chapter we will first derive the general expression for the body force equivalent to slip on a fault following Burridge and Knopoff (1964). This expression will then be used to obtain the double couple $M_{13} + M_{31}$, which in turn will be the basis for the discussion of a fault of arbitrary orientation. The chapter also includes the definition of the seismic moment tensor, a discussion of the relation between the parameters of the pair of fault planes (known as conjugate planes) that produce the same radiation pattern, and a consideration of practical aspects of focal mechanism studies. Although our emphasis is on the displacements as observed in the far field, because of the increasing importance of strong-motion studies, which require consideration of the ground displacement in the near field, we also discuss the permanent deformation, known as static displacement, generated by an earthquake source, and present examples of synthetic seismograms for points near the source. Finally, to show how the theory developed for a homogeneous medium

---

[1] Dislocation is Love's translation of the Italian word *distorsione*, which was used by Volterra and the other Italian mathematicians who developed most of the corresponding theory (Love, 1927; Steketee, 1958).

can be applied to more realistic media, an equation for the far field based on the
ray theory approximation is derived.

10.2 A representation theorem

In general, the displacement affecting the particles of an elastic solid subject to
a body force will depend on the nature of the applied force. Let $\mathbf{u}$ and $\mathbf{v}$ be the
displacements corresponding to body forces $\mathbf{f}$ and $\mathbf{g}$. Each displacement–force pair
will satisfy the elastic wave equation independently. Here we will derive a relation
(involving integrals) between these four vectors known as a reciprocity, or *reciprocal, theorem*. The importance of this result is that if one of the displacement–force
pairs, say $\mathbf{v}$ and $\mathbf{g}$, is known, then the theorem can be used to express $\mathbf{u}$ in terms
of integrals involving $\mathbf{v}$, $\mathbf{g}$, and $\mathbf{f}$. When $\mathbf{g}$ is a Dirac's delta in the space and time
domains, the integrals are significantly simplified, and the corresponding result is
known as a *representation theorem*. If Green's function (i.e., the $\mathbf{v}$ obtained when $\mathbf{g}$
is a Dirac's delta) is available, then a potentially complicated problem may become
easier to solve using the representation theorem. This is what Burridge and Knopoff
(1964) did to find the body force equivalent to slip on a fault. Their analysis will
be followed closely here.

The starting point in the derivation is the stress–strain relation (4.5.9) and the
strain–displacement relation (2.4.1). First, note that

$$\tau_{ij} = c_{ijpq}\varepsilon_{pq} = \frac{1}{2}c_{ijpq}(u_{p,q} + u_{q,p}) = \frac{1}{2}c_{ijpq}u_{p,q} + \frac{1}{2}c_{ijpq}u_{q,p} = c_{ijpq}u_{p,q},$$
(10.2.1)

where the relation $c_{ijqp} = c_{ijpq}$ has been used (see (4.5.3b)). Next, introduce this
result in the equation of motion (4.2.5) slightly rearranged

$$(c_{ijpq}(\mathbf{x})u_{p,q}(\mathbf{x}, t))_{,j} - \rho(\mathbf{x})\ddot{u}_i(\mathbf{x}, t) = -f_i(\mathbf{x}, t),$$
(10.2.2)

where f_i now indicates a force per unit volume.

For the force $\mathbf{g}$ and corresponding displacement $\mathbf{v}$ we have

$$(c_{ijpq}v_{p,q})_{,j} - \rho\ddot{v}_i = -g_i.$$
(10.2.3)

The forces are assumed to be equal to zero for $t < -T$, where T is a positive
constant, which means that to satisfy causality the displacements must also be zero
for $t < -T$. Before proceeding (10.2.3) will be rewritten with t replaced by $-t$,

$$(c_{ijpq}\bar{v}_{p,q})_{,j} - \rho\ddot{\bar{v}}_i = -\bar{g}_i,$$
(10.2.4)

where

$$\bar{v}_p(\mathbf{x}, t) = v_p(\mathbf{x}, -t); \qquad \bar{g}_p(\mathbf{x}, t) = g_p(\mathbf{x}, -t).$$
(10.2.5)

This operation will allow the significant simplifications effected below. Because $v_p(\mathbf{x}, t) = 0$ for $t < -T$, $\bar{v}_p(\mathbf{x}, t) = 0$ for $t > T$. Now contract (10.2.2) with $\bar{v}_i$ and (10.2.4) with u_i and subtract the two resulting equations:

$$\bar{v}_i\{(c_{ijpq}u_{p,q}),_j - \rho\ddot{u}_i)\} - u_i\{(c_{ijpq}\bar{v}_{p,q}),_j - \rho\ddot{\bar{v}}_i\}$$

$$= \{c_{ijpq}(\bar{v}_i u_{p,q} - u_i\bar{v}_{p,q})\},_j - c_{ijpq}(\bar{v}_{i,j}u_{p,q} - u_{i,j}\bar{v}_{p,q})$$

$$- \rho\frac{\partial}{\partial t}(\bar{v}_i\dot{u}_i - u_i\dot{\bar{v}}_i) = u_i\bar{g}_i - \bar{v}_i f_i. \tag{10.2.6}$$

The second term on the right of the first equals sign was added and subtracted to allow for the first term. Also note that the second term is the product of c_{ijpq}, which is symmetric on ij and pq, and a factor which is anti-symmetric on the same indices. Therefore, that term vanishes.

Next, assume that $\mathbf{f}$ and $\mathbf{g}$ act on a body with volume V and surface S, and integrate over t and V the expressions on both sides of the second equal sign in (10.2.6). The expression on the left-hand side gives

$$\int_{-\infty}^{\infty} dt \int_V \left\{ \{c_{ijpq}(\bar{v}_i u_{p,q} - u_i\bar{v}_{p,q})\},_j - \rho\frac{\partial}{\partial t}(\bar{v}_i\dot{u}_i - u_i\dot{\bar{v}}_i) \right\} dV. \tag{10.2.7}$$

The limits for the integral over t are actually $-T$ and T because the integrand vanishes outside of this interval. Because we have assumed that ρ does not depend on t, the second term in (10.2.7) gives

$$\rho(\bar{v}_i\dot{u}_i - u_i\dot{\bar{v}}_i)\Big|_{-\infty}^{\infty} = 0 \tag{10.2.8}$$

because $\bar{v}_i = \dot{\bar{v}}_i = 0$ for $t > T$ and $u_i = \dot{u}_i = 0$ for $t < -T$. After rewriting the remaining volume integral in (10.2.7) using Gauss' theorem (see §1.4.9) and rearranging terms, after integration equation (10.2.6) becomes

$$\int_{-\infty}^{\infty} dt \int_V (u_i\bar{g}_i - \bar{v}_i f_i)\, dV = \int_{-\infty}^{\infty} dt \int_S (\bar{v}_i c_{ijpq}u_{p,q} - u_i c_{ijpq}\bar{v}_{p,q})n_j\, dS, \tag{10.2.9}$$

where n_j indicates an outward unit vector normal to S.

Equation (10.2.9) is the integrated version of Betti's reciprocal theorem and to give some physical meaning to it note that the integrands on the left- and right-hand sides of (10.2.9) correspond to the work done by the body and surface forces, respectively (see §4.4; Love, 1927; Sokolnikoff, 1956).

Equation (10.2.9) will now be specialized to the case where $\mathbf{g}$ is a concentrated force in space and time along the x_n axis. This force will be represented as follows:

$$g_i(\mathbf{x}, t) = \delta_{in}\delta(\mathbf{x} - \boldsymbol{\xi})\delta(t + \tau) \equiv \delta_{in}\delta(\mathbf{x}, t; \boldsymbol{\xi}, -\tau). \tag{10.2.10}$$

Equation (10.2.10) represents a point force applied at point $\boldsymbol{\xi}$ at time $-\tau$ in the direction of the x_n axis. When this equation is introduced in (10.2.3) the corresponding solution is known as Green's function, indicated by $G_{in}(\mathbf{x}, t; \boldsymbol{\xi}, -\tau)$. Note the following conventions regarding the arguments and subscripts in Green's function. The first two arguments represent the observation point and time, while the last two arguments represent the source location in space and time. The first subscript indicates which component of the displacement G_{in} represents; the second gives the direction of the applied force. We have already encountered a Green's function (corresponding to an unbounded homogeneous medium) in §9.6.

Using (10.2.10), the concentrated force and Green's function corresponding to $\bar{g}_i(\mathbf{x}, t)$ are

$$\bar{g}_i(\mathbf{x}, t) = \delta_{in}\delta(\mathbf{x}, -t; \boldsymbol{\xi}, -\tau) \quad \text{and} \quad G_{in}(\mathbf{x}, -t; \boldsymbol{\xi}, -\tau). \quad (10.2.11\text{a,b})$$

Introducing (10.2.10) and (10.2.11a) in (10.2.9), using the properties of Dirac's delta, replacing $\bar{v}_i$ by (10.2.11b), and rearranging terms gives

$$u_n(\boldsymbol{\xi}, \tau) = \int_{-\infty}^{\infty} dt \int_V G_{in}(\mathbf{x}, -t; \boldsymbol{\xi}, -\tau) f_i(\mathbf{x}, t) \, dV_x$$

$$+ \int_{-\infty}^{\infty} dt \int_S \{G_{in}(\mathbf{x}, -t; \boldsymbol{\xi}, -\tau) c_{ijpq}(\mathbf{x}) u_{p,q}(\mathbf{x}, t)$$

$$- u_i(\mathbf{x}, t) c_{ijpq}(\mathbf{x}) G_{pn,q}(\mathbf{x}, -t; \boldsymbol{\xi}, -\tau)\} n_j \, dS_x. \quad (10.2.12)$$

The subscript in dV and dS indicates the integration variable (Problem 10.1).

Note that the variable $\boldsymbol{\xi}$ represents an observation point on the left-hand side of (10.2.12) and a source point in Green's functions on the right-hand side of (10.2.12). To change this situation and have $\boldsymbol{\xi}$ represent an observation point everywhere it is necessary to interchange the arguments in Green's function. However, that can be done only if the displacement and Green's function in (10.2.12) satisfy the same homogeneous boundary conditions (i.e., they are equal to zero, see §10.4) on the surface S. In such a case, if $\mathbf{f}$ is a point force applied at point $\boldsymbol{\xi}'$ at time τ' in the direction of x_m,

$$f_i(\mathbf{x}, t) = \delta_{im}\delta(\mathbf{x}, t; \boldsymbol{\xi}', \tau') \quad (10.2.13)$$

then (10.2.12) gives

$$G_{nm}(\boldsymbol{\xi}, \tau; \boldsymbol{\xi}', \tau') = G_{mn}(\boldsymbol{\xi}', -\tau'; \boldsymbol{\xi}, -\tau), \quad (10.2.14)$$

where the left-hand side of (10.2.14) is Green's function corresponding to (10.2.13). Note the different order of the subscripts m and n in the two functions. Equation (10.2.14) represents a space–time reciprocity for Green's function.

Now combining (10.2.12) and (10.2.14) under the assumption of homogeneous boundary conditions and interchanging $\boldsymbol{\xi}$ and $\mathbf{x}$ and t and τ (so that $\mathbf{x}$ and t represent the observation point and time everywhere) we obtain

$$
\begin{aligned}
u_n(\mathbf{x}, t) = &\int_{-\infty}^{\infty} d\tau \int_V G_{ni}(\mathbf{x}, t; \boldsymbol{\xi}, \tau) f_i(\boldsymbol{\xi}, \tau) dV_\xi \\
&+ \int_{-\infty}^{\infty} d\tau \int_S \{ G_{ni}(\mathbf{x}, t; \boldsymbol{\xi}, \tau) c_{ijpq}(\boldsymbol{\xi}) u_{p,q}(\boldsymbol{\xi}, \tau) \\
&- u_i(\boldsymbol{\xi}, \tau) c_{ijpq}(\boldsymbol{\xi}) G_{np,q'}(\mathbf{x}, t; \boldsymbol{\xi}, \tau) \} n_j \, dS_\xi,
\end{aligned}
\qquad (10.2.15)
$$

where the prime on q represents a derivative with respect to ξ_q.

Equation (10.2.15) is the representation theorem that will be used to find the body force equivalent to an earthquake, but before doing that it is necessary to modify the surface integral to allow for the presence of a discontinuity surface within the volume. This is necessary because, as discussed in §10.1, the model for an earthquake will be a surface (a fault) across which the displacement is discontinuous.

10.3 Gauss' theorem in the presence of a surface of discontinuity

Consider Gauss' theorem for a vectorial or tensorial quantity B,

$$
\int_V \nabla \cdot B \, dV = \int_S B \cdot \mathbf{n} \, dS
\qquad (10.3.1)
$$

(see §1.4.9) when B is discontinuous across a surface Σ within the volume V. As usual, S is the surface of V and $\mathbf{n}$ is a unit vector normal to S.

The analysis of this problem requires the following steps (based on Kraut, 1967): (a) surround Σ by a cavity having surface C (Fig. 10.1), (b) introduce a volume V' having external surface S and internal surface C, i.e., V' is equal to V with the cavity subtracted, (c) apply Gauss' theorem to V', within which B is continuous, (d) use a limiting argument to extend the theorem to the whole of V.

Let Σ^+ and Σ^- indicate the two sides of Σ and B^+ and B^- the corresponding values of B. Let C^+ and C^- indicate the portions of C on the sides of Σ^+ and Σ^- (Fig. 10.1). Using (10.3.1) with the volume V' gives

$$
\int_{V'} \nabla \cdot B \, dV = \int_S B \cdot \mathbf{n} \, ds + \int_{C^+} B \cdot \boldsymbol{v}^+ dC^+ + \int_{C^-} B \cdot \boldsymbol{v}^- dC^-.
\qquad (10.3.2)
$$

The vectors $\boldsymbol{v}^+$ and $\boldsymbol{v}^-$ are the outer normals to C^+ and C^-, respectively.

Now we let C shrink in such a way that C^+ and C^- approach Σ^+ and Σ^-, respectively. As C shrinks, $\boldsymbol{v}^+$ approaches $-\boldsymbol{v}^-$. Therefore, in the limit as C goes

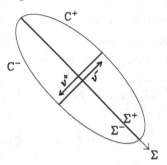

Fig. 10.1. Geometry for Gauss' theorem in the presence of a surface of a discontinuity (Σ). Σ^+ and Σ^- indicate the two sides of Σ. C^+ and C^- combined form a cavity C surrounding Σ and $\boldsymbol{v}^+$ and $\boldsymbol{v}^-$ are unit normal vectors.

into Σ,

$$\int_V \nabla \cdot B \, dV = \int_S B \cdot \mathbf{n} \, ds + \int_{\Sigma^+} B^+ \cdot \boldsymbol{v}^+ \, d\Sigma^+ + \int_{\Sigma^-} B^- \cdot \boldsymbol{v}^- \, d\Sigma^-$$

$$= \int_S B \cdot \mathbf{n} \, ds - \int_\Sigma (B^+ - B^-) \cdot \boldsymbol{v} \, d\Sigma = \int_S B \cdot \mathbf{n} \, ds$$

$$- \int_\Sigma [B] \cdot \boldsymbol{v} \, d\Sigma, \tag{10.3.3}$$

where $\boldsymbol{v}^-$ was renamed $\boldsymbol{v}$ and

$$[B] = B^+ - B^- \tag{10.3.4}$$

is the value of the discontinuity in B across Σ. The difference $[B]$ is assumed to be finite and is known as a jump discontinuity. In the following, an expression within square brackets will be interpreted as a discontinuity in the sense of (10.3.4).

10.4 The body force equivalent to slip on a fault

As noted in §10.1, an earthquake will be modeled as slip on a fault in a medium that at this point will be assumed to be inhomogeneous (its elastic properties vary with position) and anisotropic. Later, more restrictive conditions will be introduced. The surface rupture will be similar to Σ in Fig. 10.1, with the normal vector given by $\boldsymbol{v}$ as defined previously. As before, homogeneous boundary conditions on the displacement and Green's function will be assumed. Furthermore, Green's function and its derivatives will be assumed to be continuous through V, which means that only the displacement $\mathbf{u}$ and its derivatives will be allowed to be discontinuous.

Under these conditions (10.2.15) and (10.3.3) give

$$u_n(\mathbf{x}, t) = \int_{-\infty}^{\infty} d\tau \int_V G_{ni}(\mathbf{x}, t; \boldsymbol{\xi}, \tau) f_i(\boldsymbol{\xi}, t) \, dV_\xi$$

$$+ \int_{-\infty}^{\infty} d\tau \int_\Sigma \{ [u_i(\boldsymbol{\sigma}, \tau)] c_{ijpq}(\boldsymbol{\sigma}) G_{np,q'}(\mathbf{x}, t; \boldsymbol{\sigma}, \tau)$$

$$- G_{ni}(\mathbf{x}, t; \boldsymbol{\sigma}, \tau) c_{ijpq}(\boldsymbol{\sigma}) [u_{p,q}(\boldsymbol{\sigma}, \tau)] \} \nu_j \, d\Sigma. \tag{10.4.1}$$

The vector $\boldsymbol{\sigma}$ identifies points on the fault and $d\Sigma$ is the corresponding surface element. The integral over S is zero because of the homogeneous boundary conditions.

To find the body force equivalent, the second term in (10.4.1) will be written as a volume integral using the properties of Dirac's delta

$$G_{ni}(\mathbf{x}, t; \boldsymbol{\sigma}, \tau) = \int_V \delta(\boldsymbol{\xi} - \boldsymbol{\sigma}) G_{ni}(\mathbf{x}, t; \boldsymbol{\xi}, \tau) \, dV_\xi \tag{10.4.2}$$

$$G_{np,q'}(\mathbf{x}, t; \boldsymbol{\sigma}, \tau) = \frac{\partial}{\partial \sigma_q} G_{np}(\mathbf{x}, t; \boldsymbol{\sigma}, \tau) = \int_V \delta(\boldsymbol{\xi} - \boldsymbol{\sigma}) \frac{\partial}{\partial \xi_q} G_{np}(\mathbf{x}, t; \boldsymbol{\xi}, \tau) \, dV_\xi$$

$$= - \int_V \frac{\partial}{\partial \xi_q} \delta(\boldsymbol{\xi} - \boldsymbol{\sigma}) G_{np}(\mathbf{x}, t; \boldsymbol{\xi}, \tau) \, dV_\xi$$

$$\equiv - \int_V \delta_{,q}(\boldsymbol{\xi} - \boldsymbol{\sigma}) G_{np}(\mathbf{x}, t; \boldsymbol{\xi}, \tau) \, dV_\xi \tag{10.4.3}$$

(Problem 10.2).

Introducing (10.4.2) and (10.4.3) in (10.4.1), changing the order of integration and interchanging the dummy indices i and p gives

$$u_n(\mathbf{x}, t) = \int_{-\infty}^{\infty} d\tau \int_V G_{np}(\mathbf{x}, t; \boldsymbol{\xi}, \tau) \Big\{ f_p(\boldsymbol{\xi}, \tau)$$

$$- \int_\Sigma \{ [u_i(\boldsymbol{\sigma}, \tau)] c_{ijpq}(\boldsymbol{\sigma}) \delta_{,q}(\boldsymbol{\xi} - \boldsymbol{\sigma})$$

$$+ [u_{i,q}(\boldsymbol{\sigma}, \tau)] c_{pjiq}(\boldsymbol{\sigma}) \delta(\boldsymbol{\xi} - \boldsymbol{\sigma}) \} \nu_j \, d\Sigma \Big\} \, dV_\xi. \tag{10.4.4}$$

As $f_p(\boldsymbol{\xi}, \tau)$, which is a body force, and the integral over Σ enter (10.4.4) in the same way, the contribution of the discontinuities across Σ is equivalent to that of a body force given by

$$e_p(\boldsymbol{\xi}, \tau) = - \int_\Sigma \{ [u_i(\boldsymbol{\sigma}, \tau)] c_{ijpq}(\boldsymbol{\sigma}) \delta_{,q}(\boldsymbol{\xi} - \boldsymbol{\sigma}) + [u_{i,q}(\boldsymbol{\sigma}, \tau)] c_{pjiq}(\boldsymbol{\sigma}) \delta(\boldsymbol{\xi} - \boldsymbol{\sigma}) \} \nu_j \, d\Sigma$$

$$\tag{10.4.5}$$

acting on an unfaulted medium. Note that (10.4.5) involves summation over i, j, and q, which in the most general case gives rise to 27 terms. As shown below, however, in special cases this number may be as low as one.

Equation (10.4.5) will be rewritten using the fact that the contribution of $[u_{i,q}]$ can be expressed in terms of the discontinuity of the stress vector. Using (10.2.1) and (3.5.11) we obtain

$$c_{pjiq}[u_{i,q}]v_j = [\tau_{pj}]v_j = [T_p]. \tag{10.4.6}$$

Then,

$$e_p(\boldsymbol{\xi}, \tau) = -\int_\Sigma \{[u_i(\boldsymbol{\sigma}, \tau)]c_{ijpq}(\boldsymbol{\sigma})\delta_{,q}(\boldsymbol{\xi} - \boldsymbol{\sigma})v_j + [T_p(\boldsymbol{\sigma}, \tau)]\delta(\boldsymbol{\xi} - \boldsymbol{\sigma})\} \, d\Sigma. \tag{10.4.7}$$

Note that $[T_p]$ is the stress vector associated with the displacement $\mathbf{u}$ and the normal $\boldsymbol{v}$.

Some comments (Burridge and Knopoff, 1964) regarding (10.4.7) are in order. First, the equivalent force does not represent a real force acting in the real medium. Secondly, the discontinuities in the displacement and its derivatives cannot be assigned arbitrarily. For example, if Σ is in the plane (x_1, x_2), only the derivatives with respect to x_3 can be assigned arbitrary values. In the following we will ignore the discontinuities in $u_{i,q}$, and thus in T_p, because $[T_p]$ is taken as zero for the earthquake model. This assumption is reasonable because the continuity of T_p, which corresponds to Newton's law of action and reaction, is expected unless external forces are applied to Σ. Finally, when the tractions are continuous the total equivalent force and the total moment about any coordinate axis are zero (Problem 10.3).

Equation (10.4.6) without discontinuities is also useful because when used in (10.2.12) the resulting equation

$$u_n(\boldsymbol{\xi}, \tau) = \int_{-\infty}^{\infty} dt \int_V G_{in}(\mathbf{x}, -t; \boldsymbol{\xi}, -\tau) f_i(\mathbf{x}, t) \, dV_x$$

$$+ \int_{-\infty}^{\infty} dt \int_S \{G_{in}(\mathbf{x}, -t; \boldsymbol{\xi}, -\tau) T_i(\mathbf{x}, t)$$

$$- u_i(\mathbf{x}, t)c_{ijpq}(\mathbf{x})G_{pn,q}(\mathbf{x}, -t; \boldsymbol{\xi}, -\tau)n_j\} \, dS_x \tag{10.4.8}$$

can be used to make more specific statements regarding homogeneous boundary conditions for Green's function. As discussed by Aki and Richards (1980), there are two possibilities. One is that the boundary is rigid, in which case both the displacement and Green's function are zero on S. The other possibility is that the stress vector (or traction) is zero on the surface. This means that T_i and the traction $c_{ijpq}G_{pn,q}n_j$ are zero on S. In either case the integral over S in (10.4.8) is zero. As we have seen in §6.3, the surface of the Earth is nearly traction-free.

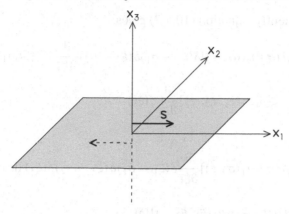

Fig. 10.2. Model for an earthquake source. The fault plane is in the (x_1, x_2) plane and plays the role of the surface Σ in Fig. 10.1, with Σ^+ on the side of $+x_3$. Σ^+ moves in the direction of $+x_1$, as indicated by the slip vector **s**. (After Aki and Richards, 1980.)

10.5 Slip on a horizontal plane. Point-source approximation. The double couple

Let us consider the case of a fault surface Σ in the horizontal plane, with the (plane) surfaces Σ^+ and Σ^- in the positive and negative directions of ξ_3 (Fig. 10.2). This means that $\boldsymbol{v}$ is in the $+\xi_3$ direction. Furthermore, it will be assumed that slip takes place in the ξ_1 direction. Therefore, the motion on Σ^+ is in the $+\xi_1$ direction and the motion on Σ^- is in the $-\xi_1$ direction. Under these conditions we can write

$$\boldsymbol{v} = (v_1, v_2, v_3) = (0, 0, 1); \qquad [u_1] \neq 0; \qquad [u_2] = [u_3] = 0. \qquad (10.5.1)$$

Two additional assumptions will be made. First, $[T_p]$ will be taken as zero, as discussed above. Secondly, the medium will be assumed to be isotropic in the vicinity of the fault, so that c_{ijpq} has the simple form

$$c_{ijpq} = \lambda \delta_{ij} \delta_{pq} + \mu(\delta_{ip}\delta_{jq} + \delta_{iq}\delta_{jp}) \qquad (10.5.2)$$

(see (4.6.2)).

From the first term in the integral in (10.4.7) and the special form of $\boldsymbol{v}$ and $\mathbf{u}$ given in (10.5.1) we can see that the only nonzero contributions to the integral are those for which c_{ijpq} is of the form c_{13pq}. Then from (10.5.2) we obtain

$$c_{13pq} = \lambda \delta_{13}\delta_{pq} + \mu(\delta_{1p}\delta_{3q} + \delta_{1q}\delta_{3p}) = \mu(\delta_{1p}\delta_{3q} + \delta_{1q}\delta_{3p}) \qquad (10.5.3)$$

because $\delta_{13} = 0$. The subscripts 1 and 3 come from $[u_1]$ and v_3, respectively, which are the only components of $\mathbf{u}$ and $\boldsymbol{v}$ different from zero. Equation (10.5.3) shows that the possible 27 terms in (10.4.7) reduce to one because the only nonzero contributions have the combinations $p = 1, q = 3$ and $p = 3, q = 1$. In both cases the result is μ. Consequently, equation (10.4.7) gives

$$e_1(\boldsymbol{\xi}, \tau) = -\int_\Sigma \mu(\boldsymbol{\sigma})[u_1(\boldsymbol{\sigma}, \tau)]\delta(\xi_1 - \sigma_1)\delta(\xi_2 - \sigma_2)\frac{\partial}{\partial\xi_3}\delta(\xi_3)\,d\sigma_1\,d\sigma_2$$

$$= -\mu(\xi_1, \xi_2)[u_1(\xi_1, \xi_2, \tau)]\frac{\partial}{\partial\xi_3}\delta(\xi_3) \tag{10.5.4}$$

$$e_2(\boldsymbol{\xi}, \tau) = 0 \tag{10.5.5}$$

$$e_3(\boldsymbol{\xi}, \tau) = -\int_\Sigma \mu(\boldsymbol{\sigma})[u_1(\boldsymbol{\sigma}, \tau)]\frac{\partial}{\partial\xi_1}\delta(\xi_1 - \sigma_1)\delta(\xi_2 - \sigma_2)\delta(\xi_3)\,d\sigma_1\,d\sigma_2$$

$$= -\frac{\partial}{\partial\xi_1}\{\mu(\xi_1, \xi_2)[u_1(\xi_1, \xi_2, \tau)]\}\delta(\xi_3). \tag{10.5.6}$$

In (10.5.4) and (10.5.6) the derivatives are with respect to variables different from the integration variables and for this reason can be taken out of the integrals.

As shown below, the derivative of Dirac's delta in (10.5.4) represents a double couple in the ξ_1 direction with an arm in the ξ_3 direction. Therefore, e_1 represents a system of couples over Σ. This force has a moment about the ξ_2 axis given by

$$M(\tau) = \int_V \xi_3 e_1 \,dV = -\int_\Sigma \mu[u_1]\,d\sigma_1\,d\sigma_2 \int \xi_3\frac{\partial}{\partial\xi_3}\delta(\xi_3)\,d\xi_3 = \int_\Sigma \mu[u_1]\,d\sigma_1\,d\sigma_2, \tag{10.5.7}$$

where (A.27) has been used (Problem 10.4). If the medium is homogeneous in the vicinity of the fault, then μ is constant and can be placed outside of the integral in (10.5.7). If, in addition, $[u_1]$ is replaced by its average value $\bar{u}$ over Σ, then

$$M(\tau) = \mu\bar{u}(\tau)A; \qquad \bar{u} = \bar{u}(\tau) = \frac{1}{A}\int_\Sigma [u_1(\sigma_1, \sigma_2, \tau)]\,d\sigma_1\,d\sigma_2, \tag{10.5.8a,b}$$

where A is the area of Σ. Equation (10.5.8a) was derived by Aki (1966) for the concentrated displacement discussed below. Aki (1990) gives an interesting historical account of the earlier work that influenced his 1966 results.

Equation (10.5.6) represents a more complicated force distribution, discussed in detail in Aki and Richards (1980). To simplify the problem we will introduce the assumption that the linear dimensions of Σ are much smaller than the wavelengths of the observed waves and that the periods are much longer than the source duration (Backus and Mulcahy, 1976; Aki and Richards, 1980). Under this approximation Σ will act as a *point source*, with the displacement represented by a concentrated distribution, which in view of (10.5.8) will be written as

$$[u_1(\xi_1, \xi_2, \tau)] = \frac{1}{\mu}M_o\delta(\xi_1)\delta(\xi_2)H(\tau), \tag{10.5.9}$$

where M_o is given by

$$M_o = \mu \bar{u} A \qquad (10.5.10)$$

and is assumed to be constant.

Equation (10.5.9) is an extension of the expression used by Burridge and Knopoff (1964), who did not include the M_o/μ factor. $H(t)$ is Heaviside's unit step function, and is introduced to indicate that the slip occurs suddenly at time $\tau = 0$ and remains constant after that.

When (10.5.9) is introduced in (10.5.4) and (10.5.6) we obtain

$$e_1(\boldsymbol{\xi}, \tau) = -M_o\delta(\xi_1)\delta(\xi_2)\frac{\partial}{\partial \xi_3}\delta(\xi_3)H(\tau) \qquad (10.5.11)$$

$$e_3(\boldsymbol{\xi}, \tau) = -M_o\frac{\partial}{\partial \xi_1}\delta(\xi_1)\delta(\xi_2)\delta(\xi_3)H(\tau). \qquad (10.5.12)$$

Now we will interpret the derivatives in (10.5.11) and (10.5.12). To do that, write the derivative in (10.5.11) (including the minus sign) as

$$-\frac{\partial}{\partial \xi_3}\delta(\xi_3) \approx \frac{1}{h}\left[\delta\left(\xi_3 - \frac{h}{2}\right) - \delta\left(\xi_3 + \frac{h}{2}\right)\right]. \qquad (10.5.13)$$

Since e_1 is in the ξ_1 direction, the derivative in (10.5.13) can be represented by a couple in the ξ_1 direction with arm in the ξ_3 direction (Fig. 10.3). Comparison with Fig. 9.8 shows that the derivative in (10.5.11) represents the couple M_{13}. In a similar way we can see that the derivative in (10.5.12) corresponds to M_{31}. Therefore, the forces e_1 and e_3 can be represented by $M_o M^{dc}$, where M^{dc} is the double couple $M_{13} + M_{31}$ introduced in §9.9.4.

M_o is known as the *scalar seismic moment* and is used to measure the size of an earthquake. It must be noted however, that the definition of M_o includes the rigidity (μ) of the medium. An alternative way to measure the size of an earthquake is to use the ratio $M_o/\mu = \bar{u}A$, which Ben-Menahem and Singh (1981) call *potency*. In principle, either definition would be perfectly acceptable if μ were well known, but since it is not, it has been suggested that potency instead of seismic moment be used to quantify the size of an earthquake (Ben-Zion, 2001). It must be noted, however, that neither of these entities is actually an observable, and that to determine each of them it is necessary to solve an inverse problem, which in turn requires choosing Earth and source models. Therefore, any errors in these models will translate into errors in the estimates of source strength.

Another comment pertinent to the definition of M_o is that the slip may not necessarily be constant, particularly in large earthquakes, in which case $M_o H(\tau)$ in (10.5.9) must be replaced by $M(\tau)$ with the slip averaged at time τ as shown in (10.5.8b). However, for large earthquakes the point-source approximation may

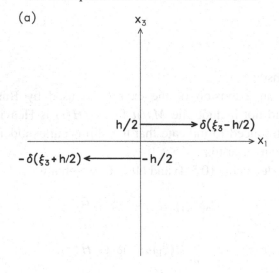

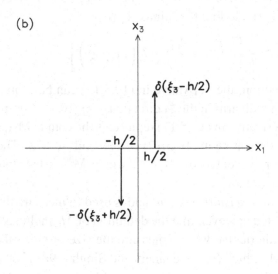

Fig. 10.3. (a) The derivative of $-\partial\delta(\xi_3)/\partial\xi_3$ that appears in the expression for $e_1(\boldsymbol{\xi}, \tau)$ (see (10.5.11)) can be represented by the couple M_{13}. (b) The derivative $-\partial\delta(\xi_1)/\partial\xi_1$ along the ξ_3 direction (see (10.5.12)) can be represented by the couple M_{31}. The combination of the two couples produces no net rotation.

not be appropriate, in which case the fault area can be subdivided into a number of smaller faults for which the approximation is valid. Then the effect of the large fault will be equal to the sum of the effects of the subfaults. An example of variable slip is given in §10.10.

10.6 The seismic moment tensor

In §9.9 we introduced the moment tensor without reference to any particular physical process occurring in the Earth. On the other hand, in §10.5 we saw that slip in the (x_1, x_2) plane can be represented by the double couple $M_o(M_{13} + M_{31})$, which has the moment tensor $M_o\mathbf{M}^{\mathrm{dc}}$ (see (9.9.4)). Therefore, we can study the displacement generated by this source using (9.9.1) without further consideration of the analysis carried out in the preceding sections. It is possible, however, to derive two equations similar to (9.9.1) using results from §10.4 after introducing new definitions of the moment tensor, which in turn will be used to complement and expand our earlier results. The new definitions arise from consideration of two alternative expressions for the displacement caused by the dislocation. One of them follows from (10.4.1),

$$u_n(\mathbf{x}, t) = \int_{-\infty}^{\infty} d\tau \int_{\Sigma} [u_i(\boldsymbol{\sigma}, \tau)]c_{ijpq}(\boldsymbol{\sigma})G_{np,q'}(\mathbf{x}, t; \boldsymbol{\sigma}, \tau)v_j \, d\Sigma, \qquad (10.6.1)$$

where we have assumed that body forces are absent and that the tractions are continuous. Now we will introduce the *moment tensor density* m_{pq}

$$m_{pq}(\boldsymbol{\sigma}, \tau) = [u_i(\boldsymbol{\sigma}, \tau)]c_{ijpq}(\boldsymbol{\sigma})v_j \qquad (10.6.2)$$

(Aki and Richards, 1980), use the relation

$$G_{np}(\mathbf{x}, t; \boldsymbol{\sigma}, \tau) = G_{np}(\mathbf{x}, t - \tau; \boldsymbol{\sigma}, 0) \qquad (10.6.3)$$

(Problem 10.5), and assume as before that the fault is a point source, so that $G_{np,q'}$ can be considered to be a constant over Σ. Under these conditions (10.6.1) becomes

$$u_n(\mathbf{x}, t) = \int_{-\infty}^{\infty} \left[\int_{\Sigma} m_{pq}(\boldsymbol{\sigma}, \tau) \, d\Sigma \right] G_{np,q'}(\mathbf{x}, t - \tau; \boldsymbol{\sigma}, 0) \, d\tau$$

$$= M_{pq}(t) * G_{np,q'}(\mathbf{x}, t; \boldsymbol{\sigma}, 0) = M_{pq} * \frac{\partial G_{np}}{\partial \sigma_q}, \qquad (10.6.4)$$

where

$$M_{pq}(t) = \int_{\Sigma} m_{pq}(\boldsymbol{\sigma}, t) \, d\Sigma \qquad (10.6.5)$$

is the *seismic moment tensor* (Aki and Richards, 1980). Equation (10.6.4) is equal to (9.9.1), although it must be noted that the moment tensor derived from (10.6.2) is symmetric (see (4.5.3b)), while that in Chapter 9 does not have any special symmetry properties.

In the case of an isotropic medium, using (10.5.2), equation (10.6.2) becomes

$$m_{pq} = \lambda[u_i]v_i\delta_{pq} + \mu\left([u_p]v_q + [u_q]v_p\right). \qquad (10.6.6)$$

The first term on the right-hand side of (10.6.6) is the scalar product of $[\mathbf{u}]$ and $\boldsymbol{v}$, which is zero when the two vectors are perpendicular to each other. This situation arises when [**u**] is on Σ, as in §10.5, in which case

$$m_{pq} = \mu\left([u_p]v_q + [u_q]v_p\right). \tag{10.6.7}$$

Moreover, when the point-source approximation is applicable, from (10.6.5) and (10.6.7) we obtain

$$M_{pq} = A\bar{m}_{pq} = A\mu\left([\bar{u}_p]v_q + [\bar{u}_q]v_p\right), \tag{10.6.8}$$

where A is the fault area and the overbar indicates an average value (Backus and Mulcahy, 1976). For the particular case discussed in §10.5 the moment tensor is

$$\mathbf{M} = M_o \begin{pmatrix} 0 & 0 & 1 \\ 0 & 0 & 0 \\ 1 & 0 & 0 \end{pmatrix} = M_o\mathbf{M}^{dc}, \tag{10.6.9}$$

in agreement with our earlier result.

A second expression for the displacement caused by the dislocation follows from consideration of the equivalent body force e_p. Under the same assumptions that led to (10.6.1) and using (10.4.4) and (10.4.5), the displacement can be written as

$$u_n(\mathbf{x}, t) = \int_{-\infty}^{\infty} d\tau \int_{V_o} G_{np}(\mathbf{x}, t; \boldsymbol{\xi}, \tau) e_p(\boldsymbol{\xi}, \tau)\, dV_\xi, \tag{10.6.10}$$

where V_o is the volume where these forces are not zero. Now we will expand Green's function in a Taylor series about a reference point $\bar{\boldsymbol{\xi}}$,

$$G_{np}(\mathbf{x}, t; \boldsymbol{\xi}, \tau) = G_{np}(\mathbf{x}, t; \bar{\boldsymbol{\xi}}, \tau) + (\xi_q - \bar{\xi}_q)G_{np,q'}(\mathbf{x}, t; \bar{\boldsymbol{\xi}}, \tau)$$

$$+ \frac{1}{2}(\xi_q - \bar{\xi}_q)(\xi_r - \bar{\xi}_r)G_{np,q'r'}(\mathbf{x}, t; \bar{\boldsymbol{\xi}}, \tau) + \cdots, \tag{10.6.11}$$

where, as before, the primes on q and r indicate derivatives with respect to ξ_q and ξ_r. The selection of the reference point depends on the nature of the fault. For small faults $\bar{\boldsymbol{\xi}}$ can be the hypocenter, while for large faults it is more convenient to choose the centroid of the seismic source, which is a weighted mean position of the source (Backus, 1977; Dziewonski *et al.*, 1981; Dziewonski and Woodhouse, 1983a). For a large fault the hypocenter corresponds to the point of rupture initiation. For small faults the first two terms in the expansion (10.6.11) provide a good approximation to G_{np} (Backus and Mulcahy, 1976; Stump and Johnson, 1977) and the volume integral in (10.6.10) becomes

$$\int_{V_o} G_{np}(\mathbf{x}, t; \boldsymbol{\xi}, \tau) e_p(\boldsymbol{\xi}, \tau) \, dV_\xi = G_{np}(\mathbf{x}, t; \bar{\boldsymbol{\xi}}, \tau) \int_{V_o} e_p(\boldsymbol{\xi}, \tau) \, dV_\xi$$

$$+ G_{np,q'}(\mathbf{x}, t; \bar{\boldsymbol{\xi}}, \tau) \int_{V_o} (\xi_q - \bar{\xi}_q) e_p(\boldsymbol{\xi}, \tau) \, dV_\xi$$

$$= G_{np,q'}(\mathbf{x}, t; \bar{\boldsymbol{\xi}}, \tau) \acute{M}_{pq}(\tau), \qquad (10.6.12)$$

where

$$\acute{M}_{pq}(\tau) = \int_{V_o} (\xi_q - \bar{\xi}_q) e_p(\boldsymbol{\xi}, \tau) \, dV_\xi. \qquad (10.6.13)$$

The integral of e_p in (10.6.12) is zero because we have assumed continuous tractions (see Problem 10.3). Consequently, $\acute{M}_{pq}$ is symmetric (Problem 10.7).

Now introducing (10.6.12) into (10.6.10) and using (10.6.3) we obtain

$$u_n(\mathbf{x}, t) = \int_{-\infty}^{\infty} G_{np,q'}(\mathbf{x}, t - \tau; \bar{\boldsymbol{\xi}}, 0) \acute{M}_{pq}(\tau) \, d\tau = \acute{M}_{pq} * \frac{\partial G_{np}}{\partial \xi_q}. \qquad (10.6.14)$$

Comparison of (10.6.4) and (10.6.14) shows that $\acute{M}_{pq}$ is an equivalent representation of the moment tensor M_{pq} (Problem 10.7). Moreover, because the integrand in (10.6.13) is the product of a force and a distance, equation (10.6.13) generalizes the definition of moment of a couple introduced in connection with (9.8.2).

The concepts of moment tensor density and seismic moment tensor were introduced by Backus and Mulcahy (1976) in a general context, in which the earthquake source is a particular case. These authors considered *indigenous* sources, which are restricted to forces inside the Earth (such as earthquakes). Meteor impacts and bodily tides are examples of nonindigenous sources. The vanishing of the total force (the integral of e_p) and the symmetry of the seismic moment tensor are general properties of indigenous sources, and correspond to the conservation of the total linear and angular momentum (see (3.2.19) and (3.2.20)) of the Earth.

10.7 Moment tensor for slip on a fault of arbitrary orientation

Before discussing the moment tensor it is necessary to introduce the parameters that will be used to describe the fault and the slip. Because we are interested in relating our results to observations in the Earth we will introduce a coordinate system based on the geographical system (Fig. 10.4). In this system the origin is the Earth's center, the x_1 axis is in the Greenwich meridian and the x_3 axis passes through the north pole. Then the source location is identified by the longitude ϕ_0 and the colatitude θ_0, and the unit vectors $\boldsymbol{\Theta}^\circ$, $\boldsymbol{\Phi}^\circ$, and $\boldsymbol{\Gamma}^\circ$ point in the south, east, and up (i.e., towards the zenith) directions. The superscript was used to avoid

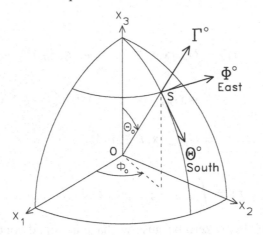

Fig. 10.4. Coordinates of the seismic source (indicated with S) in the geographic coordinate system. The point O represents the center of the Earth, ϕ_o is longitude and θ_o is colatitude. (After Ben-Menahem and Singh, 1981.)

confusion with the unit vectors in Fig. 9.10, which has the source at the center of the sphere. Now consider the following two operations. First, rotate the system (x_1, x_2, x_3) about the x_3 axis through an angle ϕ_o. After this operation the new axis x_1 is in the meridian that passes through the source. Next, rotate the resulting system about the new x_2 axis through an angle θ_o, so that the new x_3 axis passes through the source. The final system, called the *epicentral system of coordinates* by Ben-Menahem and Singh (1981), has the axes x_1, x_2, and x_3 in the south, east, and up directions, respectively. This coordinate system is convenient for global seismology studies and except for §10.9 it will be the system used here. This means that in the context of this chapter, the coordinate system (x_1, x_2, x_3) in Fig. 9.10 is the epicentral system. An alternative system with x_3 pointing down is also commonly used, and for this reason the corresponding results are included at the end of this section.

The fault parameters and related variables are shown in Fig. 10.5. The *strike* φ is the azimuth of the intersection of the fault with the horizontal plane (x_1, x_2). The strike is measured clockwise from north, with $0 \leq \varphi \leq 2\pi$. Because strikes of φ and $\varphi + \pi$ represent the same line, we will avoid this ambiguity by introducing the *strike direction*, which is chosen so that the fault dips to the right when looking in the strike direction. The *dip angle* (or dip) is the angle between the horizontal plane and the fault plane and is measured down from the (x_1, x_2) plane in a vertical plane perpendicular to the strike. In the convention used here, $0 \leq \delta \leq \pi/2$. The *dip direction* is given by the vector $\mathbf{d}$ and is equal to the strike direction plus $\pi/2$. The *hanging wall* and *foot wall* of a fault are defined as the blocks above and below of

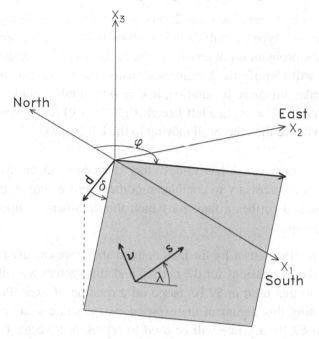

Fig. 10.5. Definition of the fault parameters. A geographical coordinate system centered at the source is used. North, south and east are in the $-x_1, x_1$, and x_2 directions, respectively, and the x_3 direction points up (i.e., towards the zenith). The strike φ is the azimuth of the intersection of the fault with the horizontal plane (x_1, x_2). The arrow indicates the strike direction. The dip angle δ is measured down from the horizontal plane in a vertical plane perpendicular to the strike. The vector $\boldsymbol{v}$ is normal to the fault. The vector $\mathbf{s}$ is the slip vector. It is in the fault plane and indicates the motion of the hanging wall with respect to the foot wall. The foot wall (hanging wall) is defined as the block below (above) the fault plane. The slip angle λ is the angle between $\mathbf{s}$ and the strike direction and is measured in the fault plane.

the fault plane, respectively. The hanging wall is to the right of an observer looking in the strike direction, which means that Fig. 10.5 corresponds to the foot wall. The *slip vector* indicates the relative motion of the hanging wall with respect to the foot wall. The *slip angle* λ is measured in the fault plane counterclockwise from the strike direction, with $0 \leq \lambda \leq 2\pi$. The unit vector normal to the fault will be indicated by $\boldsymbol{v}$.

Faults can be classified based on the values of λ and δ as follows, although it must be noted that some other authors have somewhat different definitions.

Dip-slip faults. λ is equal to $\pi/2$ or $3\pi/2$, which means that there is no horizontal displacement. If $\lambda = \pi/2$ the hanging wall moves up with respect to the foot wall and the fault is known as *reverse*. If $\lambda = 3\pi/2$ the opposite happens and the fault is said to be *normal*.

Strike-slip faults. $\delta = \pi/2$ and $\lambda = 0$ (or 2π) or $\lambda = \pi$. In this case the slip is purely horizontal. For this type of fault there is ambiguity in the definition of the strike direction, but the problem is not serious; of the two possible strike directions choose one and this will identify the hanging wall using the convention that it is to the right of the strike direction. In addition, if $\lambda = 0$ the fault is said to be *left lateral*, and *right lateral* if $\lambda = \pi$. In a left-lateral (right-lateral) fault, an observer on one of the walls will see the other wall moving to the left (right).

Oblique-slip faults. In this case λ has values other than those given above. To describe these faults it is necessary to combine modifiers. For example, the term left-lateral normal fault describes a fault for which the dominant component of motion is normal dip-slip.

To determine the moment tensor for the fault defined above we can use (10.6.8), but to avoid finding the expressions for the normal and slip vectors we will apply an argument similar to that used in §9.10, based on a rotation of axes (Pujol and Herrmann, 1990). Using this argument the expressions for $\mathbf{s}$ and $\boldsymbol{\nu}$ are obtained without effort. As in §9.10, a prime will be used to represent a vector or tensor in the rotated system. Since we already know the moment tensor for slip in the horizontal plane in the x_1 direction, the coordinate axes in Fig. 10.5 will be rotated such that $\boldsymbol{\nu}'$ is in the positive x_3' direction and $\mathbf{s}'$ is in the x_1' direction. In the rotated system the surfaces Σ^+ and Σ^- (see Fig. 10.2) are in the hanging and foot walls, respectively, and the moment tensor $\mathbf{M}'$ is $M_o \mathbf{M}^{dc}$ (see (10.6.9)). Then the moment tensor $\mathbf{M}$ in the epicentral system can be obtained using (9.10.3) with $\mathbf{R}$ a rotation matrix to be determined below. In component form we obtain

$$M_{ij} = M_o R_{ki} \mathbf{M}_{kl}^{dc} R_{lj} = M_o(R_{1i} R_{3j} + R_{3i} R_{1j}). \tag{10.7.1}$$

The second equality in (10.7.1) arises because only M_{13}^{dc} and M_{31}^{dc} are different from zero. Note that $M_{ij} = M_{ji}$, as expected, since $\mathbf{M}^{dc}$ is symmetric.

The same arguments can be applied to the normal vector and the slip vector. In the rotated system $\boldsymbol{\nu}' = (0, 0, 1)^T$. Then, from (1.3.16), $\boldsymbol{\nu} = \mathbf{R}^T \boldsymbol{\nu}'$ and

$$\nu_i = R_{ki} \nu_k' = R_{3i}. \tag{10.7.2}$$

Therefore, the components of the normal vector are given by the elements of the third row of the rotation matrix.

Similarly, $\mathbf{s}' = (1, 0, 0)^T$, $\mathbf{s} = \mathbf{R}^T \mathbf{s}'$, and

$$s_i = R_{ki} s_k' = R_{1i}, \tag{10.7.3}$$

so that the components of the slip vector are given by the elements of the first row of the rotation matrix.

Using (10.7.1)–(10.7.3) we obtain

$$M_{ij} = M_0(v_i s_j + v_j s_i),\tag{10.7.4}$$

in agreement with (10.6.8). This equation is sometimes written in dyadic form

$$\mathcal{M} = M_0(\boldsymbol{vs} + \boldsymbol{sv}).\tag{10.7.5}$$

Equations (10.7.4) and (10.7.5) show that $\boldsymbol{v}$ and $\boldsymbol{s}$ contribute to the moment tensor in a symmetric way, which means that slip in the direction of $\boldsymbol{s}$ in a plane with normal $\boldsymbol{v}$ has the same moment tensor as slip in the direction of $\boldsymbol{v}$ in a plane with normal $\boldsymbol{s}$. Therefore, in the point-source approximation it is not possible to establish which plane is the actual fault plane, and additional information is required to remove this ambiguity (i.e., the distribution of aftershocks, surface evidence). The two possible planes are called *conjugate planes*. If one of them is the actual fault plane, then the other one is known as the *auxiliary plane*. The relations between the parameters of the two planes are given in §10.8.

Now $\mathbf{R}$ will be expressed as the product of three simple rotation matrices:

$$\mathbf{R} = \mathbf{R}_3 \mathbf{R}_2 \mathbf{R}_1 \tag{10.7.6}$$

where

$\mathbf{R}_1$: counterclockwise rotation of angle $(\pi - \varphi)$ about the x_3 axis. The new co-ordinate system will be indicated with $(x_1^1, x_2^1, x_3^1 = x_3)$, with the superscripts identifying the rotation. This operation brings the strike direction and the x_1^1 axis into coincidence.

$$\mathbf{R}_1 = \begin{pmatrix} \cos(\pi - \varphi) & \sin(\pi - \varphi) & 0 \\ -\sin(\pi - \varphi) & \cos(\pi - \varphi) & 0 \\ 0 & 0 & 1 \end{pmatrix} = \begin{pmatrix} \cos\varphi & -\sin\varphi & 0 \\ \sin\varphi & \cos\varphi & 0 \\ 0 & 0 & 1 \end{pmatrix}.\tag{10.7.7}$$

$\mathbf{R}_2$: counterclockwise rotation of angle δ about the x_1^1 axis. The rotated system is $(x_1^2 = x_1^1, x_2^2, x_3^2)$. This rotation brings the fault plane and the (x_1^2, x_2^2) plane into coincidence and aligns the x_3^2 axis with the normal vector $\boldsymbol{v}$.

$$\mathbf{R}_2 = \begin{pmatrix} 1 & 0 & 0 \\ 0 & \cos\delta & \sin\delta \\ 0 & -\sin\delta & \cos\delta \end{pmatrix}.\tag{10.7.8}$$

$\mathbf{R}_3$: counterclockwise rotation of angle λ about the x_3^2 axis. The rotated system is $(x_1^3, x_2^3, x_3^3 = x_3^2)$. This rotation aligns the x_1^3 axis with the slip vector $\boldsymbol{s}$.

$$\mathbf{R}_3 = \begin{pmatrix} \cos\lambda & \sin\lambda & 0 \\ -\sin\lambda & \cos\lambda & 0 \\ 0 & 0 & 1 \end{pmatrix}.\tag{10.7.9}$$

Then

$$\mathbf{R} =$$

$$\begin{pmatrix} -\cos\lambda\cos\varphi - \sin\lambda\cos\delta\sin\varphi & \cos\lambda\sin\varphi - \sin\lambda\cos\delta\cos\varphi & \sin\lambda\sin\delta \\ \sin\lambda\cos\varphi - \cos\lambda\cos\delta\sin\varphi & -\sin\lambda\sin\varphi - \cos\lambda\cos\delta\cos\varphi & \cos\lambda\sin\delta \\ \sin\delta\sin\varphi & \sin\delta\cos\varphi & \cos\delta \end{pmatrix}.$$

$$(10.7.10)$$

After the rotation $\mathbf{R}$ has been applied, the original coordinate system (x_1, x_2, x_3) becomes (x_1', x_2', x_3'). An equation similar to (10.7.10) arises in the study of rigid-body rotations, with the rotation angles known as Euler angles. In seismology the series of rotations indicated in (10.7.6) can be found in Jarosch and Aboodi (1970) and Ben-Menahem and Singh (1981).

Now we can use (10.7.10) with (10.7.1), which gives the components of the moment tensor in the epicentral system:

$$M_{11} = -M_o(\sin\delta\cos\lambda\sin 2\varphi + \sin 2\delta\sin\lambda\sin^2\varphi)$$

$$M_{12} = M_{21} = -M_o(\sin\delta\cos\lambda\cos 2\varphi + \tfrac{1}{2}\sin 2\delta\sin\lambda\sin 2\varphi)$$

$$M_{13} = M_{31} = -M_o(\cos\delta\cos\lambda\cos\varphi + \cos 2\delta\sin\lambda\sin\varphi) \qquad (10.7.11)$$

$$M_{22} = M_o(\sin\delta\cos\lambda\sin 2\varphi - \sin 2\delta\sin\lambda\cos^2\varphi)$$

$$M_{23} = M_{32} = M_o(\cos\delta\cos\lambda\sin\varphi - \cos 2\delta\sin\lambda\cos\varphi)$$

$$M_{33} = M_o\sin 2\delta\sin\lambda.$$

These expressions, with the subscripts 1, 2, 3 replaced by θ, ϕ, r, respectively, correspond to the moment tensor components in the Harvard centroid-moment tensor catalog (Dziewonski and Woodhouse, 1983a,b). The first explicit relation between moment tensor components and fault parameters is due to Mendiguren (1977).

From (10.7.2) and (10.7.3) the components of the vectors $\mathbf{s}$ and $\mathbf{\nu}$ are given by the elements of the first and third rows of $\mathbf{R}$, respectively:

$$\mathbf{s} = \begin{pmatrix} -\cos\lambda\cos\varphi - \sin\lambda\cos\delta\sin\varphi \\ \cos\lambda\sin\varphi - \sin\lambda\cos\delta\cos\varphi \\ \sin\lambda\sin\delta \end{pmatrix}; \qquad \mathbf{\nu} = \begin{pmatrix} \sin\delta\sin\varphi \\ \sin\delta\cos\varphi \\ \cos\delta \end{pmatrix} \quad (10.7.12a,b)$$

(e.g., Ben-Menahem and Singh, 1981; Dziewonski and Woodhouse, 1983a).

Vectors $\mathbf{p}$, $\mathbf{t}$ and $\mathbf{b}$ can also be written in terms of the fault parameters with little effort. In fact, in the rotated system

$$\mathbf{p}' = \frac{1}{\sqrt{2}}(1, 0, -1)^T = \frac{1}{\sqrt{2}}(\mathbf{s}' - \mathbf{\nu}') \qquad (10.7.13)$$

$$t' = \frac{1}{\sqrt{2}}(1, 0, 1)^T = \frac{1}{\sqrt{2}}(s' + v'). \qquad (10.7.14)$$

Since the operation of rotation is linear with respect to the sum of vectors,

$$p = \frac{1}{\sqrt{2}}(s - v) \qquad (10.7.15)$$

and

$$t = \frac{1}{\sqrt{2}}(s + v). \qquad (10.7.16)$$

The components of p and t are obtained directly from (10.7.15), (10.7.16), and (10.7.12).

The components of b are obtained from $b = R^T b'$ with $b' = (0, 1, 0)^T$,

$$b_i = R_{ki} b'_k = R_{2i}, \qquad (10.7.17)$$

so that they are given by the elements of the second row of R:

$$b = \begin{pmatrix} \sin\lambda\cos\varphi - \cos\lambda\cos\delta\sin\varphi \\ -\sin\lambda\sin\varphi - \cos\lambda\cos\delta\cos\varphi \\ \cos\lambda\sin\delta \end{pmatrix} \qquad (10.7.18)$$

(e.g., Ben-Menahem and Singh, 1981; Dziewonski and Woodhouse, 1983a).

An alternative coordinate system, generally used with body waves, has the axes x_1, x_2, x_3 in the north, east and down (i.e., towards the nadir) directions. This is the coordinate system used by Aki and Richards (1980), and to obtain the corresponding moment tensor components the coordinate system of Fig. 10.5 will be rotated an angle π about the x_2 axis. After this rotation the x_1 axis points north, as opposed to south in Fig. 10.5. The matrix corresponding to this rotation is given by

$$R = \begin{pmatrix} -1 & 0 & 0 \\ 0 & 1 & 0 \\ 0 & 0 & -1 \end{pmatrix}. \qquad (10.7.19)$$

Therefore, if M_{ij} are the components of the moment tensor in the unrotated system (given by (10.7.11)), then in the rotated system we have

$$M'_{ij} = R_{ik} M_{kl} R_{jl} \qquad (10.7.20)$$

(see (1.4.17)). Since the only nonzero elements of R are $R_{11} = -1$, $R_{22} = 1$, and $R_{33} = -1$, $M'_{ij} = M_{ij}$ except for $M'_{12} = -M_{12}$ and $M'_{23} = -M_{23}$. Therefore the moment tensors components are related as follows:

$$\begin{pmatrix} M'_{11} & M'_{12} & M'_{13} \\ M'_{21} & M'_{22} & M'_{23} \\ M'_{31} & M'_{32} & M'_{33} \end{pmatrix} = \begin{pmatrix} M_{11} & -M_{12} & M_{13} \\ -M_{21} & M_{22} & -M_{23} \\ M_{31} & -M_{32} & M_{33} \end{pmatrix}, \qquad (10.7.21)$$

where the tensor on the left is that given by Aki and Richards (1980). Note $\mathbf{M}^{dc}$ has the same components in the two coordinate systems.

The corresponding relations for the normal and slip vectors are:

$$(v_1', v_2', v_3') = (-v_1, v_2, -v_3) \tag{10.7.22}$$

and

$$(s_1', s_2', s_3') = (-s_1, s_2, -s_3). \tag{10.7.23}$$

10.8 Relations between the parameters of the conjugate planes

Let $\mathbf{v}_1$ and $\mathbf{s}_1$ be the normal and slip vectors for one of the conjugate planes and $\mathbf{v}_2$ and $\mathbf{s}_2$ be the corresponding vectors for the other plane. From our discussion following (10.7.5) we know that the normal and slip vectors are related as follows:

$$\mathbf{v}_2 = \mathbf{s}_1 \tag{10.8.1}$$

$$\mathbf{s}_2 = \mathbf{v}_1. \tag{10.8.2}$$

Equations (10.8.1) and (10.8.2) will be used to find a set of relations between the parameters $\varphi_1, \delta_1, \lambda_1$, and $\varphi_2, \delta_2, \lambda_2$ that identify the two planes. From (10.8.1), using (10.7.12) and equating components we obtain

$$-\sin \delta_2 \sin \varphi_2 = \cos \lambda_1 \cos \varphi_1 + \sin \lambda_1 \cos \delta_1 \sin \varphi_1 \tag{10.8.3}$$

$$\sin \delta_2 \cos \varphi_2 = \cos \lambda_1 \sin \varphi_1 - \sin \lambda_1 \cos \delta_1 \cos \varphi_1 \tag{10.8.4}$$

$$\cos \delta_2 = \sin \lambda_1 \sin \delta_1. \tag{10.8.5}$$

Multiplying (10.8.3) by $\cos \varphi_1$, equation (10.8.4) by $\sin \varphi_1$ and adding the resulting equations gives

$$\sin \delta_2 \sin(\varphi_1 - \varphi_2) = \cos \lambda_1. \tag{10.8.6}$$

Multiplying (10.8.3) by $\sin \varphi_1$, equation (10.8.4) by $\cos \varphi_1$ and subtracting the resulting equations gives

$$-\sin \delta_2 \cos(\varphi_1 - \varphi_2) = \sin \lambda_1 \cos \delta_1. \tag{10.8.7}$$

From (10.8.2) we obtain

$$\sin \lambda_2 = \frac{\cos \delta_1}{\sin \delta_2} \tag{10.8.8}$$

$$\cos \lambda_2 = -\sin \delta_1 \sin(\varphi_1 - \varphi_2). \tag{10.8.9}$$

Equation (10.8.8) follows directly from equating the third components and (10.8.9) from steps similar to those that lead to (10.8.6). If $\varphi_1, \delta_1, \lambda_1$ are known, this set of equations can be used to determine $\varphi_2, \delta_2, \lambda_2$. First, use (10.8.5) to obtain δ_2; then use (10.8.6) and (10.8.7) to get $(\varphi_1 - \varphi_2)$ in the appropriate quadrant and to find φ_2. Finally use (10.8.8) and (10.8.9) to obtain λ_2 in the appropriate quadrant. If $\pi/2 < \delta_2 \leq \pi$ (which identifies the hanging wall), the computed values of δ_2, φ_2, and λ_2 must be replaced by $\pi - \delta_2$, $\varphi_2 + \pi$ and $2\pi - \lambda_2$. This transformation refers the parameters back to the foot wall (Jarosch and Aboodi, 1970).

10.9 Radiation patterns and focal mechanisms

The moment tensor components given in (10.7.11) can be combined with (9.9.17) to give the expressions for the P, SH, and SV displacements in terms of the fault parameters, but for computational purposes this step is not required. It is sufficient to compute the components M_{ij} first and then to use (9.9.17) for the displacements and (9.9.18) for the radiation patterns. To display the radiation patterns we can use 3-D perspective plots, as done in Chapter 9, but an alternative, more convenient, approach is to use a lower-hemisphere equal-area projection, which is common in focal mechanism studies (e.g., Aki and Richards, 1980; Lee and Stewart, 1981). In addition, because this projection is used with the (north, east, down) coordinate system, here we will follow the same convention. This requires a rotation of our system about the x_2 axis through an angle of π, as done in §10.7, after which the x_3 axis points down.

In Fig. 9.10 the set of points with $0 \leq \phi \leq 2\pi$ and $0 \leq \theta \leq \pi/2$ define the upper hemisphere (corresponding to $x_3 > 0$). After the rotation of axes the same set of points describes the lower hemisphere. In the rotated system the unit vectors defined in (9.9.16) remain unchanged, but now the angle ϕ must be measured clockwise from north (as this is the direction to which the x_1 axis points). If the radius of the sphere is taken as 1, each point on the sphere is identified by the angles ϕ and θ. To this point corresponds a value of the radiation pattern given by (9.9.18). The equal-area projection of a particular point (ϕ, θ) is the point (in the horizontal plane) with polar coordinates $(\phi, r \sin(\theta/2))$, where r is the radius of the circle used to represent the hemisphere. In this projection the center of the circle represents the x_3 axis ($\theta = 0$) and the circle represents the horizontal plane (x_1, x_2), corresponding to $\theta = \pi/2$. To plot the P-wave radiation patterns, $+$ and $-$ signs will be used to indicate polarity, with the size of the symbols indicating relative amplitudes. For the SV- and SH-wave radiation patterns two displays will be used, one similar to that for the P waves and another in which an arrow indicates the sense of motion, with the size of the arrow indicating amplitude (Kennett, 1988). For any of the three wave types a positive (negative) value of the radiation pattern means motion in the same (opposite) direction as the corresponding unit vector. Another representation of the S-wave motion will be based on the polarization angle ε (given by (9.9.21)) and the absolute value of the radiation pattern of $\mathbf{u}^S$.

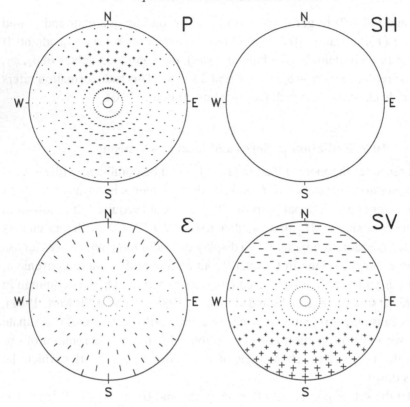

Fig. 10.6. Radiation patterns for the single couple M_{31}. A lower-hemisphere, equal-area projection was used. The polarity and relative amplitudes are indicated by the scaled plus and minus signs. The polarization angle ε is also shown. (After Pujol and Herrmann, 1990; plotting software based on Kennett, 1988.)

This will be done using a $-$ sign with size proportional to $|\mathcal{R}^S|$ (obtained from the first equality in (9.12.5)). The angle ε is measured with respect to ϕ (Herrmann, 1975).

The P and SV radiation patterns for the single couple M^{sc} (see §9.12) are shown in a lower-hemisphere projection in Fig. 10.6. These plots can be interpreted easily by consideration of those in Fig. 9.12 If in Fig. 9.12(a) we draw an upper hemisphere, label the x_1 axis north, and then rotate the axes about x_2 so that x_3 is down, then we have the geometry required for the projection. The positive and negative values in Fig. 10.6 correspond to points with positive and negative x_1 in Fig. 9.12(a), with their largest absolute values for $\theta = \pi/4$ and ϕ equal to zero and π. Along the x_2 axis (east–west direction), $\mathcal{R}^P$ is equal to zero (see Fig. 9.12c). The nodal planes, or planes where the radiation pattern is equal to zero (see §9.12), are represented by the circle ($\theta = \pi/2$) and by the line WE ($\phi = \pi/2$), and correspond

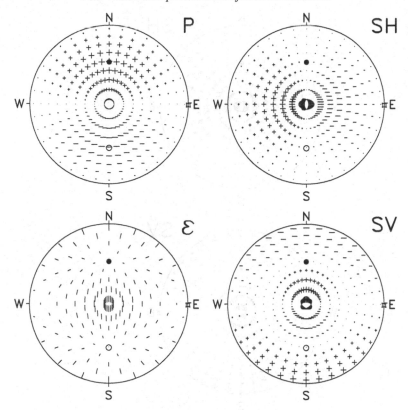

Fig. 10.7. Radiation patterns for the double couple $M^{dc} = M_{13} + M_{31}$. A lower-hemisphere, equal-area projection was used. The polarity and relative amplitudes are indicated by the scaled plus and minus signs. The polarization angle ε is also shown. The T, P, and B axes are indicated by the dot, open circle, and star respectively. (After Pujol and Herrmann, 1990; plotting software based on Kennett, 1988.)

to the planes (x_1, x_2) and (x_2, x_3), respectively. Using a similar approach it is easy to see that $\mathcal{R}^{SV}$ (Fig. 9.12b) has the lower-hemisphere projection shown in Fig. 10.6. The largest absolute values correspond to points along the x_1 axis.

The radiation patterns for the P and S waves corresponding to the double couple M^{dc} (see §9.12) are shown in Fig. 10.7 together with the position of the P, T, and B axes (open circle, dot, and star, respectively). This figure makes clear the relation between the three axes and the points of extremal and zero values of the P-wave radiation pattern (as discussed in §9.12), and the relation between the polarization angle and the P and T axes. As for the case of M^{sc}, the nodal planes for the P waves are the planes (x_1, x_2) and (x_2, x_3). The direction of motion for the SV and SH waves is shown in Fig. 10.8. Comparison of Figs 10.6 and 10.7 shows again that the single and double couples produce different S-wave radiation patterns.

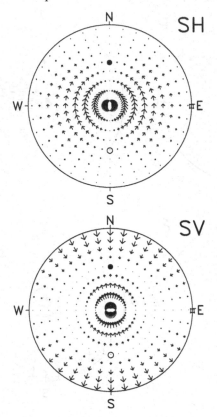

Fig. 10.8. Similar to SH and SV of Fig. 10.7 but showing the direction of motion. (Plotting software based on Kennett, 1988.)

The orientation of the double couple M^{dc} is not representative of the fault orientations observed in the Earth. A more realistic case is considered next. Figures 10.9 and 10.10 show the P, SV, and SH radiation patterns and the polarization angle for a reverse fault having strike, dip, and slip of $70°$, $60°$, and $70°$, respectively. The same patterns are obtained if the parameters for the conjugate plane (computed using the equations in §10.8) are used. Also shown are the conjugate planes, the slip vectors (squares) and the P, B, and T axes, which were obtained by diagonalizing the moment tensor with primed components in (10.7.21). For the P-wave radiation pattern the P and T axes again correspond to extremal values. For this pattern the conjugate planes are also nodal planes, and correspond to the nodal planes for the double couple M^{dc} discussed above.

Now we will give a brief description of a technique commonly used for the determination of the focal mechanism of earthquakes and will relate it to the results obtained so far. In the context of the point-source approximation, we are interested

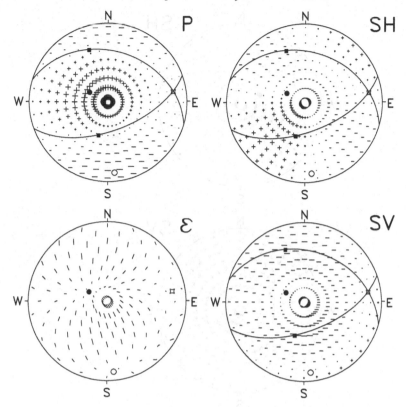

Fig. 10.9. Similar to Fig. 10.7 for a fault with strike and dip and slip angles of 70°, 60°, and 70°, respectively. Also shown are the projections of the conjugate planes and the projections of the slip vectors (squares). (Plotting software based on Kennett, 1988.)

in the determination of the three parameters (φ, δ, λ) that characterize the fault which ruptured in an earthquake. The simplest approach is based on the analysis of the polarity of the first motion of P waves recorded at a number of stations placed on the surface of the Earth. S waves and others can also be used (e.g. Stauder, 1960, 1962), but P waves are preferred because generally their polarities can be determined without ambiguity and because of the simplicity of the analysis. It is also worth noting that although the method described below appears straightforward now, it took many years of effort to develop and was intimately related to the studies on the nature of the earthquake source (Stauder, 1962). The basic idea is to consider a small homogeneous sphere centered at the hypocenter (or *focus*), known as the *focal sphere*, and to transfer the information recorded by the seismic stations to the focal sphere. Since the radius of this sphere is not important, it can be taken as 1. To identify the point in the focal sphere corresponding to a given station two angles are used. One is the *take-off angle* i_t, which is the angle with which the

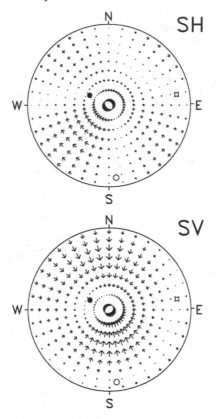

Fig. 10.10. Similar to *SH* and *SV* of Fig. 10.9 but showing the direction of motion. (Plotting software based on Kennett, 1988.)

ray reaching the station leaves the focus. This angle is measured from the positive x_3 direction. The other angle is the azimuth ϕ_s of the station with respect to the hypocenter and is measured from north. Note that the pair of angles i_t, ϕ_s is similar to the pair θ, ϕ that we have used so far. Once the observed data have been traced back to the focal sphere a lower-hemisphere equal-area projection is used to map the sphere onto its equatorial plane. The use of a lower-hemisphere projection is predicated on the fact that most rays leave the focus in a downward direction, which means that $i_t < \pi/2$. However, for small station–epicenter distances the ray may leave in the upward direction ($i_t > \pi/2$), thus intersecting the upper hemisphere, rather than the lower one. In this case the point (ϕ_s, i_t) to be projected must be replaced by ($\phi_s + \pi$, $\pi - i_t$).

In summary, for focal mechanism studies based on *P*-wave first motions the following information for each station is needed: take-off and azimuth angles and polarity. This information is then transferred directly to an equal-area (or Schmidt)

net, with different symbols to indicate compressions (positive values) or dilatations (negative values). Sometimes the stations with amplitudes close to zero are
specially marked to help in the identification of the nodal plane(s). After this is
done it is necessary to draw two planes satisfying the conditions that they separate
compressional from dilatational arrivals and are perpendicular to each other. The
diagram thus obtained is known as the *fault plane solution* or *focal mechanism*. As
the practical aspects of this process are well described in Lee and Stewart (1981),
here we will concentrate on the steps required to determine the fault parameters.

Fig. 10.11(a) has been derived from the P-wave radiation pattern in Fig. 10.9
and can be interpreted as representing actual observations, which in turn requires
that the angles ϕ and θ be interpreted as ϕ_s and i_t, respectively. The two possible
fault planes are labeled π_1 and π_2, and the corresponding parameters are identified
by the subscripts 1 and 2. Now we must determine the parameters that define the
two fault planes. If these parameters are to be used to compute radiation patterns,
they must conform strictly to the convention used to define them in §10.7. The dips
δ_1 and δ_2, indicated by the double-arrow lines, must be determined first. As dip is
measured from the horizontal plane, a point on the circle corresponds to a dip of
$0°$ and the center of the circle corresponds to $90°$. Next the dip directions, given by
lines passing through the center of the circle and containing the double-arrow lines,
must be determined. Then the strike of each plane is obtained by subtracting $\pi/2$
from the corresponding dip direction. This is how the strikes φ_1 and φ_2 (identified
by the triangles) have been determined. The slip vectors are obtained under the
condition that s_1 is perpendicular to π_2 and s_2 is perpendicular to π_1. To determine
λ_1, imagine that the plane π_1 can be rotated independently of the rest of the figure
in such a way that φ_1 is brought into coincidence with the north direction. Then λ_1
is the angle measured from the point marked S (south) to s_1 along π_1. To find λ_2
repeat the process using π_2. The T axis can be determined using (10.7.16), (10.8.1)
and (10.8.2), which show that the T axis bisects the vectors s_1 and s_2. Then the P
axis is obtained from the T axis because they are $90°$ apart. The B axis is in the
intersection of the two conjugate planes. The three axes shown in the figure were
actually drawn using the components obtained from (10.7.15) and (10.7.16) with
(10.7.12), and (10.7.18), which verifies the consistency of the equations used to
generate the P-wave radiation patterns in Figs 10.9 and 10.11(a).

Fig. 10.11(a) corresponds to a reverse fault. To show the differences involved
when a normal fault is considered, the previous value of λ_1 was increased to
$250°$ (i.e., a π difference) (Fig. 10.11b). The effect of this change in the radiation
patterns is to change positive values to negative and vice versa, without changing
the absolute values. Comparison of Figs 10.11(a) and (b) shows that the positions
of the P and T axes have been interchanged, which means that the central portion
of the circle between the two planes now corresponds to negative values. Also note

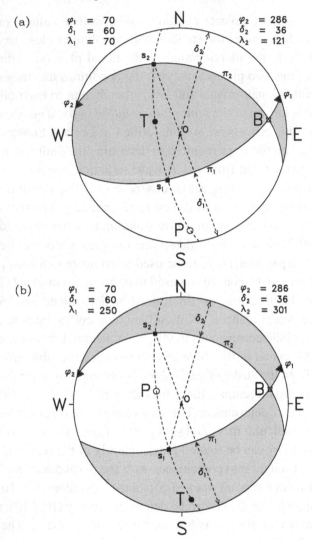

Fig. 10.11. Plot of two conjugate planes (π_1 and π_2) corresponding to the fault parameters ϕ_1, δ_1, λ_1 and ϕ_2, δ_2, λ_2. Shaded areas represent positive values of the P-wave radiation pattern (i.e., compressional arrivals). (a) Reverse fault. (b) Normal fault. The only differ- ence between (a) and (b) is in the values of the slip angles λ_1 and λ_2, which differ by π. In (a) the shaded area corresponds to the area of positive P-wave arrivals in Fig. 10.9. The projections of the P, B, and T axes and the slip vectors s_1 and s_2 are also shown. In focal mechanism studies the planes π_1 and π_2 are determined from the polarity of the arrivals and all other parameters must be determined as explained in the text.

that the value of λ_2 differs from the previous value by π, and that to obtain the correct values of λ_1 and λ_2 it is necessary to add π to the values obtained when applying the procedure described in connection with Fig. 10.11(a). This applies to any normal fault, which is recognized by the fact that the dilatations occupy the central portion of the circle.

10.10 The total field. Static displacement

The equations for the total field have been derived in §9.13, which should be used with the moment tensor expressions given in (10.7.11). Knowledge of the total field is essential in strong motion seismology, which deals with waves recorded in the near-field of large earthquakes. In this case the point-source approximation introduced in §10.5 is no longer valid, and the finite nature of the fault must be taken into account, and although we will not address this problem here, a brief overview will be given to show its connection to the rest of this section.

The most basic approach to calculating the displacement field owing to a finite fault is to divide the fault into a number of subfaults for which the point-source approximation is valid and then to sum the displacements at a given observation point caused by all the subfaults. This, in turn, requires a model for the rupture of the fault. A simple model is one in which the rupture begins at some point within the fault and then propagates with constant velocity and constant slip in all directions. Finally, a source time function is also needed. Because the rupture velocity is finite, the contributions of the different subfaults to the displacement at a given observation point will arrive at different times, so that the shape and duration of the recorded signal will depend on the dimensions of the fault and the position of the observation point with respect to the fault. Another point to consider is that when trying to compare observed and synthetic seismograms, the Earth's free surface must be taken into account. The correct way to do that is to use Green's function for a half-space (Johnson, 1974; Anderson, 1976). An alternative, approximate approach is to account for the free-surface effects using the equations derived in §6.5.2.1 for the P waves and the corresponding equations for the SV waves (Anderson, 1976).

With the preceding caveats we can now proceed with the analysis of the total field corresponding to a point source. As in §9.13, we will assume that all the components of the moment tensor have the same time dependence, so that the total field is given by (9.13.3). In that equation $D(t)$ plays the role of the displacement $\bar{u}(t)$ introduced in (10.5.8). $D(t)$ is usually not a unit step; rather, it reaches its maximum, constant, value, in a finite amount of time (known as the *rise time*). This constant value will be indicated by D_o. Under these conditions the scalar moment can be written as

$$M_o = \mu D_o A \tag{10.10.1}$$

(Harkrider, 1976). In the following we will ignore the factor of μA, but it should be included when comparing actual and synthetic data.

Now we will determine the *static displacement*, which is the permanent displacement that affects the points of the medium after the passage of the wave with velocity β. This displacement, indicated by $\mathbf{u}^s$, is obtained using (9.13.3) with $D(t)$ replaced by D_0. Under this condition the integral becomes

$$\int_{r/\alpha}^{r/\beta} \tau D(t - \tau)\,d\tau = D_0 \int_{r/\alpha}^{r/\beta} \tau\,d\tau = \frac{1}{2}D_0 r^2 \left(\frac{1}{\beta^2} - \frac{1}{\alpha^2} \right), \qquad (10.10.2)$$

and $\dot{D}(t) = 0$, so that the far field does not contribute to $\mathbf{u}^s$. Therefore,

$$\mathbf{u}^s(\mathbf{x}) = \frac{1}{4\pi\rho}\frac{D_0}{r^2}\left[\frac{1}{2}\left(\frac{1}{\beta^2} - \frac{1}{\alpha^2} \right)\mathbf{A}^N + \frac{1}{\alpha^2}\mathbf{A}^{I\alpha} - \frac{1}{\beta^2}\mathbf{A}^{I\beta} \right]; \qquad t \gg \frac{r}{\beta}.$$
$$(10.10.3)$$

Thus, aside from radiation pattern effects, the static displacement decays as r^{-2}. Equation (10.10.3) gives the components of the static displacement in Cartesian coordinates. To obtain the expressions in spherical coordinates we must project $\mathbf{u}^s$ along the axes $\mathbf{\Gamma}$, $\mathbf{\Theta}$, and $\mathbf{\Phi}$. Doing that and assuming that the moment tensor is symmetric and has zero trace we find

$$u_\Gamma^s(\mathbf{x}) = \frac{1}{8\pi\rho}\frac{D_0}{r^2}\left(\frac{3}{\beta^2} - \frac{1}{\alpha^2} \right)\mathbf{\Gamma}^{\mathrm{T}}\bar{\mathbf{M}}\mathbf{\Gamma} \qquad (10.10.4)$$

$$u_\Theta^s(\mathbf{x}) = \frac{1}{4\pi\rho}\frac{D_0}{r^2}\frac{1}{\alpha^2}\mathbf{\Theta}^{\mathrm{T}}\bar{\mathbf{M}}\mathbf{\Gamma} \qquad (10.10.5)$$

$$u_\Phi^s(\mathbf{x}) = \frac{1}{4\pi\rho}\frac{D_0}{r^2}\frac{1}{\alpha^2}\mathbf{\Phi}^{\mathrm{T}}\bar{\mathbf{M}}\mathbf{\Gamma} \qquad (10.10.6)$$

(Harkrider, 1976; Aki and Richards, 1980; Problem 10.9).

Now we will illustrate the effect of distance and source time function on the displacement, which will be computed using (9.13.11). The displacements obtained with that equation are in Cartesian coordinates, which will be converted to spherical coordinates using the following transformation:

$$\begin{pmatrix} u_\Theta \\ u_\Phi \\ u_\Gamma \end{pmatrix} = \mathbf{R}\begin{pmatrix} u_1 \\ u_2 \\ u_3 \end{pmatrix}, \qquad (10.10.7)$$

where

$$\mathbf{R} = \begin{pmatrix} \cos\theta\cos\phi & \cos\theta\sin\phi & -\sin\theta \\ -\sin\phi & \cos\phi & 0 \\ \sin\theta\cos\phi & \sin\theta\sin\phi & \cos\theta \end{pmatrix} = \begin{pmatrix} \mathbf{\Theta}^{\mathrm{T}} \\ \mathbf{\Phi}^{\mathrm{T}} \\ \mathbf{\Gamma}^{\mathrm{T}} \end{pmatrix} \qquad (10.10.8)$$

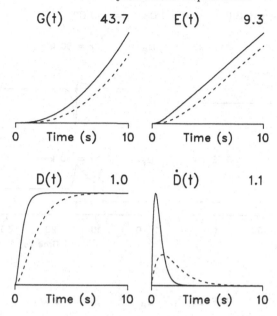

Fig. 10.12. Plots of the functions $G(t)$, $E(t)$, $D(t)$, and $\dot{D}(t)$ given in (10.10.12) and (10.10.9)–(10.10.11), respectively. Solid and dashed lines correspond to $a = 3$ and 1, respectively. In each plot the two curves are normalized to the maximum value of the solid line, shown by the number on top.

(Problem 10.10). Equation (10.10.7) will also be applied to the components of $\mathbf{u}^s(\mathbf{x})$ computed using (10.10.3).

The functions $E(t)$, $D(t)$, and $\dot{D}(t)$ used in (9.13.11) are given by

$$E(t) = D_o J(t) \qquad (10.10.9)$$

$$D(t) = D_o J'(t) \qquad (10.10.10)$$

$$\dot{D}(t) = D_o J''(t) \qquad (10.10.11)$$

with J, J', and J'' as in (9.5.21)–(9.5.23), respectively, while

$$G(t) = \int_0^t E(\tau)\, d\tau = H(t)\frac{D_o}{2a^2}\left[6\left(1 - e^{-at}\right) + at\left(at - 4 - 2e^{-at}\right)\right]$$
$$(10.10.12)$$

(Harkrider, 1976; Problem 10.11). The function $D(t)$ was selected for convenience, not for physical significance. These four functions have been plotted in Fig. 10.12 for $a = 1$ and 3.

Figure 10.13(a) shows the displacement in the Γ and Θ directions caused by the double couple $M_{13} + M_{31}$ (see §9.8) at a point with $\theta = 60°$, $\phi = 30°$, and several values of r. The velocity and density models were those used to generate

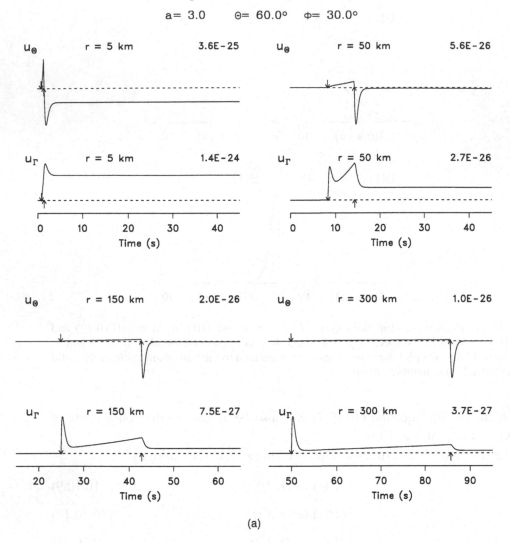

Fig. 10.13. (a) Total-field synthetic seismograms for the double couple $M^{dc} = M_{13} + M_{31}$ computed using (9.13.11) and (10.10.7) with the functions G, E, D and $\dot{D}$ shown in Fig. 10.12. The P and S velocities and the density are given by $\alpha = 6$ km s^{-1}, $\beta = 3.5$ km s^{-1}, and $\rho = 2.8$ g m cm^{-3}. The observation point has $\theta = 60°$, $\phi = 30°$ and several values of r, as noted in each plot. The values of a and D_o are equal to 3 and 1, respectively. For each value of r the components u_Γ and u_Θ of the displacement vector are shown. The u_Φ component is similar to u_Θ. The down and up arrows show the times r/α and r/β, respectively. The number in exponential notation is the largest amplitude for each curve; its dimension is cm when the force is given in dynes.

the seismograms in Fig. 9.4 and the values of a and D_o are 3 and 1, respectively. The Φ component of the displacement is not shown because it is similar in shape

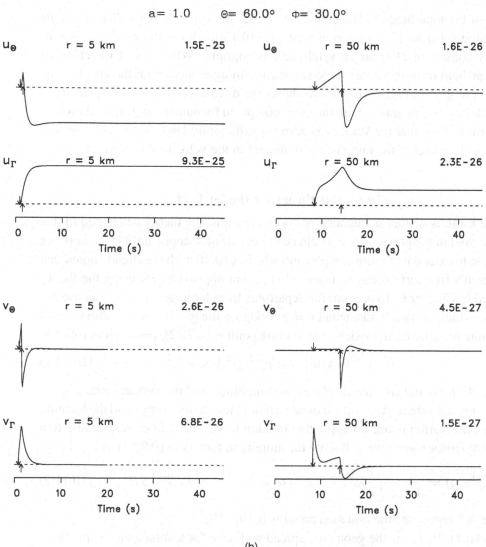

(b)

Fig. 10.13. (b) The four upper plots are similar to the corresponding plots in (a) except for a different value of the parameter a, set equal to 1. The four lower plots show the components v_r and v_Θ of the velocity vector, obtained from the upper plots by numerical differentiation.

to that in the Θ direction. This figure shows that the near- and intermediate-field terms are significant even for large distances (e.g., $r = 300$ km), as demonstrated by the ramp that precedes the arrival with the velocity of the S waves. In fact, the long-period wave between the P- and S-wave arrival times observed in some actual displacement seismograms recorded at teleseismic distances may be explained in terms of the total field (Vidale *et al.*, 1995). Figure 10.13(b) shows displacements computed for $a = 1$. Comparison with Fig. 10.13(a) shows the effect of the frequency content of $D(t)$ on the synthetic seismograms. When $a = 1$ the effect of the near-field terms becomes more significant, in agreement with the conclusions reached in §9.5.2. Fig. 10.13(b) also shows the differences between displacement and velocity seismograms, with the latter computed by numerical differentiation of the former. Note that the velocity is zero for sufficiently large times (as expected) and that the effect of the ramp is less dominant in the velocity seismograms.

10.11 Ray theory for the far field

As the Earth is neither unbounded nor homogeneous, the theory developed in this chapter and in the previous one would be of very limited applicability if it were not possible to extend it to more complex models. In §10.10 we have already noted that the Earth's free surface can be taken into account approximately using the theory derived in Chapter 6. To account for departures from homogeneity one can use ray theory, which plays a fundamental role in bridging the gap between homogeneous and more realistic Earth models. The starting point is (8.7.22), rewritten as follows:

$$\rho_s^{1/2} c_s^{1/2} A(\mathbf{x}_s) \sigma_s^{1/2} = \rho_o^{1/2} c_o^{1/2} A(\mathbf{x}_o) \sigma_o^{1/2}, \tag{10.11.1}$$

where A and σ indicate the displacement amplitude and the surface area, respectively, and the subscripts s and o indicate a point close to the source and the location of the observation point. To apply this relation to the far field corresponding to a moment tensor source we will write the amplitude factors in (9.9.17) as

$$A^R(\mathbf{x}_s) = \frac{1}{4\pi} \frac{1}{\rho_s c_s^3} \frac{1}{r} \mathcal{R}^R; \qquad c = \alpha, \beta; \qquad R = P, SV, SH, \tag{10.11.2}$$

where $\mathcal{R}^R$ represents the radiation patterns in (9.9.18).

In (10.11.2), $1/r$ is the geometric spreading factor for a homogeneous medium. When this factor is removed, the remainder of $A^R(\mathbf{x}_s)$ is equivalent to $A(\mathbf{x}_s)$ in (10.11.1). Then, taking this equivalence into account, equation (10.11.1) becomes

$$\frac{1}{4\pi} \frac{1}{\rho_s^{1/2} c_s^{5/2}} \mathcal{R}^R \sigma_s^{1/2} = \rho_o^{1/2} c_o^{1/2} A(\mathbf{x}_o) \sigma_o^{1/2} \tag{10.11.3}$$

so that

$$A(\mathbf{x}_o) = \frac{1}{4\pi} \frac{1}{\rho_s^{1/2} c_s^{5/2}} \frac{1}{\rho_o^{1/2} c_o^{1/2}} \left(\frac{\sigma_s}{\sigma_o}\right)^{1/2} \mathcal{R}^R. \tag{10.11.4}$$

Next, we assume that the source area is homogeneous within a sphere of unit radius centered at $\mathbf{x}_s$ and choose σ_s to be on the surface of this sphere. Recall that in a homogeneous medium the rays are straight lines through the origin and the wave fronts are spherical (see §8.4). With this choice of σ_s, if the whole medium is homogeneous equations (10.11.2) for the far field are recovered. This can be seen from (10.11.4) using the fact that the element of area in spherical coordinates is

$$d\sigma = r^2 \sin\theta \, d\theta \, d\phi \tag{10.11.5}$$

(Problem 10.12).

An equation similar to (10.11.4) was derived by Aki and Richards (1980). The factor involving the ratio of surface areas is the geometric spreading factor. The inverse of this factor measures the spread of the wave front as it moves away from the source. Because of its importance in applications, an explicit expression for the geometric spreading factor for a spherically symmetric Earth model will be derived. Fig. 10.14 shows a ray tube associated with the ray from the source (located at point E) to the observation point B. To carry out the derivation it is convenient to introduce a spherical coordinate system centered at the source. In this system the angle i_h corresponds to colatitude, while $\delta\phi$ is an increment in the longitude of the vertical plane that contains the ray EB (see §8.6).

The intersection of the ray tube, assumed to be very narrow, with the surface of the Earth is the element $BCFD$. We are interested in the area σ_o of the wave front, represented by $BCIG$, which can be obtained using the following relations:

$$\overline{BD} = r_o \delta\Delta \tag{10.11.6}$$

$$\overline{HB} = r_o \sin\Delta \tag{10.11.7}$$

$$\overline{BC} = \overline{HB}\delta\phi = r_o\delta\phi \sin\Delta \tag{10.11.8}$$

$$\overline{BG} = \overline{BD} \cos i_o = r_o\delta\Delta \cos i_o \tag{10.11.9}$$

where the overbar indicates distance. Then

$$\sigma_o = \overline{BC}\,\overline{BG} = r_o^2\delta\Delta \, \delta\phi \sin\Delta \cos i_o. \tag{10.11.10}$$

To find σ_s use (10.11.5) with $\theta = i_h$ and $r = 1$,

$$\sigma_s = \delta\phi \, \delta i_h \sin i_h. \tag{10.11.11}$$

Therefore,

$$\left(\frac{\sigma_s}{\sigma_o}\right)^{1/2} = \frac{1}{r_o}\left(\frac{\sin i_h}{\sin\Delta \cos i_o}\frac{\delta i_h}{\delta\Delta}\right)^{1/2}. \tag{10.11.12}$$

Now (10.11.12) will be rewritten in terms of the difference δt between the travel times along the rays EB and ED. From Fig. 10.14,

$$\delta t = \frac{\overline{GD}}{c_o} = \frac{\overline{BD}\sin i_o}{c_o} = \frac{r_o\delta\Delta \sin i_o}{c_o} = p\delta\Delta, \tag{10.11.13}$$

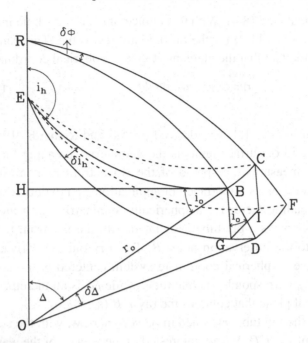

Fig. 10.14. Geometry for the determination of the geometric spreading factor for a spherically symmetric Earth model. Points E and B indicate source and observation locations and are in the plane passing through the points O, R, and D. The arc RD is part of a great circle. The arcs emanating from E represent a very narrow ray tube, which has widths δi_h and $\delta \phi$ at the source. The intersection of the ray tube with the surface of the Earth is the surface element $BCFD$, while the surface element $BCIG$ is the portion of the wave front bounded by the ray tube. The segments BH and EH are perpendicular to each other, and r_o is the Earth's radius.

where c_o indicates the velocity at the surface and p is ray parameter introduced in (8.4.44). Therefore

$$\frac{\delta t}{\delta \Delta} = p = \frac{r \sin i_h}{c} \qquad (10.11.14)$$

where r is the distance $\overline{OE}$ and c is the velocity at the source. In addition, in the limit as δi_h goes to zero, the ratios of increments in (10.11.12) and (10.11.14) can be replaced by the corresponding derivatives. Then taking the derivative of (10.11.14) with respect to Δ we obtain

$$\frac{d^2 t}{d\Delta^2} = \frac{r \cos i_h}{c} \frac{d i_h}{d\Delta} \qquad (10.11.15)$$

so that

$$\frac{d i_h}{d\Delta} = \frac{c}{r \cos i_h} \frac{d^2 t}{d\Delta^2}. \qquad (10.11.16)$$

Introducing (10.11.16) in (10.11.12) we obtain the following expression for the geometric spreading factor:

$$\left(\frac{\sigma_s}{\sigma_o}\right)^{1/2} = \frac{1}{r_o}\left(\frac{c\tan i_h}{r\sin\Delta\cos i_o}\frac{d^2t}{d\Delta^2}\right)^{1/2} \tag{10.11.17}$$

(e.g., Ben-Menahem and Singh, 1981). Given $c(r)$, all the information on the right-hand side of (10.11.17) can be computed (e.g., Gubbins, 1990).

An application of ray theory to the study of the radiation from an earthquake source is presented by Spudich and Frazer (1984), who also discuss the conditions for the validity of the ray theory approach (see also Spudich and Archuleta, 1987).

Problems

10.1   Verify (10.2.12).

10.2   Verify (10.4.3).

10.3   Consider the equivalent body force given by (10.4.7). Show that when the tractions are continuous the total equivalent force and the total moment about any coordinate axis are zero.

10.4   Verify (10.5.7).

10.5   Verify (10.6.3).

10.6   Consider a crack in the (x_1, x_2) plane having area A. The opening of the crack can be represented by $[u_1] = [u_2] = 0$; $[u_3] \neq 0$. Show that for an isotropic medium the corresponding moment tensor is given by

$$M_{ij} = A\overline{[u_3]}(\lambda\delta_{ij} + 2\mu\delta_{i3}\delta_{j3})$$

(after Kennett, 1983). This is a diagonal tensor with a graphical representation similar to that shown in Fig. 9.9(d).

10.7   Refer to §10.6. Show that the moment tensor $\acute{M}_{pq}$ is symmetric and that $\acute{M}_{pq} = M_{pq}$.

10.8   Refer to §10.7.

(a) Let $\mathbf{b}$ be the vector introduced in (9.11.6). Show that $\mathbf{b} = \boldsymbol{\nu} \times \mathbf{s}$.

(b) Verify that vectors $\mathbf{p}$ and $\mathbf{t}$ given by (10.7.15) and (10.7.16) and the vector $\mathbf{b}$ of (a) are eigenvectors of M_{ij} (see (10.7.4)) with eigenvalues -1, 1, and 0, respectively.

10.9   Verify (10.10.4)–(10.10.6).

10.10 Refer to §10.10.

(a) Verify (10.10.8).

(b) Apply $\mathbf{R}$ to $\boldsymbol{\Theta}$, $\boldsymbol{\Phi}$, and $\boldsymbol{\Gamma}$.

(c) Find $\mathbf{R}^{-1}$ and verify that $\mathbf{RR}^{-1} = \mathbf{I}$.

10.11 Verify (10.10.12).

10.12 Verify (10.11.5).

11

Anelastic attenuation

11.1 Introduction

The theory developed so far is not completely realistic because it does not account for the observed fact that the elastic energy always undergoes an irreversible conversion to other forms of energy. If this were not the case, a body excited elastically would oscillate for ever. The Earth, in particular, would still be oscillating from the effect of past earthquakes (Knopoff, 1964). The process by which elastic energy is lost is known as *anelastic attenuation*, and its study is important for several reasons. For example, because attenuation affects wave amplitudes and shapes, it is necessary to account for their variations when computing synthetic seismograms for comparison with observations. Properly accounting for the reduction in wave amplitude was particularly important during the cold-war period because of the use of seismic methods to estimate the yield of nuclear explosions in the context of nuclear test-ban treaties. In addition, because attenuation depends on temperature and the presence of fluids, among other factors, its study has the potential for shedding light on the internal constitution of the Earth. The study of attenuation may also help us understand the Earth's rheology, although the relation between the two is not clear. For a discussion of these and related matters see Der (1998), Karato (1998), Minster (1980), and Romanowicz and Durek (2000).

From a phenomenological point of view, the effect of attenuation is a relative loss of the high-frequency components of a propagating wave. As a result, a seismic wave propagating in an attenuating medium will become richer in lower frequencies, thus changing shape, in addition to suffering an overall amplitude reduction. However, as discussed below, this is not the only process that affects wave forms and amplitudes. Attenuation is also referred to as internal friction, but this term should not be interpreted in the usual macroscopic sense of loss of energy owing to friction between grains or across cracks. At the small strains (less than 10^{-6})

357

typical of seismology this process does not contribute to attenuation (Winkler *et al.*, 1979). The mechanisms actually responsible for attenuation are varied. In crystals, attenuation is explained in terms of microscopic effects involving crystal imperfections (such as point defects and line defects, or dislocations), the presence of interstitial impurities, and thermoelastic effects. In polycrystalline substances, the presence of grain boundaries also contributes to attenuation. See Nowick and Berry (1972) for a detailed discussion of these matters and Anderson (1989) and Lakes (1999) for useful summaries. All of these mechanisms (as well as others) are frequency dependent, with characteristic peaks at which the attenuation is a maximum. The frequencies of these peaks range between about 10^{-13} and 10^8 Hz, with the peak for grain boundary slip at about 10^{-13} Hz. Representative figures can be found in Liu *et al.* (1976) and Lakes (1999).

In the Earth, the attenuation mechanisms referred to above must be present, but relating them to the observations is a difficult task (Karato and Spetzler, 1990). This fact should not be surprising, as the determination of attenuation relies on the measurement of amplitude variations, which are highly dependent on factors such as geometric spreading, the effect of heterogeneities, and the presence of cracks, fluids and volatiles.

The importance of volatiles was initially recognized in the context of the lunar seismic studies carried out as part of the Apollo experiments during 1969–77. The seismograms recorded in the Moon showed very long durations and extremely low attenuation, with estimated values of the quality factor Q (whose inverse is a measure of attenuation) of at least 3000 for P and S waves in the crust and upper mantle (Nakamura and Koyama, 1982). As the Moon is almost devoid of volatiles (it has no atmosphere), a likely explanation for these extremely high values of Q was the absence of volatiles. As discussed in Tittmann *et al.* (1980), this hypothesis was confirmed with laboratory measurements, which showed that Q could change dramatically when the volatiles were almost completely removed, which in turn required repeated heating and application of a high vacuum. In an extreme case, a sample with a Q of 60 changed to 4800 after this outgassing procedure. Laboratory studies also showed that only polar[1] volatiles (such as water and alcohol) are capable of producing drastic increases in attenuation. Moreover, this effect is associated with extremely small amounts of the volatile (one or two monolayers) bound to the internal surfaces of the sample. When the concentration of the volatile increases to the point that the cracks fill with liquid, a different attenuation mechanism takes place (e.g., Bourbié *et al.*, 1987; Winkler and Murphy, 1995).

An example of the effect of fluids and/or cracks on attenuation is provided by results obtained using data recorded in the German continental deep borehole

---

[1] In simple terms, polar molecules have net positive or negative electric charges.

(KTB) drilled to a depth of 9 km. From vertical seismic profiling data (see §11.11 for a definition) it was found that Q for P waves between 3.5 and 4.5 km depth does not exceed 32 (Pujol *et al.*, 1998). This value is much smaller than that expected for crystalline crust and was interpreted as being the result of the presence of fluids or cracks known to exist in that depth range. Independent confirmation for a low value of Q was provided by the analysis of microearthquakes induced by fluid injection at 9 km depth and recorded at 4 km (in an adjacent borehole) and at the surface. For the 0–4 km depth interval the Q for P waves is 38, while for the 4–9 km interval Q is much larger. For example, it is about 2000 for a frequency of 20 Hz (Jia and Harjes, 1997).

Scattering owing to the presence of heterogeneities causes a redistribution of the seismic energy and may produce attenuation-like effects. Thus, distinguishing the *intrinsic* attenuation from the *scattering attenuation* is difficult, although some methods have been proposed to separate the two (Fehler *et al.*, 1992). An example of scattering will be provided in §11.11, where the effect of finely layered media will be considered. Aside from that section, we will be concerned with intrinsic attenuation only.

The study of attenuation has proceeded along several directions, namely, observational (using seismic data), experimental (using data acquired in the laboratory), and theoretical. The latter, in turn, has developed along different lines. One is based on the theory of *viscoelasticity*, which differs from elasticity in that the relation between stress and strain does not follow Hooke's law, which corresponds to an instantaneous response of the elastic body to an applied stress. In contrast, in a viscoelastic solid the strain depends on the past history of stress, and the constitutive law includes an instantaneous elastic response plus an integral over past values of the stress. Viscoelastic behavior (in solid or liquids) is frequently modeled with systems involving springs and dashpots arranged in such a way that they reproduce certain observed features. A *dashpot* provides viscous damping and can be represented by a cylinder containing oil and a piston with a diameter such that the oil can flow around it. A combination of a spring and a dashpot in parallel connected in series to a second spring reproduces the most basic features of an anelastic solid, namely it responds elastically for very fast and very slow deformations (Hunter, 1976). This spring–dashpot arrangement represents the so-called *standard linear solid*, although a more appropriate term would be a *standard anelastic solid* (Nowick and Berry, 1972).

Another approach to the analysis of attenuation is based on the constraints imposed by causality on wave propagation. This approach is independent of the mechanisms that cause attenuation, and applies to viscoelastic solids as well. However, unlike viscoelasticity, which is well discussed in a number of texts, there is not a unified treatment of causality. For this reason, in this chapter the latter is discussed in detail, while for the former the reader is referred to Hunter (1976), Mase (1970) and Bourbié *et al.* (1987).

Our treatment of attenuation begins with the analysis of a spring–dashpot system. Although this is an extremely simple system, it is very valuable because it allows the introduction of the basic concepts of attenuation and a temporal Q. A spatial Q and an attenuation coefficient, denoted by α, are then introduced by consideration of the one-dimensional wave equation with complex velocity. These preliminary results are used to establish an attenuation model for the Earth, but as causality considerations show, this model cannot be established arbitrarily. The main conclusions are that α, Q, and the wave velocity must be frequency-dependent, that α and the wavenumber k cannot be chosen independently, and that α cannot grow as fast as ω as ω goes to infinity. Reaching these conclusions requires an elaborate mathematical framework, which constitutes the bulk of the chapter. Other topics included here are t^*, which is a factor used to quantify attenuation along a raypath when the properties of the medium vary with position, the spectral ratio method and the possible bias introduced by the use of tapering windows, and the attenuation-like effects introduced by a sequence of thin layers.

11.2 Harmonic motion. Free and damped oscillations

Consider a spring with constant κ and a body of mass m on one end of the spring (Fig. 11.1). This system is known as a linear harmonic oscillator. The differential equation for the displacement $y(t)$ from a position of equilibrium is

$$m\ddot{y} + \kappa y = 0, \qquad (11.2.1)$$

where a dot represents a derivative with respect to time. Dividing by m (11.2.1) becomes

$$\ddot{y} + \omega_o^2 y = 0; \qquad \omega_o = \sqrt{\frac{\kappa}{m}} \qquad (11.2.2a,b)$$

(e.g., Arya, 1990). The most general solution of (11.2.2a) is of the form

$$y(t) = A_1 \cos \omega_o t + A_2 \sin \omega_o t, \qquad (11.2.3)$$

where the constants A_1 and A_2 are determined from the initial conditions. Letting

$$A_1 = A \cos \phi; \qquad A_2 = A \sin \phi \qquad (11.2.4)$$

equation (11.2.3) becomes

$$y(t) = A \cos(\omega_o t - \phi), \qquad (11.2.5)$$

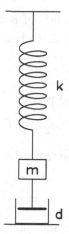

Fig. 11.1. Spring–dashpot system used to introduce the concept of damped oscillations. The damping is due to friction in the dashpot.

where

$$A = \sqrt{A_1^2 + A_2^2}; \qquad \tan\phi = \frac{A_2}{A_1}. \qquad (11.2.6)$$

Equation (11.2.5) corresponds to a harmonic motion that will never die out. However, if the spring–mass system includes frictional forces, then the motion will eventually vanish. The introduction of friction can be accomplished with a dashpot (Fig. 11.1). This system introduces a dissipative or damping force proportional to the velocity $\dot{y}(t)$. Letting d (positive) be the proportionality constant, the differential equation for the system becomes

$$\ddot{y} + 2\beta\dot{y} + \omega_o^2 y = 0; \qquad \beta = \frac{d}{2m}. \qquad (11.2.7a,b)$$

where the factor of 2 has been introduced for convenience. To solve (11.2.7a) we will use a trial solution of the form

$$y(t) = e^{\gamma t}, \qquad (11.2.8)$$

where γ must be determined under the condition that (11.2.7a) is satisfied. Introducing (11.2.8) in (11.2.7a) and after cancellation of a common factor $\exp(\gamma t)$ we obtain

$$\gamma^2 + 2\beta\gamma + \omega_o^2 = 0 \qquad (11.2.9)$$

for which the solution is

$$\gamma_{1,2} = -\beta \pm \sqrt{\beta^2 - \omega_o^2}, \qquad (11.2.10)$$

where the subscripts 1 and 2 correspond to the + and − signs. The type of solution represented by (11.2.10) depends on the values of β and ω_o. The case of interest to us is $\omega_o > \beta$, so that the square root in (11.2.10) becomes imaginary. Let

$$\omega = \sqrt{\omega_o^2 - \beta^2}. \tag{11.2.11}$$

Then the solutions corresponding to γ_1 and γ_2 are

$$y_1 = e^{-\beta t} e^{i\omega t}; \qquad y_2 = e^{-\beta t} e^{-i\omega t}. \tag{11.2.12}$$

Because at this point we are interested in real solutions we will replace (11.2.12) by

$$y_1 = e^{-\beta t} \sin \omega t; \qquad y_2 = e^{-\beta t} \cos \omega t. \tag{11.2.13}$$

In analogy with (11.2.5), we can write the most general solution of (11.2.7a) (with $\omega_o > \beta$) as

$$y(t) = A e^{-\beta t} \cos(\omega t - \delta). \tag{11.2.14}$$

Equation (11.2.14) shows that the effect of damping is to introduce an exponential decay that increases with time. The cosine term in (11.2.14) corresponds to an oscillatory motion with a frequency ω, and although the whole motion is not strictly periodic, we may introduce a "quasi-period" (Boyce and Di Prima, 1977) T equal to

$$T = \frac{2\pi}{\omega}. \tag{11.2.15}$$

Furthermore, because of (11.2.11), $\omega < \omega_o$. Solutions (11.2.5) and (11.2.14) are plotted in Fig. 11.2.

11.2.1 Temporal Q

Here we will investigate the loss of energy in the damped oscillator discussed above and will introduce the quality factor Q, whose inverse is used to quantify the damping. The energy $E(t)$ of the system is equal to the sum of the kinetic and potential energies, and is given by

$$E(t) = \frac{1}{2} m \dot{y}^2 + \frac{1}{2} \kappa y^2 \tag{11.2.16}$$

with $y(t)$ given by (11.2.14) (e.g., Arya, 1990). To simplify the derivations it is assumed that the damping is so small that the exponential in (11.2.14) is almost constant over one oscillation cycle and that $\omega_o \approx \omega$. Under these conditions and using (11.2.2) we have

$$E(t) = \frac{1}{2} \kappa A^2 e^{-2\beta t} = E_o e^{-2\beta t}, \tag{11.2.17}$$

where $E_o = E(0)$.

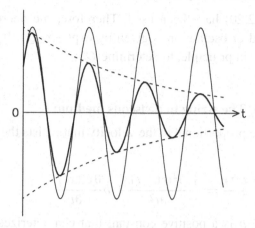

Fig. 11.2. Free and damped oscillations (thin and bold curves, respectively) obtained by solving the differential equation corresponding to the system shown in Fig. 11.1. See (11.2.5) and (11.2.14) for the equations of the curves. The damped motion shown here requires $\omega_o > \beta$ (defined in (11.2.2b) and (11.2.7b)). The dashed curves correspond to the exponential decay factor in (11.2.14).

We are now ready to introduce the quality factor Q, defined as 2π times the ratio of the energy stored in one cycle to the energy lost in one cycle. From (11.2.17) and (11.2.15), the energy lost in one cycle, corresponding to one period, is

$$|\Delta E| = T \left|\frac{dE}{dt}\right| = \frac{4\pi\beta}{\omega}E \qquad (11.2.18)$$

and

$$Q = 2\pi \frac{E}{\Delta E} = \frac{\omega}{2\beta} \qquad (11.2.19)$$

so that

$$\beta = \frac{\omega}{2Q}. \qquad (11.2.20)$$

Using (11.2.20) we can write (11.2.14) and (11.2.17) in terms of Q

$$y(t) = Ae^{-\omega t/2Q} \cos(\omega t - \delta) \qquad (11.2.21)$$

$$E(t) = E_o e^{-\omega t/Q}. \qquad (11.2.22)$$

Because of its association with t, the Q in (11.2.21) and (11.2.22) will be referred to as a *temporal Q*.

Using (11.2.21) we can give a more useful interpretation of Q. First note that if t_m corresponds to a peak of (11.2.21), then the next peak will occur at $t_m + T$. Then, forming the ratios of peak values at these two times gives

$$\frac{y(t_m + T)}{y(t_m)} = \frac{Ae^{-\beta(t_m+T)}}{Ae^{-\beta t_m}} = e^{-\beta T} = e^{-\pi/Q}, \qquad (11.2.23)$$

where (11.2.15) and (11.2.20) have been used. Therefore, the decrease in peak amplitude over one period of oscillation is given by $\exp(-\pi/Q)$. It also follows that (11.2.23) can be used, in principle, to determine Q.

11.3 The string in a viscous medium

In this case there is a force proportional to the velocity that resists the motion. The representative equation is

$$\frac{\partial^2 \psi(x,t)}{\partial x^2} = \frac{1}{c^2}\frac{\partial^2 \psi(x,t)}{\partial t^2} + b\frac{\partial \psi(x,t)}{\partial t} \tag{11.3.1}$$

(after Graff, 1975) where b is a positive constant that characterizes the viscous damping. When b is equal to zero (11.3.1) becomes the familiar wave equation. To solve (11.3.1) we will use a trial solution of the form

$$\psi(x,t) = Ae^{i(\omega t - \gamma x)}, \tag{11.3.2}$$

where γ must be determined under the condition that (11.3.1) is satisfied, which requires that

$$\gamma^2 = c^{-2}\omega^2 - ib\omega \equiv Be^{-i\phi} \tag{11.3.3}$$

with

$$B = \sqrt{c^{-4}\omega^4 + b^2\omega^2}; \qquad \tan\phi = \frac{c^2 b}{\omega}. \tag{11.3.4a,b}$$

Equation (11.3.3) has two solutions

$$\gamma_1 = B^{1/2}e^{-i\phi/2} \equiv k - i\alpha \tag{11.3.5}$$

$$\gamma_2 = B^{1/2}e^{-i(\phi+2\pi)/2} = -(k - i\alpha) \tag{11.3.6}$$

where, from (11.3.5),

$$k = B^{1/2}\cos\frac{\phi}{2}; \qquad \alpha = B^{1/2}\sin\frac{\phi}{2} \tag{11.3.7a,b}$$

(Graff, 1975). Introducing (11.3.5) and (11.3.6) in (11.3.2) gives

$$\psi(x,t) = Ae^{i[\omega t \mp (k-i\alpha)x]} = Ae^{\mp\alpha x}e^{i(\omega t \mp kx)}. \tag{11.3.8}$$

Assuming $x \geq 0$, a bounded solution is given by

$$\psi(x,t) = Ae^{-\alpha x}e^{i(\omega t - kx)}; \qquad \alpha > 0. \tag{11.3.9}$$

Equation (11.3.9) represents a wave propagating in the positive x direction with an amplitude that decreases with distance.

11.4 The scalar wave equation with complex velocity

A convenient way to introduce attenuation in a wave propagation problem is to allow some of its parameters to be complex quantities (e.g., Ewing *et al.*, 1957; Schwab and Knopoff, 1971). To see that consider the scalar wave equation,

$$v^2 \frac{\partial^2 \psi(x, t)}{\partial x^2} = \frac{\partial^2 \psi(x, t)}{\partial t^2}, \qquad (11.4.1)$$

where

$$v = \sqrt{\frac{M}{\rho}}, \qquad (11.4.2)$$

ρ is the density and M depends on the type of problem represented by (11.4.1). In the case of a vibrating string, M corresponds to the tangential tension applied to the string (e.g., Boyce and Di Prima, 1977). For the special cases of P or S waves described in Problems 4.11 and 4.12 M is equal to either $\lambda + 2\mu$ or μ and is sometimes referred to as the elastic modulus. Now we will let M be a complex function of frequency, and will search for solutions of the form

$$\psi(x, t) = A e^{i(\omega t - Kx)}, \qquad (11.4.3)$$

where

$$K(\omega) = k(\omega) - i\alpha(\omega); \qquad \alpha > 0 \qquad (11.4.4)$$

is the *complex wavenumber*. From (11.3.9) we know that (11.4.3) represents a wave with amplitude attenuated in the positive x direction. In addition, let

$$M = M_R + iM_I \qquad (11.4.5)$$

and introduce the *complex velocity*

$$v = \sqrt{\frac{M}{\rho}} \equiv v_R + iv_I = |v| e^{i\phi}; \qquad \tan\phi = \frac{v_I}{v_R}. \qquad (11.4.6a,b)$$

When M is real v is the usual wave velocity. Introducing (11.4.3) in (11.4.1) gives

$$\omega^2 = v^2 K^2. \qquad (11.4.7)$$

Therefore

$$K = k - i\alpha = \frac{\omega}{|v|} e^{-i\phi} = \frac{\omega}{|v|} (\cos\phi - i\sin\phi) = \frac{\omega}{|v|^2} (v_R - iv_I) \qquad (11.4.8)$$

and

$$k = \omega \frac{v_R}{|v|^2}; \qquad \alpha = \omega \frac{v_I}{|v|^2} \qquad (11.4.9a,b)$$

(O'Connell and Budiansky, 1978).

To write v_R and v_I in terms of M start with (11.4.5) and (11.4.6a), which gives

$$\frac{M_R}{\rho} = v_R^2 - v_I^2; \qquad \frac{M_I}{\rho} = 2v_R v_I. \qquad (11.4.10\text{a,b})$$

Then, solving for v_R and v_I we obtain

$$v_R = \sqrt{\frac{|M| + M_R}{2\rho}}; \qquad v_I = \sqrt{\frac{|M| - M_R}{2\rho}} \qquad (11.4.11\text{a,b})$$

(Problem 11.1).

11.4.1 Spatial Q

In analogy with (11.2.23), we will introduce a *spatial Q* defined in such a way that the decay of the peak amplitude over one wavelength is $\exp(-\pi/Q)$. Introducing (11.4.4) in (11.4.3) gives

$$\psi(x, t) = Ae^{-\alpha x} e^{i(\omega t - kx)}, \qquad (11.4.12)$$

where α is the *attenuation coefficient*. If x_m indicates the value of x corresponding to a peak of ψ and λ is the wavelength, for a fixed value of t we have

$$\frac{\psi(x_m + \lambda, t)}{\psi(x_m, t)} = \frac{Ae^{-\alpha(x_m + \lambda)}}{Ae^{-\alpha x_m}} = e^{-\alpha\lambda} = e^{-\pi/Q}, \qquad (11.4.13)$$

which means that

$$\alpha\lambda = \alpha\frac{2\pi}{k} = \frac{\pi}{Q} \qquad (11.4.14)$$

where (5.4.18) has been used. Therefore,

$$Q = \frac{k}{2\alpha} = -\frac{\mathcal{R}\{K\}}{2\mathcal{I}\{K\}} = \frac{v_R}{2v_I}, \qquad (11.4.15)$$

where $\mathcal{R}$ and $\mathcal{I}$ indicate real and imaginary parts. The second equality in (11.4.15) makes more explicit the relation between Q and K, while the right-hand side follows from (11.4.9a,b). In addition, from (11.4.14) and using $k = \omega/c$ (see (6.9.2b)) we have

$$\alpha = \frac{k}{2Q} = \frac{\omega}{2Qc}. \qquad (11.4.16)$$

Note that, for fixed x, ψ is a function of t. In this case instead of a complex wavenumber we can introduce a *complex frequency* $\omega + i\beta$ and apply the definition of temporal Q given in (11.2.19). Furthermore, because $c = x/t$, from (11.4.16) we see that $\alpha x = \beta t$. Consequently, the temporal and spatial definitions of Q lead to the same exponential decay, so that it is not necessary to distinguish between them (within the simplified context considered here). Therefore, Q can, in principle, be determined from the temporal or spatial decay of ψ.

Now we will write the v introduced in (11.4.6) in terms of Q. To do that rewrite v as

$$v = v_R \left(1 + i\frac{v_I}{v_R}\right) = v_R \left(1 + i\frac{1}{2Q}\right) \qquad (11.4.17)$$

where (11.4.15) was used. Clearly, when there is no attenuation ($Q = \infty$), $v = v_R$, which is the usual velocity. Equation (11.4.17) provides a practical way to introduce attenuation in the computation of synthetic seismograms (e.g., Ganley, 1981), although it does not account for dispersion (see §11.8).

Finally, we note that Q has also been defined as follows:

$$\bar{Q} = \frac{M_R}{M_I} \qquad (11.4.18)$$

(White, 1965; O'Connell and Budiansky, 1978). This definition is based on the consideration of viscoelastic solids. From (11.4.10a,b) and (11.4.15) we obtain

$$\bar{Q} = \frac{v_R^2 - v_I^2}{2v_R v_I} = \frac{v_R}{2v_I} - \frac{v_I}{2v_R} = Q - \frac{1}{4Q}. \qquad (11.4.19)$$

This equation shows that $\bar{Q} \approx Q$ if $Q \gg 1$ or $\alpha \ll k$.

11.5 Attenuation of seismic waves in the Earth

Although the precise mechanisms responsible for attenuation are not well known, an exponential decay similar to that studied in the previous section provides a convenient phenomenological framework for the description of seismic attenuation in the Earth, as well as in solids in general. Specifically, we will write a plane harmonic wave as

$$u(x, t) = u_o e^{-\alpha x} e^{i(\omega t - kx)} = u_o e^{i(\omega t - Kx)} \qquad (11.5.1)$$

with α and K defined by (11.4.16) and (11.4.4) and u_o a constant. For the case of waves other than plane waves, the treatment of attenuation requires consideration of the geometric spreading factor (see §10.11).

Equation (11.5.1), however, is not enough to completely describe attenuation. One of the most important questions is the relation between Q and frequency. This matter has been studied for a long time, and early work seemed to indicate that Q was independent of frequency (e.g., Knopoff, 1964), although as noted in §11.1, the analysis of the possible attenuation mechanisms showed that a frequency dependence should be expected. In view of this discrepancy it was suggested that attenuation was the result of the superposition of a variety of mechanisms

with different time scales that resulted in a smearing of the frequency dependence
(Stacey *et al.*, 1975). More recent work, however, shows that Q actually depends
on frequency, although this dependence seems to be weak, with Q increasing with
frequency (e.g., Anderson, 1989; Der, 1998). In any case, the early assumption
that Q was frequency independent was useful because it simplified the analysis of
attenuation considerably.

A second question that was the matter of some debate was whether attenuation is
a linear process. In this context, linearity means that if a particular harmonic wave
is affected by attenuation in any specific way, it will be equally affected when the
wave is part of a superposition of harmonic waves. The mathematical equivalent of
this statement is the following. If $u(x, t)$ represents a seismic wave, then

$$u(x, t) \equiv \frac{1}{2\pi} \int_{-\infty}^{\infty} u(x, \omega) e^{i\omega t} \, d\omega = \frac{1}{2\pi} \int_{-\infty}^{\infty} u(0, \omega) e^{i[\omega t - K(\omega)x]} \, d\omega$$

$$= \frac{1}{2\pi} \int_{-\infty}^{\infty} u(0, \omega) e^{-\alpha x} e^{i[\omega t - k(\omega)x]} \, d\omega \qquad (11.5.2)$$

(after Futterman, 1962) with K as in (11.5.1) and $u(x, \omega)$ and $u(0, \omega)$ the Fourier
transforms of $u(x, t)$ and $u(0, t)$, respectively (see (5.4.24)). Therefore, if linearity
applies, taking care of attenuation is very simple. First, decompose a wave into
its Fourier components, multiply each component by $\exp(-\alpha x)$, and then go back
to the time domain by applying the inverse Fourier transform to the attenuated
components. Linearity simplifies the analysis of attenuation significantly and has
been found to be applicable for strains of about 10^{-6} or smaller, which are typical
in the far field of earthquakes (Stacey, 1992).

Now we will see that the combination of an attenuation model given by (11.5.1)
and linearity produces a serious unwanted result. Using (11.4.4), (11.4.16) and
$k = \omega/c$ we have

$$K(\omega)x = [k(\omega) - i\alpha(\omega)]x = \omega T - i\frac{\omega T}{2Q}; \qquad T = \frac{x}{c}. \qquad (11.5.3a,b)$$

Introducing (11.5.3a) in (11.5.2) and writing $A(\omega)$ for $u(0, \omega)$ we obtain

$$u(x, t) = \frac{1}{2\pi} \int_{-\infty}^{\infty} \left[A(\omega) \left(e^{-i\omega T} e^{-|\omega|T/2Q} \right) \right] e^{i\omega t} \, d\omega, \qquad (11.5.4)$$

where the absolute value of ω is needed to ensure an exponential decay when ω is
negative.

Equation (11.5.4) shows that $u(x, t)$ is the inverse Fourier transform of the
function in square brackets, which in turn is equal to the product of two functions.

Therefore, equation (11.5.4) can be written as the following convolution:

$$u(x,t) = \mathcal{F}^{-1}\{A(\omega)\} * \mathcal{F}^{-1}\left(e^{-i\omega T}e^{-|\omega|T/2Q}\right) = A(t) * \frac{1}{\pi}\frac{T/2Q}{(T/2Q)^2 + (t-T)^2}$$
$$(11.5.5)$$

(Problem 11.2). Because of causality, $A(t) = 0$ for $t < 0$ and we expect $u(x,t)$ to be zero before the arrival time $T = x/c$, but this is not the case because $A(t)$ is convolved with a function that is nonzero for $t < T$. For example, if $A(t) = \delta(t)$, because Dirac's delta is the unit element with respect to convolution (Appendix A), equation (11.5.5) becomes

$$u(x,t) = \frac{1}{\pi}\frac{x/2cQ}{(x/2cQ)^2 + (t - x/c)^2} \qquad (11.5.6)$$

(Stacey *et al.*, 1975), which has nonzero amplitudes for $t < x/c$.

So far we have assumed that linearity applies, and that Q and c are independent of ω, but the noncausal result obtained above indicates that these three assumptions cannot hold simultaneously. Therefore, now we will allow c to be a function of ω (so that wave propagation becomes dispersive, see §7.6), while keeping Q constant (i.e., independent of frequency) and will investigate whether this new assumption is consistent with causality. To do that let us write

$$e^{i(\omega t - Kx)} = e^{i\omega t}e^{-i[\omega/c(\omega)]x - \alpha(\omega)x} \equiv e^{i\omega t}F(\omega), \qquad (11.5.7)$$

where $F(\omega)$ is defined by the identity. To satisfy the causality condition the inverse Fourier transform $f(t)$ of $F(\omega)$ must be zero for

$$t < \tau = x/c_\infty; \qquad c_\infty = \lim_{\omega \to \infty} c(\omega). \qquad (11.5.8a,b)$$

The existence of the *limiting velocity* c_∞ can be proved analytically and will be discussed further in §11.7. Because for attenuating media c_∞ is unknown, it follows that the arrival time τ is also unknown, so that the measured arrival time of a wave will depend on its frequency content.

Now we want to solve the following problem. Given the function $F(\omega)$, find whether there is a relation between $c(\omega)$ and $\alpha(\omega)$ that makes its inverse Fourier transform $f(t)$ equal to zero for $t < \tau$. To analyze this problem we will apply a translation of the axis such that $t = \tau$ in the old system becomes $t = 0$ in the new system. Then the function $f(t)$ becomes $f(t + \tau)$, with a Fourier transform given by

$$\mathcal{F}\{f(t + \tau)\} = e^{i\omega x/c_\infty}F(\omega) = e^{-i[\omega/c(\omega) - \omega/c_\infty - i\alpha(\omega)]x} \equiv F_\tau(\omega). \qquad (11.5.9)$$

$F_\tau(\omega)$ is the function whose inverse must be causal, and to find the relation between $c(\omega)$ and $\alpha(\omega)$ it is necessary to be familiar with a number of mathematical results concerning causality that will be discussed in §11.6.

Finally, we will see that the fact that $u(x, t)$ is a real function has important consequences because it imposes certain symmetry properties on $k(\omega)$ and $\alpha(\omega)$. To see that consider a real function $s(t)$ with a Fourier transform given by

$$S(\omega) = \int_{-\infty}^{\infty} s(t)e^{-i\omega t} \, dt. \tag{11.5.10}$$

The complex conjugate of $S(\omega)$ is given by

$$S^*(\omega) = \int_{-\infty}^{\infty} s^*(t)e^{i\omega t} \, dt = \int_{-\infty}^{\infty} s(t)e^{i\omega t} \, dt = S(-\omega). \tag{11.5.11}$$

Here we have used the fact that because $s(t)$ is real, it is equal to its complex conjugate (Byron and Fuller, 1970). Therefore

$$S^*(\omega) = S(-\omega). \tag{11.5.12}$$

Equation (11.5.12) can be applied to $u(0, \omega)$ and $u(x, \omega)$ (see (11.5.2)), which are the Fourier transforms of real functions:

$$u^*(0, \omega) = u(0, -\omega) \tag{11.5.13}$$

$$u^*(x, \omega) = u(x, -\omega). \tag{11.5.14}$$

Equation (11.5.14) can be written as

$$u^*(0, \omega)e^{iK^*(\omega)x} = u(0, -\omega)e^{-iK(-\omega)x}, \tag{11.5.15}$$

where $u(x, \omega) = u(0, \omega) \exp(-iKx)$ was used. Combining (11.5.13) and (11.5.15) gives

$$K^*(\omega) = -K(-\omega) \tag{11.5.16}$$

which, in turn, implies

$$k(\omega) = -k(-\omega) \qquad \alpha(\omega) = \alpha(-\omega) \tag{11.5.17a,b}$$

(Futterman, 1962). Therefore, k and α are odd and even functions of ω, respectively. Furthermore, from (11.5.17a) and $k = \omega/c$ we obtain

$$c(\omega) = c(-\omega) \tag{11.5.18}$$

(Ben-Menahem and Singh, 1981).

11.6 Mathematical aspects of causality and applications

Causality is an important concept in mathematics, physics, and signal processing, and consequently there is a well-established body of knowledge. Here we will introduce the most relevant aspects of it, but because the study of causality requires heavy use of analytic functions, which is beyond the scope of this book, a number of basic results are just quoted. Causality can be investigated using two different, but complementary points of view, both of which have been used in the study of attenuation.

11.6.1 The Hilbert transform. Dispersion relations

As shown in Appendix B, if $s(t)$ is a causal function, then the real and imaginary parts of its Fourier transform $S(\omega)$ constitute a Hilbert transform pair. Let

$$S(\omega) = R(\omega) + iI(\omega). \qquad (11.6.1)$$

Then

$$R(\omega) = \frac{1}{\pi\omega} * I(\omega) = -\frac{1}{\pi}P\int_{-\infty}^{\infty}\frac{I(\omega')}{\omega' - \omega}\,d\omega', \qquad (11.6.2)$$

$$I(\omega) = -\frac{1}{\pi\omega} * R(\omega) = \frac{1}{\pi}P\int_{-\infty}^{\infty}\frac{R(\omega')}{\omega' - \omega}\,d\omega'. \qquad (11.6.3)$$

Equations (11.6.2) and (11.6.3) were obtained for ω a real variable, but it is much more fruitful to carry out the analysis in the complex domain, with $\omega = \mathcal{R}\{\omega\} + i\mathcal{I}\{\omega\}$. However, the relations of interest to us are those for which $\omega = \mathcal{R}\{\omega\}$. When working in the complex domain it is found that (11.6.2) and (11.6.3) are valid as long as $S(\omega)$ is analytic and does not have poles in the lower half-plane (i.e., $\mathcal{I}\{\omega\} < 0$). If $S(\omega)$ is analytic and does not have poles in the upper half-plane, then $s(t) = 0$ for $t \geq 0$ (Solodovnikov, 1960). If the function $S(\omega)$ has a pole at $\omega = \omega_p$, then $S(\omega_p) = \infty$. For example, the function $1/[(\omega - \omega_1)(\omega - \omega_2)]$ has poles at ω_1 and ω_2. The fact that $S(\omega)$ must be analytic and bounded in the lower half-plane is related to the way the Fourier transform has been defined. Had the alternative definition been used (see §5.4), then the upper plane should replace the lower plane.

As noted in Appendix B, the proof of (11.6.2) and (11.6.3) in the complex domain also requires that

$$\lim_{|\omega|\to\infty} |S(\omega)| = 0 \qquad (11.6.4)$$

in the lower half-plane (including the real axis). If this condition is not satisfied but the limit in (11.6.4) has a finite value (known or unknown), the Hilbert relations have to be modified as follows. If ω_o is a point on the real axis for which $S(\omega)$ is analytic, then $S(\omega)$ is differentiable there and the function

$$\frac{S(\omega) - S(\omega_o)}{\omega - \omega_o} \qquad (11.6.5)$$

is not singular at $\omega = \omega_o$, is analytic, and satisfies (11.6.4). Then, an analysis in the complex domain similar to that leading to (11.6.2) and (11.6.3) gives the following pair:

$$R(\omega) = R(\omega_o) - \frac{1}{\pi}(\omega - \omega_o)P\int_{-\infty}^{\infty} \frac{I(\omega')}{(\omega' - \omega)(\omega' - \omega_o)}\,d\omega', \quad (11.6.6)$$

$$I(\omega) = I(\omega_o) + \frac{1}{\pi}(\omega - \omega_o)P\int_{-\infty}^{\infty} \frac{R(\omega')}{(\omega' - \omega)(\omega' - \omega_o)}\,d\omega' \quad (11.6.7)$$

(Ben-Menahem and Singh, 1981). A clear exposition of these matters is provided by Byron and Fuller (1970) (some sign differences are due to the use of the alternative definition of the Fourier transform).

Equations (11.6.6) and (11.6.7) are known as *dispersion relations with one subtraction*. The so-called *Kramers–Kronig relations*, named after the two physicists that introduced them in the 1920s, are examples of dispersion relations. These relations were initially derived in the context of electromagnetic theory (e.g., Toll, 1956; Arfken, 1985), but now the name is used with similar relations that arise in other branches of physics. In particular, they play a critical role in attenuation studies, as will be shown in §11.7.

11.6.2 Minimum-phase-shift functions

Let us consider a causal function $g(t)$ having Fourier transform $G(\omega)$ given by

$$G(\omega) = A(\omega)e^{i\varphi(\omega)}; \qquad A \geq 0. \quad (11.6.8)$$

We want to address the following question: given $A(\omega)$, what can we say about $\varphi(\omega)$? To discuss this matter take the logarithm of both sides of (11.6.8). This gives

$$\ln G(\omega) = \ln A(\omega) + i\varphi(\omega). \quad (11.6.9)$$

Equation (11.6.9) is similar to (11.6.1) with $R(\omega)$ and $I(\omega)$ replaced by $\ln A(\omega)$ and $\varphi(\omega)$, respectively, so that we can use (11.6.3), which gives

$$\varphi(\omega) = -\frac{1}{\pi}P\int_{-\infty}^{\infty} \frac{\ln A(u)}{\omega - u}\,du \quad (11.6.10)$$

provided that $\ln G(\omega)$ satisfies the conditions that $S(\omega)$ must satisfy for (11.6.3) to be valid, namely, $\ln G(\omega)$ is bounded in the lower half-plane. This means that in addition to not having poles, $G(\omega)$ cannot have zeros, because otherwise $\ln G(\omega)$ would not be bounded. Clearly, this condition limits the class of functions $G(\omega)$ for which the phase can be determined from knowledge of the amplitude. Functions $G(\omega)$ for which (11.6.10) applies are known as *minimum-phase-shift* functions (or minimum-phase, for short), and they have the property that of all the functions with the same amplitude, they have the least phase and phase difference (Toll, 1956; Solodovnikov, 1960; Papoulis, 1962).

The concept of a minimum-phase function will be illustrated with the following example (Papoulis, 1962). Let

$$g_1(t) = e^{-2t} H(t); \qquad g_2(t) = \left(3e^{-2t} - 2e^{-t}\right) H(t), \qquad (11.6.11a,b)$$

where $H(t)$ is the unit step. These functions are plotted in Fig. 11.3. The corresponding Fourier transforms are

$$G_1(\omega) = \frac{1}{2 + i\omega}; \qquad G_2(\omega) = \frac{1}{2 + i\omega} \frac{i\omega - 1}{i\omega + 1}. \qquad (11.6.12a,b)$$

Note that G_2 is equal to G_1 times a function of ω having unit amplitude, which means that G_1 and G_2 have the same amplitude, given by

$$|G_1(\omega)| = |G_2(\omega)| = \frac{1}{\sqrt{4 + \omega^2}} \qquad (11.6.13)$$

but different phases (Fig. 11.3). Therefore, the amplitude spectrum does not constrain the phase spectrum. On the other hand, $G_1(\omega)$ has minimum phase, while $G_2(\omega)$ does not because it has a zero for $\omega_o = -i$, which is in the lower half-plane. Consequently, equation (11.6.10) can be used to determine the phase of $G_1(\omega)$ but not the phase of $G_2(\omega)$.

Minimum-phase functions have the property that they are *minimum-delay* functions in the time domain. Of all the functions with the same amplitude spectrum, the minimum delay function has the shortest time duration (Robinson, 1967). Given a function $g(t)$, its *duration D* can be estimated using

$$D = \int_0^\infty t^2 |g(t)|^2 \, dt. \qquad (11.6.14)$$

The factor t^2 can be considered to be a weighting function that increases its value as the width of $g(t)$ increases. Therefore, qualitatively, the narrower the function the smaller the value of D (as long as the amplitude spectra are the same). To make this statement more precise and to show the relation between duration and phase we will work in the frequency domain. The starting point is the following Fourier transform pair (Papoulis, 1962) and the expression (11.6.8) for $G(\omega)$

$$-itg(t) \longleftrightarrow \frac{dG(\omega)}{d\omega} = \left(\frac{dA}{d\omega} + iA\frac{d\varphi}{d\omega}\right) e^{i\varphi}. \qquad (11.6.15)$$

We also need Parseval's formula, which relates a Fourier transform pair $b(t)$, $B(\omega)$ as follows:

$$\int_{-\infty}^{\infty} |b(t)|^2 \, dt = \frac{1}{2\pi} \int_{-\infty}^{\infty} |B(\omega)|^2 \, d\omega \qquad (11.6.16)$$

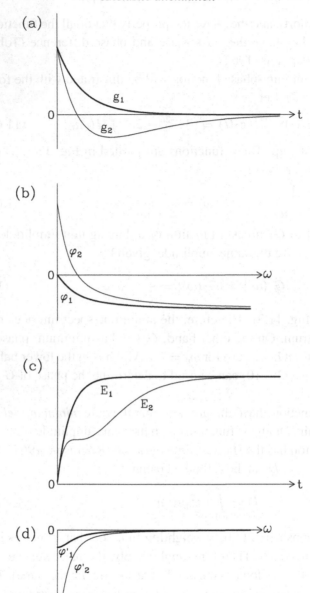

Fig. 11.3. Comparison between a minimum-phase (or minimum-delay) function (bold curves) and a nonminimum-phase function (thin curves) with the same amplitude spectra. (a) Plot of the functions $g_1(t)$ and $g_2(t)$ given by equations (11.6.11a,b). (b), (d) Phase and derivative of phase with respect to ω of the Fourier transforms of $g_1(t)$ and $g_2(t)$ (see (11.6.12a,b)). (c) Partial energy of $g_1(t)$ and $g_2(t)$, computed using (11.6.18).

(Papoulis, 1962). If $b(t)$ is causal the lower boundary of the first integral actually is zero. Then, using (11.6.15) and (11.6.16), equation (11.6.14) gives

$$D = \int_0^\infty t^2 |g(t)|^2 \, dt = \frac{1}{2\pi} \int_{-\infty}^\infty \left| \frac{dG(\omega)}{d\omega} \right|^2 \, d\omega$$

$$= \frac{1}{2\pi} \int_{-\infty}^\infty \left[\left(\frac{dA}{d\omega} \right)^2 + A^2 \left(\frac{d\varphi}{d\omega} \right)^2 \right] d\omega. \qquad (11.6.17)$$

Therefore, under the assumption that $A(\omega)$ is the same for a collection of functions, the function with the shortest duration is that for which $d\varphi/d\omega$ is minimum. The minimum-delay function also has the property that for a given value of t it has the largest partial energy among all the functions with the same amplitude spectrum (Robinson, 1967). For an arbitrary causal signal $b(t)$, the partial energy is defined by

$$E^P(t) = \int_0^t |b(\tau)|^2 \, d\tau. \qquad (11.6.18)$$

These two properties of minimum-delay functions are demonstrated in Fig. 11.3 for the functions $g_1(t)$ and $g_2(t)$ defined by (11.6.11a,b). Note that the two functions have the same total energy (given by $E^P(\infty)$), as can be expected from (11.6.18) and (11.6.16).

11.6.3 The Paley–Wiener theorem. Applications

Equations (11.6.2) and (11.6.3) are a basic result in the study of causality, but not the only one. Another very important result is a theorem due to Paley and Wiener (1934, Theorem XII), which will be used to establish a class of allowable attenuation factors $\alpha(\omega)$. The original version of the theorem refers to time functions that are zero for positive times, but it also applies to causal functions (Wiener, 1949). The theorem (also known as a criterion or a condition) says that if $A(\omega)$ is a real nonnegative (i.e., $A(\omega) \geq 0$) square-integrable function, i.e., if

$$\int_{-\infty}^\infty |A(\omega)|^2 \, d\omega < \infty \qquad (11.6.19)$$

then a necessary and sufficient condition for the existence of a causal function $f(t)$ having a Fourier transform $F(\omega)$ such that $|F(\omega)| = A(\omega)$ is that

$$\int_{-\infty}^\infty \frac{|\ln A(\omega)|}{1 + \omega^2} \, d\omega < \infty \qquad (11.6.20)$$

(see, e.g., Zadeh and Desoer, 1963; Papoulis, 1962, 1977). An important consequence of the theorem is that $A(\omega)$ cannot be identically zero on any finite interval, because if it were then the integrand would equal infinity on the interval and the integral would diverge. However, isolated zeros are acceptable.

Let us apply the Paley–Wiener theorem to $A(\omega) = |F_\tau(\omega)|$, with $F_\tau(\omega)$ given by (11.5.9). In this case $|\ln A(\omega)|$ is equal to $\alpha(\omega)x$ and (11.6.20) becomes

$$\int_{-\infty}^{\infty} \frac{\alpha(\omega)}{1 + \omega^2}\, d\omega < \infty. \qquad (11.6.21)$$

This condition is satisfied if

$$\alpha(\omega) \propto \omega^s; \qquad s < 1 \qquad (11.6.22\text{a,b})$$

(Guillemin, 1963; see also Problem 11.3) or, equivalently, if

$$\lim_{\omega \to \infty} \frac{\alpha(\omega)}{\omega} = 0. \qquad (11.6.23)$$

Equations (11.6.22) and (11.6.23) must be satisfied regardless of the mechanism postulated to represent attenuation as long as it is modeled by (11.5.1) and have been used as the starting point of the analysis in §11.8 and to derive general results regarding the dependence of Q on frequency. First of all note that (11.6.22b) is a strict inequality, so that from (11.4.16) we see that the product Qc cannot be constant. Next we show that Q itself cannot be constant when $c = c(\omega)$. To see that, take logarithms on both sides of the identity in (11.5.9), which gives

$$\ln[F_\tau(\omega)] = -\alpha(\omega)x - i \left[\frac{\omega}{c(\omega)} - \frac{\omega}{c_\infty} \right] x \qquad (11.6.24)$$

to which we can apply (11.6.3) under the conditions discussed in connection with (11.6.10). Doing that gives

$$\frac{\omega}{c(\omega)} = \frac{\omega}{c_\infty} - \frac{1}{\pi \omega} * \alpha(\omega) = \frac{\omega}{c_\infty} + \breve{\alpha}(\omega), \qquad (11.6.25)$$

where the ˘ indicates the Hilbert transform (see (B.10)). Using (11.4.16), equation (11.6.25) can be rewritten as

$$2Q = \frac{\omega}{c_\infty \alpha(\omega)} + \frac{\breve{\alpha}(\omega)}{\alpha(\omega)}. \qquad (11.6.26)$$

This equation shows that Q cannot be constant because if it were, $\alpha(\omega)$ (as well as $\breve{\alpha}(\omega)$) would be linear in ω, which would violate the condition (11.6.22). This result is due to Strick (1970).

Finally, because $\alpha(\omega)$ is an even function (see (11.5.17b)) its Hilbert transform can be written as follows:

$$\breve{\alpha}(\omega) = -\frac{1}{\pi} \mathcal{P} \int_{-\infty}^{\infty} \frac{\alpha(\omega')}{\omega - \omega'}\, d\omega' = -\frac{2}{\pi} \omega \mathcal{P} \int_{0}^{\infty} \frac{\alpha(\omega')}{\omega^2 - \omega'^2}\, d\omega' \qquad (11.6.27)$$

(Problem 11.4). Written in this way, $\breve{\alpha}$ depends on positive frequencies, which are the only ones available in an experiment. When (11.6.27) is introduced in (11.6.25) we obtain

$$\frac{1}{c(\omega)} = \frac{1}{c_\infty} - \frac{2}{\pi}\mathcal{P}\int_0^\infty \frac{\alpha(\omega')}{\omega^2 - \omega'^2}\,d\omega'. \tag{11.6.28}$$

Equation (11.6.28) will be used in §11.8.

11.7 Futterman's relations

Futterman (1962) and Lamb (1962) were the first to apply causality to the analysis of seismic attenuation, although Lamb's analysis was less comprehensive. Futterman (1962) assumed that $\alpha(\omega)$ was strictly a linear function of frequency in the range of measurement and introduced as a basic variable the *index of refraction* of the medium, defined by

$$n(\omega) = \frac{K(\omega)}{K_o(\omega)} \equiv \mathcal{R}\{n(\omega)\} - i\mathcal{I}\{n(\omega)\} \tag{11.7.1}$$

where K_o defines a nondispersive K at the same frequency. Because Futterman defined the Fourier transform with $i\omega t$ in the exponent, his $K(\omega)$ is the complex conjugate of the $K(\omega)$ used here. Futterman also assumed the existence of a low-frequency cut-off ω_o, characteristic of the material, below which there was no dispersion and such that

$$K(\omega) = \frac{\omega}{C}; \qquad \omega < \omega_o, \tag{11.7.2}$$

where C is the nondispersive low-frequency limit of $c(\omega)$. As shown below, $\omega_o > 0$. Then, by definition,

$$K_o(\omega) = \frac{\omega}{C}; \qquad \text{all } \omega. \tag{11.7.3}$$

Also, from (11.7.1)–(11.7.3) we see that

$$n(0) = 1. \tag{11.7.4}$$

Using (11.7.1) and (11.7.3), $K(\omega)$ can be written as

$$K(\omega) = \frac{\omega}{C}n(\omega) = \frac{\omega}{C}[\mathcal{R}\{n(\omega)\} - i\mathcal{I}\{n(\omega)\}]. \tag{11.7.5}$$

Then, by comparison with (11.4.4) we find that $k(\omega)$ and $\alpha(\omega)$ are given by

$$k(\omega) = \frac{\omega}{C}\mathcal{R}\{n(\omega)\}; \qquad \alpha(\omega) = \frac{\omega}{C}\mathcal{I}\{n(\omega)\}. \tag{11.7.6a,b}$$

In addition, from the properties of dispersive waves we know that $c(\omega)$ is the phase velocity, equal to

$$c(\omega) = \frac{\omega}{k(\omega)} = \frac{C}{\mathcal{R}\{n(\omega)\}} \qquad (11.7.7)$$

(see §7.6.2).

Futterman (1962) also introduced the reduced quality factor Q_o, equal to

$$Q_o = \frac{\omega}{2\alpha(\omega)C}, \qquad (11.7.8)$$

which is independent of frequency in the range where $\alpha(\omega)$ is proportional to ω.

With the previous definitions and using (11.4.16) we can write Q, which as we showed must be a function of ω, as

$$Q(\omega) = \frac{k(\omega)}{2\alpha(\omega)} = \frac{\omega}{2\alpha(\omega)c(\omega)} = \frac{CQ_o}{c(\omega)} = \frac{\mathcal{R}\{n(\omega)\}}{2\mathcal{I}\{n(\omega)\}}. \qquad (11.7.9)$$

A basic result in Futterman's (1962) work are the Kramers–Kronig relations, which involve the real and imaginary parts of $K(\omega)$. However, because $K(\omega)$ is not the Fourier transform of a causal function (Weaver and Pao, 1981), the derivation of the Kramers–Kronig relations requires showing that $K(\omega)$ is analytic in the lower half-plane (for a proof see Futterman, 1962; Aki and Richards, 1980) and that the limit c_∞ (11.5.8b) exists. Using the first equality in (11.7.7), c_∞ can be written as

$$c_\infty = \lim_{\omega \to \infty} \frac{\omega}{k(\omega)}. \qquad (11.7.10)$$

The existence of this limit is essential in the following derivation, and although it can be proved for electromagnetic media from an analysis of the physical mechanisms involved, it is not clear whether it must also be valid for all possible attenuation mechanisms in solids (or liquids). These questions are discussed by Weaver and Pao (1981), who show that (11.7.10) applies to any medium, regardless of the physical processes involved, as long as linearity, causality, and passivity (i.e., absence of energy sources within the medium) hold. In addition, proving that c_∞ exists requires that $\mathcal{I}\{n(\infty)\} = 0$ (Ben-Menahem and Singh, 1981). This assumption is also critical in Futterman's (1962) analysis, who justified it with the argument that "it is difficult to envision Earth structures of such small dimensions that they resonate to the infinite frequency component of the incident displacement wave". However, as shown below, this assumption is not essential as long as $\mathcal{I}\{n(\infty)\}$ is bounded. With this caveat, there is no attenuation at infinite frequency and

$$\mathcal{I}\{n(\infty)\} = 0; \qquad \mathcal{R}\{n(\infty)\} = n(\infty). \qquad (11.7.11)$$

Now, using (11.7.6), (11.7.7), (11.7.11), and (11.7.1) the frequency-dependent part of the exponent of (11.5.9) becomes

$$h(\omega) \equiv \left[\frac{\omega}{c(\omega)} - \frac{\omega}{c_\infty} - i\alpha(\omega) \right]$$

$$= \frac{\omega}{C} \left[\mathcal{R}\{n(\omega)\} - \mathcal{R}\{n(\infty)\} - i\left(\mathcal{I}\{n(\omega)\} - \mathcal{I}\{n(\infty)\} \right) \right]$$

$$= \frac{\omega}{C} [n(\omega) - n(\infty)] \equiv \frac{\omega}{C} \Delta n(\omega) \qquad (11.7.12)$$

so that

$$\Delta n(\omega) \equiv \mathcal{R}\{\Delta n(\omega)\} - i\mathcal{I}\{\Delta n(\omega)\} = C\frac{h(\omega)}{\omega}. \qquad (11.7.13)$$

As $h(\omega)$ is an analytic function (Futterman, 1962), equation (11.7.13) shows that $\Delta n(\omega)$ is an analytic function in the lower half-plane (because it is the ratio of two such functions). Therefore, we can apply the Hilbert transforms to $\Delta n(\omega)$, which gives

$$\mathcal{R}\{\Delta n(\omega)\} = \frac{1}{\pi} \mathcal{P} \int_{-\infty}^{\infty} \frac{\mathcal{I}\{\Delta n(\omega')\}}{\omega' - \omega} \, d\omega' \qquad (11.7.14)$$

and

$$\mathcal{I}\{\Delta n(\omega)\} = -\frac{1}{\pi} \mathcal{P} \int_{-\infty}^{\infty} \frac{\mathcal{R}\{\Delta n(\omega')\}}{\omega' - \omega} \, d\omega'. \qquad (11.7.15)$$

Next, using

$$\mathcal{P} \int_{-\infty}^{\infty} \frac{d\omega'}{\omega' - \omega} = 0 \qquad (11.7.16)$$

and (11.7.14) we obtain

$$\mathcal{R}\{n(\omega) - n(\infty)\} = \frac{1}{\pi} \mathcal{P} \int_{-\infty}^{\infty} \frac{\mathcal{I}\{n(\omega')\}}{\omega' - \omega} \, d\omega' \qquad (11.7.17)$$

which is valid as long as $\mathcal{I}\{n(\infty)\}$ is bounded (Problem 11.5). Applying this relation to the zero frequency gives

$$\mathcal{R}\{n(0) - n(\infty)\} = \frac{1}{\pi} \mathcal{P} \int_{-\infty}^{\infty} \frac{\mathcal{I}\{n(\omega')\}}{\omega'} \, d\omega'. \qquad (11.7.18)$$

Upon subtraction of (11.7.18) from (11.7.17) we obtain

$$\mathcal{R}\{n(\omega) - n(0)\} = \frac{\omega}{\pi} \mathcal{P} \int_{-\infty}^{\infty} \frac{\mathcal{I}\{n(\omega')\}}{\omega'(\omega' - \omega)} \, d\omega'. \qquad (11.7.19)$$

Note that (11.7.19) can be obtained from (11.6.6) and (11.7.1) with $\omega_o = 0$, which means that this basic result does not depend on the details of the behavior of $\mathcal{I}\{n(\infty)\}$ as long as it is bounded. Finally, using

$$\mathcal{I}\{n(\omega)\} = -\mathcal{I}\{n(-\omega)\} \qquad (11.7.20)$$

we obtain

$$\mathcal{R}\{n(\omega) - n(0)\} = \mathcal{R}\{n(\omega)\} - 1 = \frac{2\omega^2}{\pi} \mathcal{P} \int_0^\infty \frac{\mathcal{I}\{n(\omega')\}}{\omega'(\omega'^2 - \omega^2)} \, d\omega' \quad (11.7.21)$$

(Problem 11.6). Equation (11.7.21), or variations thereof, and the corresponding expressions for $\mathcal{I}\{n(\omega)\}$ are known as the Kramers–Kronig relations.

Futterman (1962) proposed three expressions for $\mathcal{I}\{n(\omega)\}$ and computed $\mathcal{R}\{n(\omega)\}$ using (11.7.21). One of the expressions is given by

$$\mathcal{I}\{n(\omega)\} = \frac{1}{2Q_o} \left(1 - e^{-\omega/\omega_o}\right); \qquad \frac{\omega}{\omega_o} \geq 0. \quad (11.7.22)$$

Introducing this expression in (11.7.21) gives the corresponding value of $\mathcal{R}\{n(\omega)\}$, which involves two exponential terms that can be neglected for $\omega > 6\omega_o$, in which case

$$\mathcal{R}\{n(\omega)\} = 1 - \frac{1}{\pi Q_o} \ln \gamma \frac{\omega}{\omega_o} \quad (11.7.23)$$

where $\ln \gamma$ is equal to 0.5772. This equation shows that ω_o must be larger than zero, because $\ln(\omega/\omega_o)$ goes to infinity as ω_o goes to zero.

Using (11.7.7) and (11.7.23) we obtain

$$c(\omega) = C \left(1 - \frac{1}{\pi Q_o} \ln \gamma \frac{\omega}{\omega_o}\right)^{-1}. \quad (11.7.24)$$

When $\omega/\omega_o \gg \gamma$,

$$\ln \gamma \frac{\omega}{\omega_o} = \ln \frac{\omega}{\omega_o} + \ln \gamma \approx \ln \frac{\omega}{\omega_o} \quad (11.7.25)$$

and

$$c(\omega) = C \left(1 - \frac{1}{\pi Q_o} \ln \frac{\omega}{\omega_o}\right)^{-1}. \quad (11.7.26)$$

Moreover, when the logarithmic term in (11.7.26) satisfies

$$\left|\frac{1}{\pi Q_o} \ln \frac{\omega}{\omega_o}\right|^2 \ll 1 \quad (11.7.27)$$

we can expand $c(\omega)$ in a Taylor series about the origin keeping the first two terms only. In this way we obtain

$$c(\omega) = C \left(1 + \frac{1}{\pi Q_o} \ln \frac{\omega}{\omega_o}\right). \quad (11.7.28)$$

Finally, from (11.7.9) and (11.7.26),

$$Q = \frac{C Q_o}{c(\omega)} = Q_o \left(1 - \frac{1}{\pi Q_o} \ln \gamma \frac{\omega}{\omega_o}\right). \quad (11.7.29)$$

Inspection of (11.7.29) shows that Q will become negative when

$$\omega > \frac{\omega_o}{\gamma} e^{\pi Q_o}, \tag{11.7.30}$$

which requires an upper cut-off frequency. In addition, equation (11.7.29) shows that Q is a function of frequency, although its variation is small over a wide range of frequencies (see Problem 11.7), and for this reason Futterman's model is known as a *nearly constant Q* model. The selection of ω_o, on the other hand, is somewhat arbitrary and should be chosen so that it is small compared with the lowest measured frequency (Futterman, 1962).

Equation (11.7.24) has been widely quoted, and was the subject of some debate. To describe it, note that from (11.7.24) and (11.7.29) we obtain

$$c(\omega) = C \quad \text{if} \quad Q = Q_o = \infty \tag{11.7.31}$$

$$c(\omega) > C \quad \text{if} \quad Q, Q_o < \infty. \tag{11.7.32}$$

Equations (11.7.31) and (11.7.32) seem to imply that the velocity in an attenuating medium is larger than in a perfectly elastic medium, with the attenuated wave arriving earlier than expected (Stacey *et al.*, 1975). The problem with this interpretation is that C is the low-frequency limit of $c(\omega)$, while the elastic wave velocity, say c_E, is actually the high-frequency limit of $c(\omega)$, corresponding to the immediate response of the medium (Savage, 1976). Because of the upper cut-off frequency required by (11.7.30), c_E cannot be determined, but the following argument (due to Savage, 1976) removes this difficulty. Let ω_1 be a frequency larger than the frequencies in the bandpass of the recording instrument. Then the instrument will not distinguish between $\omega = \omega_1$ and $\omega = \infty$. Therefore $c_E = c(\omega_1)$ and using (11.7.24) we find that

$$\frac{c(\omega)}{c_E} = \frac{\pi Q_o - \ln(\gamma \omega_1/\omega_o)}{\pi Q_o - \ln(\gamma \omega/\omega_o)} < 1 \tag{11.7.33}$$

as long as $\omega_o < \omega < \omega_1$. Therefore, the attenuated wave will always arrive later than a wave propagating with the elastic wave velocity.

11.8 Kalinin and Azimi's relation. The complex wave velocity

Kalinin *et al.* (1967) approached the analysis of attenuation using the Paley–Wiener theorem. Three expressions for $\alpha(\omega)$ were discussed, one of them being

$$\alpha(\omega) = \frac{\alpha_o \omega}{1 + \alpha_1 \omega}, \tag{11.8.1}$$

where α_o and α_1 are positive constants that, in principle, should be derived from attenuation measurements. This $\alpha(\omega)$ satisfies the condition (11.6.23) and is approximately linear when $\alpha_1\omega \ll 1$. Then, using (11.8.1) and (11.6.28), Kalinin *et al.* (1967) derived the following relation:

$$c(\omega) = c_\infty \left[1 + \frac{2}{\pi} \frac{c_\infty \alpha_o \ln(1/\alpha_1\omega)}{1 - \alpha_1^2\omega^2}\right]^{-1}, \qquad (11.8.2)$$

which is usually attributed to Azimi *et al.* (1968). Note that $c(\infty) = c_\infty$ and $c(0) = 0$.

Now we will introduce a number of approximations that will lead to an expression for the phase velocity that does not include c_∞, α_o, or α_1. Under the assumption that $\alpha_1\omega \ll 1$, equation (11.8.2) becomes

$$c(\omega) = c_\infty \left(1 + \frac{2}{\pi}c_\infty\alpha_o \ln\frac{1}{\alpha_1\omega}\right)^{-1}. \qquad (11.8.3)$$

In addition, equation (11.8.3) will be rewritten as

$$\frac{1}{c(\omega)} = \frac{1}{c_\infty}\left(1 + \frac{2}{\pi}c_\infty\alpha_o \ln\frac{1}{\alpha_1\omega}\right). \qquad (11.8.4)$$

To eliminate the factor $c_\infty\alpha_o$ from (11.8.3) and (11.8.4) we use (11.6.26) rewritten as

$$2Q\alpha(\omega) = \frac{\omega}{c_\infty} + \breve{\alpha}(\omega). \qquad (11.8.5)$$

The expression for $\breve{\alpha}(\omega)$ can be found by comparison of (11.8.2) with (11.6.25). Introducing it in (11.8.5), using (11.8.1), assuming $\alpha_1\omega \ll 1$, and canceling a common factor ω gives

$$2Q\alpha_o = \frac{1}{c_\infty} + \frac{2\alpha_o}{\pi}\ln\frac{1}{\alpha_1\omega} \qquad (11.8.6)$$

(Problem 11.8). If we now assume that the second term on the right of (11.8.6) can be neglected (see below) we obtain

$$2\alpha_o c_\infty = \frac{1}{Q} \qquad (11.8.7)$$

(Aki and Richards, 1980).

To eliminate c_∞ and α_1 we will consider two frequencies ω_1 and ω_2, and use (11.8.3), (11.8.4), and (11.8.7), with (11.8.3) expanded in a Taylor series about the origin keeping the first two terms only. In this way we obtain

$$c(\omega_1) = c_\infty \left(1 + \frac{1}{\pi Q}\ln\alpha_1\omega_1\right) \qquad (11.8.8)$$

$$\frac{1}{c(\omega_2)} = \frac{1}{c_\infty}\left(1 + \frac{1}{\pi Q}\ln\frac{1}{\alpha_1\omega_2}\right). \qquad (11.8.9)$$

Equation (11.8.8) is valid for

$$\left| \frac{1}{\pi Q} \ln \alpha_1 \omega_1 \right|^2 \ll 1. \tag{11.8.10}$$

Multiplying (11.8.8) and (11.8.9) and neglecting the term with $1/\pi^2 Q^2$ gives

$$\frac{c(\omega_1)}{c(\omega_2)} = 1 + \frac{1}{\pi Q} \ln \frac{\omega_1}{\omega_2} \tag{11.8.11}$$

(Aki and Richards, 1980). Equations similar to (11.8.11) have been derived for viscoelastic solids using different approaches (e.g., Liu *et al.*, 1976; Kjartansson, 1979).

To give an idea of the approximations behind (11.8.11) let us consider the following values: $Q = 30$, $c_\infty = 5$ km s^{-1}, and $\alpha_1 \omega = 0.001$. Then, from (11.8.7), $\alpha_o = 0.0033$ and the second term on the right-hand side of (11.8.6) is equal to 0.015. This number is considerably smaller than $1/c_\infty$, which is equal to 0.2, and the approximation that leads to (11.8.7) is justified in this case. The condition (11.8.10) is also satisfied because the left-hand side is equal to 0.005. Also note that for ω_1/ω_2 equal to 1000 and 10 000, $c(\omega_1)/c(\omega_2)$ is equal to 1.07 and 1.10, respectively, so that the dispersion over a wide range of frequencies is quite small. These results explain why the dispersion of body waves is difficult to detect.

Finally we will derive an expression for the complex velocity that generalizes (11.4.17). First write $K(\omega)$, given by (11.4.4), as

$$K(\omega) = \frac{\omega}{c(\omega)} - i\alpha(\omega) = \frac{\omega}{c(\omega)} \left(1 - \frac{i}{2Q} \right), \tag{11.8.12}$$

where (11.4.16) was used. The last term in (11.8.12) can be considered the Taylor expansion about the origin of $1/(1+i/2Q)$ keeping the first two terms only, so that

$$K(\omega) = \frac{\omega}{c(\omega)(1 + i/2Q)} \equiv \frac{\omega}{\mathcal{C}(\omega)}. \tag{11.8.13}$$

Equation (11.8.13) defines the complex velocity $\mathcal{C}(\omega)$, which will be rewritten using (11.8.11) with ω_1, ω_2, and $c(\omega_2)$ replaced by ω, ω_r, and c_r, where the subscript r stands for reference. With these changes we get

$$\mathcal{C}(\omega) = c(\omega) \left(1 + \frac{i}{2Q} \right) = c_r \left(1 + \frac{1}{\pi Q} \ln \frac{\omega}{\omega_r} \right) \left(1 + \frac{i}{2Q} \right)$$

$$= c_r \left(1 + \frac{1}{\pi Q} \ln \frac{\omega}{\omega_r} + \frac{i}{2Q} \right). \tag{11.8.14}$$

In the last step a term containing the factor $1/Q^2$ was neglected. The approximations involved in (11.8.13) are (11.8.14) are similar to those introduced above.

Equation (11.8.14) is given by Aki and Richards (1980) for a reference frequency of 1 Hz (i.e., $\omega_r = 2\pi$). The difference in the sign of the imaginary part is due to the difference in the definition of the Fourier transform. The complex wave velocity is extremely important because it provides a way to account for attenuation and dispersion. To solve a wave propagation problem when attenuation is present, formulate the equivalent problem for the elastic case and then replace the wave velocity with the complex velocity (e.g., Aki and Richards, 1980; Ganley, 1981; Kennett, 1983).

11.9 t^*

Our discussion of attenuation so far has assumed that waves propagate in a homogeneous medium. If the properties of the medium vary with position, then the variation of amplitude due to attenuation is represented by

$$A \propto A_o e^{-\omega t^*/2} \tag{11.9.1}$$

where t^* (t star) is given by

$$t^* = \int_{\text{raypath}} \frac{ds}{Q(s)c(s)} \tag{11.9.2}$$

and A_o and A are the amplitudes at the starting and ending points of the raypath, respectively. Equation (11.9.1) is an extension of the exponential decay introduced in (11.5.1) with α (see (11.4.16)) generalized to the case of a heterogeneous medium, and is applicable as long as ray theory applies and Q and c are frequency independent within the frequency band of interest. For teleseismic data, values of t^* of about 1 and 4 s for P and S waves, respectively, are generally used. A review of values of t^* in the Earth for different frequency bands is provided by Der (1998). An application of (11.9.1) is given below.

11.10 The spectral ratio method. Window bias

The *spectral ratio method* is widely used to estimate attenuation using seismic data. The method is applied in the frequency domain and is based on the following decomposition of the amplitude spectrum of a seismogram:

$$A(\omega, r, \theta, \phi) = G(r)S(\omega, \theta, \phi)|R(\theta, \phi)|I(\omega)P(\omega)e^{-\omega t^*/2} \equiv F(\omega, r)e^{-\omega t^*/2}, \tag{11.10.1}$$

where r indicates distance, θ and ϕ are the direction angles introduced in §9.9.1, the functions of ω correspond to amplitude spectra, t^* is given by (11.9.1), and the factors in the middle expression represent the contributions from geometric spreading (G), the source time function (S), the radiation pattern (R), the instrument response

(I), propagation effects (P), and attenuation, respectively (e.g. Teng, 1968; Pilant, 1979). For point sources S is a function of ω only. The decomposition (11.10.1) is valid as long as ray theory is applicable (Ben-Menahem *et al.*, 1965).

Equation (11.10.1) can be applied to two or more stations (variable r) or frequencies (variable ω), and to indicate either two values of r or ω, we will use (11.10.1) with subscripts 1 and 2 in the functions, not in the variables. Then forming the ratio A_1/A_2 and taking logarithms we find

$$\ln \frac{A_1(r, \omega, \theta, \phi)}{A_2(r, \omega, \theta, \phi)} = \ln \frac{F_1(r, \omega, \theta, \phi)}{F_2(r, \omega, \theta, \phi)} - \frac{1}{2}\left[(\omega t^*)_1 - (\omega t^*)_2\right]. \qquad (11.10.2)$$

Equation (11.10.2) is the basis of the spectral ratio method. In practice, the seismic events and stations are chosen in such a way that there is a cancelation (usually only approximate) of factors in the first ratio on the right-hand side of (11.10.2) that results in a linear relation in r or ω between observations (left ratio) and the unknown t^*. Of course, whether the observations follow a linear trend or not depends to a large extent on the assumptions made to arrive at the linear relation. The spectral ratio method has been applied to surface and body waves and a comprehensive review is provided by Båth (1974).

To apply the spectral ratio method it is necessary to isolate the wave of interest, which may be surrounded by other waves which have to be removed to avoid interference. This isolation is done with tapering windows (e.g., Båth, 1974), which also have the purpose of reducing leakage in the computation of the spectra, which in turn is the result of the truncation of a signal. Windowing, however, may bias the results obtained with the spectral ratio method. This will be demonstrated with the following simple attenuation model:

$$\frac{A_z(\omega)}{A_o(\omega)} = Ge^{-\beta\omega}; \qquad \beta = \frac{z - z_o}{2Qv}, \qquad (11.10.3\text{a,b})$$

which is appropriate for attenuation studies using data recorded in a borehole. Here A_z and A_o represent the amplitude spectra of the waves recorded at depths z and z_o, G is a ratio of geometric spreading factors, Q is assumed to be independent of frequency, and v is velocity. To investigate the effect of windowing, let

$$\mathcal{A}_z(\omega) = A_z(\omega)e^{i\phi(\omega)} \qquad (11.10.4)$$

and

$$\mathcal{A}_o(\omega) = A_o(\omega)e^{i\phi_o(\omega)} \qquad (11.10.5)$$

be the Fourier transforms of the waves at depths z and z_o, respectively, and $\mathcal{W}(\omega)$ the Fourier transform of the windowing function. Since the window is applied in the time domain, the ratio of amplitudes in the frequency domain is obtained from

$$\frac{|\mathcal{A}_z(\omega) * \mathcal{W}(\omega)|}{|\mathcal{A}_o(\omega) * \mathcal{W}(\omega)|} = \frac{\left|\int_{-\infty}^{\infty} A_z(\omega - s)e^{i\phi(\omega - s)}\mathcal{W}(s)\,ds\right|}{|\mathcal{A}_o(\omega) * \mathcal{W}(\omega)|}$$

$$= \frac{\left|G \int_{-\infty}^{\infty} e^{-\beta(\omega - s)}A_o(\omega - s)e^{i\phi(\omega - s)}\mathcal{W}(s)\,ds\right|}{|\mathcal{A}_o(\omega) * \mathcal{W}(\omega)|}$$

$$= \frac{A_z(\omega)}{A_o(\omega)}\frac{\left|[A_o(\omega)e^{i\phi(\omega)}] * [e^{\beta\omega}\mathcal{W}(\omega)]\right|}{|\mathcal{A}_o(\omega) * \mathcal{W}(\omega)|} \qquad (11.10.6)$$

(Pujol and Smithson, 1991). For the last two steps equation (11.10.3a) was used. Equation (11.10.6) shows that the ratio of amplitudes that should be used in equation (11.10.3a) is, in fact, multiplied by a function that includes the product of the spectrum of the window and a function that grows exponentially with frequency. This means that if the side lobes of $\mathcal{W}(\omega)$ are relatively large their contribution to the spectral ratio may be important, thus introducing a bias in the results. Experiments with synthetic data show that windowing using a rectangular function or a combination of cosine functions can introduce significant spurious and undetectable dependences of Q on frequency (Pujol and Smithson, 1991). To minimize the window bias Harris *et al.* (1997) recommend the use of a multitaper approach.

11.11 Finely layered media and scattering attenuation

The effect of very thin layers on wave propagation was first studied in the context of exploration seismology, and to introduce it we will consider synthetic seismograms corresponding to the propagation of waves through a stack of thin layers having the acoustic impedance (equal to *velocity* × *density*, see §6.6.2.1) shown in Fig. 11.4(a). The source of waves is assumed to be at the top of the stack, with the receivers inside it along a vertical line that includes the source. This arrangement is the idealized version of a vertical seismic profile (VSP), which in practice is conducted with a receiver placed at different depths in a borehole. For the computation of the synthetic data an acoustic medium is assumed (Burridge *et al.*, 1988). The computations can be carried out without and with reflections included (Figs 11.4b,c and d,e, respectively). Figures 11.4(b,c) show the *impulse response* (the source time function is an impulse) and the seismogram computed for a nonimpulsive time function. Note the drastic change in amplitude as a function of depth. Because the synthetic seismograms are computed for an incident plane wave, the amplitude reduction is not the result of geometric spreading. Also, because the acoustic impedance used is roughly representative of actual data, from the synthetic seismograms we would have to conclude that the seismic energy

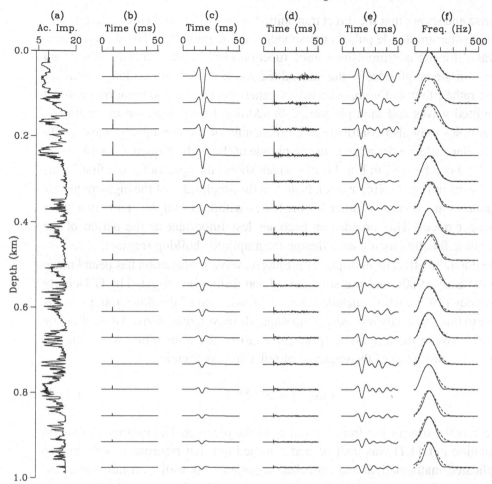

Fig. 11.4. (a) Actual acoustic impedance (*velocity × density*) as a function of depth generated for the velocity (km s^{-1}) and density (g cm^{-3}) shown in Fig. 3 of Pujol and Smithson (1991). (b) Vertical-incidence synthetic seismograms for a 1-D layered medium generated by sampling the acoustic impedance model in such a way that the two-way travel time in each layer is 0.2 ms. The source time function is an impulse and only transmitted waves are included. (c) Similar to (b) for a nonimpulsive source function. (d), (e) Similar to (b) and (c) but including transmitted and reflected waves. All the seismic traces are normalized to the same maximum amplitude. Software provided by R. Burridge and described in Burridge *et al.* (1988). (f) Amplitude spectra for the first 28 ms of the traces in (e) (solid lines) and (c) (dashed lines). All the spectra are normalized to their corresponding maximum values.

would not be able reach the depths that it actually does. The problem with this conclusion is that we have ignored the effect of multiple reflections (multiples, for short) within each of the layers, similar to those described in §8.8.1. This effect is evident in Figs 11.4(d,e). The first of these figures corresponds to the impulse response and shows that the direct transmitted impulse is followed by a large number of smaller-amplitude pulses corresponding to the multiples. The second figure, generated for the nonimpulsive source function used before, shows a significant increase in the amplitudes of the transmitted waves, which are no longer primary waves; rather, they are composite waves generated by the combination of direct transmitted waves and multiple waves. In addition to this increase in amplitude, the multiples may also contribute to a broadening of the downgoing waves, with the attendant relative decrease in the amplitude of the high-frequency components. This effect can be seen in Fig. 11.4(f), which shows the spectra for the first 28 ms of the traces in Figs 11.4(c) and (e). Note that the amplitudes of the high-frequency components are generally larger for the traces without multiples, particularly for the deeper traces. This broadening becomes less important as the period of the source time function increases, although the amplitude buildup remains.

The potential effect of multiple reflections on wave propagation has been known since the early 1960s, and was summarized and further investigated by O'Doherty and Anstey (1971), who concluded that the broadening of the downgoing wave is similar to that caused by intrinsic attenuation. Moreover, they derived the following relation between the amplitude spectrum $T(\omega)$ of the transmitted wave and the power spectrum $R(\omega)$ of the sequence of reflection coefficients:

$$T(\omega, \tau) = e^{-\tau R(\omega)}, \qquad\qquad (11.11.1)$$

where τ is the travel time from the source to the receiver. The original derivation of equation (11.11.1) was unclear, and it turned out that rigorous proofs involve complicated mathematical and statistical arguments, as well as a number of assumptions on the nature of the reflection coefficients. Burridge *et al.* (1988) provide one of those proofs and Resnick (1990) summarizes some of the work done on this subject.

The attenuation-like effect of the multiples was confirmed by Schoenberger and Levin (1978), who used actual velocity and density logs to generate synthetic seismograms similar to those shown in Fig. 11.4 Another conclusion reached by these authors is that if the total observed attenuation is the sum of the intrinsic attenuation and that owing to the multiples, then the latter could be as important as the former. The hypothesis that the two types of attenuations are additive is used to estimate the intrinsic attenuation when using VSP data (Hauge, 1981; Pujol and Smithson, 1991) and was investigated with numerical experiments by Richards and Menke (1983), who found that the additivity is approximately satisfied. Finally, we note that in a medium with one-dimensional variations in elastic properties, the multiple reflections constitute the scattered wave field and for this reason the attenuation caused by the multiples is an example of scattering attenuation.

Problems

11.1 Verify (11.4.10) and (11.4.11).

11.2 Verify (11.5.5).

11.3 Show that

$$I(s) = \int_0^\infty \frac{\omega^s}{1 + \omega^2} \, d\omega$$

has a finite value for $s < 1$ but not for $s = 1$. Why is the lower limit not equal to $-\infty$?

11.4 Verify (11.6.27).

11.5 Verify (11.7.16) and show that (11.7.17) applies as long as $\mathcal{I}\{n(\infty)\}$ is bounded.

11.6 Verify (11.7.20) and (11.7.21).

11.7 Plot (11.7.29) for $Q_o = 100$, $\omega_o = 0.001$ and $0.01 \le \omega \le 10$ Hz.

11.8 Verify (11.8.6).

Hints

Chapter 1

General references: McConnell (1957), Mase (1970), and Segel (1977).

1.1 See §1.3.

1.2 Use the definition (1.3.1). Clockwise rotation:

$$A = \begin{pmatrix} \cos\alpha & 0 & \cos(\pi/2+\alpha) \\ 0 & 1 & 0 \\ \cos(\pi/2-\alpha) & 0 & \cos\alpha \end{pmatrix} = \begin{pmatrix} \cos\alpha & 0 & -\sin\alpha \\ 0 & 1 & 0 \\ \sin\alpha & 0 & \cos\alpha \end{pmatrix}.$$

1.3 (a) Multiply (1.3.2) scalarly with e_k and use (1.2.7).

(b) $\mathbf{e}_j = a_{ij}\mathbf{e}'_i$.

1.4 For Fig. 1.2(a), $v_1 = 0.55$, $v_2 = 0.75$, $v'_1 = 0.33$, $v'_2 = 0.85$. For Fig. 1.2(b), $v_1 = v'_1 = 0.39$, $v_2 = v'_2 = 0.65$.

1.5 Start with the following expressions:

(a) $\nabla \cdot (\mathbf{a} \times \mathbf{b}) = \left(\epsilon_{ijk}a_j b_k\right)_{,i} = \epsilon_{ijk}a_{j,i}b_k + \epsilon_{ijk}a_j b_{k,i}$.

(b) $\nabla \cdot (f\mathbf{a}) = (f\mathbf{a})_{i,i} = f_{,i}a_i + fa_{i,i}$.

(c) $(\nabla \times (f\mathbf{a}))_{,i} = \epsilon_{ijk}(f\mathbf{a})_{k,j}$.

(d) $(\nabla \times \mathbf{r})_i = \epsilon_{ijk}x_{k,j} = \epsilon_{ijk}\delta_{kj}$.

(e) $((\mathbf{a} \cdot \nabla)\mathbf{r})_j = a_i x_{j,i}$.

(f) $(\nabla|\mathbf{r}|)_i = (\sqrt{x_j x_j})_{,i} = (1/2|\mathbf{r}|)(x_j x_j)_{,i}$.

1.6 $|\mathbf{v}|^2 = v_i v_i = a_{ji}v'_j a_{ki}v'_k = |\mathbf{v}'|^2$.

1.7 Let $\mathbf{n}'$ be equal to $(1, 0, 0)$, $(0,1,0)$, $(0, 0, 1)$.

1.8 The x_i are the components of a vector.

1.9 Intermediate step: $(\nabla(\nabla \cdot \mathbf{u}) - \nabla \times \nabla \times \mathbf{u})_i = u_{j,ji} - \epsilon_{ijk}\epsilon_{klm}u_{m,lj}$.

1.10 Let i be equal to 1, 2, 3, and write the ensuing expressions in full.

1.11 (a) Let $|B| = \epsilon_{ijk}b_{i1}b_{j2}b_{k3}$ (column expansion). Let C be the matrix obtained by interchanging the first and second columns of B. Then $|C| = \epsilon_{ijk}b_{i2}b_{j1}b_{k3} = -|B|$ (after a rearrangement of indices). A similar argument shows that this result applies when any other pair of columns or rows is interchanged.

(b) Let the first two columns of B be equal to each other. Then $|B| = \epsilon_{ijk}b_{i1}b_{j2}b_{k3} = -\epsilon_{jik}b_{j1}b_{i2}b_{k3} = -|B|$.

(c) Let $d_1 = \epsilon_{lmn}|B|$, $d_2 = \epsilon_{ijk}b_{il}b_{jm}b_{kn}$, $d_3 = \epsilon_{ijk}b_{li}b_{mj}b_{nk}$. Unless l, m and n are all different, $d_2 = d_3 = 0$ because of (b) and d_1 is also equal to zero. If $(l, m, n) = (1, 2, 3)$, $d_2 = d_3$ from the definition of the determinant and $d_1 = |B|$. If (l, m, n) is an even permutation of $(1, 2, 3)$, d_1 and d_2 have an even number of permutations and $d_2 = d_3 = |B| = d_1$. If (l, m, n) is an odd permutation, then $d_2 = d_3 = -|B| = d_1$.

1.12 Apply the result of Problem 1.11(c) with $B = A$, use $|A| = 1$, and contract with a_{pn}.

1.13 Let $\mathbf{w} = \mathbf{u} \times \mathbf{v}$. Show that $w'_r = a_{rn}w_n$. Start with $w'_r = \epsilon_{rst}u'_s v'_t = \epsilon_{rst}a_{sp}u_p a_{tq}v_q$ and use the result of Problem 1.12.

1.14 (a) Start with $t_{ij}v_j = \lambda v_i$. Assume λ is complex. Then, $t_{ij}v_j^* = \lambda^* v_i^*$, where the $*$ indicates complex conjugation. Contract the first equation with v_i^* and the second with v_i. Subtract one of the resulting equations from the other.

(b) Start with $t_{ij}u_j = \lambda u_i$ and $t_{ij}v_j = \mu v_i$. Contract the first equation with v_i and the second with u_i. Subtract one of the resulting equations from the other.

1.15 Straightforward.

1.16 (a) Start with (1.4.10) and contract i and j.

(b) Start with (1.4.80) and write the tensor and vector using (1.3.9) and (1.4.12).

1.17 Introduce (1.4.107) in (1.4.113) and use (1.4.65) and $\delta_{kk} = 3$.

1.18 Start with $w'_i = \frac{1}{2}\epsilon_{ijk}W'_{jk} = \frac{1}{2}\epsilon_{ijk}a_{jm}a_{kn}W_{mn}$ and use the result of Problem 1.12 to show that $w'_i = a_{ip}w_p$.

1.19 (a)

$$\begin{vmatrix} -\lambda & w_3 & -w_2 \\ -w_3 & -\lambda & w_1 \\ w_2 & -w_1 & -\lambda \end{vmatrix} = -\lambda\left(\lambda^2 + |\mathbf{w}|^2\right) = 0.$$

(b) $W_{ij}w_j = \epsilon_{ijk}w_k w_j = 0$.

1.20 Use (1.4.113) with $W_{ij} = a_i b_j - b_i a_j$ and (1.4.56).

1.21

$$A \approx \begin{pmatrix} 1 & 0 & -\alpha \\ 0 & 1 & 0 \\ \alpha & 0 & 1 \end{pmatrix}; \qquad \alpha \le 8°.$$

1.22 $\mathcal{T}_c \cdot \mathbf{v} = \mathbf{v} \cdot \mathcal{T} = t_{ij}v_i \mathbf{e}_j$.

Chapter 2

2.1 Use relations of the form $u_{i,jkl} = u_{i,kjl} = u_{i,ljk} = \cdots$.

2.2 Use $x_{i,1} = u_{i,1} + \delta_{i1}$; $(1 + 2a)^{1/2} \approx 1 + a$ when $a \ll 1$.

2.3 The only nonzero component of the vector product corresponds to $i = 3$ and is equal to $\epsilon_{321} u_{1,2} = -\alpha$.

2.4 The eigenvalues of $\mathcal{E}$ are 0 and $\pm\alpha$, and the corresponding eigenvectors are $(0, 0, \pm1)$, $(1/\sqrt{2})(1, 1, 0)$, and $(1/\sqrt{2})(1, -1, 0)$.

2.5

$$\frac{\rho - \rho_o}{\rho_o} = \frac{V_o - V}{V} = -\frac{dV}{V_o + dV} = -\frac{\varepsilon_{ii}}{1 + \varepsilon_{ii}} \approx -\varepsilon_{ii}.$$

2.6 Let $\mathbf{b} = \frac{1}{2}\nabla \times \mathbf{u}$. Start with $(\mathbf{e}_p \mathbf{e}_p \times \mathbf{b})_{ij} = \delta_{pi}\epsilon_{jkl}\delta_{pk}b_l = -\epsilon_{ijl}b_l$.

2.7 If $\mathbf{v}$ is an eigenvector of $\mathbf{T}$ and λ is the corresponding eigenvalue, $\mathbf{x}^T\mathbf{T}\mathbf{x} = \lambda\mathbf{x}^T\mathbf{x}$.

2.8 (a) Start with an expression similar to (2.5.1). Use (2.5.4) and (1.6.32).

  (b) Before computing the dot product apply (1.6.32) to $d\mathbf{R}$ on the right.

  (c) Use (2.6.4) and $dx'_i \mathbf{e}'_i \cdot (1 - 2\epsilon_J)\mathbf{e}'_j\mathbf{e}'_j \cdot dx'_k \mathbf{e}'_k = (1 - 2\epsilon_J)(dx'_J)^2$; $J = 1, 2, 3$.

  (d) The volume of the ellipsoid $(x/a)^2 + (y/a)^2 + (z/a)^2 = 1$ is $\frac{4}{3}\pi abc$. Use $(1 - 2d)^{-1/2} \approx 1 + d$ and the result of Problem 1.16(a).

Chapter 3

General references: Atkin and Fox (1980), Mase (1970).

3.1 Start with (3.2.4) with p replaced by fg.

3.2 Because J is a function of $\mathbf{R}$ and t, $D/Dt = \partial/\partial t$. Start with $J = \epsilon_{ijk}x_{1,i}x_{2,j}x_{3,k}$. After taking the derivative there are three similar terms. One of them is $C = \epsilon_{ijk}(\partial x_{1,i}/\partial t)x_{2,j}x_{3,k}$, with $\partial x_{1,i}/\partial t = \partial v_1/\partial X_i$. Because $v_i = v_i(\mathbf{r}, t)$ and $x_j = x_j(\mathbf{R}, t)$, $\partial v_1/\partial X_i = (\partial v_1/\partial x_l)(\partial x_l/\partial X_i)$. Introducing this expression in that for C gives $(\partial v_1/\partial x_i)J$ (two other terms are zero because we are in the case of Problem 1.11(b)).

3.3 Change the variables of integration x_i to X_i in the second integral. This changes the integration volume to V_o. When changing variables the Jacobian (2.2.3) is needed. Then subtract one integral from the other assuming continuity of the integrand.

3.4 (a) From Problem 3.3, because ρ_o does not depend on time, $D(\rho J)/Dt = 0$. Then use the results of Problems 3.1 and 3.2.

  (b) Use the result in (a) and (3.2.4) and (3.2.6).

3.5 Convert the integral over V into an integral over V_o (the volume before deformation) and note that V_o is fixed in time, so that differentiation and integration commute. The result is

$$\frac{d}{dt}\int_V \rho\phi\,dV = \int_{V_o}\frac{D}{Dt}(J\rho\phi)\,dV_o.$$

Then use the result of Problem 3.1 and see Problem 3.4.

3.6 Let r and h be the radius and thickness of the disk. The volume of the disk is $\pi r^2 h$ and the area of the lateral surface is $2\pi rh$. See Problem 3.7(b).

3.7 (a) Equation of the plane through points A, B, C: $ax_1 + bx_2 + cx_3 = d$. The unit vector normal to the plane is $\mathbf{n} = \mathbf{p}/|\mathbf{p}|$, with $\mathbf{p} = (a, b, c)$. The area dS_n of the triangle ABC is one-half of the absolute value of the vector product of the vectors BA and CA. Then, $dS_n = h^2/2n_1 n_2 n_3$. The volume of the tetrahedron is given by

$$V = \frac{1}{2}\int_0^h dS_n(h')\,dh' = \frac{1}{3}h\,dS_n$$

where $dS_n(h')$ is the area of the triangle parallel to ABC a distance h' from the origin.

(b) Use the following result:

$$\int_V f(x_1, x_2, x_3)\,dV \le \max\{|f|\}V.$$

(c) Consider the triangle APB. Following the procedure outlined in (a), the area of the triangle is $n_3\,dS_n$.

3.8 (a) Equation of the plane: $x_1 + 3x_2 + 3x_3 = 3$. Normal vector: $\mathbf{n} = (1, 3, 3)/\sqrt{19}$; $\mathbf{T} = (1, 0, 0)/\sqrt{19}$.

(b) $\mathbf{T}^N = (1/19)\mathbf{n}$; $\mathbf{T}^S = (18, -3, -3)/(19\sqrt{19})$; $\mathbf{T}^S \cdot \mathbf{n} = 0$.

(c) C_1: $(\tau_n + 1)^2 + \tau_s^2 \ge 1$. C_2: $(\tau_n + \frac{1}{2})^2 + \tau_s^2 \le (\frac{3}{2})^2$. C_3: $(\tau_n - \frac{1}{2})^2 + \tau_s^2 \ge (\frac{1}{2})^2$.

Chapter 4

4.1 Use an argument similar to that used in connection with (1.4.68). Start with $\tau'_{ij} = c'_{ijkl}\varepsilon'_{kl} = a_{im}a_{jn}\tau_{mn} = c'_{ijkl}a_{kp}a_{lq}\varepsilon_{pq}$. Use $\tau_{mn} = c_{mnpq}\varepsilon_{pq}$. Show that $a_{im}a_{jn}c_{mnpq} = a_{kp}a_{lq}c'_{ijkl}$. Contract with a_{rp} and a_{sq}.

4.2 Use $c_{ijkl} = c_{jikl}$ to obtain $2\nu\epsilon_{mij}\epsilon_{mkl} = 0$. When $i = j$ or $k = l$, (4.6.1) gives (4.6.2).

4.3 Start with $\varepsilon_{ij}x_j = \nu x_i$ and use (4.6.8).

4.4 From (4.6.13), $\lambda + \mu = \lambda/2\sigma$ and $\mu = \lambda(1 - 2\sigma)/2\sigma$. Introduce in (4.6.12).

4.5 Divide the numerator and denominator on the right of (4.6.12) and (4.6.13) by λ and let λ go to ∞.

4.6 Introduce (4.6.14) and (4.6.15) in (4.6.20).

4.7 Use the symmetry of ε_{ij}.

4.8 Using (4.7.3), $\varepsilon_{ij}\varepsilon_{ij} = \bar{\varepsilon}_{ij}\bar{\varepsilon}_{ij} + \frac{2}{3}\bar{\varepsilon}_{ii}\varepsilon_{kk} + (\varepsilon_{kk})^2\delta_{ii}/9 = \bar{\varepsilon}_{ij}\bar{\varepsilon}_{ij} + (\varepsilon_{kk})^2/3$. Introduce this expression in the first equality of (4.7.1) and use (4.6.20).

4.9 Start with

$$\nabla \cdot [\nabla(\nabla \cdot \mathbf{u})] = [\nabla(\nabla \cdot \mathbf{u})]_{i,i} = u_{j,jii}$$

$$\nabla \cdot (\nabla \times \nabla \times \mathbf{u}) = [\nabla \times (\nabla \times \mathbf{u})]_{i,i} = u_{j,iji} - u_{i,jji}$$

$$\{\nabla \times [\nabla(\nabla \cdot \mathbf{u})]\}_i = \epsilon_{ijk}u_{l,lkj}$$

$$[\nabla \times (\nabla \times \nabla \times \mathbf{u})]_i = \epsilon_{ijk}(u_{l,klj} - u_{k,llj}).$$

4.10 Use (4.8.5) to write λ and μ in terms of α and β. Introduce the resulting expressions in (4.6.13).

4.11 Use (4.8.4) with $\nabla \cdot \mathbf{u} = u_{3,3}$, $\nabla(\nabla \cdot \mathbf{u}) = u_{3,33}\mathbf{e}_3$, $\nabla \times \mathbf{u} = \mathbf{0}$, and $\ddot{\mathbf{u}} = \ddot{u}_3\mathbf{e}_3$, where the two dots indicate a second derivative with respect to time.

4.12 Use (4.8.4) with $\nabla \cdot \mathbf{u} = 0$, $\nabla \times \mathbf{u} = -u_{2,3}\mathbf{e}_1$, and $\nabla \times \nabla \times \mathbf{u} = -u_{2,33}\mathbf{e}_2$.

Chapter 5

5.1 Application of the boundary conditions and integration of the equation involving derivatives gives

$$h(-x/c) + g(x/c) = F(x)$$

and

$$-h(-x/c) + g(x/c) = \frac{1}{c}\int_0^x G(s)\,ds + k.$$

Solve for $h(-x/c)$ and $g(x/c)$, and then proceed as in §5.2.1.

5.2 Replace ϕ in the expression for the Laplacian in spherical coordinates.

5.3 Start with $\nabla \cdot \mathbf{M} = M_{i,i} = \epsilon_{ijk}\phi_{,ji}a_k$.

5.4 Start with $\psi_{,p} = -i(\mathbf{k} \cdot \mathbf{r})_{,p}\psi = -i(k_j r_j)_{,p}\psi = -ik_j r_{j,p}\psi$.

5.5 $[(\mathbf{k} \times \mathbf{a})_q\psi]_{,j} = -i(\mathbf{k} \times \mathbf{a})_q k_j\psi$.

5.6 The spatial derivatives are the same as those in §5.6. The only difference is the replacement of $-\ddot{\psi}/c^2$ by $k_c^2\psi$.

5.7 From (5.8.60), $u_1 = c_1 \cos(\omega t - k_1 x_1)$ and $u_3 = -c_3 \sin(\omega t - k_1 x_1)$.

5.8 From (5.8.53)–(5.8.55), the real parts of the displacements are

$$\mathbf{u}_P = A(l, 0, n) f(\alpha),$$

$$\mathbf{u}_{SV} = B(-n, 0, l) f(\beta),$$

$$\mathbf{u}_{SH} = C(0, 1, 0) f(\beta),$$

where $f(c) = \cos[\omega(t - \mathbf{p} \cdot \mathbf{r}/c)]$. To apply (5.9.5) the following results are needed. For P waves (nonzero terms)

$$u_{1,1} \propto l^2 A\omega/\alpha; \qquad u_{3,3} \propto n^2 A\omega/\alpha;$$

$$u_{k,k} \propto A\omega/\alpha; \qquad (u_{1,3} + u_{3,1}) \propto 2ln A\omega/\alpha.$$

For SV waves (nonzero terms)

$$u_{1,1} \propto -nl B\omega/\beta; \qquad u_{3,3} = -u_{1,1}; \qquad (u_{1,3} + u_{3,1}) \propto (l^2 - n^2) B\omega/\beta.$$

For SH waves (nonzero terms)

$$u_{2,1} \propto lC\omega/\beta; \qquad u_{2,3} \propto nC\omega/\beta.$$

In all cases a factor $\sin[\omega(t - \mathbf{p} \cdot \mathbf{r}/c)]$ $(c = \alpha, \beta)$ has been omitted.

5.9 The velocity vectors to be multiplied by the matrices (5.9.6)–(5.9.8) are

$$-A\omega(l, 0, n)^{\mathrm{T}}, \qquad -B\omega(-n, 0, l)^{\mathrm{T}}, \qquad -C\omega(0, 1, 0)^{\mathrm{T}},$$

respectively. The corresponding sine factors have been omitted.

5.10 Use $\mathbf{a}_3 \cdot \mathbf{p} = n$.

5.11 $T = 2\pi/\omega$ and $\langle \sin^2(\omega t - b) \rangle = \frac{1}{2}$.

Chapter 6

6.1 Use $\tau_{3i} = \lambda\delta_{3i}u_{k,k} + \mu(u_{3,i} + u_{i,3})$. For P and SV waves $\tau_{32} = 0$ because $u_2 = 0$ and u_3 has no dependence on x_2. Similarly, $\tau_{31} = \tau_{33} = 0$ for SH waves.

6.2 The factor $b_j - b_k$ is a wavenumber, so that

$$\int_{-\infty}^{\infty} e^{i(b_j - b_k)x_1} \, dx_1 = \mathcal{F}\{1\}.$$

Use (A.62) and the comments following (A.14).

6.3 (a) Apply an argument similar to that used for the general case to show that $e = e_1$.

(b) Use the results obtained in (a) and apply the argument again. Recall that $\exp(0) = 1$.

(c) The equation without solution is $\sin e/\alpha = 1/\beta$.

6.4 B_1/A is negative or zero if $f \leq \pi/4$. Let $e = \pi/2$ (for largest value of f). Use Snell's law. If $f \geq \pi/4$, $\beta/\alpha \geq \sqrt{2}/2$. Use this result with the expression for σ given in Problem 4.10.

6.5 Aside from the exponential factor, the component in the $\mathbf{a}_1$ direction is given by $\sin 2e[2 \sin e \sin 2f + 2(\alpha/\beta) \cos f \cos 2f]$. Use $2(\alpha/\beta) \cos f \cos 2f = 2(\alpha/\beta) \cos f \cos^2 f - \sin e \sin 2f$. Aside from the exponential factor, the component in the $\mathbf{a}_3$ direction is given by $-2(\alpha/\beta)^2 \cos e(\cos^2 2f + \sin^2 f \cos 2f)$.

6.6 Straight application of Snell's law.

6.7 Equations (5.9.9)–(5.9.11) give the energy flux in the direction of propagation per unit area. Multiply by the cross-sectional areas of the beams, take absolute values, and average as done in Problem 5.11.

6.8 Use Snell's law to write the numerator of (6.5.36) as

$$2 \sin f \sin 2f \left(\frac{\beta^2}{\alpha^2} - \sin^2 f \right)^{1/2} - \cos^2 2f = 0.$$

Solve this equation using a computer.

6.9 Use the method described after (B.14). For the Fourier transforms of $\cos at$ and $\sin at$ see Problem A.3 of Appendix A. For the δ see (A.60) and (A.75).

6.10 For the reflection coefficient start with the expression to the left of the last equality in (6.6.11) and use $\mu/\beta = \rho\beta$ and an equivalent relation for μ'. For the transmission coefficient start with (6.6.9). To compute the impedance for each of the three waves use $\tau_{32}/\dot{u}_2$ with τ_{32} given by (6.4.8).

6.11 Start with

$$\mathbf{u} = \mathbf{a}_2 C \exp[i\omega(t - x_1 \sin f/\beta)]\{\exp(i\omega x_3 \cos f/\beta)$$

$$- \exp[-i(\omega x_3 \cos f/\beta) - 2i\chi]\}$$

and multiply and divide by $\exp i\chi$, and

$$\mathbf{u}' = \mathbf{a}_2 2C \sin \chi \exp(i\omega x_3 \cos f'/\beta') \exp[i\omega(t - x_1 \sin f'/\beta')]$$

$$\times \exp[i(\pi/2 - \chi)]$$

and use Snell's law to modify the exponent involving $\cos f'$.

6.12 Start with

$$\mathcal{R}\{\mathbf{u}'\} = -2\mathbf{a}_2 C \sin \chi \sin[\omega(t - x_1/c) - \chi)]$$

$$\times \exp[\omega x_3 (\sin^2 f - \beta^2/\beta'^2)^{1/2}/\beta].$$

6.13 $\sin 2e'/\sin 2e = \alpha'/\alpha$ as e goes to zero.

6.14 Similar to the problem discussed in §6.5.1.

Chapter 7

7.1 Refer to §7.3.2. Multiply (7.3.11) by $i\tan K$, add the result to (7.3.10) and use (7.3.19).

7.2 Start with (6.9.16a), use $c = \beta$, and rearrange the equation into (7.3.32) with $m = N - 1$.

7.3 Start with (4.2.5) and (4.6.3). If λ and μ are not constant,

$$\tau_{ij,j} = \mu(\nabla^2\mathbf{u})_i + (\lambda+\mu)[\nabla(\nabla\cdot\mathbf{u})]_i + (\nabla\lambda)_i(\nabla\cdot\mathbf{u}) + (\nabla\mu)_j\cdot(\nabla\mathbf{u}+\mathbf{u}\nabla)_{ij}.$$

7.4 Use $\lambda + 2\mu = \mu\alpha^2/\beta^2$ and (7.2.14).

7.5 Straightforward.

7.6 Straightforward.

7.7 Use $1 + \gamma_\beta^2 = 2 - c^2/\beta^2$.

7.8 After the operations indicated the determinant must be multiplied by i^4.

7.9 Straightforward.

7.10

$$\mathcal{F}\{g(ax)\} = \int_{-\infty}^{\infty} g(ax)e^{ikx}\,dx.$$

If $a > 0$, introduce the change of variable $ax = u$. If $a < 0$, use $ax = -|a|x = u$. The Fourier transform of $\exp(-at^2)$ is equal to $(\pi/a)^{1/2}\exp(-\omega^2/4a)$ (e.g., Papoulis, 1962).

7.11 Change variables so that the integral in (7.6.31) becomes

$$I = \int_{-\infty}^{\infty} e^{iau^2}\,du.$$

Separate into real and imaginary parts. Introduce changes of variables similar to those in Problem 7.10. Use the fact that the sine and cosine are odd and even functions. Use a table of integrals.

7.12 Start with

$$\phi(k) = \omega(k) + \frac{1}{t}\psi(k) - k\frac{x}{t}.$$

7.13 The derivation is straightforward but lengthy. An important intermediate result is

$$\frac{dc}{dk} = -\frac{1}{c}A$$

where

$$A = (\mu_1^2\eta_1^2 + \mu_2^2\eta_2^2)\eta_1^2\beta_1^2\beta_2^2\gamma_2 H$$

$$\times \frac{1}{\mu_1\eta_1^2\beta_1^2\mu_2 + \mu_1\mu_2\beta_2^2\gamma_2^2 + kH(\mu_1^2\eta_1^2 + \mu_2^2\eta_2^2)\gamma_2\beta_2^2}.$$

Use this expression with (7.6.15).

7.14 Start with

$$f(x,t) = \frac{2}{\sqrt{2\pi t |\omega_o''|}} \cos\left(\omega_o t - k_o x + \frac{\pi}{4}\right)$$

and use (7.6.49) and (7.6.50).

7.15 Show that after the change of variable the integral has the same expression for c positive or negative.

Chapter 8

8.1 Refer to §8.3. The six terms on the right-hand sides of (8.3.4) and (8.3.6) will be labeled I, II, ..., VI, VII, VIII. Then the coefficient of $\ddot{f}$ comes from VI, VII and (8.3.7), the coefficient of $\dot{f}$ from II, IV, V, VIII and (8.3.5), and the coefficient of f from I and III.

8.2 Interchange j and l and use the symmetry of c_{ijkl}.

8.3 The determinant is

$$D = \begin{vmatrix} B + CT_{11} & CT_{12} & CT_{13} \\ CT_{12} & B + CT_{22} & CT_{23} \\ CT_{13} & CT_{23} & B + CT_{33} \end{vmatrix}$$

where $T_{ij} = T_{,i} T_{,j}$. Use $T_{IJ} T_{JL} T_{IL} = T_{,1}^2 T_{,2}^2 T_{,3}^2$ (no summation over upper-case indices) and similar relations.

8.4 Take the derivative of $\mathbf{t} \cdot \mathbf{t} = t_i t_i$

8.5 Refer to Fig. 8.2. Replace $\sin \Delta\theta/2$ by its expression for small Δs.

8.6 From (8.5.7), $\mathbf{t}$ is a constant vector. Use (8.5.4) and (8.4.24).

8.7 From (8.5.12), $\mathbf{b}$ is a constant vector. Show that $d(\mathbf{r} \cdot \mathbf{b})/ds = 0$, so that $\mathbf{r} \cdot \mathbf{b}$ is a constant, which is the equation of a plane.

8.8 Refer to §8.5. Let $\mathbf{r} = (a\cos u, a\sin u, b\,u)$. Use (8.5.3) and (8.5.5) to obtain $\dot{\mathbf{r}}$ and $\mathbf{t}$. Use $d\mathbf{t}/ds = (d\mathbf{t}/du)(du/ds)$ and a similar relation for $\mathbf{b}$, and (8.5.2). For τ use (8.5.12).

8.9 Use $\mathbf{a} \times \mathbf{a} = \mathbf{0}$ ($\mathbf{a}$ arbitrary), $\mathbf{b} \times \mathbf{t} = \mathbf{n}$, and similar relations.

8.10 Use (10.11.5) with radii r_1 and r_o.

8.11 In the first term change the dummy index j to l, interchange i and k and use the symmetry of c_{ijkl}.

8.12 Start with (8.7.3) and use (8.7.25).

8.13 Multiply (8.7.43) scalarly with $\mathbf{t}$.

8.14 Refer to §8.7.2.2. Multiply (8.7.53a,b) by $\cos\theta$ and $\sin\theta$, respectively, add the corresponding results, and use (8.7.47). This gives (8.5.28).

8.15 Apply an inverse rotation.

8.16 $t = 2\sqrt{(H^2 + x^2/4)/\alpha}$.

Chapter 9

9.1 Introduce a change of variable similar to that in Problem 7.10.

9.2 Consider

$$I = \int_{-\infty}^{\infty} f(x - x_o)\varphi(x)\,dx = \int_{-\infty}^{\infty} f(x)\varphi(x + x_o)\,dx$$

where the second integral is obtained by a change of variables. If the δ were a regular function, (9.2.12) would follow immediately (see (A.3)), but in the context of distribution theory it is a definition (see (A.10)).

9.3 Use $r - c(t - t_o) = c[t_o - (t - r/c)]$.

9.4 Use (9.3.5), (1.4.61), (1.4.53), and (9.4.15).

9.5 $\nabla\phi = (\phi_{,1}, 0, \phi_{,3})$ and $\nabla \times \boldsymbol{\psi} = (-\psi_{,3}, 0, \psi_{,1})$. Apply (9.4.1).

9.6 Start with (9.4.2). $\nabla \times (\nabla \times \mathbf{u}) = -(0, u_{2,11} + u_{2,33}, 0)$.

9.7 Let $\mathbf{v} = (r^{-1}, 0, 0)$. $\nabla \times \mathbf{v} = (0, (r^{-1})_{,3}, -(r^{-1})_{,2})$.

9.8 Write (9.5.6) in component form. For the first component use (9.2.14). The other two components satisfy equations similar to (9.5.5).

9.9 Use $\partial(x_j - \xi_j)/\partial x_i = \delta_{ij}$ and (9.5.13).

9.10 γ_j is a constant. See the next problem.

9.11 Let $v_i = 3\gamma_i\gamma_j - \delta_{ij}$ (j fixed) and θ be the angle between v_i and γ_i. Show that $\cos\theta \neq 0$. Also show that $\mathbf{v} \times \boldsymbol{\Gamma} \neq \mathbf{0}$.

9.12 Integrate by parts as indicated below

$$\int_{r/\alpha}^{r/\beta} \underbrace{\tau}_{u}\ \underbrace{J''(t - \tau)\,d\tau}_{dv}\ .$$

9.13 Given

$$f(t) = \frac{1}{2\pi}\int_{-\infty}^{\infty} A(\omega)e^{i\omega t}\,d\omega$$

we know from (5.4.25) that $A(\omega)$ is the Fourier transform of $f(t)$. Apply this fact to the expressions obtained by taking the first and second derivatives of both sides of the integral with respect to time. Use (11.6.16). The energy of $f(t)$ and its derivatives depends on the frequency content of $f(t)$.

9.14 Start with

$$\mathcal{F}\{h'(t)\} = \int_{-\infty}^{\infty} h'(t)e^{-i\omega t}\,dt.$$

Integrate by parts and assume that $h(\pm\infty) = 0$. This gives

$$\mathcal{F}\{h'(t)\} = i\omega\mathcal{F}\{h(t)\}.$$

Also use (6.5.68).

9.15 To get $J'(t)$ integrate $J''(\tau)$ between 0 and t. To justify the $H(t)$, take the derivative of $J'(t)$ and use $f(t)\delta(t) = f(0)\delta(t)$ (see (A.20)) and $0\delta(t) = 0$.

9.16 Separate the Fourier integral into real and imaginary parts. The final result is $|\mathcal{F}\{J''(t)\}| = a^2/(a^2 + \omega^2)$.

9.17 Refer to Fig. 9.5(a). For $x_3 > 0$, $|\cos\theta| = \cos\theta$, and

$$x_Q = |\cos\theta|\sin\theta = \cos\theta\sin\theta; \qquad y_Q = |\cos\theta|\cos\theta = \cos^2\theta$$

and

$$x_Q^2 + \left(y_Q - \frac{1}{2}\right)^2 = \frac{1}{4}.$$

For $x_3 < 0$, $|\cos\theta| = -\cos\theta$, and the center of the circle is at $(0, -\frac{1}{2})$. For the plot of $|\sin\theta|$ consider $x_1 > 0$ and $x_1 < 0$.

9.18 Let $t_{ij} = \gamma_i\gamma_j$. Show that after a rotation of coordinates $t'_{ij} = \gamma'_i\gamma'_j$. The time dependence comes from scalar quantities.

9.19 To avoid dealing with the convolution apply the Fourier transform in the time domain to both sides of (9.7.2):

$$u_i(\mathbf{x}, \omega) \equiv u_i = F_j(\omega)G_{ij}(\mathbf{x}, \omega; \boldsymbol{\xi}, 0) \equiv G_{ij}F_j.$$

Here u_i and F_j are vectors. Start with $u'_i = G'_{ij}F'_j$, write u'_i and F'_j in terms of u_m and F_n, and use the equation above to eliminate u_m. Contract with an appropriate component of the rotation matrix to get an equation similar to (1.4.10).

9.20 As done in the previous problem, start with (9.9.1) in the frequency domain

$$u_k = M_{ij}G_{ki,j} = M_{ij}s_{kij}.$$

Here u_k is a vector and $s_{kij} = G_{ki,j}$ is a tensor (see the two previous problems and §1.4.3). Start with $u'_k = M'_{ij}s'_{kij}$ and use a procedure similar to that used in the previous example to end up with a relation similar to (1.4.12).

9.21 Start with (9.13.3)–(9.13.8). The only nonzero terms correspond to $\mathbf{A}^{I\beta}$ and $\mathbf{A}^{FS}$, which are both equal to $(-\gamma_3, 0, \gamma_1)^T$ and thus, perpendicular to $\boldsymbol{\Gamma}$.

9.22 Start with (9.13.3)–(9.13.8). The only nonzero terms correspond to $\mathbf{A}^{I\alpha}$ and $\mathbf{A}^{FP}$, which are both equal to $\boldsymbol{\Gamma}$.

9.23 Use (9.2.6), (9.5.13) and $\partial(x_j - \xi_j)/\partial\xi_i = -\delta_{ij}$.

9.24 Refer to Fig. 9.10. Given $\boldsymbol{\Gamma}$, the angles between $\boldsymbol{\Theta}$ and the x_1 and x_3 axes are ϕ and $\theta + \pi/2$, respectively. For $\boldsymbol{\Phi}$ the corresponding angles are $\phi + \pi/2$ and $\pi/2$. To verify that the three vectors form a right-handed coordinate system perform the appropriate vector products.

9.25 Refer to (9.9.18). For a symmetric moment tensor, the conditions for extremal values are

$$\frac{\partial \mathcal{R}^P}{\partial \theta} = 2\frac{\partial \mathbf{\Gamma}^T}{\partial \theta}\mathbf{M}\mathbf{\Gamma} = 0; \qquad \frac{\partial \mathcal{R}^P}{\partial \phi} = 2\frac{\partial \mathbf{\Gamma}^T}{\partial \phi}\mathbf{M}\mathbf{\Gamma} = 0.$$

In addition, $\partial \mathbf{\Gamma}/\partial \theta = \mathbf{\Theta}$ and $\partial \mathbf{\Gamma}/\partial \phi = \sin\theta\, \mathbf{\Phi}$ (Harkrider, 1976; R. Herrmann, personal communication, 1994).

9.28 The pairs (θ, ϕ) that satisfy the conditions of Problem 9.25 are $(\pi/4, 0)$, $(-\pi/4, 0)$ and $(0, \pi/2)$. The first two pairs define the directions of $\mathbf{t}$ and $\mathbf{p}$.

9.29 Use

$$\gamma_i \bar{M}_{ij}\gamma_j \gamma_k = (\mathbf{\Gamma}^T\bar{\mathbf{M}}\mathbf{\Gamma})(\mathbf{\Gamma})_k; \qquad \delta_{ij}\bar{M}_{ij}\gamma_k = \bar{M}_{ii}\gamma_k = \text{tr}(\bar{\mathbf{M}})(\mathbf{\Gamma})_k$$

$$\delta_{kj}\bar{M}_{ij}\gamma_i = (\bar{\mathbf{M}}^T\mathbf{\Gamma})_k; \qquad \delta_{ki}\bar{M}_{ij}\gamma_j = (\bar{\mathbf{M}}\mathbf{\Gamma})_k.$$

9.30 Apply integration by parts as indicated in Problem 9.12. Write the ensuing integral between r/α and r/β as the sum of two integrals between 0 and $t - r/\alpha$ and 0 and $t - r/\beta$.

9.31 Verifying the equations is straightforward using $\bar{\mathbf{M}}^T = \bar{\mathbf{M}}, \mathbf{\Gamma}^T\mathbf{\Gamma} = 1$, and the orthogonality of $\mathbf{\Gamma}, \mathbf{\Theta}$ and $\mathbf{\Phi}$. Applying the equations to $\bar{\mathbf{M}}^{dc}$ gives $\text{tr}(\bar{\mathbf{M}}) = 0, \mathbf{\Gamma}^T\bar{\mathbf{M}}\mathbf{\Gamma} = \sin 2\theta \cos\phi, \mathbf{\Theta}^T\bar{\mathbf{M}}\mathbf{\Gamma} = \cos 2\theta \cos\phi, \mathbf{\Phi}^T\bar{\mathbf{M}}\mathbf{\Gamma} = -\cos\theta \sin\phi$.

Chapter 10

10.1 Intermediate step

$$\int_{-\infty}^{\infty} dt \int_V [u_i\delta_{in}\delta(\mathbf{x}, -t; \boldsymbol{\xi}, -\tau) - G_{in}(\mathbf{x}, -t; \boldsymbol{\xi}, -\tau)f_i]\, dV_x$$

$$= u_n(\boldsymbol{\xi}, \tau) - \int_{-\infty}^{\infty} dt \int_V G_{in}(\mathbf{x}, -t; \boldsymbol{\xi}, -\tau)f_i\, dV_x.$$

10.2 Write the δ as indicated in (9.2.3). Then, using (A.31) we can write

$$\int \delta(\xi_q - \sigma_q)\frac{\partial}{\partial \xi_q}G_{np}\, d\xi_q = -\int \frac{\partial}{\partial \xi_q}\delta(\xi_q - \sigma_q)G_{np}\, d\xi_q$$

$$= -\int \delta_{,q}(\xi_q - \sigma_q)G_{np}\, d\xi_q.$$

10.3 Total body force: is given by the integral over V of $e_p(\boldsymbol{\xi}, t)$. The volume integral involves the δ only, which using Gauss' theorem can be written as

$$\int_V \delta_{,q}(\boldsymbol{\xi} - \boldsymbol{\sigma})\, dV_\xi = \int_S \delta(\boldsymbol{\xi} - \boldsymbol{\sigma})n_q\, dS_\xi = 0$$

where S and n_q are as in (10.2.9). The integral on the right is zero because the δ is nonzero over Σ only and because V and S do not have a common point (Burridge and Knopoff, 1964).

Total moment: here we are interested in the torque caused by a force, which is given by the vector product of the vector $\mathbf{r} = (x_1, x_2, x_3)$ and the force (e.g., Arya, 1990). The qth component of the torque, τ_q, is given by $\tau_q = \epsilon_{qrp} x_r f_p$. In our case the force is the body force. To get the total moment integrate over V. Use ξ_r instead of x_r. The volume integral involves the factors that depend on $\boldsymbol{\xi}$ and gives

$$\int_V \xi_r \delta_{,q}(\boldsymbol{\xi} - \boldsymbol{\sigma})\, \mathrm{d}V_\xi = -\sigma_{r,q} = -\delta_{rq}$$

where δ_{rq} is Kronecker's delta. Then the surface integral is equal to zero because it involves the product of a symmetric and an antisymmetric tensor (after Burridge and Knopoff, 1964).

10.4 In the integral involving the delta there is a factor $-\xi_{3,3} = -1$.

10.5 Green's function $G_{np}(\mathbf{x}, t; \boldsymbol{\sigma}, \tau)$ satisfies (10.2.2) with f_i given by the product of deltas in (10.2.10). One of the terms in the resulting equation is $\rho \partial^2 G_{np} / \partial t^2$. Introduce the change of variable $t' = t - \tau$ and verify that the equation is satisfied for $G_{np}(\mathbf{x}, t - \tau; \boldsymbol{\sigma}, 0)$.

10.6 To get the moment tensor density use (10.6.2) and (10.5.2). The only nonzero contributions come from $[u_3]$ and $\nu_3 = 1$. To get the moment tensor use (10.6.5) and replace the integral by the area A times the average value of $[u_3]$.

10.7 Introduce (10.4.7) in (10.6.13) (assuming $[T_p] = 0$). Proceeding as in Problem 10.3 (total moment), there will be a Kronecker delta. The final result is an integral over Σ with the integrand equal to the moment tensor density. This shows that $\acute{M}_{pq}$ is symmetric and equal to M_{pq}.

10.8 (a) Start with $\mathbf{b} = \mathbf{t} \times \mathbf{p}$ and use (10.7.15) and (10.7.16).
   (b) Show that $M_{ij} v_j = \lambda v_i$ with M_{ij} given by (10.7.4) and $\mathbf{v}$ equal to $\mathbf{t}$, $\mathbf{p}$, or $\mathbf{b}$.

10.9 Start with (10.10.3) and use (9.13.4), (9.13.5) and (9.13.7) to get the projections along $\boldsymbol{\Gamma}$, $\boldsymbol{\Theta}$ and $\boldsymbol{\Phi}$ (this step is similar to that used to get (9.13.15)–(9.13.17)).

10.10 (a) $\mathbf{R}$ is the product of two rotation matrices. One corresponds to a counterclockwise rotation of angle ϕ about the x_3 axis. The second is a clockwise rotation of angle θ about the new x_2 axis.
   (b) $(1, 0, 0)^{\mathrm{T}}$, $(0, 1, 0)^{\mathrm{T}}$ $(0, 0, 1)^{\mathrm{T}}$.
   (c) $\mathbf{R}^{-1} = \mathbf{R}^{\mathrm{T}}$.

10.11 Straightforward integration. It may be easier to start with $J(\tau) = (1/a)\{H(\tau)[a\tau + \exp(-a\tau) - 1] - J'(\tau)\}$.

10.12 Refer to Fig. 10.14. Let $\Delta = \theta$ and $r_o = r$. We are interested in the area $d\sigma$ of the surface element with corners B, C, D, F, equal to $\overline{BC}\,\overline{BD}$.

Chapter 11

11.1 Solve (11.4.10b) for v_R; introduce it in (11.4.10a). Solve for v_I^2 first and then for v_I (choose signs so that it is real). Use (10.4.10a) to get v_R.

11.2 $\mathcal{F}^{-1}\{\exp(-\alpha|\omega|)\} = (1/\pi)\alpha/(t^2 + \alpha^2)$. Use (6.5.68).

11.3 $I(s) = \pi/[2\cos(\pi s/2)]$. See (11.5.17b).

11.4 Write the integral as the sum of two integrals, one between $-\infty$ and 0 and the other between 0 and ∞. In the first one introduce the change of variable $\omega' = -u$, operate and replace u with ω' (dummy variable). The new integral is between 0 and ∞. Combine the two integrals into one.

11.5 Show that

$$\mathcal{P}\int_{-R}^{R}\frac{d\omega'}{\omega' - \omega} = \lim_{\delta \to 0}\left(\int_{-R}^{\omega - \delta}\frac{d\omega'}{\omega' - \omega} + \int_{\omega + \delta}^{R}\frac{d\omega'}{\omega' - \omega}\right)$$

$$= \ln\frac{R - \omega}{R + \omega}; \qquad -R < \omega < R.$$

The first equality corresponds to the definition of the principal value. The final integral is elementary. In the second one change ω' to $-u$. Integrate. Adding the two results gives the result on the second line. Finally, take the limit as R goes to ∞ (Byron and Fuller, 1970). If $\mathcal{I}\{n(\infty)\}$ is bounded, it can be taken out of the integral in (11.7.14). Then, by (11.7.16), the contribution from this integral vanishes.

11.6 To verify (11.7.20) solve (11.7.6b) for $\mathcal{I}\{n(\omega)\}$ and use (11.5.17b). To verify (11.7.21) proceed as in Problem 11.4.

11.8 From (11.6.25), $c = c_\infty[1 + (c_\infty/\omega)\check{\alpha}(\omega)]^{-1}$. Comparison with (11.8.2) shows that

$$\check{\alpha}(\omega) = \frac{2}{\pi}\omega\alpha_o\frac{\ln(1/\alpha_1\omega)}{1 - \alpha_1^2\omega^2} \approx \frac{2}{\pi}\omega\alpha_o\ln\frac{1}{\alpha_1\omega}.$$

Appendix A

A.1 Show that

$$\int_{-\infty}^{\infty}\operatorname{sgn}x\,\varphi(-x)\,dx = -\int_{-\infty}^{\infty}\operatorname{sgn}x\,\varphi(x)\,dx.$$

A.2 Let $T(x) = \operatorname{sgn}x$ and $D(\omega) = \hat{T} = 2/(i\omega)$. Use an equation similar to (A.61). This gives $\mathcal{F}\{D(x)\} = -2\pi T(\omega)$.

A.3 Use an argument similar to that in the previous problem to show that $\mathcal{F}\{\exp(-iat)\} = 2\pi\,\delta_{-a} = \delta(\omega + a)$ (symbolic notation). Then

$$\mathcal{F}\{\cos at\} = \pi[\delta(\omega + a) + \delta(\omega - a)];$$

$$\mathcal{F}\{\sin at\} = i\pi[\delta(\omega + a) - \delta(\omega - a)].$$

Graphical representation of the transforms. For the cosine, a pair of spikes in the up direction located at $\omega = \pm a$. For the sine, a pair of imaginary spikes, one in the up direction at $\omega = -a$ and one in the down direction at $\omega = a$.

Appendix C

C.1 Use (9.2.3) and (A.58) modified to be consistent with (5.4.26).

C.2 See Problem 9.13.

Appendix D

D.1 Straightforward. As an exercise in the use of indicial notation, let $r = (x_i x_i)^{1/2}$. For (D.5) show that $(1/r)_{,jk} = (3r^{-2}x_j x_k - \delta_{jk})r^{-3}$ and contract indices. For (D.6) use Problem 1.5(f).

Appendix A

Introduction to the theory of distributions

Dirac's delta plays a critical role in mathematical physics, and was heavily used in Chapters 9 and 10, where seismic sources were introduced. It is common to refer to the delta as a function, but from a mathematical point of view, the typical equations used to define it, namely

$$\delta(x - a) = 0 \qquad \text{if } x \neq a, \tag{A.1}$$

$$\int_{-\infty}^{\infty} \delta(x - a) \, dx = 1, \tag{A.2}$$

do not make sense within the classical theory of functions, as the integral of a function equal to zero everywhere except one point is equal to zero. If equations (A.1) and (A.2) are accepted as valid, then they are consistent with

$$\int_{-\infty}^{\infty} g(x)\delta(x - a) \, dx = g(a), \tag{A.3}$$

where $g(x)$ is a function continuous at $x = a$. In addition, because of (A.1), the integration interval in (A.2) and (A.3) can be made arbitrarily small as long as $x = a$ is included.

The theory of distributions (or generalized functions), introduced by L. Schwartz in the 1940s, deals with a broad class of mathematical entities, including the delta, that cannot be described within the context of classical mathematical analysis, and has widespread application in the solution of partial differential equations (among other applications). The following simplified treatment is based on Hörmander (1983), Al-Gwaiz (1992) and Friedlander and Joshi (1998). Other useful references are Schwartz (1966), Zemanian (1965) and Beltrami and Wohlers (1966). For simplicity, only the one-dimensional case will be considered here, but the extension to higher dimensions is not difficult.

A basic element in the theory of distributions is the concept of a *test function*, defined by the property of being equal to zero outside of some finite interval and

407

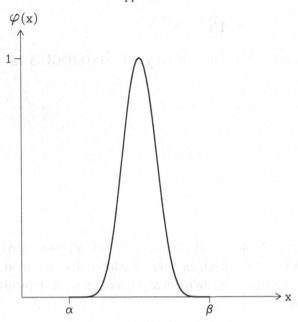

Fig. A1.1. Plot of the test function described by equation (A.4).

having derivatives of all orders. An example is given by

$$\varphi(x) = \begin{cases} e^{8(\alpha-\beta)^{-2}} e^{-[(x-\alpha)^{-2}+(x-\beta)^{-2}]}; & \alpha < x < \beta, \\ 0; & \text{elsewhere} \end{cases} \qquad \text{(A.4a,b)}$$

(see Strichartz, 1994). The constant factor in (A.4a) was chosen to make the maximum value of φ equal to 1 (Fig. A1.1). In general, the closed interval[1] $[\alpha, \beta]$ outside which $\varphi(x)$ is zero is known as the *support* of φ. Because the interval is finite, the support is said to be *compact*.

To motivate the definition of distribution let us introduce the following operation. For a locally integrable[2] function $f(x)$ and an arbitrary test function $\varphi(x)$ let

$$\langle f, \varphi \rangle = \int_{-\infty}^{\infty} f(x)\varphi(x)\,dx. \qquad \text{(A.5)}$$

Given φ, $\langle f, \varphi \rangle$ is a number. Any relation (not necessarily an integral) or "rule" that assigns a number to a function is known as a *functional*, which should be distinguished from a *function* (of a single variable), which assigns a number to a number. A different kind of functional is the $I[x]$ introduced in (8.6.5). The functional defined by (A.5) has the property that it is linear, which means that

---

[1] An interval is closed if it includes its limiting points; otherwise it is open. Square brackets and parentheses will be used to indicate them.

[2] A function is *locally integrable* if it is integrable over any finite interval.

given two test functions φ_1 and φ_2, then

$$\langle f, c_1\varphi_1 + c_2\varphi_2\rangle = c_1\langle f, \varphi_1\rangle + c_2\langle f, \varphi_2\rangle, \tag{A.6}$$

where c_1 and c_2 are constants.

Another property of (A.5) is that it is continuous. This means that if φ is the limit of a sequence of test functions, i.e., if

$$\lim_{n\to\infty} \varphi_n(x) = \varphi(x), \tag{A.7}$$

then (ignoring certain technical details)

$$\lim_{n\to\infty} \langle f, \varphi_n(x)\rangle = \langle f, \varphi(x)\rangle. \tag{A.8}$$

Equations (A.7) and (A.8) are also written as

$$\varphi_n(x) \to \varphi(x); \qquad \langle f, \varphi_n(x)\rangle \to \langle f, \varphi(x)\rangle. \tag{A.9a,b}$$

We are now ready to introduce the following definition. A *distribution* is a linear and continuous functional. The functionals of the form (A.5) are known as *regular distributions*. Usually no distinction is made between the function $f(t)$ and the corresponding distribution f, but sometimes it is convenient to use a different symbol to identify the latter. Examples are given below (see (A.40) and (A.47)). All other functionals that cannot be represented by (A.5) are known as *singular distributions*. For example, Dirac's delta is defined by the rule

$$\langle \delta_a, \varphi\rangle = \varphi(a), \tag{A.10}$$

which is a linear functional, as

$$\langle \delta_a, c_1\varphi_1 + c_2\varphi_2\rangle = c_1\varphi_1(a) + c_2\varphi_2(a) = c_1\langle \delta_a, \varphi_1\rangle + c_2\langle \delta_a, \varphi_2\rangle, \tag{A.11}$$

and

$$\langle \delta_a, \varphi_n\rangle = \varphi_n(a) \to \varphi(a) = \langle \delta_a, \varphi\rangle. \tag{A.12}$$

When $a = 0$ the subscript a will be omitted. Equation (A.10) is commonly written in a symbolic way as shown in (A.3). It is worth emphasizing, however, that (A.3) should not be understood as an actual integral, although in applications it is as if (A.3) actually holds. Another important point to note is that distributions cannot be evaluated at a point, so that, for example, the expression $\delta(x - a)$ should be interpreted in a symbolic way. It is possible, however, to state that the δ is equal to zero under certain conditions, but before doing that it is necessary to introduce two new definitions.

A distribution T is said to be *zero* in an open interval (γ, ε) if

$$\langle T, \varphi\rangle = 0 \tag{A.13}$$

for all φ with support contained in (γ, ε).

The *support* of a distribution T is the smallest closed interval outside of which T is equal to zero (Schwartz, 1966). For example,

$$\langle \delta_a, \varphi \rangle = 0 \qquad\qquad\qquad\qquad (A.14)$$

if the support of φ does not include the point a. Therefore $\delta_a = 0$ in the interval $(-\infty, \infty)$ with the point a removed. This result justifies (A.1).

Given two distributions S and T, they are said to be *equal* if

$$\langle S, \varphi \rangle = \langle T, \varphi \rangle \qquad\qquad\qquad\qquad (A.15)$$

for all φ.

As with functions, *even* and *odd* distribution can be defined. Let

$$\check{\varphi}(x) = \varphi(-x). \qquad\qquad\qquad\qquad (A.16)$$

Then, a distribution is even if

$$\langle T, \varphi(x) \rangle = \langle T, \check{\varphi}(x) \rangle, \qquad\qquad\qquad\qquad (A.17)$$

and odd if

$$\langle T, \varphi(x) \rangle = -\langle T, \check{\varphi}(x) \rangle. \qquad\qquad\qquad\qquad (A.18)$$

Because

$$\langle \delta, \varphi(x) \rangle = \langle \delta, \varphi(-x) \rangle = \varphi(0), \qquad\qquad\qquad\qquad (A.19)$$

the delta is even. An example of an odd distribution is sgn x (see (6.5.43)) (Problem A.1).

In general, the product of two distributions is not defined, but if $f(x)$ is a function with derivatives of all orders and T is a distribution, then the product fT is defined by

$$\langle fT, \varphi \rangle = \langle T, f\varphi \rangle. \qquad\qquad\qquad\qquad (A.20)$$

For example,

$$\langle x\delta, \varphi \rangle = \langle \delta, x\varphi \rangle = x\varphi \Big|_{x=0} = 0, \qquad\qquad\qquad\qquad (A.21)$$

which implies that

$$x\delta = 0. \qquad\qquad\qquad\qquad (A.22)$$

Conversely,

$$xT = 0, \qquad\qquad\qquad\qquad (A.23)$$

implies

$$T = c\delta, \qquad\qquad\qquad\qquad (A.24)$$

where c is a constant (Schwartz, 1966).

As another example,

$$\langle x\delta_a, \varphi \rangle = \langle \delta_a, x\varphi \rangle = a\varphi(a) = \langle a\delta_a, \varphi \rangle, \tag{A.25}$$

so that

$$x\delta_a = a\delta_a. \tag{A.26}$$

The *derivative* of a distribution is defined by

$$\left\langle \frac{dT}{dx}, \varphi \right\rangle = -\left\langle T, \frac{d\varphi}{dx} \right\rangle. \tag{A.27}$$

This definition generalizes the following relation that applies to a differentiable function $f(x)$:

$$\int_{-\infty}^{\infty} f'(x)\varphi(x)\,dx = -\int_{-\infty}^{\infty} f(x)\varphi'(x)\,dx, \tag{A.28}$$

where the primes indicate derivatives with respect to x. To verify (A.28) apply integration by parts to the left-hand side and use $\varphi(\pm\infty) = 0$. When $f(x)$ is not differentiable the left-hand side of (A.28) is not defined but the right-hand side may still exist. For example, if $f(x)$ is Heaviside's unit step $H(x)$, from the right-hand side of (A.28) we obtain

$$-\int_{-\infty}^{\infty} H(x)\varphi'(x)\,dx = -\int_{0}^{\infty} \varphi'(x)\,dx = -\varphi(x)\Big|_{0}^{\infty} = \varphi(0). \tag{A.29}$$

Comparison of (A.29) and (A.10) with $a = 0$ gives

$$\frac{d}{dx}H = \delta. \tag{A.30}$$

Therefore, by introducing distributions it is possible to make sense out of differentiating functions such as $H(x)$, which do not have a derivative at the origin. More generally, if a function has a derivative everywhere except at a point $x = a$ where it has a jump discontinuity, then the derivative at a is equal to the jump of the function at the discontinuity times δ_a (Schwartz, 1966). Simple examples of functions with jump discontinuities are $H(x)$ and $\operatorname{sgn} x$, with jumps of one and two, respectively. Another example is the derivative of the function $h(x)$ defined in (A.39) (see below). In this case the jump is equal to b.

Another useful example involving derivatives is

$$\langle \delta', \varphi \rangle = -\langle \delta, \varphi' \rangle = -\varphi'(0). \tag{A.31}$$

Equation (A.31) defines a dipole (Schwartz, 1966), which is consistent with the interpretation of δ' in our discussion of the double couple (see §10.5). Recall that

$$\varphi'(x) = \lim_{\epsilon \to 0} \frac{\varphi(x + \epsilon/2) - \varphi(x - \epsilon/2)}{\epsilon}, \tag{A.32}$$

so that

$$-\varphi'(0) \approx \frac{\varphi(-\epsilon/2) - \varphi(\epsilon/2)}{\epsilon}. \tag{A.33}$$

The second derivative of a distribution is obtained using (A.27) twice

$$\left\langle \frac{d^2 T}{dx^2}, \varphi \right\rangle = - \left\langle \frac{dT}{dx}, \frac{d\varphi}{dx} \right\rangle = \left\langle T, \frac{d^2 \varphi}{dx^2} \right\rangle. \tag{A.34}$$

An important result involving differential operators is the following. Let L be the operator

$$L = \frac{d^2}{dx^2} + c_1 \frac{d}{dx} + c_2, \tag{A.35}$$

where c_1 and c_2 are constants. Further, let $f_1(x)$ and $f_2(x)$ be solutions of the differential equation

$$Lf = 0 \tag{A.36}$$

satisfying the following conditions:

$$f_1(a) = f_2(a), \tag{A.37}$$

$$f_2'(a) - f_1'(a) = b, \tag{A.38}$$

where b is a constant, known as the jump of $f'(x)$ at the point a. Finally, let $h(x)$ be the function defined by

$$h(x) = \begin{cases} f_1(x); & x \leq a, \\ f_2(x); & x > a. \end{cases} \tag{A.39}$$

Because of (A.37), $h(x)$ is continuous at $x = 0$. Under these conditions, if $\tilde{h}$ is the regular distribution associated with $h(x)$, then

$$L\tilde{h} = b\delta_a \tag{A.40}$$

(Al-Gwaiz, 1992). The tilde ($\tilde{}$) was used to identify the distribution.

To verify (A.40) consider the three terms of (A.35) one at a time. The first term gives

$$\left\langle \frac{d^2 \tilde{h}}{dx^2}, \varphi \right\rangle = \left\langle \tilde{h}, \frac{d^2 \varphi}{dx^2} \right\rangle = \int_{-\infty}^{\infty} h(x)\varphi''(x)\,dx$$

$$= h\varphi' \Big|_{-\infty}^{\infty} - h'\varphi \Big|_{-\infty}^{\infty} + \int_{-\infty}^{\infty} h''(x)\varphi(x)\,dx$$

$$= f_1\varphi'\Big|_{-\infty}^{a} + f_2\varphi'\Big|_{a}^{\infty} - f_1'\varphi\Big|_{-\infty}^{a} - f_2'\varphi\Big|_{a}^{\infty} + \int_{-\infty}^{\infty} h''(x)\varphi(x)\,\mathrm{d}x$$

$$= [f_2'(a) - f_1'(a)]\varphi(a) + \int_{-\infty}^{\infty} h''(x)\varphi(x)\,\mathrm{d}x$$

$$= b\varphi(a) + \int_{-\infty}^{\infty} h''(x)\varphi(x)\,\mathrm{d}x. \tag{A.41}$$

The third equality follows from integration by parts twice, the fourth from (A.39), the fifth from (A.37) and $\varphi'(\pm\infty) = 0$, and the last one from (A.38).

The second term of (A.35) gives

$$\left\langle \frac{\mathrm{d}\tilde{h}}{\mathrm{d}x}, \varphi \right\rangle = -\left\langle \tilde{h}, \frac{\mathrm{d}\varphi}{\mathrm{d}x} \right\rangle = -\int_{-\infty}^{\infty} h(x)\varphi'(x)\,\mathrm{d}x = \int_{-\infty}^{\infty} h'(x)\varphi(x)\,\mathrm{d}x. \tag{A.42}$$

Here integration by parts and (A.37) were used.

Finally, the third term gives

$$\langle \tilde{h}, \varphi \rangle = \int_{-\infty}^{\infty} h(x)\varphi(x)\,\mathrm{d}x. \tag{A.43}$$

Using (A.41)–(A.43) we obtain

$$\langle L\tilde{h}, \varphi \rangle = b\varphi(a) + \int_{-\infty}^{\infty} [Lh(x)]\varphi(x)\,\mathrm{d}x = b\varphi(a) = \langle b\delta_a, \varphi \rangle. \tag{A.44}$$

The integral is equal to zero because $Lh = 0$ by hypothesis (see (A.36)). Equation (A.40) follows from (A.44) and (A.15). This result justifies the arguments used to solve (C.4).

Of particular importance to us is the *Fourier transform* of distributions. Given the distribution T, its Fourier transform is defined by the relation

$$\langle \hat{T}, \psi \rangle = \langle T, \hat{\psi} \rangle, \tag{A.45}$$

where the circumflex accent (ˆ) indicates a Fourier transform and

$$\hat{\psi}(\omega) = \int_{-\infty}^{\infty} \psi(x)\mathrm{e}^{-\mathrm{i}\omega x}\,\mathrm{d}x. \tag{A.46}$$

Here we do not make the distinction between space and time variables introduced in §5.4. The test functions ψ associated with the Fourier transform of distributions are not the same functions that were used to define distributions. The reason for this is that the functions $\hat{\varphi}$ do not have compact support (Schwartz, 1966). Roughly speaking, the functions $\psi(x)$ and their derivatives go to zero as $|x|$ goes to infinity faster than any power of $1/|x|$. An example is e^{-x^2}, which is not a function of compact support.

When T is the regular distribution (A.5), the definition (A.45) agrees with the classical Fourier transform

$$\langle \hat{\tilde{f}}, \psi(\omega) \rangle = \langle \tilde{f}, \hat{\psi}(x) \rangle = \int_{-\infty}^{\infty} f(x) \left(\int_{-\infty}^{\infty} \psi(\omega) e^{-i\omega x} \, d\omega \right) dx$$

$$= \int_{-\infty}^{\infty} \psi(\omega) \left(\int_{-\infty}^{\infty} f(x) e^{-i\omega x} \, dx \right) d\omega = \int_{-\infty}^{\infty} \psi(\omega) \hat{f}(\omega) \, d\omega = \langle \tilde{\hat{f}}, \psi \rangle.$$

$$(A.47)$$

As before, the tilde was used to identify the distributions. It must be noted that the functional on the right-hand side of the first equal sign is defined on the variable x, which requires writing the second integral as shown. In this way, the integral is a function of x.

For any function $f(x)$ for which $\hat{f}(\omega)$ exists, the following inversion relation applies

$$f(x) = \frac{1}{2\pi} \int_{-\infty}^{\infty} \hat{f}(\omega) e^{i\omega x} \, d\omega. \qquad (A.48)$$

This equation will be used to derive the following important property. If $\hat{f}(\omega)$ is the Fourier transform of $f(x)$, then the Fourier transform of $\hat{f}(x)$ is $2\pi f(-\omega)$. To show this first replace x by $-x$ in (A.48)

$$f(-x) = \frac{1}{2\pi} \int_{-\infty}^{\infty} \hat{f}(\omega) e^{-i\omega x} \, d\omega, \qquad (A.49)$$

and then replace x by ω

$$f(-\omega) = \frac{1}{2\pi} \int_{-\infty}^{\infty} \hat{f}(x) e^{-i\omega x} \, dx. \qquad (A.50)$$

The integral in the right-hand side of (A.50) if the Fourier transform of $\hat{f}(x)$, to be indicated by $\hat{\hat{f}}$. Therefore we will write

$$\hat{\hat{f}} = 2\pi \check{f}(\omega) \qquad (A.51)$$

(Hörmander, 1983), where $\check{f}(\omega)$ is defined as in (A.16). Equation (A.51) should be interpreted as follows. Given a function $f(x)$ its Fourier transform is $\hat{f}(\omega)$. Replace ω by x and take the Fourier transform of $\hat{f}(x)$. The result is a function $\hat{\hat{f}}(\omega)$ equal to 2π times the original function f with x replaced by $-\omega$.

For distributions, the equation equivalent to (A.51) is derived using the definition (A.45)

$$\langle \hat{\hat{T}}, \psi \rangle = \langle \hat{T}, \hat{\psi} \rangle = \langle T, \hat{\hat{\psi}} \rangle = 2\pi \langle T, \check{\psi} \rangle. \qquad (A.52)$$

Examples of the use of (A.52) are given below.

A result needed below is that the Fourier transform of an odd distribution is odd. This follows from

$$\langle \hat{T}, \check{\psi} \rangle = \langle T, \hat{\check{\psi}} \rangle = \langle T, \check{\hat{\psi}} \rangle = -\langle T, \hat{\psi} \rangle = -\langle \hat{T}, \psi \rangle. \qquad (A.53)$$

The relation $\hat{\check{\psi}} = \check{\hat{\psi}}$ is easy to verify. The third equality follows from the definition of an odd distribution (see (A.18)). In a similar way it can be shown that the Fourier transform of an even distribution is even.

Finally, we will consider relations that involve derivatives. Using the definition (A.46), the Fourier transform of $d\psi/dx$ is given by

$$\widehat{\frac{d\psi}{dx}}(\omega) = \int_{-\infty}^{\infty} \frac{d\psi(x)}{dx} e^{-i\omega x}\, dx = i\omega \int_{-\infty}^{\infty} \psi(x) e^{-i\omega x}\, dx = i\omega\hat{\psi}(\omega). \qquad (A.54)$$

The second equality is obtained by solving the first integral by integration by parts and using $\psi(\pm\infty) = 0$ (see Problem 9.14).

Next take the derivative of $\hat{\psi}(\omega)$. Using (A.46) again,

$$\frac{d\hat{\psi}(\omega)}{d\omega} = -i \int_{-\infty}^{\infty} x\psi(x) e^{-i\omega x}\, dx = -i\widehat{x\psi}(\omega). \qquad (A.55)$$

Exchanging the order of differentiation and integration is possible because of the properties of the Fourier integral.

Now we will derive an equation equivalent to (A.54) for distributions.

$$\left\langle \widehat{\frac{dT}{dx}}, \psi(\omega) \right\rangle = \left\langle \frac{dT}{dx}, \hat{\psi}(x) \right\rangle = -\left\langle T, \frac{d\hat{\psi}(x)}{dx} \right\rangle = \langle T, i\widehat{\omega\psi}(x) \rangle$$

$$= \langle \hat{T}, i\omega\psi(\omega) \rangle = \langle i\omega\hat{T}, \psi(\omega) \rangle. \qquad (A.56)$$

Here equations (A.20), (A.27), (A.45) (twice), and (A.55) have been used. To use (A.55), recall the comment made in connection with (A.47), which when applied to (A.56) requires the interchange of x and ω in (A.55). From (A.56) we obtain

$$\widehat{\frac{dT}{dx}} = i\omega\hat{T}. \qquad (A.57)$$

Finding the Fourier transform of distributions using the definition may be difficult, so in general it may be more convenient to find the transform by other means and then verify that it satisfies (A.45) (Strichartz, 1994). However, a number of transforms needed elsewhere in the text can be derived with little effort.

(1) Fourier transform of δ_a. Apply the definition (A.45) and (A.10)

$$\langle \hat{\delta}_a, \psi(\omega) \rangle = \langle \delta_a, \hat{\psi}(x) \rangle = \int_{-\infty}^{\infty} \psi(\omega) e^{-i\omega x} \, d\omega \bigg|_{x=a}$$

$$= \int_{-\infty}^{\infty} \psi(\omega) e^{-i\omega a} \, d\omega = \langle e^{-i\omega a}, \psi(\omega) \rangle. \qquad (A.58)$$

Therefore,

$$\hat{\delta}_a = e^{-i\omega a}. \qquad (A.59)$$

In particular, if $a = 0$,

$$\hat{\delta} = 1. \qquad (A.60)$$

(2) Fourier transform of $f(x) = 1$. This function does not have a transform in a classical sense because it is not integrable. However, it has a transform in the sense of distributions. To find it use (A.52) with $T = \delta$, equation (A.60), and the fact that the delta is even (see (A.19)), which gives

$$\langle \hat{\hat{\delta}}, \psi \rangle = \langle \hat{1}, \psi \rangle = 2\pi \langle \delta, \check{\psi} \rangle = 2\pi \langle \delta, \psi \rangle. \qquad (A.61)$$

Using (A.15), from (A.61) we obtain

$$\hat{1} = 2\pi \delta. \qquad (A.62)$$

(3) Fourier transform of $\operatorname{sgn} x$ in the sense of distributions. To compute the transform use the fact that

$$\operatorname{sgn} x = 2H(x) - 1. \qquad (A.63)$$

From (A.63) and (A.30) we obtain

$$\frac{d}{dx} \operatorname{sgn} x = 2\delta. \qquad (A.64)$$

Applying the Fourier transform to (A.64) and using (A.57) and (A.60) gives

$$i\omega \, \widehat{\operatorname{sgn} x} = 2, \qquad (A.65)$$

so that

$$\widehat{\operatorname{sgn} x} = \frac{2}{i\omega} + c\delta, \qquad (A.66)$$

where c is a constant. To verify that the second term on the right-hand side is necessary, multiply (A.66) by $i\omega$ and then use (A.22) with x replaced by ω. The final result is (A.65). To determine c we use the fact that $\operatorname{sgn} x$ as well as its Fourier transform (see (A.53)) and $1/\omega$ are odd, while δ is even. Therefore, c must be zero and

$$\widehat{\operatorname{sgn} x} = \frac{2}{i\omega}. \qquad (A.67)$$

The distribution $1/\omega$ must be interpreted as a *Cauchy principal value*, defined by

$$\left\langle \frac{1}{\omega}, \psi \right\rangle = \lim_{\epsilon \to 0} \left(\int_{-\infty}^{-\epsilon} \frac{\psi(\omega)}{\omega}\, d\omega + \int_{\epsilon}^{\infty} \frac{\psi(\omega)}{\omega}\, d\omega \right). \qquad (A.68)$$

(4) Fourier transform of $H(x)$ in the sense of distributions. Because

$$H(x) = \frac{1}{2}(\operatorname{sgn} x + 1), \qquad (A.69)$$

$$\hat{H}(\omega) = \frac{1}{2}(\widehat{\operatorname{sgn} x} + \hat{1}) = \frac{1}{i\omega} + \pi\delta. \qquad (A.70)$$

In the last step (A.67) and (A.62) were used.

Equation (A.70) will be used to find expressions for two integrals that have meaning in a distribution sense only. Writing $\hat{H}(\omega)$ as

$$\hat{H}(\omega) = \int_{-\infty}^{\infty} H(t)e^{-i\omega x}\, dx = \int_{0}^{\infty} e^{-i\omega x}\, dx$$

$$= \int_{0}^{\infty} \cos \omega x\, dx - i \int_{0}^{\infty} \sin \omega x\, dx = \pi\delta - \frac{i}{\omega} \qquad (A.71)$$

and equating real and imaginary parts gives

$$\int_{0}^{\infty} \cos \omega x\, dx = \pi\delta, \qquad (A.72)$$

and

$$\int_{0}^{\infty} \sin \omega x\, dx = \frac{1}{\omega}. \qquad (A.73)$$

Equation (A.72) is used in Appendix C.

(5) Fourier transform of $1/x$ in the sense of distributions. To obtain this transform use (A.52), (A.67) and the result of Problem 1, which gives

$$\widehat{\frac{2}{ix}} = -2\pi \operatorname{sgn} \omega \qquad (A.74)$$

so that

$$\widehat{\frac{1}{x}} = -i\pi \operatorname{sgn} \omega \qquad (A.75)$$

(Problem A.2)

At this point the convolution of distributions and its Fourier transform should be discussed, but because they have been used little in the text and require the introduction of additional concepts, these matters will not be discussed here. Relatively simple treatments can be found in Al-Gwaiz (1992) and Friedlander and Joshi (1998). For our purposes, the most important results are that the Fourier transform of the convolution of two distributions is equal to the product of their Fourier transforms and that the delta is the unit element with respect to the convolution operation. This means that the convolution of any distribution with the delta is equal to the same distribution. This result was used in §9.6 and (B.4).

Problems

A.1　Verify that sgn x is an odd distribution.

A.2　Verify (A.75).

A.3　Find the Fourier transforms of $\cos at$ and $\sin at$, where a is a constant, and give their graphical representations.

Appendix B

The Hilbert transform

A convenient way to introduce the Hilbert transform is to consider a real causal function, i.e., a function $s(t)$ with the property that

$$s(t) = 0; \qquad t < 0. \tag{B.1}$$

Using the Heaviside's unit step, $s(t)$ can be written as

$$s(t) = H(t)s(t), \qquad \text{all } t \tag{B.2}$$

(Berkhout, 1985).

Applying the Fourier transform to (B.2) (see (5.4.24)) and using the frequency convolution theorem (e.g. Papoulis, 1962) gives

$$S(\omega) = \frac{1}{2\pi} [H(\omega) * S(\omega)], \tag{B.3}$$

where $S(\omega)$ and $\mathcal{H}(\omega)$ are the transforms of $s(t)$ and $H(t)$.

After introducing the expression for $\mathcal{H}(\omega)$ (see (A.70)), equation (B.3) becomes

$$S(\omega) = \frac{1}{2\pi} \left[\pi \delta(\omega) - \frac{i}{\omega} \right] * S(\omega) = \frac{1}{2} \delta(\omega) * S(\omega) - \frac{i}{2\pi\omega} * S(\omega)$$

$$= \frac{1}{2} S(\omega) - \frac{i}{2\pi\omega} * S(\omega), \tag{B.4}$$

where in the last step we used the fact that the delta is the unit element with respect to the operation of convolution (Appendix A).

From (B.4) we find that

$$S(\omega) = -\frac{i}{\pi\omega} * S(\omega). \tag{B.5}$$

In general, $S(\omega)$ is a complex function, so that it can be written as the sum of its real and imaginary parts

$$S(\omega) = R(\omega) + iI(\omega). \tag{B.6}$$

419

Introducing (B.6) in (B.5) gives

$$R(\omega) + \mathrm{i}I(\omega) = -\frac{\mathrm{i}}{\pi\omega} * R(\omega) + \frac{1}{\pi\omega} * I(\omega). \tag{B.7}$$

Equating real and imaginary parts, equation (B.7) gives

$$R(\omega) = \frac{1}{\pi\omega} * I(\omega), \tag{B.8}$$

$$I(\omega) = -\frac{1}{\pi\omega} * R(\omega). \tag{B.9}$$

Equations (B.8) and (B.9) show that the real and imaginary parts of the Fourier transform of a causal function are related to each other in a very specific way. These equations constitute a Hilbert transform–inverse transform pair.

Now we introduce the Hilbert transform of an arbitrary real function $y(t)$, indicated by $\breve{y}(t)$

$$\breve{y}(t) = -\frac{1}{\pi t} * y(t) = \frac{1}{\pi}\mathcal{P}\int_{-\infty}^{\infty}\frac{y(\tau)}{\tau - t}\,d\tau, \tag{B.10}$$

where $\mathcal{P}$ indicates Cauchy's principal value of the integral (see (A.68)). Definition (B.10) is general, and can be introduced by consideration of the integral on the right-hand side of (B.10) when the integration is carried out in the complex plane (e.g., Jeffrey, 1992; Arfken, 1985).

Further insight into the Hilbert transform is gained when the Fourier transform is applied to $\breve{y}(t)$. Because the transform of a convolution is equal to the product of the transforms (e.g., Papoulis, 1962; Appendix A), from the left equality in (B.10) we obtain

$$\mathcal{F}\{\breve{y}(t)\} = \mathcal{F}\left\{-\frac{1}{\pi t}\right\}Y(\omega) = \mathrm{i}\,\mathrm{sgn}\,\omega\,Y(\omega), \tag{B.11}$$

where

$$\mathcal{F}\{y(t)\} = Y(\omega) \tag{B.12}$$

and (A.75) has been used.

Comparison of (B.11) and (B.12) shows that the Hilbert transform does not change the amplitude spectrum of $y(t)$. However, its phase spectrum is modified because

$$\mathrm{i}\,\mathrm{sgn}\,\omega = \begin{cases} \mathrm{i} = \mathrm{e}^{\mathrm{i}\pi/2}; & \omega > 0, \\ 0; & \omega = 0, \\ -\mathrm{i} = \mathrm{e}^{-\mathrm{i}\pi/2}; & \omega < 0. \end{cases} \tag{B.13}$$

(see (6.5.43)). In addition, from (B.11) we see that

$$\breve{y}(t) = \mathcal{F}^{-1}\{\mathrm{i}\,\mathrm{sgn}\,\omega\,Y(\omega)\}. \tag{B.14}$$

Equations (B.11)–(B.14) are extremely useful for the numerical computation of the Hilbert transform. Given a function $y(t)$, compute $Y(\omega)$, change the phase of $Y(\omega)$ as indicated in (B.13), and then compute the inverse Fourier transform. The result of these operations is $\breve{y}(t)$.

Another consequence of (B.11) is that since

$$(\mathrm{i}\,\mathrm{sgn}\,\omega)^2 = -1, \tag{B.15}$$

the result of two successive Hilbert transforms is a phase change of π. Therefore,

$$-\frac{1}{\pi t} * \left[-\frac{1}{\pi t} * y(t) \right] = -\frac{1}{\pi t} * \breve{y}(t) = -y(t), \tag{B.16}$$

which implies that

$$y(t) = \frac{1}{\pi t} * \breve{y}(t) = -\frac{1}{\pi} P \int_{-\infty}^{\infty} \frac{\breve{y}(\tau)}{\tau - t} \, d\tau \tag{B.17}$$

(Meskó, 1984). Equation (B.17) is the expression for the inverse Hilbert transform.

The Hilbert transform is of importance in physics and signal processing (e.g., Arfken, 1985; Bose, 1985) and in the study of anelastic media (see §11.6–§11.8). In the latter context the following additional information is needed (e.g., Jeffrey, 1992). Let

$$z = x + \mathrm{i}y \tag{B.18}$$

indicate a complex variable. If $y = 0$, z becomes the real axis x. Also let

$$f(z) = u(x, y) + \mathrm{i}v(x, y), \tag{B.19}$$

where $u(x, y)$ and $v(x, y)$ are real functions, indicate a function of the complex variable z. Then, if $f(z)$ is an analytic function in the lower half-plane (i.e., $y \leq 0$) and

$$\lim_{|z| \to \infty} |f(z)| = 0 \tag{B.20}$$

then $u(x, 0)$ and $v(x, 0)$ constitute a Hilbert transform pair. The argument 0 in u and v indicates that these functions are specified on the real axis, so that we can write, for example,

$$u(x, 0) = \breve{\phi}(x) \tag{B.21}$$

$$v(x, 0) = \phi(x) \tag{B.22}$$

with $\breve{\phi}(x)$ and $\phi(x)$ satisfying equations similar to (B.10) and (B.17). These equations involve real variables and functions only. Therefore, in the context of Chapter 11 the most important consideration is the existence of the condition (B.20).

Appendix C
Green's function for the 3-D scalar wave equation

We want to find the function $G(\mathbf{x}, t; \mathbf{x}_o, t_o)$ that solves

$$\frac{\partial^2 G}{\partial t^2} = c^2 \nabla^2 G + \delta(\mathbf{x} - \mathbf{x}_o)\delta(t - t_o) \tag{C.1}$$

under the causality condition

$$G(\mathbf{x}, t; \mathbf{x}_o, t_o) = 0; \qquad t < t_o. \tag{C.2}$$

The following derivation follows Haberman (1983). The first step is to apply to (C.1) the triple Fourier transform in the space domain (represented by $\mathbf{x}$). Let

$$\hat{G}(\mathbf{k}, t; \mathbf{x}_o, t_o) = \mathcal{F}\{G(\mathbf{x}, t; \mathbf{x}_o, t_o)\}, \tag{C.3}$$

where $\mathcal{F}$ indicates the Fourier transform introduced in (5.4.26). Then (C.1) and (C.2) become

$$\frac{\partial^2 \hat{G}}{\partial t^2} + c^2 k^2 \hat{G} = e^{i\mathbf{k} \cdot \mathbf{x}_o}\delta(t - t_o) \tag{C.4}$$

$$\hat{G}(\mathbf{k}, t; \mathbf{x}_o, t_o) = 0; \qquad t < t_o, \tag{C.5}$$

(Problems C.1 and C.2) where k is the absolute value of the vector $\mathbf{k}$ given in (5.4.11).

To solve (C.4) we use the fact that $\delta(t - t_o)$ is zero everywhere except at $t = t_o$ (see also Appendix A). Therefore,

$$\frac{\partial^2 \hat{G}}{\partial t^2} + c^2 k^2 \hat{G} = 0; \qquad t > t_o. \tag{C.6}$$

For fixed $\mathbf{k}$ and $\mathbf{x}_o$, this is an ordinary differential equation in $\hat{G}$ with solution

$$\hat{G} = A \cos ck(t - t_o) + B \sin ck(t - t_o); \qquad t > t_o, \tag{C.7}$$

where A and B may depend on $\mathbf{x}_o$, t_o, and $\mathbf{k}$. To determine A and B we must examine what happens at $t = t_o$. First, because $\hat{G}$ is continuous at $t = t_o$ (see below), from (C.5) and (C.7) we have

$$\hat{G}(\mathbf{k}, t_o; \mathbf{x}_o, t_o) = 0 = A. \tag{C.8}$$

To obtain B, integrate (C.4) between t_{o-} and t_{o+}, with (t_{o-}, t_{o+}) a small interval around t_o. As the integral of the delta is equal to one we obtain

$$\int_{t_{o-}}^{t_{o+}} \frac{\partial^2 \hat{G}}{\partial t^2} dt + c^2 k^2 \int_{t_{o-}}^{t_{o+}} \hat{G}\, dt = \frac{\partial \hat{G}}{\partial t}\bigg|_{t_{o-}}^{t_{o+}} + c^2 k^2 \int_{t_{o-}}^{t_{o+}} \hat{G}\, dt = e^{i\mathbf{k}\cdot\mathbf{x}_o}. \tag{C.9}$$

The next step is to let t_{o-} and t_{o+} go to t_o. Because $\hat{G}$ is continuous, the last integral in (C.9) vanishes and (C.9) becomes

$$\lim_{t_{o-}, t_{o+} \to t_o} \frac{\partial \hat{G}}{\partial t}\bigg|_{t_{o-}}^{t_{o+}} = e^{i\mathbf{k}\cdot\mathbf{x}_o}. \tag{C.10}$$

Equation (C.10) shows that $\partial\hat{G}/\partial t$ is discontinuous at $t = t_o$ with a jump equal to the right-hand side of (C.10). The discontinuity of $\partial\hat{G}/\partial t$ is consistent with (C.4) and with the discussion in Appendix A of the derivative of a function with a jump discontinuity. In a qualitative way, $\hat{G}$ is a continuous function and its first derivative is discontinuous. Then the second derivative is equal to the jump times the delta. Therefore, both sides of (C.4) have the same type of singularity (represented by the delta). To apply (C.10), use (C.5) for $t < t_o$, which gives $\partial\hat{G}/\partial t = 0$, and (C.7) for $t > t_o$, which, after taking the limit, gives

$$ck B = e^{i\mathbf{k}\cdot\mathbf{x}_o}. \tag{C.11}$$

Using (C.8) and (C.11), equation (C.7) becomes

$$\hat{G}(\mathbf{k}, t; \mathbf{x}_o, t_o) = \frac{e^{i\mathbf{k}\cdot\mathbf{x}_o}}{ck} \sin ck(t - t_o); \qquad t > t_o. \tag{C.12}$$

Now we need to go back to the space domain. To do that use (5.4.27)

$$G(\mathbf{x}, t; \mathbf{x}_o, t_o) = \frac{1}{(2\pi)^3} \iiint \frac{\sin ck(t - t_o)}{ck} e^{-i\mathbf{k}\cdot(\mathbf{x}-\mathbf{x}_o)}\, dk_x\, dk_y\, dk_z. \tag{C.13}$$

To solve the integral in (C.13) introduce spherical coordinates centered at $|\mathbf{k}| = 0$ and write the dot product in the exponent in terms of the lengths k and r of the two vectors and the angle θ between them

$$\mathbf{k} \cdot (\mathbf{x} - \mathbf{x}_o) = kr\cos\theta. \tag{C.14}$$

The angle θ is similar to the angle θ in Fig. 9.10, with the difference that it is measured with respect to an axis that is not necessarily vertical. Also note that the integration is over $\mathbf{k}$, so that θ ranges between 0 and π. The volume element in spherical coordinates is

$$dV = k^2 \sin\theta \, dk \, d\theta \, d\phi, \tag{C.15}$$

where ϕ ranges between 0 and 2π and is similar to the ϕ in Fig. 9.10, and k ranges between 0 and ∞. With these changes, equation (C.13) becomes

$$G(\mathbf{x}, t; \mathbf{x}_o, t_o) = \frac{1}{(2\pi)^3} \int_0^{2\pi} \int_0^{\infty} \int_0^{\pi} k \frac{\sin ck(t - t_o)}{c} e^{-ikr\cos\theta} \sin\theta \, d\theta \, dk \, d\phi. \tag{C.16}$$

The integral over ϕ gives 2π, while

$$\int_0^{\pi} k e^{-ikr\cos\theta} \sin\theta \, d\theta = \left. \frac{e^{-ikr\cos\theta}}{ir} \right|_0^{\pi} = \frac{2}{r} \sin kr \tag{C.17}$$

(use $\sin\theta \, d\theta = -d(\cos\theta)$). Introducing these results in (C.16) and rewriting the product of the sine functions as a sum of cosines we obtain

$$G(\mathbf{x}, t; \mathbf{x}_o, t_o) = \frac{1}{(2\pi)^2 cr} \int_0^{\infty} \cos k[r - c(t - t_o)] - \cos k[r + c(t - t_o)] \, dk$$

$$= \frac{1}{4\pi cr} \{\delta[r - c(t - t_o)] - \delta[r + c(t - t_o)]\} \tag{C.18}$$

(see (A.72)). Finally, because $r > 0$ and $t > t_o$ the argument of the second delta is positive and the delta is zero (Appendix A). Therefore,

$$G(\mathbf{x}, t; \mathbf{x}_o, t_o) = \frac{1}{4\pi cr} \delta[r - c(t - t_o)]. \tag{C.19}$$

Problems

C.1 Verify that the Fourier transform of $\delta(\mathbf{x} - \mathbf{x}_o)$ is $\exp(i\mathbf{k} \cdot \mathbf{x}_o)$.

C.2 Verify that $\mathcal{F}\{\nabla^2 G\} = -k^2 \mathcal{F}\{G\}$.

Appendix D
Proof of (9.5.12)

Using index notation and after a slight rearrangement, equation (9.5.12) becomes

$$4\pi \rho u_{i1}(\mathbf{x}, t; \boldsymbol{\xi}) = \left(r^{-1}\right)_{,1i} \int_{r/\beta}^{r/\alpha} \tau T(t - \tau)\, d\tau + \frac{1}{\alpha^2 r} r_{,1} r_{,i} T(t - r/\alpha)$$

$$+ \frac{1}{\beta^2 r} \left(\delta_{i1} - r_{,1} r_{,i}\right) T(t - r/\beta). \tag{D.1}$$

To prove (D.1) we will introduce the notation

$$I^c = \int_0^{r/c} \tau T(t - \tau)\, d\tau; \qquad c = \alpha, \beta, \tag{D.2}$$

$$I_\alpha^\beta = \int_{r/\alpha}^{r/\beta} \tau T(t - \tau)\, d\tau, \tag{D.3}$$

$$I^\alpha = I^\beta - I_\alpha^\beta, \tag{D.4}$$

obtained using (D.2) and (D.3), and will use the following relations:

$$\nabla^2 \frac{1}{r} = 0; \qquad r \neq 0 \tag{D.5}$$

$$r_{,i} r_{,i} = 1 \tag{D.6}$$

(Problem D.1)

$$\left(r_{,2}\right)^2 + \left(r_{,3}\right)^2 = 1 - \left(r_{,1}\right)^2 \equiv 1 - r_{,1} r_{,1}, \tag{D.7}$$

obtained using (D.6),

$$\left(r^{-1}\right)_{,i} = -r^{-2} r_{,i}, \tag{D.8}$$

$$\left(I^c\right)_{,i} = \frac{r}{c} T\left(t - \frac{r}{c}\right) \left(\frac{r}{c}\right)_{,i} = \frac{r}{c^2} T\left(t - \frac{r}{c}\right) r_{,i}. \tag{D.9}$$

425

Equation (D.9) results from application of the Leibnitz formula for the differentiation of an integral (e.g., Arfken, 1985).

Next write (9.4.1) in detail using (9.5.10) and (9.5.11)

$$4\pi\rho u_{i1}(\mathbf{x}, t; \boldsymbol{\xi}) = 4\pi\rho(\nabla\phi)_i + (\nabla \times \boldsymbol{\Psi})_i$$

$$= -\left[I^\alpha \left(r^{-1}\right)_{,1}\right]_{,1} \mathbf{e}_1 - \left[I^\alpha \left(r^{-1}\right)_{,1}\right]_{,2} \mathbf{e}_2$$

$$- \left[I^\alpha \left(r^{-1}\right)_{,1}\right]_{,3} \mathbf{e}_3 - \left\{\left[I^\beta \left(r^{-1}\right)_{,2}\right]_{,2} + \left[I^\beta \left(r^{-1}\right)_{,3}\right]_{,3}\right\} \mathbf{e}_1$$

$$+ \left[I^\beta \left(r^{-1}\right)_{,2}\right]_{,1} \mathbf{e}_2 + \left[I^\beta \left(r^{-1}\right)_{,3}\right]_{,1} \mathbf{e}_3$$

$$= -\left[\underbrace{I^\alpha \left(r^{-1}\right)_{,11}}_{\text{I}} + \underbrace{(I^\alpha)_{,1} \left(r^{-1}\right)_{,1}}_{\text{II}}\right] \mathbf{e}_1$$

$$- \left[\underbrace{I^\alpha \left(r^{-1}\right)_{,12}}_{\text{III}} + \underbrace{(I^\alpha)_{,2} \left(r^{-1}\right)_{,1}}_{\text{IV}}\right] \mathbf{e}_2$$

$$- \left[\underbrace{I^\alpha \left(r^{-1}\right)_{,13}}_{\text{V}} + \underbrace{(I^\alpha)_{,3} \left(r^{-1}\right)_{,1}}_{\text{VI}}\right] \mathbf{e}_3$$

$$- \left[\underbrace{I^\beta \left(r^{-1}\right)_{,22}}_{\text{VII}} + \underbrace{(I^\beta)_{,2} \left(r^{-1}\right)_{,2}}_{\text{VIII}} + \underbrace{I^\beta \left(r^{-1}\right)_{,33}}_{\text{IX}} + \underbrace{(I^\beta)_{,3} \left(r^{-1}\right)_{,3}}_{\text{X}}\right] \mathbf{e}_1$$

$$+ \left[\underbrace{I^\beta \left(r^{-1}\right)_{,12}}_{\text{XI}} + \underbrace{(I^\beta)_{,1} \left(r^{-1}\right)_{,2}}_{\text{XII}}\right] \mathbf{e}_2$$

$$+ \left[\underbrace{I^\beta \left(r^{-1}\right)_{,13}}_{\text{XIII}} + \underbrace{(I^\beta)_{,1} \left(r^{-1}\right)_{,3}}_{\text{XIV}}\right] \mathbf{e}_3. \tag{D.10}$$

From I, VII, IX, (D.4) and (D.5) we obtain

$$\left[I^\beta_\alpha \left(r^{-1}\right)_{,11} - I^\beta \nabla^2 \frac{1}{r}\right] \mathbf{e}_1 = I^\beta_\alpha \left(r^{-1}\right)_{,11} \mathbf{e}_1. \tag{D.11}$$

From III, XI and (D.4) and from V, XIII and (D.4) we obtain

$$\left(-I^\beta + I^\beta_\alpha + I^\beta\right) \left(r^{-1}\right)_{,1J} \mathbf{e}_J = I^\beta_\alpha \left(r^{-1}\right)_{,1J} \mathbf{e}_J; \qquad J = 2, 3 \tag{D.12}$$

(no summation over J).

Equations (D.11) and (D.12) account for the first term in (D.1).

From II, IV, VI, (D.8) and (D.9) we obtain

$$\frac{1}{r^2} r_{,1} \frac{r}{\alpha^2} T\left(t - \frac{r}{c}\right) r_{,i} = \frac{1}{\alpha^2 r} T\left(t - \frac{r}{\alpha}\right) r_{,1} r_{,i}; \qquad i = 1, 2, 3. \tag{D.13}$$

Equation (D.13) accounts for the second term of (D.1).
From XII, XIV, (D.8) and (D.9) we obtain

$$-\frac{1}{\beta^2 r} T\left(t - \frac{r}{\beta}\right) r_{,1} r_{,i}; \qquad i = 2, 3. \tag{D.14}$$

From VIII, X, (D.7)–(D.9) we obtain

$$\frac{1}{\beta^2 r}\left[(r_{,2})^2 + (r_{,3})^2\right] T\left(t - \frac{r}{\beta}\right) e_1 = \frac{1}{\beta^2 r}\left(1 - r_{,1} r_{,1}\right) T\left(t - \frac{r}{\beta}\right) e_1. \tag{D.15}$$

Equations (D.14) and (D.15) account for the last term of (D.1).

Problems

D.1   Verify (D.5) and (D.6).

Appendix E

Proof of (9.13.1)

We will start with (9.9.1), repeated below,

$$u_k = M_{ij} * G_{ki,j} \tag{E.1}$$

and (9.6.1), rewritten as

$$4\pi\rho G_{ki}(\mathbf{x}, t; \boldsymbol{\xi}, 0) = \mathrm{I} + \mathrm{II} + \mathrm{III}, \tag{E.2}$$

where

$$\mathrm{I} = \underbrace{(3\gamma_k\gamma_i - \delta_{ki})}_{A} \underbrace{\frac{1}{r^3}}_{B} \underbrace{\left[H\left(t - \frac{r}{\alpha}\right) - H\left(t - \frac{r}{\beta}\right) \right] t}_{C}, \tag{E.3}$$

$$\mathrm{II} = \frac{1}{\alpha^2} \gamma_k\gamma_i \frac{1}{r} \delta\left(t - \frac{r}{\alpha}\right), \tag{E.4}$$

$$\mathrm{III} = -\frac{1}{\beta^2}(\gamma_k\gamma_i - \delta_{ki}) \frac{1}{r} \delta\left(t - \frac{r}{\beta}\right). \tag{E.5}$$

We need the partial derivatives $\mathrm{I}_{,j}$, $\mathrm{II}_{,j}$ and $\mathrm{III}_{,j}$. For $\mathrm{I}_{,j}$ we have

$$\mathrm{I}_{,j} = A_{,j}BC + AB_{,j}C + ABC_{,j}, \tag{E.6}$$

where $A_{,j}$ and $B_{,j}$ are given by

$$A_{,j} = \frac{1}{r} \left[3\left(\gamma_k\gamma_j - \delta_{kj}\right)\gamma_i + 3\gamma_k\left(\gamma_i\gamma_j - \delta_{ij}\right) \right] = \frac{1}{r}\left(6\gamma_k\gamma_i\gamma_j - 3\gamma_i\delta_{kj} - 3\gamma_k\delta_{ij}\right), \tag{E.7}$$

where (9.9.9) was used, and

$$B_{,j} = \frac{3}{r^4}\gamma_j, \tag{E.8}$$

where (9.9.8) was used.

428

To obtain $C_{,j}$ use

$$\left[tH\left(t-\frac{r}{c}\right)\right]_{,j} = -t\delta\left(t-\frac{r}{c}\right)\left(\frac{r}{c}\right)_{,j} = \frac{r}{c^2}\delta\left(t-\frac{r}{c}\right)\gamma_j; \qquad c = \alpha, \beta, \quad \text{(E.9)}$$

with the second equality coming from

$$t\delta(t-t_o) = t_o\delta(t-t_o) \qquad \text{(E.10)}$$

(see (A.26) and (A.30)).

Then

$$C_{,j} = \underbrace{\frac{r}{\alpha^2}\delta\left(t-\frac{r}{\alpha}\right)\gamma_j}_{D} - \underbrace{\frac{r}{\beta^2}\delta\left(t-\frac{r}{\beta}\right)\gamma_j}_{E}. \qquad \text{(E.11)}$$

From the first two terms of (E.6) we obtain

$$(A_{,j}B + AB_{,j})C = \underbrace{\frac{1}{r^4}\left(15\gamma_k\gamma_i\gamma_j - 3\gamma_k\delta_{ij} - 3\gamma_i\delta_{kj} - 3\gamma_j\delta_{ki}\right)}_{F}$$

$$\times \underbrace{\left[H\left(t-\frac{r}{\alpha}\right) - H\left(t-\frac{r}{\beta}\right)\right]t}_{C}. \qquad \text{(E.12)}$$

The factor F gives the factors preceding the integral in (9.13.1) while the convolution of $M_{ij}(t)$ with C gives the integral in the same equation (see (9.5.18)).

Next consider $\text{II}_{,j}$ and $\text{III}_{,j}$

$$\text{II}_{,j} = \underbrace{\frac{1}{\alpha^2 r^2}\left(3\gamma_k\gamma_i\gamma_j - \gamma_k\delta_{ij} - \gamma_i\delta_{kj}\right)\delta\left(t-\frac{r}{\alpha}\right)}_{G} + \underbrace{\frac{1}{\alpha^2}\gamma_k\gamma_i\frac{1}{r}\left[\delta\left(t-\frac{r}{\alpha}\right)\right]_{,j}}_{J},$$

$$\text{(E.13)}$$

$$\text{III}_{,j} = \underbrace{-\frac{1}{\beta^2 r^2}\left(3\gamma_k\gamma_i\gamma_j - \gamma_k\delta_{ij} - \gamma_i\delta_{kj} - \gamma_j\delta_{ki}\right)\delta\left(t-\frac{r}{\beta}\right)}_{K}$$

$$\underbrace{-\frac{1}{\beta^2}(\gamma_k\gamma_i - \delta_{ki})\frac{1}{r}\left[\delta\left(t-\frac{r}{\beta}\right)\right]_{,j}}_{L}. \qquad \text{(E.14)}$$

The term $\gamma_j\delta_{ki}$ in K comes from δ_{ki}/r in (E.5).

Now we are ready to derive the other terms in (9.13.1). The second term comes from terms in (E.6), (E.11), and (E.13)

$$ABD + G = \frac{1}{\alpha^2 r^2}\left(6\gamma_k\gamma_i\gamma_j - \gamma_k\delta_{ij} - \gamma_i\delta_{kj} - \gamma_j\delta_{ki}\right)\delta\left(t-\frac{r}{\alpha}\right) \qquad \text{(E.15)}$$

after convolving $M_{ij}(t)$ with the delta.

Similarly, the third term comes from terms in (E.6), (E.11), and (E.14)

$$-ABE - K = -\frac{1}{\beta^2 r^2} \left(6\gamma_k\gamma_i\gamma_j - \gamma_k\delta_{ij} - \gamma_i\delta_{kj} - 2\gamma_j\delta_{ki}\right) \delta\left(t - \frac{r}{\beta}\right). \quad \text{(E.16)}$$

Finally, the last two terms in (9.13.1) come from $J - L$, as derived in §9.9.

Bibliography

Achenbach, J., 1973. *Wave propagation in elastic solids*, North-Holland, Amsterdam.

Aki, K., 1966. Generation and propagation of G waves from the Niigata earthquake of June 16, 1964. Part 2. Estimation of earthquake moment, released energy, and stress–strain drop from the G wave spectrum, *Bull. Earthq. Res. Inst.* **44**, 73–88.

Aki, K., 1990. Haskell's source mechanism papers and their impact on modern seismology, *in* A. Ben-Menahem, ed., *Vincit Veritas: a portrait of the life and work of Norman Abraham Haskell*, 1905–1970, American Geophysical Union, 42–45.

Aki, K. and P. Richards, 1980. *Quantitative seismology*, vol. I, Freeman, San Francisco, CA.

Al-Gwaiz, M., 1992. *Theory of distributions*, Dekker, New York.

Anderson, D., 1989. *Theory of the earth*, Blackwell, London.

Anderson, J., 1976. Motions near a shallow rupturing fault: evaluation of effects due to the free surface, *Geophys. J. R. Astr. Soc.* **46**, 575–593.

Arfken, G., 1985. *Mathematical methods for physicists*, Academic, New York.

Arya, A., 1990. *Introduction to classical mechanics*, Prentice-Hall, Englewood Cliffs, NJ.

Atkin, R. and N. Fox, 1980. *An introduction to the theory of elasticity*, Longman, London.

Auld, B., 1990. *Acoustic fields and waves in solids*, vol. I, Krieger, Malabar, FL.

Azimi, Sh., A. Kalinin, V. Kalinin, and B. Pivovarov, 1968. Impulse and transient characteristics of media with linear and quadratic absorption laws, *Izv., Phys. Solid Earth*, **2**, 88–93.

Backus, G., 1977. Interpreting the seismic glut moments of total degree two or less, *Geophys. J. R. Astr. Soc.* **51**, 1–25.

Backus, G. and M. Mulcahy, 1976. Moment tensor and other phenomenological descriptions of seismic sources – I. Continuous displacements, *Geophys. J. R. Astr. Soc.* **46**, 341–361.

Bard, P.-Y. and M. Bouchon, 1985. The two-dimensional resonance of sediment-filled valleys, *Bull. Seism. Soc. Am.* **75**, 519–541.

Båth, M., 1968. *Mathematical aspects of seismology*, Elsevier, Amsterdam.

Båth, M., 1974. *Spectral analysis in geophysics*, Elsevier, Amsterdam.

Beltrami, E. and M. Wohlers, 1966. *Distributions and the boundary values of analytic functions*, Academic, New York.

Ben-Menahem, A., 1995. A concise history of mainstream seismology: origins, legacy, and perspective, *Bull. Seism. Soc. Am.* **85**, 1202–1225.

Ben-Menahem, A. and W. Beydoun, 1985. Range of validity of seismic ray and beam methods in general inhomogeneous media – I. General theory, *Geophys. J. R. Astr.*

Soc. **82**, 207–234.

Ben-Menahem, A. and S. Singh 1981. *Seismic waves and sources*, Springer, Berlin.

Ben-Menahem, A., S. Smith, and T.-L. Teng, 1965. A procedure for source studies from spectrums of long-period seismic body waves, *Bull. Seism. Soc. Am.* **55**, 203–235.

Ben-Zion, Y., 2001. On quantification of the earthquake source, *Seism. Res. Lett.* **72**, 151–152.

Berkhout, A., 1985. *Seismic migration*, Developments in solid earth geophysics **14A**, Elsevier, Amsterdam.

Biot, M., 1957. General theorems on the equivalence of group velocity and energy transport, *Phys. Rev.* **105**, 1129–1137.

Bird, R., W. Stewart, and E. Lightfoot, 1960. *Transport phenomena*, Wiley, New York.

Bleistein, N., 1984. *Mathematical methods for wave phenomena*, Academic, New York.

Born, M. and E. Wolf, 1975. *Principles of optics*, Pergamon, Oxford.

Bose, N., 1985. *Digital filters*, North-Holland, Amsterdam.

Bourbié, T., O. Coussy, and B. Zinszner, 1987. *Acoustics of porous media*, Gulf and Editions Technip.

Boyce, W. and R. Di Prima, 1977. *Elementary differential equations and boundary value problems*, Wiley, New York.

Boyles, C., 1984. *Acoustic wave guides. Applications to oceanic science*, Wiley, New York.

Brillouin, L., 1964. *Tensors in mechanics and elasticity*, Academic, New York.

Burns, S., 1987. Negative Poisson's ratio materials, *Science* **238**, 551.

Burridge, R., 1976. *Some mathematical topics in seismology*, Courant Institute of Mathematical Sciences, New York University, New York.

Burridge, R. and L. Knopoff, 1964. Body force equivalents for seismic dislocations, *Bull. Seism. Soc. Am.* **54**, 1875–1888.

Burridge, R., G. Papanicolaou, and B. White, 1988. One-dimensional wave propagation in a highly discontinuous medium, *Wave Motion* **10**, 19–44.

Byron, F. and R. Fuller, 1970. *Mathematics of classical and quantum physics*, vol. 2, Addison-Wesley, Reading, MA. (Reprinted by Dover, New York, 1992.)

Cerveny, V., 1985. The application of ray tracing to the numerical modeling of seismic wavefields in complex structures, *in* G. Dohr, ed., *Seismic shear waves. Part A: theory*, Geophysical Press, 1–124.

Cerveny, V., 2001. *Seismic ray theory*, Cambridge University Press, Cambridge.

Cerveny, V. and F. Hron, 1980. The ray series method and dynamic ray tracing system for three-dimensional inhomogeneous media, *Bull. Seism. Soc. Am.* **70**, 47–77.

Cerveny V. and R. Ravindra, 1971. *Theory of seismic head waves*, Toronto University Press, Toronto.

Cerveny V., I. Moloktov, and I. Psencik, 1977. *Ray method in seismology*, Charles University Press, Prague.

Chou, P. and N. Pagano, 1967. *Elasticity, tensor, dyadic and engineering approaches*, Van Nostrand, Princeton, NJ.

Choy, G. and P. Richards, 1975. Pulse distortion and Hilbert transformation in multiply reflected and refracted body waves, *Bull. Seism. Soc. Am.* **65**, 55–70.

Cornbleet, S., 1983. Geometrical optics reviewed: a new light on an old subject, *Proc. IEEE* **71**, 471–502.

Dahlen, F. and J. Tromp, 1998. *Theoretical global seismology*, Princeton University Press, Princeton, NJ.

Davis, H. and A. Snider, 1991. *Introduction to vector analysis*, Brown, Dubuque, IA.

Der, Z., 1998. High-frequency P- and S-wave attenuation in the earth, *Pure Appl. Geophys.* **153**, 273–310.

Dziewonski, A. and J. Woodhouse, 1983a. Studies of the seismic source using normal-mode theory, *in* H. Kanamori and E. Boschi, eds., *Earthquakes: observation, theory and interpretation*, North-Holland, Amsterdam, 45–137.

Dziewonski, A. and J. Woodhouse, 1983b. An experiment in systematic study of global seismicity: centroid-moment tensor solutions for 201 moderate and large earthquakes of 1981, *J. Geophys. Res.* **88**, 3247–3271.

Dziewonski, A., T.-A. Chou, and J. Woodhouse, 1981. Determination of earthquake source parameters from waveform data for studies of global and regional seismicity, *J. Geophys. Res.* **86**, 2825–2852.

Eringen, A., 1967. *Mechanics of continua*, Wiley, New York.

Eringen, A. and E. Suhubi, 1975. *Elastodynamics*, vol. II, Academic, New York.

Eu, B., 1992. *Kinetic theory and irreversible thermodynamics*, Wiley, New York.

Ewing, W., W. Jardetzky, and F. Press, 1957. *Elastic waves in layered media*, McGraw-Hill, New York.

Fehler, M., M. Oshiba, H. Sato, and K. Obara, 1992. Separation of scattering and intrinsic attenuation for the Kanto-Tokai region, Japan, using measurements of S-wave energy versus hypocentral distance, *Geophys. J. Int.* **108**, 787–800.

Fermi, E., 1937. *Thermodynamics*, Prentice-Hall, Englewood Cliffs, NJ. (Reprinted by Dover, New York, 1956.)

Friedlander, G. and M. Joshi, 1998. *Introduction to the theory of distributions*, Cambridge University Press, Cambridge.

Futterman, W., 1962. Dispersive body waves, *J. Geophys. Res.* **67**, 5279–5291.

Ganley, D., 1981. A method for calculating synthetic seismograms which include the effects of absorption and dispersion, *Geophysics* **46**, 1100–1107.

Goetz, A., 1970. *Introduction to differential geometry*, Addison-Wesley, Reading, MA.

Goodbody A., 1982. *Cartesian tensors*, Horwood, Chichester.

Graff, K., 1975. *Wave motion in elastic solids*, Clarendon Press, Oxford.

Green, G., 1838. On the laws of the reflection and refraction of light at the common surface of two non-crystallized media, *Trans. Cambridge Philosophical Soc.* (Reprinted in *Mathematical papers of George Green*, Chelsea, Bronx, NY, 1970.)

Green, G., 1839. On the propagation of light in crystallized media, *Trans. Cambridge Philosophical Soc.* (Reprinted in *Mathematical papers of George Green*, Chelsea, Bronx, NY, 1970.)

Gregory, A., 1976. Fluid saturation effects on dynamic elastic properties of sedimentary rocks, *Geophysics* **41**, 895–921.

Gubbins, D., 1990. *Seismology and plate tectonics*, Cambridge University Press, Cambridge.

Guillemin, E., 1963. *Theory of linear physical systems*, Wiley, New York.

Gutenberg, B., 1957. Effects of ground on earthquake motion, *Bull. Seism. Soc. Am.* **47**, 221–250.

Haberman, R., 1983. *Elementary applied partial differential equations*, Prentice-Hall, Englewood Cliffs, NJ.

Hansen, W., 1935. A new type of expansion in radiation problems, *Phys. Rev.* **47**, 139–143.

Hanyga, A., 1985. Asymptotic theory of wave propagation, *in* A. Hanyga, ed., *Seismic wave propagation in the earth*, Elsevier, Amsterdam, 35–168.

Harkrider, D., 1976. Potentials and displacements for two theoretical seismic sources, *Geophys. J. R. Astr. Soc.* **47**, 97–133.

Harris, P., C. Kerner, and R. White, 1997. Multichannel estimation of frequency-dependent Q from VSP data, *Geophys. Prosp.* **45**, 87–109.

Haskell, N., 1953. The dispersion of surface waves on multilayered media, *Bull. Seism. Soc. Am.* **43**, 17–34.

Haskell, N., 1960. Crustal reflections of plane SH waves, *J. Geophys. Res.* **65**, 4147–4150.

Haskell, N., 1962. Crustal reflections of plane P and SV waves, *J. Geophys. Res.* **67**, 4751–4767.

Haskell, N., 1963. Radiation pattern of Rayleigh waves from a fault of arbitrary dip and direction of motion in a homogeneous medium, *Bull. Seism. Soc. Am.* **53**, 619–642.

Haskell, N., 1964. Total energy and energy spectral density of elastic wave radiation from propagating faults, *Bull. Seism. Soc. Am.* **54**, 1811–1841.

Hauge, P., 1981. Measurements of attenuation from vertical seismic profiles, *Geophysics* **46**, 1548–1558.

Havelock, T., 1914. *The propagation of disturbances in dispersive media*, Cambridge University Press, Cambridge. (Reprinted by Stechert-Hafner Service Agency, New York, 1964.)

Herrmann, R., 1975, A student's guide to the use of P and S wave data for focal mechanism determination, *Earthquake Notes* **46**, no 4, 29–39.

Herrmann, R., 1998. *Computer programs in seismology (3.0)*, Dept. of Earth and Atmospheric Sciences, St Louis University, St Louis, MO.

Hill, D., 1974. Phase shift and pulse distortion in body waves due to internal caustics, *Bull. Seism. Soc. Am.* **64**, 1733–1742.

Hörmander, L., 1983. *The analysis of linear partial differential operators*, vol. 1, Springer, Berlin.

Hudson, J., 1980. *The excitation and propagation of elastic waves*, Cambridge University Press, Cambridge.

Hunter, S., 1976. *Mechanics of continuous media*, Horwood, Chichester.

Jarosch, H. and E. Aboodi, 1970. Towards a unified notation of source parameters, *Geophys. J. R. Astr. Soc.* **21**, 513–529.

Jeffrey, A., 1992, *Complex analysis and applications*, Chemical Rubber Company, Boca Raton, FL.

Jeffreys, H. and B. Jeffreys, 1956. *Methods of mathematical physics*, Cambridge University Press, Cambridge.

Jia, Y. and H.-P. Harjes, 1997. Seismische Q-Werte als Ausdruck von intrinsischer Dämpfung und Streudämpfung der kristallinen Kruste um die KTB-lokation, ICDP/KTB Kolloqium, Bochum, Germany.

Johnson, L., 1974. Green's function for Lamb's problem, *Geophys. J. R. Astr. Soc.* **37**, 99–131.

Kalinin, A., Sh. Azimi, and V. Kalinin, 1967. Estimate of the phase-velocity dispersion in absorbing media, *Izv., Phys. Solid Earth*, **4**, 249–251.

Kanai, K., 1957. The requisite conditions for the predominant vibration of ground, *Bull. Earthq. Res. Inst.* **35**, 457–471.

Karal, F. and J. Keller, 1959. Elastic wave propagation in homogeneous and inhomogeneous media, *J. Acoust. Soc. Am.* **31**, 694–705.

Karato, S., 1998. A dislocation model of seismic wave attenuation and micro-creep in the earth: Harold Jeffreys and the rheology of the solid earth, *Pure Appl. Geophys.* **153**, 239–256.

Karato, S. and H. Spetzler, 1990. Defect microdynamics in minerals and solid-state mechanisms of seismic wave attenuation and velocity dispersion in the mantle, *Rev.*

Geophys. **28**, 399–421.

Keller, J., R. Lewis, and B. Seckler, 1956. Asymptotic solutions of some diffraction problems, *Comm. Pure Appl. Math.* **9**, 207–265.

Kennett, B., 1983. *Seismic wave propagation in stratified media*, Cambridge University Press, Cambridge.

Kennett, B., 1988, Radiation from a moment-tensor source, *in* D. Doornbos, ed., *Seismological algorithms*, Academic, New York, 427–441.

Kjartansson, E., 1979. Constant Q – wave propagation and attenuation, *J. Geophys. Res.* **84**, 4737–4748.

Kline, M. and I. Kay, 1965. *Electromagnetic theory and geometrical optics*, Interscience, New York.

Knopoff, L., 1964. *Q. Rev. Geophys.* **2**, 625–660.

Kraut, E., 1967. *Fundamentals of mathematical physics*, McGraw-Hill, New York.

Kulhanek, O., 1990. *Anatomy of seismograms*, Elsevier, Amsterdam.

Lakes, R., 1987a. Foam structures with a negative Poisson's ratio, *Science* **235**, 1038–1040.

Lakes, R., 1987b. Negative Poisson's ratio materials, *Science* **238**, 551.

Lakes, R., 1999. *Viscoelastic solids*, Chemical Rubber Company, Boca Raton, FL.

Lamb, G., 1962. The attenuation of waves in a dispersive medium, *J. Geophys. Res.* **67**, 5273–5277.

Lanczos, C., 1970. *The variational principles of mechanics*, University of Toronto Press, Toronto.

Lass, H., 1950. *Vector and tensor analysis*, McGraw-Hill, New York.

Lee, W. and S. Stewart, 1981. *Principles and applications of microearthquake networks*, Academic, New York.

Lewis, R., 1966. Geometrical optics and the polarization vectors, *IEEE Trans. Antennas Propag.* **AP-14**, 100–101.

Lindberg, W., 1983. Continuum mechanics class notes (unpublished), University of Wyoming, Laramie, WY.

Liu, H.-P., D. Anderson, and H. Kanamori, 1976. Velocity dispersion due to anelasticity; implications for seismology and mantle composition, *Geophys. J. R. Astr. Soc.* **47**, 41–58.

Love, A., 1911. *Some problems of geodynamics*, Cambridge University Press, Cambridge.

Love, A., 1927. *A treatise on the mathematical theory of elasticity*, Cambridge University Press, Cambridge. (Reprinted by Dover, New York, 1944.)

Luneburg, R., 1964. *Mathematical theory of optics*, University of California Press, Berkeley, CA.

Maruyama. T., 1963. On the force equivalents of dynamical elastic dislocations with reference to the earthquake mechanism, *Bull. Earthq. Res. Inst.* **41**, 467–486.

Mase, G., 1970. *Theory and problems of continuum mechanics, Schaum's outline series*, McGraw-Hill, New York.

McConnell, A., 1957. *Applications of tensor analysis*, Dover, New York.

Mendiguren, J., 1977. Inversion of surface wave data in source mechanism studies, *J. Geophys. Res.* **82**, 889–894.

Meskó, A., 1984. *Digital filtering: applications in geophysical exploration for oil*, Halsted, New York.

Miklowitz, J., 1984. *The theory of elastic waves and waveguides*, North-Holland, Amsterdam.

Minster, J., 1980. Anelasticity and attenuation, *in* Dziewonski, A. and E. Boschi, eds., *Physics of the earth's interior*, North-Holland, Amsterdam, 152–212.

Mooney, H. and B. Bolt, 1966. Dispersive characteristics of the first three Rayleigh modes for a single surface layer, *Bull. Seism. Soc. Am.* **56**, 43–67.

Morse, P. and H. Feshbach, 1953. *Methods of theoretical physics* (2 vols), McGraw-Hill, New York.

Munk, W., 1949. Note on period increase of waves, *Bull. Seism. Soc. Am.* **39**, 41–45

Murphy, J., A. Davis, and N. Weaver, 1971. Amplification of seismic body waves by low-velocity surface layers, *Bull. Seism. Soc. Am.* **61**, 109–145.

Nadeau, G., 1964. *Introduction to elasticity*, Holt, Rinehart and Winston, New York.

Nakamura, Y. and J. Koyama, 1982. Seismic Q of the lunar upper mantle, *J. Geophys. Res.* **87**, 4855–4861.

Noble, B. and J. Daniel, 1977. *Applied linear algebra*, Prentice-Hall, Englewood Cliffs, NJ.

Nowick, A. and B. Berry, 1972. *Anelastic relaxation in crystalline solids*, Academic, New York.

O'Connell, R. and B. Budiansky, 1978. Measures of dissipation in viscoelastic media, *Geophys. Res. Lett.* **5**, 5–8.

O'Doherty, R. and A. Anstey, 1971. Reflections on amplitudes, *Geophys. Prosp.* **19**, 430–458.

Officer, C., 1974. *Introduction to theoretical geophysics*, Springer, Berlin.

Paley, W. and N. Wiener, 1934. Fourier transforms in the complex domain, *Am. Math. Soc. Colloquium Publications* **XIX**.

Papoulis, A., 1962. *The Fourier integral and its applications*, McGraw-Hill, New York.

Papoulis, A., 1977. *Signal analysis*, McGraw-Hill, New York.

Pekeris, C., 1948. Theory of propagation of explosive sound in shallow water, *in Propagation of sound in the ocean*, Memoir 27, The Geological Society of America.

Pilant, W., 1979. *Elastic waves in the Earth*, Elsevier, Amsterdam.

Psencik, I., 1979. Ray amplitudes of compressional, shear, and converted seismic body waves in 3D laterally inhomogeneous media with curved interfaces, *J. Geophys.* **45**, 381–390.

Pujol, J. and R. Herrmann, 1990. A student's guide to point sources in homogeneous media, *Seism. Res. Lett.* **61**, 209–224.

Pujol, J. and S. Smithson, 1991. Seismic wave attenuation in volcanic rocks from VSP experiments, *Geophysics* **56**, 1441–1455.

Pujol, J., E. Lüschen, and Y. Hu, 1998. Seismic wave attenuation in metamorphic rocks from VP data recorded in Germany's continental super-deep borehole, *Geophysics* **63**, 354–365.

Pujol, J., S. Pezeshk, Y. Zhang, and C. Zhao, 2002. Unexpected values of Q_s in the unconsolidated sediments of the Mississippi embayment, *Bull. Seism. Soc. Am.* **92**, 1117–1128.

Ramsey, J., 1967. *Folding and fracturing of rocks*, McGraw-Hill, New York.

Resnick, J., 1990. Stratigraphic filtering, *Pure Appl. Geophys.* **132**, 49–65.

Rey Pastor, J., P. Pi Calleja, and C. Trejo, 1957. *Analisis Matematico*, vol. II, Editorial Kapelusz, Buenos Aires.

Richards, P. and W. Menke, 1983. The apparent attenuation of a scattering medium, *Bull. Seism. Soc. Am.* **73**, 1005–1021.

Robinson, E., 1967. *Statistical communication and detection*, Hafner, New York.

Romanowicz, B. and J. Durek, 2000. Seismological constraints on attenuation in the earth: a review, *in* S. Karato, A. Forte, R. Liebermann, G. Masters and L. Stixrude, eds., *Earth's deep interior: mineral physics and tomography from the atomic to the global scale*, Geophysical Monograph 117, American Geophysical Union, 161–179.

Santalo, L., 1969. *Vectores y tensores*, Editorial Universitaria de Buenos Aires.

Savage, J., 1969. A new method of analyzing the dispersion of oceanic Rayleigh waves, *J. Geophys. Res.* **74**, 2608–2617.

Savage, J., 1976. Anelastic degradation of acoustic pulses in rock – comments, *Phys. Earth Plan. Int.* **11**, 284–285.

Savarenskii, Y., 1975. *Seismic waves*. Translated from the Russian by the Israel Program for Scientific Translations, Keter, Jerusalem.

Schoenberger, M. and F. Levin, 1978. Apparent attenuation due to intrabed multiples, II, *Geophysics* **43**, 730–737.

Scholz, C., 1990. *The mechanics of earthquakes and faulting*, Cambridge University Press, Cambridge.

Schwab, F. and L. Knopoff, 1971. Surface waves on multilayered anelastic media, *Bull. Seism. Soc. Am.* **61**, 893–912.

Schwartz, L., 1966. *Mathematics for the physical sciences*, Addison-Wesley, Reading, MA.

Segel, L., 1977. *Mathematics applied to continuum mechanics*, Macmillan, New York.

Sokolnikoff, I., 1956. *Mathematical theory of elasticity*, McGraw-Hill, New York.

Solodovnikov, V., 1960. *Introduction to the statistical dynamics of automatic control systems*, Dover, New York.

Spiegel, M., 1959. *Vector analysis, Schaum's outline series*, McGraw-Hill, New York.

Spudich, P. and R. Archuleta, 1987. Techniques for earthquake ground-motion calculation with applications to source parameterization of finite faults, *in Seismic strong motion synthetics*, B. Bolt, ed., Academic, New York, 205–265.

Spudich, P. and N. Frazer, 1984. Use of ray theory to calculate high-frequency radiation from earthquake sources having spatially variable rupture velocity and stress drop, *Bull. Seism. Soc. Am.* **74**, 2061–2082.

Stacey, F., 1992. *Physics of the earth*, Brookfield, Brisbane.

Stacey, F., M. Gladwin, B. McKavanagh, A. Linde, and L. Hastie, 1975. Anelastic damping of acoustic and seismic pulses, *Geophys. Surv.* **2**, 133–151.

Stauder, W., 1960. *S* waves and focal mechanisms: the state of the question, *Bull. Seism. Soc. Am.* **50**, 333–346.

Stauder, W., 1962. The focal mechanism of earthquakes, *in* H. Landsberg and J. Van Mieghem, eds., *Advances in Geophysics* vol 9, Academic, New York, 1–76.

Stavroudis, O., 1972. *The optics of rays, wavefronts and caustics*, Academic, New York.

Steketee, J., 1958. Some geophysical applications of the elasticity theory of dislocations, *Can. J. Phys.* **36**, 1168–1198.

Stratton, J., 1941. *Electromagnetic theory*, McGraw-Hill, New York.

Strichartz, R., 1994. *A guide to distribution theory and Fourier transforms*, Chemical Rubber Company, Boca Raton, FL.

Strick, E., 1970. A predicted pedestal effect for pulse propagation in constant-Q solids, *Geophysics* **35**, 387–403.

Struik, D., 1950. *Lectures on classical differential geometry*, Addison-Wesley, Reading, MA.

Stump, B. and L. Johnson, 1977. The determination of source properties by the linear inversion of seismograms, *Bull. Seism. Soc. Am.* **67**, 1489–1502.

Teng, T.-L., 1968. Attenuation of body waves and the Q structure of the mantle, *J. Geophys. Res.* **73**, 2195–2208.

Thomson, W., 1950. Transmission of elastic waves through a stratified solid medium, *J. Appl. Phys.* **21**, 89–93.

Timoshenko, S., 1953. *History of the strength of materials*, McGraw-Hill, New York.

Tittmann, B., A. Clark, J. Richardson, and T. Spencer, 1980. Possible mechanism for seismic attenuation in rocks containing small amounts of volatiles *J. Geophys. Res.* **85**, 5199–5208.

Toll, J., 1956. Causality and the dispersion relation: logical foundations, *Phys. Rev.* **104**, 1760–1770.

Tolstoy, I., 1973. *Wave propagation*, McGraw-Hill, New York.

Tolstoy, I. and E. Usdin, 1953. Dispersive properties of stratified elastic and liquid media, a ray theory, *Geophysics* **18**, 844–870.

Truesdell, C. and W. Noll, 1965. The non-linear field theories of mechanics, *in* S. Flügge, ed., *Handbuch der Physik*, Springer, Berlin, 1–602.

Truesdell, C. and R. Toupin, 1960. The classical field theories, *in* S. Flügge, ed., *Handbuch der Physik*, Springer, Berlin, 226–793.

Vidale, J., S. Goes, and P. Richards, 1995. Near-field deformation seen on distant broadband seismograms, *Geophys. Res. Lett.* **22**, 1–4.

Weaver, R. and Y.-H. Pao, 1981. Dispersion relations for linear wave propagation in homogeneous and inhomogeneous media, *J. Math. Phys.* **22**, 1909–1918.

White, J., 1965. *Seismic waves*, McGraw-Hill, New York.

Whitham, G., 1974. *Linear and nonlinear waves*, Wiley, New York.

Wiechert, E. and K. Zoeppritz, 1907. Über Erdbebenwellen, *Nachrichten Königl. Gesell. Wissenschaften zu Göttingen*, 427–469.

Wiener, N., 1949. *Extrapolation, interpolation, and smoothing of stationary time series*, The Technology Press of the Massachusetts Institute of Technology and Wiley, New York.

Wilson, E., 1901. *Vector analysis, founded upon the lectures of J. Willards Gibbs*, Yale University Press, New Haven, CT.

Winkler, K. and W. Murphy, 1995. Acoustic velocity and attenuation in porous rocks, *in* T. Ahrens, ed., *Rock physics and phase relations*, Reference Shelf 3, American Geophysical Union, 20–34.

Winkler, K., A. Nur, and M. Gladwin, 1979. Friction and seismic attenuation in rocks, *Nature* **277**, 528–531.

Yeats, R., K. Sieh, and C. Allen, 1997. *The geology of earthquakes*, Oxford University Press, Oxford.

Young, G. and L. Braile, 1976. A computer program for the application of Zoeppritz's amplitude equations and Knott's energy equations, *Bull. Seism. Soc. Am.* **66**, 1881–1885.

Zadeh, L. and C. Desoer, 1963. *Linear system theory*, McGraw-Hill, New York.

Zemanian, A., 1965. *Distribution theory and transform analysis*, McGraw-Hill, New York. (Reprinted by Dover, New York, 1987.)

Zoeppritz, K., 1919. Über Reflexion und Durchgang seismischer Wellen durch Unstetigkeitsflächen, *Nachrichten Königl. Gesell. Wissenschaften zu Göttingen* **1**, 66–84.

Index

439

Printed in the United States
by Baker & Taylor Publisher Services